T0318648

Safe Water in Healthcare

Safe Water in Healthcare

A Practical and Clinical Guide

James T. Walker
Walker on Water, Salisbury, United Kingdom

Susanne Surman-Lee
Leegionella, Ringwood, United Kingdom

Paul J. McDermott
PJM-HS Consulting Ltd, Alsager, United Kingdom

Michael J. Weinbren
Consultant Medical Microbiologist, Chesterfield, United Kingdom

Academic Press is an imprint of Elsevier
125 London Wall, London EC2Y 5AS, United Kingdom
525 B Street, Suite 1650, San Diego, CA 92101, United States
50 Hampshire Street, 5th Floor, Cambridge, MA 02139, United States
The Boulevard, Langford Lane, Kidlington, Oxford OX5 1GB, United Kingdom

ISBN: 978-0-323-90492-6

For information on all Academic Press publications
visit our website at https://www.elsevier.com/books-and-journals

Publisher: Stacy Masucci
Acquisitions Editor: Elizabeth Brown
Editorial Project Manager: Pat Gonzalez
Production Project Manager: Swapna Srinivasan
Cover Designer: Miles Hitchen

Typeset by STRAIVE, India

JW would like to dedicate this book to all his teachers and educators who never gave up on him and who assisted along the way at improving his handwriting and English.

My loving sister Catherine passed away while we were writing the book, and I dedicate the content to Cathy, who instilled in me a love of literature and the importance of education. Taken from us too soon my dear sister.

My mother-in-law Joyce Anlezark had a long life. Joyce was lovely lady, who was always described as the best dressed lady in the village and who kept us all on our toes. Joyce, you were a great example of how we should all live our lives, and you were wise enough to know that there are things that you could have done earlier, perhaps a lesson that we all need to think about.

In addition, I would like to thank my family who have had to put up with me working on this book at all hours and everywhere we have traveled. To Gill, who has had to listen to me talk about this book over and over again. To our Philip, Rebecca, Emme, and Ben for allowing me to hide away in their house and patio in New Mexico and beaver away in the background. To Richard, Leanda, and Florence for letting us cat sit in your lovely home by the sea.

Thank you to you all.

Contents

Preface

The overarching principle of this book was to produce a publication suitable for individuals whose work involves water and wastewater in hospitals and healthcare buildings but who may have limited knowledge on the transmission risks and subsequent water-related infections. Over the years, the authors have struggled finding literature and publications that have presented the microbial risks related to water systems in an understandable format for a wide range of staff in healthcare. Consequently, the book is primarily aimed at covering the basics of water systems and transmission risks using diagrams and case studies to understand the risk that water can pose to vulnerable patients.

Understanding a subject is easier if one can get a clear high-level overview of the main basic concepts. Once these are understood, then other information can be layered and built upon. Additionally, understanding the basic concepts of water systems provides an insight into how to apply that learning to prevent, investigate, and remediate issues.

We have often been asked who the intended audience was for the book and various chapters. Getting the balance right, not making the book too simplistic or at the other extreme overwhelming the reader with information has been more difficult than it sounds. As such we have tried to achieve a balance where readers who may be expert in one area can learn from other chapters where their knowledge may be limited. This book will not make the reader "an expert" but will enable them to gain an understanding of the many different areas in which water is used in healthcare and to further understand the risks to vulnerable patients.

This is not an academic book, nor is it a reproduction of the guidance or standard documents that are in circulation. There are many other resources, for example, published reviews and guidance documents where the reader can obtain more information to add to your knowledge.

We thank you for purchasing the book as you, the reader, will be in the best position to judge whether we have achieved our goal, and we welcome feedback on the contents and how it can be improved.

Any errors are the responsibility of the authors, and we would value your contribution in ensuring that any further updates to the book are improved.

Acknowledgment

First of all, we would like to thank you for purchasing a copy of "Safe Water in Healthcare—a practical and clinical guide." Like many projects, this one started as an inkling of an idea, which took some time to bring together, and we hope that the end results will be informative for many of you that work in healthcare.

We are extremely grateful to a large number of friends and colleagues who read and edited early versions and took the time to point out issues, errors, and particularly where more explanation was required. One of the many questions was "who is this chapter/section aimed at?" Well that was a question we discussed at length as with the subtitle, the book is aimed at any staff who are responsible for water, use water, interact with water in any way that could potentially lead to transmission of waterborne infections. After all, water safety is everyone's responsibility. We hope that different aspects of the book will appeal to different types of staff. But what we want everyone to realize is that water and waste services are a risk to patients in healthcare, and as such we can no longer accept that waterborne infections are part of a patient's stay in hospital.

The authors would like to thank Kattie Washington, Pat Gonzalez, Elizabeth Brown, and Swapna Srinivasan at Elsevier for assisting us through the publication process. For their enthusiasm for this book, for their encouragement and assistance with the editorial process. Pat deserves a special thank you for her patience with us over many aspects of the book as we deleted sections and changed others through the process and particularly for our inability to decide on the book cover.

Without a doubt, any errors in the book are the sole responsibility of the authors, and in due course we hope that you the reader will assist us to improve on the book by identifying errors and making suggestions for future editions.

There are many colleagues and friends who have helped us improve the content, who gave up their own time to read section and make comments.

We would like to thank the following for reviewing and editing sections of the book:

Mary Henderson for comments on the dental section,

Harry Evans for his contribution to many different parts including the sources of drinking water, cold water, vending machines section, and the microbiology section. Thank you Harry for your many, many comments to ensure that sections were up to date, and we are very thankful for your detailed editorial style and complements.

Lauren Gray (Great Western Hospitals NHS Foundation Trust) Speech and Language Therapy (SALT) for her additions to the ice for patients section and for her enthusiasm for our approach.

Jackie Hook (Head Chemist at JLA) for editorial comments and allowing us to use her images in the laundry section as this section would not have read so fluently without your comments.

Dianne Ridout and Adam MacLean for their comments on the birthing pool section and for ensuring that we did not stray too far from the main remit and topic. Thanks also to Mel Burden for suggesting some colleagues to look at this section.

Val O'Brien for her comments on the endoscopy section—thanks Val, great to have precise comments that were straight to the point!

Cathy Whapman, who has been generous with her knowledge, expertise, and use of the red pen to improve out content. Cathy, you are a lady to whom I am deeply indebted. You identified so many gaps and weaknesses in the text, and I am very grateful that you took time out of managing your homestead of managing your sheep, apple picking, and rescuing baby hedgehogs during the 2022 heatwave!

Teresa Inkster for not only allowing us to use your images related to drain contents and issues with ceramic tap fittings but also for commenting on a number of the sections, e.g., surveillance, which has vastly improved the content. Teresa, you were an absolute pleasure to work with, and we thank you for your contribution.

Thank you also to Teresa Inkster for enabling us to use the images of *Cupriavidus pauculus* taken by BMS Gareth Wilson.

Sarah Wratten thank you for your detailed comments on the hydrotherapy section. Sarah, we really appreciated the speed with which you responded particularly over the Platinum Jubilee bank holiday weekend.

Emma Boldock thank you for your comments on the patient drinking water and ice for patient use section, which really assisted in us ensuring that the content was applicable to patient care and the terminology was appropriate.

Sarah Mortor and colleagues in the Infection Prevention and Control Department at Norfolk and Norwich University Hospital NHS Foundation Trust NNUHFT including Prof. Fontaine, who is the DIPC and Chief nurse at NNUH, who supported the project to install screens on the hand wash basins and sinks and allowed us to use their images.

Peter Brown for supplying image and comments on the reverse osmosis section.

Becky Hill for your comments on the sources of patient drinking water and the ecology section. Thanks Becky for letting us know that the sections were enjoyable and informative—that was like being given a comfort blanket knowing that we were heading in the right direction with the content!

Alyson Prince for your comments in the patient drinking water, dirty utility, control, and microbiology chapters. Alyson, we cannot thank you enough for your in-depth and real-world insights that made a major contribution as to why we restructured different sections to ensure that the content was applicable.

Dr. Chloe Keane, who time and again sent us revisions on the ice section to ensure that the content was accurate and reflected what actually happens in clinical practice.

Steve and Ross Finch, who provided the first draft of the remote monitoring chapter. Gentlemen, we are indebted to you for your enthusiasm and expertise that has resulted in technology that can assist healthcare trusts reduce the risks to vulnerable patients.

Jonathan and Elaine Waggott for their ability to demonstrate that innovation and technology can be applied to the provision of water and hand hygiene in the healthcare sector. You have allowed us to use your technologies within the control and monitoring sections, and we thank you for your enthusiasm for the book.

For access to images, we would like to thank:

Peter Hoffman for the use of his images from ice machines.

Daniel Pitcher and the Water Hygiene Center for allowing us access to their library of pictures and for searching for specialist images for us—forever grateful!

Zak Prior for supplying an image of *Pseudomonas aeruginosa* during the writing of HTM 01-04.

We are sure that there are some people who we have forgotten to thank, and we hope that you will forgive us for this mishap.

Personally, last but not least, thanks go to my coauthors:

Susanne Surman-Lee, Michael J. Weinbren, and Paul J. McDermott.

All of you put up my constant emails, almost weekly meetings and badgering for content and comments. While at times, we struggled to agree on the appropriate content in terms of detail and depth, writing the book with you has been fantastic.

Sue, I want to thank you for your contributions to the book. You have skill sets, knowledge, and technical information that is almost encyclopaedic, and I thank you for those contributions when you have so many demands on your time.

Mike, it has been a pleasure to work with you over the last 18 months, and I thank you for many different aspects of the book not least our regular meetings including your generosity when I visited you to work on the content. Your ability to take an idea and produce graphics within a very short timescale is amazing, and your skills have livened up what would otherwise have been very dry sections. Thanks also for your multitude of images that have been used in the book and for your very own clinical perspective.

Paul, there is no other person whom I would trust to edit content. Your eye for detail and ability to identify the salient points to get across to the reader are uncanny. You put up with a lot of badgering from me particularly those phone calls on a Friday morning when you were otherwise busy. Thanks also for ensuring the content was spot on and to the point.

Importance of leadership and governance

1

Overview

This book sets out to provide an understanding of water systems for those whose work involves water and wastewater in hospitals and healthcare buildings but may have had limited access to training. One of the important concepts of the book is the incorporation of diagrams and case studies for the reader to understand the implication of unsafe water in healthcare and how ease it is to contaminate water that will then come into contact with vulnerable patients.

Water as delivered by the water supplier has to meet particular standards to ensure that the product delivered to you the customer is wholesome (DWI, 2015).

Once the water is delivered to the your site, the quality of that water becomes your responsibility, and it is once the water enters your building that the quality of the water will start to deteriorate and present a risk to vulnerable patients.

Through the book, we have described a wide range of scenarios where the quality of the water may pose a risk to vulnerable patients. The engineering aspect of water systems is clearly important in controlling the growth or microorganisms and biofilm. There are numerous guidance and standard documents that provide advice on how water systems should be built, commissioned, and operated and these are cited through the book.

Cooling towers have not been included in this edition, and further information is available from the HSE (2013a).

Outbreaks still occur, and there is no doubt that waterborne infections such Legionnaires' disease are seen by the Health and safety Executive as being preventable (HSE, 2013b, 2014).

Leadership

The consequence of microbially contaminated water can result in infections that may be fatal. Where fatalities that have involved water occur, there may be subsequent investigations and even court proceedings. Such investigations will want to understand the contamination of the water, the transmission route, the nature of the infection and how the organization has managed the water system, its staff, and how it has responded to waterborne acquired infections. Does an organization have leadership and policies in place to manage waterborne outbreaks and to implement strategies that will ensure the safety of vulnerable, such as those in oncology units (Inkster and Cuddihy, 2021)?

Safe Water in Healthcare. https://doi.org/10.1016/B978-0-323-90492-6.00006-9

Such policies and leadership could be referred to as clinical governance, i.e., "a system through which NHS organizations are accountable for continuously improving the quality of their services and safeguarding high standards of care by creating an environment in which excellence in clinical care will flourish" (NHS England, 2022).

Governance

Governance is defined variously but, in the healthcare setting, a useful definition is provided in the English Health Technical Memorandum (HTM) 04-01 Part B (DHSC, 2016). Here, it is explained that governance is concerned with how an organization directs, manages, and monitors its activities to ensure compliance with legislative requirements while ensuring that the safety of its patients, visitors, and staff is not compromised. It goes on to highlight the importance of ensuring safe processes, working practices, and risk management strategies are in place and supported by sufficient resources and suitably qualified and competent staff.

This can be achieved by appointing a water safety group with sufficient expertise and developing and implementing a suitable water safety plan. Key to successful governance is effective communication with and active support from senior management, for example, the Chief Executive Officer (CEO) and the Board of Directors. This requires senior managers to delegate responsibility for delivering the water safety plan to the water safety group, while at the same time acknowledging that the duty to ensure the safe provision of water remains with them.

CEO's and Board members are busy people, so the means by which important water safety issues are communicated to them needs to be clearly defined and efficient. There must be effective communication pathways between the water safety group and Board level, e.g., via other forums such as the Infection Prevention and Control Committees and Health and Safety Committee, and it is important that key water safety concerns are communicated in an agreed and unambiguous way (Fig. 1). In other high-risk industry sectors (e.g., chemical, oil, and gas), safety performance indicators have proved to be useful in this respect (HSE, 2011). Where they work best is in organizations where these key indicators have been agreed and set by the senior managers themselves and reports on their current status are provided regularly.

Safety performance indicators can vary from organization to organization but might include metrics on specific elements of risk management, such as planned preventive maintenance, reactive maintenance activities, out-of-specification monitoring data, positive microbiological testing results, progress with risk assessments, staffing levels, and training requirements. Senior management should set the tolerance parameters for each safety performance indicator and performance can be communicated by means of a red/amber/green dashboard or similar. Safety performance indicators that are red require immediate action at Board level, those that are amber provide an alert that should prompt some form of further investigation by the Board, and green safety performance indicators provide assurances that safety is being successfully maintained.

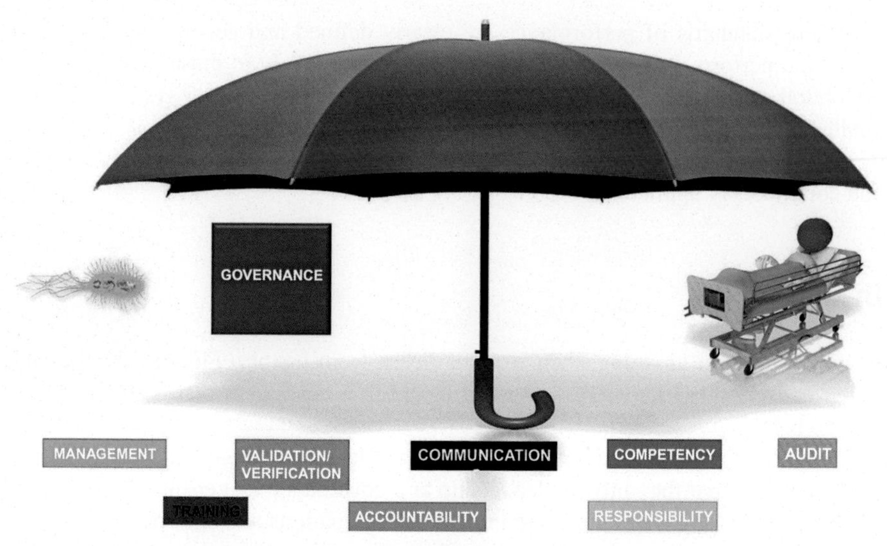

Fig. 1 Importance of communication in governance.

Ultimately, good governance is evidenced by its outcomes. It is best achieved where there is effective upward communication to Board level, so that senior managers can be assured when water safety is being controlled, and so that they can be informed when it is not and take appropriate measures to regain control. There are responsibilities at all levels, but important decisions regarding allocation of budgets and other resources are made at the top of the management hierarchy, and these decisions need to be supported by evidence-based information (Fig. 2).

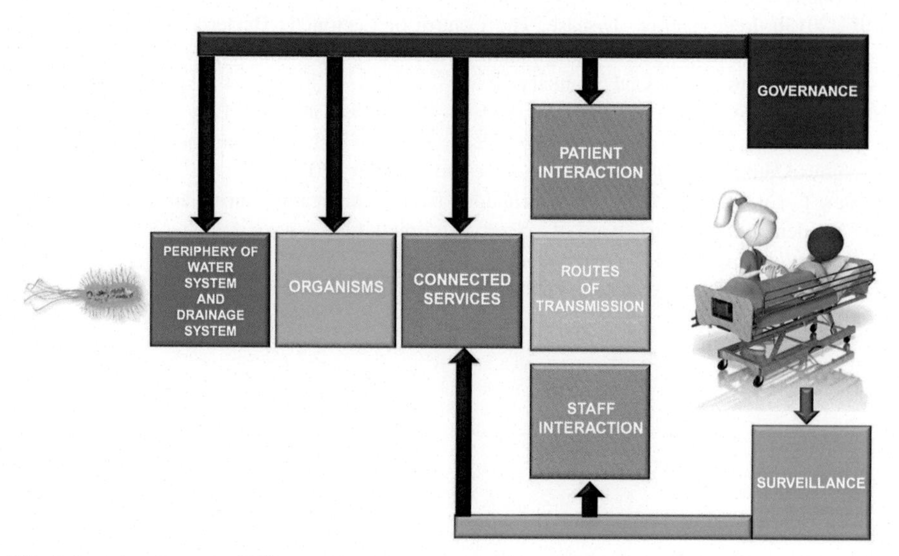

Fig. 2 Levels of responsibility required for appropriate and effective governance.

Where standards of performance are clearly defined and set by senior management, and performance levels are communicated effectively to those individuals, the likelihood of good governance and the continued operation of water systems safely in healthcare premises is increased.

However, good leadership must ensure that governance is enshrined at all levels and continuously applied.

In summary

If we are to learn anything, then we must learn from history. Florence Nightingale stated in 1859 that "The very first requirement in a hospital is that it should do the sick no harm."

Those who work in healthcare are dedicated to their patients and organizations need to ensure that they provide leadership and governance that ensures that staff are educated, supported, and empowered to act when situations and environments are not safe for patients.

References

DHSC, 2016. HTM 04-01: Safe Water in Healthcare Premises.

DWI, 2015. The Water Supply (Water Quality) Regulations 2016.

HSE, 2011. Development of Suitable Safety Performance Indicators for Level 4 Bio-containment Facilities: Phase 2.

HSE, 2013a. Legionnaires' Disease: Technical Guidance Part 1: The Control of Legionella Bacteria in Evaporative Cooling Systems.

HSE, 2013b. Legionnaires' Disease. The Control of Legionella Bacteria in Water Systems. Approved Code of Practice Legionnaires' Disease. ACOP https://www.hse.gov.uk/pubns/books/l8.htm. (Accessed 5 January 2021).

HSE, 2014. HSG 274 Legionnaires' Disease—Technical Guidance Part 2: The Control of Legionella Bacteria in Hot and Cold Water Systems Technical Guidance. http://www.hse.gov.uk/pubns/books/hsg274.htm. (Accessed 5 January 2021).

Inkster, T., Cuddihy, J., 2021. Duty of candour and communication during an infection control incident in a paediatric ward of a Scottish hospital: how can we do better? J. Med. Ethics. https://doi.org/10.1136/medethics-2020-106862.

NHS England, 2022. Governance, Patient Safety and Quality.

Design, construction, and commissioning of healthcare water systems

2

Introduction to designing safe water systems

The requirement to protect patients from any harm caused by water should be the primary aim of those designing and specifying water systems and associated equipment. National guidance in England states that "*Where new healthcare premises are planned or existing premises are to be altered or refurbished, the water safety group should be consulted at the earliest possible opportunity and water risk assessments be completed for all projects. This will enable the total water hygiene requirements to be assessed in the planning stages, and appropriate action taken, including ensuring that any pressure testing, flushing and cleaning does not lead to stagnation or contamination before being placed into service. The risk assessment should be reviewed once the system is operational.* At all stages of the design, installation and commissioning of new or extended water systems, the design team should liaise and consult with the local water safety group in a timely manner, *give consideration to HTM 04-01 Parts B and C and incorporate all operational managements requirement into their design. requirement into their design.*"

Choosing the right contractors for each stage of the project is fundamental to ensuring the project delivers safe systems. Requiring evidence that contractors have fulfilled their role successfully before and been prepared to work with the client at each stage including water safety groups. This evidence that can help make an informed choice may include reviewing examples of previous project consultations, how the commissioning was undertaken, handover process, sampling strategy, plans, and ongoing safety of the water.

At the concept stage of each capital build or major refurbishment, before the brief for the design architect is completed, the multidisciplinary WSG should carry out an assessment to identify the intended patient groups and the quality of water needed for each group. This should take account of all potential routes of exposure to all types of hazards and their associated risks to ensure all water systems in healthcare buildings are designed, constructed, installed, commissioned, operated, and maintained. The water supplied should be safe for all uses and all users from all potential water-associated hazards based on the susceptibilities of the intended population.

For the most neutropenic patients, it may be necessary to ensure they are protected from any exposure to water and rooms should then be designed without any water installations; taking account of data from the Netherlands, which has shown that removing the potential for patient exposure to water in ICUs reduced the overall levels of Gram-negative infections (Hopman et al., 2017).

Safe Water in Healthcare. https://doi.org/10.1016/B978-0-323-90492-6.00021-5

It is important to understand that once design and build contracts have gone out to tender, it is too late to make substantive changes as each contractor will be basing their costings on the information provided. Hence, the team must ensure that all those who will be involved in the project development as well as those who will be providing or managing water-related services have input before the tender stage is essential for a successful outcome.

Project water safety plan

Developing a project plan (see BS 8680) that takes account of all uses of water and the vulnerabilities and susceptibilities of intended users will help the architect/design engineer/construction/installation/commissioning teams deliver safe water (BSI, 2020, p. 86). Relevant standards and guidance need to be incorporated with the WSG agreeing derogations where these are out of date or not appropriate.

No contractor wishes to deliver a project that is going to cause harm so involving preferred/shortlisted contractors at an early stage and providing a supported learning (training) environment will help to ensure contractors understand the potential risks of causing harm. The death of patients from poorly designed, installed, and/or commissioned systems and any associated equipment has major impacts on hospitals and staff (Francis, n.d.; Inkster and Cuddihy, 2021). Potential risks from poor system design and construction could include litigation (including from corporate manslaughter or corporate killing depending on location), financial losses as well as losses of organizational and contractor reputation (HSE, 2014).

Project duty holders and water safety groups should hold workshops with all those involved in the design, engineering, and purchasing prior to the tender stage so everyone is aware of the absolute requirement to put patient safety first. There needs to be project oversight and adherence to governance and policies together with a lessons learnt exercise from previous projects and includes the time built into the project for risk assessment and review at each stage of the system development to handover and occupation (RIBA, 2020). In a large development, where it is anticipated that the building will be occupied in stages, the water safety plan should include processes for monitored staged filling and commissioning with a sampling strategy to minimize the risk of stagnation such that wholesome water is delivered.

Project governance and accountability

To avoid the many historical errors that have led to outbreaks and cases following poor design and construction of new healthcare premises, it is imperative that the water safety plan ensures there is effective governance and oversight at Trust or Board level (i.e., the client) throughout the project. This is necessary to ensure that the safety of the patients remains at the core of every decision made regarding water systems design through all stages to handover and normal operation and maintenance, for any new healthcare building, or major refurbishment. A designated Project lead needs to be appointed to take day-to-day responsibility for the safe delivery of the project on behalf of the Duty Holder supported by a project-specific water safety group. The water safety plan (BS 8680:2020) needs to be developed to ensure that the water will be delivered at each point of use so that the water and its delivery mechanism are safe (reduced splashing and aerosols) for its intended uses and users. The water safety plan also has to include the associated waste water systems, which should be appropriately sized and compartmentalized to avoid backflow and spread of waterborne pathogens (Jamal et al., 2019).

Where there is an existing water safety group, they should ensure that their existing water safety plan contains sufficient organizational governance and policies, which have been agreed at Board level and implemented to make sure that there is a chain of accountability responsibility, communication, competency, and training from the top down. It is recommended that WSGs include the processes for how new projects should be managed within the WSP, whether or not they are aware of any plans. Leaving it until a decision has been communicated could well mean that it is too late to ensure there are the right competencies and governance structures in place.

Project WSG skills and competencies

The skills needed for the project water safety groups to function effectively include personnel with the skills needed to understand architects and design engineers' drawings, specifications for materials and components, and the impact of poor design. The water safety group needs to ensure the architect/design engineers, commissioning engineers, contractors understand and carry out all the requirements relating to safe water provision taking account of the intended range of uses, quality requirements for each use, and the susceptibilities and vulnerabilities of the intended patient groups. It is essential for the completion and handover of safe water systems, and any associated equipment, that the project water safety group is empowered to make decisions. Where skills are not available in-house, then adequate checks should be made to ensure the competencies of consultants/(sub)contractors to deliver a safe project and ensure water safety (HSE, 2014).

For a large healthcare builds such as general hospitals, the project water safety group could include:

- Board representative with responsibility and accountability for project water safety.
- Infection prevention and control with appropriate knowledge of the built environment.
- Estates and facilities management personnel responsible for managing and maintaining the systems.
- Relevant specialist clinical service providers and those with water quality requirements including or ex aquatic physiotherapists, dialysis providers, decontamination, intensive care, transplant and hematology oncology units.
- Water specialist Advisor/Authorizing engineer (water) (DHSC, 2016).
- Architects and design engineers.
- Microbiologist with expertise in water microbiology.
- Contractors
- Principal contractor and principal designer (GOV.UK, 2015).

The range of competencies required includes:

1. Knowledge of regulatory, standards, and guidelines relating to water chemistry and water microbiology.
2. Experience of previous new build projects to understand the potential impact on patient health and well-being from poor water system design; specification, construction, installation, commissioning; operation, usage, and maintenance taking into account the susceptibility of the users.
3. How to develop and implement a comprehensive water safety plan to BS 8680:2020, based on risk assessments for *Legionella* as well as *Pseudomonas aeruginosa* and other waterborne pathogens as appropriate including (BSI, 2019, 2020, 2022):
   - **(a)** How to interpret proposed water system and drainage plans and schematics and their impact on patients, staff, and visitors.
   - **(b)** Knowledge of all applicable hazards, hazardous events, and risks associated with water systems and equipment within the proposed healthcare environment to be able to critically review risk assessments at each stage of the project.
   - **(c)** Knowledge of different water system methods used to control microbial risks and the relative pros and cons and contraindications.
   - **(d)** Knowledge of health and safety legislation, approved codes of practice and best practice guidance applicable to construction projects (including in the United Kingdom: The Construction (Design and Management) Regulations 2015, the Health and Safety at Work Act 1974 and associated regulations (as amended), e.g., The Management of Health and Safety at Work Regulations 1999, The Control of Substances Hazardous to Health Regulations 2002) (GOV.UK, 2015; HSE, 1974, 1999, 2020).
   - **(e)** Be able to develop and manage a team for implementation of a soft landings approach.
   - **(f)** A knowledge of all applicable legislation, guidance, and standards applicable to the project and where derogations may be required.
4. Effective communication, motivating and negotiating skills.
5. Ability to interpret architects and design engineers' drawings.
6. An understanding of the governance and accountability required to ensure the project proceeds and is delivered as required for each use and each user.
7. Ability to determine and manage key stage assurance markers.
8. Ability to audit and risk assess each stage of the project for compliance with the project key stage assurance markers.
9. Be able to manage the development and implementation of quality systems to ensure appropriate documentation and best practice is implemented through all stages of the project.

An understanding of the stages of a new build project may help the project water safety group be prepared (BSRIA, 2018; RIBA, 2020). There are a number of examples where stages are important for project water safety groups (Table 1).

Consultation

It is essential for the safety of patients, particularly those at increased risk of infection, that those responsible for designing, engineering, and advising on the quality of water required within healthcare buildings should make sure that there is adequate consultation with the building owners and all stakeholders. This will ensure that when the building is completed, water delivered at each point of use meets the required water

Table 1 Examples of key stages for consideration by the water safety group.

Stage	Key factors for consideration
Develop a brief with the client	Special water quality needs should be identified based on the intended uses and intended user groups, e.g., dialysis, decontamination and aquatic therapy as well as clinicians and ward managers of specialist units such as transplant, cystic fibrosis, oncology, and hematology
Create concept designs options	For specialist needs regulatory and best practice guidance should be considered, for example, a building with a new hydrotherapy pool should comply with relevant legislation (ATACP, 2022; GOV.UK, 2015; HSE, 2013)
Develop a set of construction information	Collate all the important information that needs to be taken into account when the tender specification is prepared. This information will be used for costing the project. The water safety group should ensure risk assessment and review at each gateway before progression to the next design and construction phase is built into the tender specification. Ensure pipework, components, and fittings are of the required quality and are wrapped and capped to prevent contamination. Ensure that pressure leak testing has been carried out and that the water is wholesome by the manufacturer with unsafe water. Any special considerations should be clearly spelt out at this point
Prepare a tender	Ensure the tender contains all the detail to deliver safe water systems as well as time for key stage risk assessments and reviews, and involves a soft landings approach (BSRIA, 2018; RIBA, 2020)
Award a building contract	Ensure there are the policies and governance in place to ensure the project will proceed with the required skills and competencies, supervision and in a supported learning environment
Construct the building	Ensure that there is ongoing supervision, assurance checks and key stage risk assessments before progression to the next stage
Inspect the construction as it progresses	Those inspecting work should have competence and to ensure the inspection is robust and would identify any matters that would affect ongoing water safety. It is essential that those inspecting have the authority and resources to intervene in a timely manner when necessary to ensure safe systems are delivered
Hand over the building	All commissioning should be witnessed and agreed as satisfactory with acceptable commissioning sampling results, all disinfection certificates, sampling plans and results, logbooks instruction manuals are handed over and the soft landings team facilitate and remain on site to trouble shoot

quality standards, i.e., wholesome; conforms to all applicable legislation, national guidance and standards to protect patients at increased risk of harm from hazards related to water use or damp environments (DHSC, 2016; HSE, 2014).

Those preparing the brief for design teams should identify all potential uses and the quality of water within the planned building and ensure risk assessments are carried out by the multidisciplinary project water safety group. This will identify where additional measures are needed to protect patients, for example, where specific water quality requirements are required for water systems and any associated equipment, such as in decontamination units, intensive care units, cystic fibrosis and transplant units, and hydrotherapy pools (Table 2). This risk assessment should be carried out before the design brief is given to the design team and include input from relevant clinical and IPC teams, ward managers, and reviewed and approved by the multidisciplinary WSG. It is important that all potential hazards are considered and not just microbiological hazards. For example, the impact on vulnerable patients of chemical contamination such as from corrosion products and water treatment chemicals, e.g., in dialysis units as well as physical risks from scalding, slips, and trips because of, for example, staff having to carry water used for personal hygiene for long distances to sluice rooms, located at long distances from patient care. For patients, such as those in mental health units, design teams need to consider where protection from physical harm is needed by installing antiligature fittings and fixtures and designing in remote access for maintenance personal.

Supply water

Water supplied into the building by a public utility at low temperatures (<20°C) and low levels of available nutrients rarely causes harm to the general public. (An exception is when people rinse their contact lens with tap water, which can lead to *Acanthamoeba* keratitis as a result of these amebae sticking to the lens and attacking the eye when reinserted).

The microbiological parameters set within drinking water regulations for water supplied to premises are based on protecting the general population from fecal contamination from humans, animals, and sewage. The absence of indicators of fecal contamination such as *Escherichia coli*, coliforms, enterococci, and *Clostridium perfringens* is used as a measure of safe water for the general population. However, it is important that those involved in the development of new healthcare buildings, or major refurbishments, which affect water provision, understand that even in the absence of faecal indicators and detectable disinfection residuals, drinking water supplies into a building are not sterile and contain communities of microorganisms of nonfecal origin, called heterotrophic bacteria, attached to surfaces throughout the distribution network and growing within biofilms. Heterotrophs are a diverse range of bacteria that can use organic nutrients for growth including all potential waterborne opportunistic pathogens such as legionellae, environmental mycobacteria, *Pseudomonas aeruginosa,* and other Gram-negatives. Once within buildings, if water systems are not well designed, engineered, and/or quality not managed effectively throughout the system, from the point of supply to all potential points of use, waterborne

Table 2 Examples of hazards, at risk groups and protections required (Bartram et al., 2003; Engelhart et al., 2001).

Patient risk group	Hazards	Risk factors	Examples of protection needed
Severe: immunosuppression protection level 3	Microbiological (Engelhart et al., 2001; WHO, 2011)	Acute or chronic leukaemia, malignant lymphoma, childhood histiocystosis Solid tumors under intensive treatment (expected duration of neutropenia <500/μL for >10 days) Solid organ transplantation under intensive treatment phase (induction or rejection therapy) Allogeneic stem cell transplantation (first 6–12 months after engraftment) AIDS patients with a count of CD4+ cells less than 200/μL and an additional factor of immunosuppression (e.g., neutropenia, corticosteroids)	Any water for human use should have a very low bacterial count • Use water filters/controlled carbonated water) • Strict control of bath installation and water for showering (showering to be avoided if no control possible)
Extreme immunosuppression protection level 4		Allogeneic stem cell transplantation (until engraftment)	Only sterile fluids for drinking, mouth care and washing allowed
Dialysis patients	Microbiological, endotoxins, toxins, chemicals including aluminium and water treatment chemicals	All patients receiving dialysis	Separate incoming drinking water supplies are recommended so that there is protection from any water treatment required for the main site Water treatment plant including RO, carbon filters
Mental health patients	Physical (ligature risk)	Those at risk of self-harm and harming others	All fixtures and fittings should be designed so that they cannot be used to affix ligatures or as offensive weapons

opportunistic pathogens can gain virulence and multiply to levels that can cause serious harm and sometimes death to those at increased susceptibility to microbiological hazards.

Patient risk factors

The WSG needs to identify all patients at high risk of waterborne infections where additional measures may need to be factored into the design, engineering, and specification of materials, components, and fittings. High-risk patient areas include those with patients with little or no immunity to infections, those likely to have with breaches in their skin's integrity allowing direct access directly into tissues and organs such as those with burns, accidental wounds, as well as those with surgical wounds, indwelling venous and urinary catheters. Patients at higher risk of waterborne respiratory infections including those from *Legionella, Pseudomonas aeruginosa,* and other Gram-negatives, for example, also are at increased risk, i.e., those with underlying medical conditions such as suppressed immune system response due to illness or treatment, diabetes, heart and existing respiratory illnesses, and those at increased risk of aspiration such as those with conditions that affect their ability to swallow, including from neurological conditions such as motor neurone disease and stroke patients, but increasing age, having an anesthetic, taking narcotics, drinking while prone can also increase the risk of aspiration pneumonia (Blatt et al., 1993; Carratala et al., 1994; Loeb et al., 1999).

Drains

The location, sizing, and design of drains should be considered as part of the water system design, adequate backflow protection should be considered to prevent backflow from drains contaminating the sinks and sluices. Predictable adverse events such as wipes, being inappropriately disposed of should be considered in the design risk assessment, and a WSG brain storming session could help to identify other factors that need to be taken into account to reduce the risk from water and sanitation provision for the design team to consider. As an example: where there are critically ill patients needing constant care, the risk of leaving the patient to dispose of water used for patient hygiene is likely to be perceived as higher than the disposing of the waste down the sinks in patients' rooms. However, disposing of inappropriate material into sinks in the patient environment has long-term consequences. There is much evidence that introducing nutrients into the sink drain increases the risk of pathogen growth and provides an environment for the transfer of antibiotic resistance between different microorganisms. The impact of this can affect not just the whole building water system but also may extend to pose a risk to the wider community as the transfer of waterborne opportunistic pathogens, including those with antibiotic resistance genes, has been shown to occur in drains outside the immediate environment and can also cause subsequent infection in patients from direct and indirect splash contamination, i.e., the patient being splashed directly from a sink or toilet contaminated by backflow by poor drainage or staff, equipment personal belongings splashed and resulting in infection from cross-contamination causing infection.

Supported learning

The development of a preselection process for interested parties, which assesses and supports them to ensure they have the knowledge and understanding of the organizational culture of patient safe engineering for all aspects of the project, could take the form of a workshop prior to tender. All those involved including the Duty holder, water safety group members, and all contractors should understand the way that water systems and associated equipment can harm patients during design, specification, construction, installation, and commissioning. This supported learning will help to ensure the project water safety plan includes the policies and processes to keep water systems safe throughout the project. This leaning could be achieved by holding workshops to include, for example:

- The importance of project governance
- How and why patients come can be harmed as a result of design and construction failures
- Why projects fail to deliver safe systems
- The cost (human, reputational, and financial)
- The need for and to facilitate culture change
- The importance of specification in ensuring patient safety (front loading the project in detail)
- Importance of the contract to deliver a safe system
- Understanding individual roles on ensuring patient safety

The development of supported learning and supported accountability for contractors and subcontractors throughout all phases of the project can be facilitated by ensuring there are experienced and competent team leaders. This could include, for example, trained healthcare plumbers with sufficient time to supervise and mentor on site.

Commissioning

The commissioning of a building water system (and associated equipment) is carried out after the main construction and installation phases. This is a process that follows a documented progression of steps to verify that, for example, the installed systems, equipment, valves, plant and control systems function as intended and are fit for purpose. Once water systems and associated equipment have been installed, the commissioning process and checks are intended to ensure that the system produces safe water for each use as intended by both the client and design team.

It can be argued that the commissioning stage is one of the most important in the whole life cycle of the system. If errors are made during the filling and commissioning stage, which allows ingress of contaminants, which can lead to colonization and growth of opportunistic pathogens, then the system can remain at risk for the remainder of its life cycle, which can be expensive to remediate.

Scanlon et al. published a systematic review that categorized the risks in community and healthcare construction projects (Scanlon et al., 2020). They identified 31 studies (7 community, 24 in healthcare) associated with 894 disease cases of Legionnaires' disease including 112 deaths. Of the construction activity risks that were identified, commissioning was the second most common in 30% of publications.

Commissioning plans

It is therefore important that commissioning is effectively managed. A bespoke documented commissioning plan should be prepared at the design stage for each system, which includes the written procedures to be followed and agreed by all involved. It is recommended that the plan includes a precommissioning risk assessment using up-to-date and accurate as-fitted drawings and identifies what documentation and checklists with detailed governance, responsibilities, and accountabilities for each stage should be available for the commissioning process. These should be reviewed and agreed by the project water safety group before the commissioning process begins. The commissioning plan should also identify the skills and competencies of the personnel required, how and when the system should be filled to ensure it is not contaminated during the commissioning process together with approved testing plans and method statements for each stage of the commissioning process and may include pressure testing with gas or where water is used, for filling, flushing, and disinfection to ensure the water is safe.

Why is commissioning such a high-risk event? During commissioning, water will be used to test the system and the natural microbial flora will proliferate and form biofilms within the water system. If the hot water system has not been balanced, there will be zones where water temperatures will be conducive to microbial growth. Similar situation will occur on the cold water system where the pipes may gain heat. Flushing regimen should be in place that replicates the occupation of the building. If flushing is to be carried out correctly, it should be based on science, which requires sufficient movement of water to maintain a temperature control regimen and to prevent stagnation. Where contamination and biofilm formation have occurred during the commission phase, the use of biocidal shock dosing may not only result in safe water being delivered for a short period of time.

System filling

Prior to system filling, all pipework should be flushed with potable water to remove metal working fluids, potential contaminants, and debris that have contaminated the pipes during the installation and construction stages. Systems should then be filled with potable water as late in the project as possible to minimize the length time that a system is filled with stagnant water prior to handover. The longer the system is filled with water and not used, the more likely that it will become colonized with opportunistic pathogens (Leslie et al., 2021). For water systems, this means leak/pressure testing the system, preferably using an inert gas rather than water to minimize the time stagnant water is held within the system before the installation of fittings (DHSC, 2016). Once the system is appropriately labeled and filled, the monitoring, disinfection, flushing, and sampling procedures, as detailed in the commissioning plan and agreed by the water safety group, should be followed. It is important to ensure that any equipment brought on site by the commissioning team has been decontaminated beforehand so that it does not pose a risk of contaminating the system. For example, the equipment may have been used previously for different systems and left with stagnant microbially contaminated water inside. As a consequence the water safety plan should include processes to ensure contractors' equipment is safe and does not pose a risk to water systems and associated equipment, fittings, and components.

The commissioning plan should also detail the requirements for supervision and client witnessing. This should include processes and communication channels to be followed including where there are predictable untoward events. For example,

- where a microbial monitoring plan has identified adverse sampling results,
- unexpected delay between filling and disinfection and/or handover, for example,
- equipment failures, leaks, and water damage.

For the commissioning of equipment, which contains, uses, or stores water, the UK Health and Safety Executive emphasizes that it is essential the commissioning process is carried out by competent people in a logical defined manner and full compliance with suppliers and installers' instructions following documented operating procedures (HSE, 2015).

Case study: Failures in filling the water system

The project water safety group had been proactive and approached the new build project armed with learning from other projects. Early on a commissioning plan had been drawn up and agreed by all parties including the contractor. The commissioning plan was detailed delineating responsibilities, timelines, and agreed criteria for moving through the individual key stages.

The agreed date for pressurizing the water systems with water, provided all the parameters necessary to proceed had been agreed, happened to be a Monday. When the parties met on the Monday to give the final go-ahead, it turned out that the contractor had started filling the water systems on the preceding Friday because the project was running behind schedule. Additionally, to speed up the process, water from a fire hydrant (likely to be heavily contaminated) was used and the point of entry filtration system was bypassed.

This flagrant breach of the commissioning plan raises a number of issues. The obvious one is the risk to the water system. Of more concern is the view that the contractors must have been sufficiently confident that no punitive action would be taken. It requires commissioning to be linked to penalty clauses in the contract if healthcare facilities are going to have a chance to prevent such actions happening again.

Case study: Failures due to partial occupation of a hospital build program

Commissioning may be further complicated by partial occupation of a new building, as happened in Coventry. In 2005, the West Wing of the new hospital was opened but only partially occupied. The remainder of the building was due to become occupied more than a year later. The flushing regimen put in place for the unoccupied areas was insufficient to maintain water temperatures. Consequently, a *Legionella* spp. were detected in the water system. There had been a Legionella risk assessment, which had highlighted the issues of lack of temperature control but did not appear to have been followed. From the Trust perspective, the unoccupied area as it had not been handed over was seen as a responsibility of the construction company. However, as the water system supplying the occupied and unoccupied areas was the same, the risks incurred in the unoccupied areas could and indeed did affect patient occupied areas. Thus, the water safety group needs to review the whole of the water system not just the occupied areas. Failure to utilize a Legionella risk assessment was also a factor in the contamination at the new hospital in Glasgow.

Handover

The project handover at the end of the project should only occur when all water systems and associated equipment within the project are delivering safe water and preforming as designed for normal operation and maintenance. As with the commissioning plan, there should be a documented plan for handover, which identifies what is required by the client for formal acceptance and handover in terms of the system performance, compliance with set targets for all regulatory and agreed control parameters, and the information and documentation required. For example, the client might include with their specification, evidence that water system that is safe for all users and uses is required and set out the criteria for how that has been determined.

Partial handover

Where there is a planned staged handover and occupation of a healthcare building, how the systems will be filled with water and managed to avoid stagnation in the unoccupied part of the building should be factored in at the design stage. The commissioning and handover plans should also reflect the additional measures to ensure both handed over and occupied and parts still to be competed are managed to ensure systems remain safe. The contractors should supply all relevant documentation that the water in the portion handed over is safe and what assurance will be supplied to ensure the systems not yet handed over will not compromise water safety.

It is important that once the water system is handed over that, there is appropriate governance and oversight of the systems and the water safety plan includes supporting documentation such as asset registers, risk assessments, maintenance schedules, training, and competency checks.

Examples of checks for commissioning and handover within the project WSP

- Are the roles and responsibilities of individuals involved in commissioning operations clearly defined?
- Have all those involved in the commissioning and ongoing management of the systems and equipment (control measures, maintenance, and monitoring schedules) had appropriate competence checks and training?
- Have all pipework and connections been checked to confirm the system and associated materials, joints, equipment, fittings, components have been fitted as designed?
- Has the commissioning plan been agreed by the water safety group as fit for purpose? This would include the sampling and monitoring plans, identification of outlets to be sampled,

sampling and monitoring targets, and the timescales for carrying these out, e.g., does sampling after disinfection allow time for recovery of sublethally damaged microorganisms?
- All systems and equipment to be commissioned are clean with no residual water and are fit for purpose before leak tests are performed?
- Is there an agreement about when systems and associated equipment will be filled and who will be present to witness key stages on behalf of the client?
- Is there a documented procedure for flushing, filling, disinfection, and biocide handling as well as neutralizing the chemicals as appropriate for all systems and related equipment?
- Has the water safety group assessed the disinfection protocol to ensure guarantees of fittings, equipment, and components are not compromised (some fittings and materials may be adversely affected by some biocides at the levels needed for shock disinfection) and that there is adequate dosing, residence time, and neutralization of products as specified by the manufacturer?
- Does the water safety group have the skill and training to understand the appropriateness of the disinfection procedures and what evidence they should be looking for?
- Is access for commissioning, future monitoring, and maintenance, safe and appropriate?
- Are there agreed documented procedures, which include COSHH, manual handling, permit to work, and PPE requirements for each stage of the commissioning process?
- Has all equipment involved in system measurement and control as well as performance management (including building management systems, artificial intelligence systems, alarms, dosing equipment, temperature and flow measurement) been calibrated and performance checked?
- Has the process for recording all relevant commissioning data and the handover process been agreed?
- Is there a checklist agreed by the project water safety group of the information and documentation, including health and safety information, required and the format this should be supplied in (electronic or paper) for commissioning and handover?
  - Design and engineering plans and specifications;
  - As-fitted drawings to reflect changes during the commissioning, handover, and problem-solving phases;
  - Plant and equipment operation and maintenance manuals and log books;
  - Competence checks and training;
  - Documented risk assessments for all relevant hazards and schemes of control;
  - Commissioning test and sampling and monitoring results;
  - Records of checks, inspections, supervision, and witnessing.
- Is there an agreed documented procedure for addressing failures identified during the commissioning procedure, e.g., adverse chemical/microbiological results with agreed timescales?
- Have all those involved in the ongoing maintenance and operation of the system been identified and trained to safely operate and maintain all relevant systems and equipment?
- Have all instruction and maintenance manuals been identified for handover with details of the recommended maintenance regimes and frequency for checks and monitoring and spare parts?
- What guarantees, log-books for systems and equipment are required at handover?
- Who is responsible for completing and verifying the asset register?

References

ATACP, 2022. Aquatic Therapy Association of Chartered Physiotherapists. https://atacp.csp.org.uk/.

Bartram, J., Cotruvo, J.A., Exner, M., Fricker, C., Glasmacher, A., 2003. Heterotrophic Plate Counts and Drinking-Water Safety: The Significance of HPCs for Water Quality and Human Health. World Health Organization. https://apps.who.int/iris/handle/10665/42612.

Blatt, S.P., Parkinson, M.D., Pace, E., Hoffman, P., Dolan, D., Lauderdale, P., Zajac, R.A., Melcher, G.P., 1993. Nosocomial legionnaires' disease: aspiration as a primary mode of disease acquisition. Am. J. Med. 95, 16–22. https://doi.org/10.1016/0002-9343(93)90227-g.

BSI, 2019. BS 8580-1 Water Quality—Risk Assessments for Legionella Control. Code of Practice [WWW Document]. URL: https://knowledge.bsigroup.com/products/water-quality-risk-assessments-for-legionella-control-code-of-practice-1/tracked-changes. (Accessed 9 August 2022).

BSI, 2020. BS 8680—Water Quality. Water Safety Plans. Code of Practice. https://shop.bsigroup.com/ProductDetail?pid=000000000030364472. (Accessed 5 January 2021).

BSI, 2022. BS 8580-2:2022—Risk Assessments for *Pseudomonas aeruginosa* and Other Waterborne Pathogens. Code of Practice https://standardsdevelopment.bsigroup.com.

BSRIA, 2018. Soft Landings Framework. [WWW Document]. URL: https://www.bsria.com/uk/product/QnPd6n/soft_landings_framework_2018_bg_542018_a15d25e1/. (Accessed 9 August 2022).

Carratala, J., Gudiol, F., Pallares, R., Dorca, J., Verdaguer, R., Ariza, J., Manresa, F., 1994. Risk factors for nosocomial *Legionella pneumophila*. Am. J. Respir. Crit. Care Med. 149, 625–629. https://doi.org/10.1164/ajrccm.149.3.8118629.

DHSC, 2016. HTM 04-01: Safe Water in Healthcare Premises.

Engelhart, S., Glasmacher, A., Kaufmann, F., Exner, M., 2001. Protecting vulnerable groups in the home: the interface between institutions and the domestic setting. J. Infect. 43, 57–59. discussion 59-60 https://doi.org/10.1053/jinf.2001.0851.

Francis, R., n.d. Report of the Mid Staffordshire NHS Foundation Trust Public Inquiry GOV.UK. [WWW Document]. URL: https://www.gov.uk/government/publications/report-of-the-mid-staffordshire-nhs-foundation-trust-public-inquiry (Accessed 12 February 2022).

GOV.UK, 2015. The Construction (Design and Management) Regulations. [WWW Document]. URL: https://www.legislation.gov.uk/uksi/2015/51/contents/made. (Accessed 9 August 2022).

Hopman, J., Tostmann, A., Wertheim, H., Bos, M., Kolwijck, E., Akkermans, R., Sturm, P., Voss, A., Pickkers, P., vd Hoeven, H., 2017. Reduced rate of intensive care unit acquired gram-negative bacilli after removal of sinks and introduction of 'water-free' patient care. Antimicrob. Resist. Infect. Control 6, 59. https://doi.org/10.1186/s13756-017-0213-0.

HSE, 1974. Health and Safety at Work etc Act 1974—Legislation Explained.

HSE, 1999. The Management of Health and Safety at Work Regulations 1999. [WWW Document]. URL: https://www.legislation.gov.uk/uksi/1999/3242/contents/made. (Accessed 9 August 2022).

HSE, 2013. Legionnaires' Disease. The Control of Legionella Bacteria in Water Systems. Approved Code of Practice Legionnaires' Disease. ACOP. https://www.hse.gov.uk/pubns/books/l8.htm. (Accessed 5 January 2021).

HSE, 2014. HSG 274 Legionnaires' Disease—Technical Guidance Part 2: The Control of Legionella Bacteria in Hot and Cold Water Systems Technical Guidance. http://www.hse.gov.uk/pubns/books/hsg274.htm. (Accessed 5 January 2021).

HSE, 2015. Operating Procedures. [WWW Document]. URL: https://www.hse.gov.uk/comah/sragtech/techmeasoperatio.htm. (Accessed 9 August 2022).

HSE, 2020. Control of Substances Hazardous to Health (COSHH). [WWW Document]. URL: https://www.hse.gov.uk/coshh/. (Accessed 1 June 2021).

Inkster, T., Cuddihy, J., 2021. Duty of candour and communication during an infection control incident in a paediatric ward of a Scottish hospital: how can we do better? J. Med. Ethics. https://doi.org/10.1136/medethics-2020-106862.

Jamal, A., Brown, K.A., Katz, K., Johnstone, J., Muller, M.P., Allen, V., Borgia, S., Boyd, D.A., Ciccotelli, W., Delibasic, K., Fisman, D., Leis, J., Li, A., Mataseje, L., Mehta, M., Mulvey, M., Ng, W., Pantelidis, R., Paterson, A., McGeer, A., 2019. Risk factors for contamination with carbapenemase-producing enterobacteriales (CPE) in exposed hospital drains in Ontario, Canada. Open Forum Infect. Dis. 6, S441. https://doi.org/10.1093/ofid/ofz360.1091.

Leslie, E., Hinds, J., Hai, F.I., 2021. Causes, factors, and control measures of opportunistic premise plumbing pathogens—a critical review. Appl. Sci. 11, 4474. https://doi.org/10.3390/app11104474.

Loeb, M., Simor, A.E., Mandell, L., Krueger, P., McArthur, M., James, M., Walter, S., Richardson, E., Lingley, M., Stout, J., Stronach, D., McGeer, A., 1999. Two nursing home outbreaks of respiratory infection with *Legionella sainthelensi*. J. Am. Geriatr. Soc. 47, 547–552. https://doi.org/10.1111/j.1532-5415.1999.tb02568.x.

RIBA, 2020. The Royal Institute of British Architects (RIBA) Plan of Work. [WWW Document]. URL: https://www.architecture.com/knowledge-and-resources/resources-landing-page/riba-plan-of-work. (Accessed 9 August 2022).

Scanlon, M.M., Gordon, J.L., McCoy, W.F., Cain, M.F., 2020. Water management for construction: evidence for risk characterization in community and healthcare settings: a systematic review. Int. J. Environ. Res. Public Health 17, E2168. https://doi.org/10.3390/ijerph17062168.

WHO (Ed.), 2011. Water Safety in Buildings.

Overview and introduction to safe water in healthcare

Background

How do bacteria get into buildings' water systems?

The story begins in the water treatment plant where microorganisms exist in abundance and are used purposely in the water treatment process. There are a wide range of organisms that naturally occur in the aquatic environment, that thrive once they are located within a building. Water provides a transport system for the bacteria, and buildings provide warmth, nutrients, surfaces, other microorganisms, and stagnant conditions in which the biofilms will form. These organisms either possess or gain a number of characteristics including relative tolerance to disinfectants, ability to grow within amebae, biofilm formation, and survival in low numbers. Some researchers have termed these microorganisms as "opportunistic premise plumbing pathogens" (OPPPS), which is a term that describes them very well (Falkinham et al., 2015).

The first hurdle for bacteria is the water treatment plant where chemical disinfectants are used to control the low numbers of microorganisms. Once discharged from the treatment plant, small numbers of microorganisms enter large bore water distribution pipes. Water temperatures tend to be below 20°C, and the presence of the residual chemicals controls microbial growth. Additionally, the high rates of flow, significant shearing forces, and relatively low pipe surface area to water volume make this arena an inhospitable place for bacterial growth. The further the water travels from the treatment plant, then the lower the concentration of chemicals and the more likely the microorganisms will proliferate.

Once water enters a building, the environmental parameters change markedly and, in the United Kingdom, the responsibility for water safety moves from the water supplier to the landlord/owner of the building. At this point, there is a shift from the inhospitable environment of large bore water supply pipes to smaller bore pipes inside the building, where conditions are conducive to bacterial growth and the establishment of biofilms. It is within the building water system that any residual disinfectant concentrations will degrade and given appropriate conditions, small numbers of bacteria will use the water system as a breeding ground. This will result in the dispersal of large numbers of organisms including pathogens that could result in patient infections. Water temperatures (>20°C and <50°C), sluggish flow or stagnation of water, high pipe surface area to water volume ratio, and materials more likely to support biofilm formation are some of the many factors that need to be stringently controlled to prevent the water system presenting a risk to vulnerable patients (Collier et al., 2021; Gómez-Gómez et al., 2020; Hayward et al., 2020; Li et al., 2017).

Safe Water in Healthcare. https://doi.org/10.1016/B978-0-323-90492-6.00010-0

There may be a general perception by healthcare staff using handwashing, showers, drinking fountains, and ice machines that water is safe for all purposes and for all users (Kanwar et al., 2017; Marshall et al., 2020; Nakamura et al., 2019; Schreiber et al., 2021). In addition, healthcare staff may consider that the microbial safety of that water is someone else's responsibility. Therefore, all staff working in healthcare premises need to understand the importance of water as a vehicle for the transmission of infections, how and why waterborne infections occur, and the importance of controlling microbial growth and transmission from water systems.

Due to the inherent microbial risks from potable water, a range of guidance and standards have been produced in the United Kingdom to prevent waterborne infections (BSI, 2000, 2015, 2020; DHSC Safe water in healthcare premises (HTM 04-01), 2017; HSE, 2014). However, the content of guidance and standards documents can often be difficult to interpret and implement even for a specialist in this area.

Legionella and Legionnaires' disease remain a high priority for those managing healthcare building water systems (HSE, 1974). Once a water system has become colonized with legionellae and other waterborne pathogens are major control challenges if the water system is not managed appropriately (Fig. 1) (Anaissie et al., 2002).

Water supply to healthcare buildings

Water systems in healthcare buildings are inherently large and complex with multiple outlets and many additional specialist systems (e.g., including endoscopy, renal, dental) and uses for patient treatment and diagnosis. In healthcare buildings, there are a high number of patients who are susceptible to infections. Therefore, ensuring the quality of the water is safe for all purposes and all users must be taken into account when designing water systems within such buildings.

Drinking water quality problems in healthcare buildings tend to reflect the deterioration of water quality within buildings due to the design, management, and condition of the building water systems, including the associated components and fittings, rather than the water supplied through the public supply. However, human behavior also has to be taken into consideration.

Mains water supply

Drinking water supplied to a building must be wholesome at the time of supply (UK Government, 1991). Wholesome water is defined as being fit to use for drinking, cooking, food preparation, or washing without any potential danger to human health (DWI, 2015). There is also the European Drinking Water Directive on the quality of water intended for human consumption (EU, 2021).

To comply with relevant national regulations, water companies collect and test water samples to ensure that the water released from the water treatment plant and supplied to buildings is wholesome, and the results of these are available on request.

Additionally, water companies must adhere to stringent hygiene procedures to ensure that their employees or contractors do not work in restricted water supply areas if they are suffering from an infectious disease.

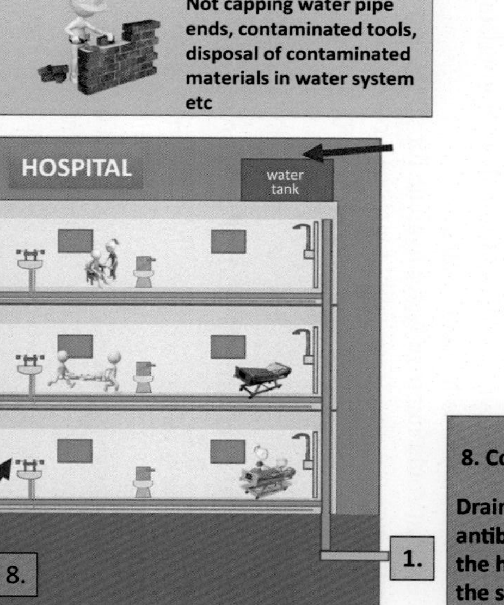

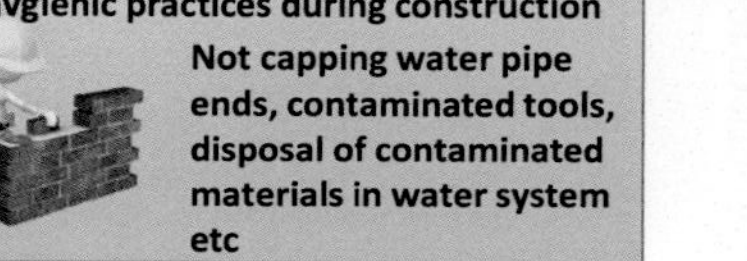

Fig. 1 Sources of microbial contamination in a hospital water system.

During distribution, the risk of contamination is increased where drinking water supply pipes and wastewater pipework are in proximity. While ingress of sewage/fecal contamination is rare in public supplies, contamination events have occurred when there have been repairs or modifications, e.g., extensions to existing infrastructure or pipe breakages resulting in cross-contamination from wastewater systems.

There is a national water hygiene scheme, known as the "Blue Card Scheme," which was developed in collaboration with all UK water companies. This scheme has a vital role in ensuring good water hygiene practices and that all operators working on the drinking water network do so with an appropriate level of knowledge and awareness with regard to hygiene issues with all individuals working on the water system.

A number of hospitals have adopted the "Blue Card Scheme" into local policies to ensure that all engineering staff working on water within the Trust, whether directly or indirectly employed, have undergone a certified training covering water hygiene practices. This includes participating in water hygiene training courses to emphasize individual responsibility with regard to the microbiological safety of the potable water supply and verifies that the employee has demonstrated an appropriate level of knowledge and awareness with regard to hygiene issues.

This is a concept that could improve water hygiene for all those working on water systems within healthcare premises especially for those working during construction, installation, and commissioning of both new and existing systems and reduce risks to vulnerable users.

Case study: Microbial contamination of water supply

The Flint water disaster was a public health crisis that started in 2014 and lasted until 2019, after the drinking water for the city of Flint, Michigan, was contaminated with lead and possibly Legionella bacteria (Pauli, 2020). The city of Flint, Michigan, switched its drinking water supply from Lake Huron to Flint River. Failure to treat the water properly at the Flint Water Treatment Plant led to a variety of problems with water quality and public health. Low concentration of disinfectants in the water system led to bacterial growth and contributed to an outbreak of Legionnaires' disease as well as high concentrations of trihalomethanes and lead.

Private water supplies

There may also be a private water supply from springs, wells, or boreholes, which are usually on or in proximity to the hospital site. As the private water supply is not supplied by a water supplier, the responsibility for maintaining quality rests with the relevant person within the organization and not a statutorily appointed water company (UK Government, 2018). In the United Kingdom, under the private water supplies regulations, there is a duty for local authorities to regulate and carry out a risk assessment and to undertake monitoring to determine compliance with drinking water standards.

The private water supply regulations also apply to public supplies, which are further distributed as a private distribution network where water is distributed to other buildings, for example, where a hospital leases a building to an outside third party.

Private water supplies used for drinking water must be treated at the point of building entry to ensure they comply with regulations before being introduced into the healthcare facility's water storage system. This water must meet the statutory limits for prescribed parameters (UK Government, 2018).

Where private water supplies are used in addition to the public supply, there is a legal requirement to ensure that the public supply is protected by ensuring there is effective backflow protection. Water companies ensure that the public supplies are protected by backflow devices to prevent backflow contamination from water used in buildings entering into the supply water.

Note: Backflow is where there is unwanted flow of water in the reverse direction and can be the source of serious health risks if the mains water supply is contaminated with contaminated water. This may happen if the mains pressure becomes lower than the pressure in the Private Water Supply pipework due to a supply failure such as a burst main or drought. Backflow devices must comply with the Water Fittings Regulations (WRAS, 1999). The regulatory body for the Private Water Supply will check the backflow measures when carrying out their risk assessments.

Monitoring water supplies

Regular or routine sampling mains water, private well, or borehole for pathogens is not recommended as the current culture methods are inherently insensitive due to the small volumes that are inherently tested. At cold water temperatures below 20°C, bacterial numbers will be low and the microorganisms may be in a viable but nonculturable state (D'Alessandro et al., 2015). If sampling for investigative purposes, follow local public health guidance and seek advice from a microbiologist.

Maintaining water supply in an emergency

UK public water suppliers have a duty to have up-to-date plans for the provision of essential water supply at all times (DEFRA, 1998). This is a statutory requirement giving priority to the domestic needs of the sick, older people, disabled people, hospitals, schools, and other vulnerable sectors of the population (UK Government, 1991). As such, NHS-funded providers should establish a formal working relationship with their water supplier, which will include registering as a priority user and have contingency plans in place in the event of a failure in water supply. (DHSC, 2014; NHS England, 2021) This should include a memorandum of understanding (MOU) for the water provider to inform the healthcare facility when there are planned maintenance works, which may affect the supply and maintenance of essential supplies of water in the event of major incidents.

The water supplier and the hospital should work together to establish, as a minimum:

- potential mains water supply risks and how these risks can be mitigated such as when the water provider may be undertaking major works;
- provision of water supply in the event of mains water failure (e.g., provision of water tankers/bowsers, bulk bottled water);
- ensure alternative supplies meet the regulatory requirements;

- ensure there is a documented process within the Water Supply Provider for what supervision is needed and where water supply is compromised;
- provision should be made for patients who are severely immunocompromised patients where bottled water is not of sufficient quality and can pose a risk of waterborne infections (Eckmanns et al., 2008);
- drinking water regulations apply to drinking water in whatever form it is delivered whether, supplied by tanker bowser, etc.

Bottled water

Bottled drinking water is categorized as a food and regulated separately in the devolved nations (UK Government, 2007). Care must be taken if considering bottled water as a substitute for mains water as the water quality may not be suitable for some patients, e.g., those who are severely immunocompromised (Caskey et al., 2018). There have been outbreaks of *P. aeruginosa* associated with bottled water use in ICU patients (Eckmanns et al., 2008). Salt concentrations may also be unsuitable for some patients including dialysis patients and for use in making up infant feeds. There is a need to risk assess the suitability of alternative supplies in association with the relevant service provider specialist technicians and consultants.

Water distribution within healthcare buildings

In the following chapters, we will establish that hospital water systems are very large and complex and as a consequence are prone to microbial colonization and growth. Minimizing the risk of microbial growth within building water systems begins at the design and specification stage. It is vital that policies are in place and that those involved in the water safety group are involved when hospitals are being designed and constructed. This will include new capital builds as well as extensions and refurbishments to reduce potential of transmission of waterborne pathogens to patients, visitors, and staff. Fundamentally, a water system has to be designed, specified, built, and commissioned with microbial contamination at the forefront of everybody's minds. Maintaining appropriate water temperatures has been the conventional mode of control. However, with increasing global temperatures, it is becomingly increasingly difficult to maintain the cold water supply temperatures below 20°C year round. Mitigation for warmer weather must be taken into account when designing systems, including resilience for further increases in incoming temperature rises due to climate change. The need for point of entry treatment (e.g., including point of entry filtration, chilling, UV, and or biocide treatment) should to be considered at the design stage and before distribution rather than having treatment systems bolted on subsequently to the water system being operated(WHO, 2011, 2017).

Infection Prevention and Control Practitioners need to have a greater understanding of the role of the built environment and facilities engineers will need to have a greater understanding of water hygiene and microbiology. There will be more competing priorities for hospital designers and managers to reduce the carbon footprint (i.e., energy costs) of the healthcare premises as well as minimizing the risk of waterborne infection. These priorities are not always complementary.

It is vital that all those who are involved in the design and build have an understanding of the basic principles of microbiologically safe water. Some of these principles will be discussed in the following sections and will include keeping the water moving throughout all parts of the system, maintaining optimal cold water and hot water system temperatures as well as using appropriate and approved materials to minimize microbial colonization and growth. Record keeping is also high on the priority list, e.g., if an action has not been completed, there is no evidence that it has been completed. Once a water system is out of control from a microbiological perspective, it is not possible to completely eradicate microbial biofilms and the system remains a continuous risk unless isolated and replaced.

As soon as the mains water enters the building, the safety of that water is the Duty Holders (previously described as the owner) responsibility (HSE, 2014). From the point of entry in most buildings, there is likely to be cold water storage, usually providing at least 12h (no more than 24h) of supply in case of a break in supply. The water safety plan must include the process for risk assessing the storage requirements to ensure there is sufficient volume for patient needs without over storage, which would result in stagnation and temperature gain. Some areas may be more prone to water supply interruptions and the cold water storage may need to be increased with measures put in place to ensure this does not stagnate and increase in temperature.

Cold water systems are described in greater detail in later chapters. Briefly, cold water is distributed around the building and supplied to the hot water heaters (usually calorifiers or plate heat exchangers) and then supplied usually via hot water storage tanks (these may also be referred to as cisterns) including for auxiliary services throughout the hospital (Fig. 2). The cold water system is not recirculated, and ideally heat gain through the system should results in no more than a 2°C heat gain. To reduce microbial growth in the cold water at the outlets, these should be flushed regularly.

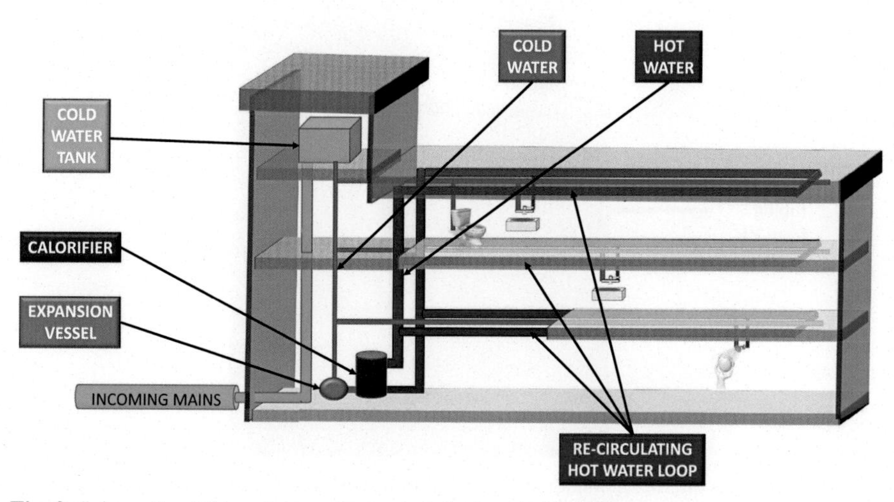

Fig. 2 Schematic of the water supply through the hospital building.

Introduction to biofilms

Biofilm is the naturally occurring residence of bacteria in healthcare buildings and has been developed and honed over billions of years (Costerton, 1995; Flemming and Wingender, 2010; Weinbren and Inkster, 2021). For biofilm to develop, the initial critical step is for bacteria to adhere to a surface within the water system. Without adherence, the organisms would pass through the water system and drains and would present little risk users. The ability of bacteria to adhere to a surface varies between different materials including plastics, copper, steel, and rubbers (Papciak et al., 2019; Rogers et al., 1994; van der Kooij et al., 2020; Waines et al., 2011).

Water systems are particularly at risk of encouraging microbial growth during the construction of new builds and refurbishments. Ingress of dirt (acting as nutrients and a source of bacteria) during manufacture, transport, storage, and fitting. During construction, there will be prolonged stasis of the water system as each floor is built, filled with water to check for leaks (to sign off that stage of the build) then left to stagnate until new sections are built and added on. (DHSC Safe water in healthcare premises (HTM 04-01), 2017).

Therefore, given the appropriate nutrients, even bacteria at undetectable numbers will begin to proliferate and attached cells will form biofilms through the water system and at peripheral outlets and in the drains.

Plumbing

Nonmetallic materials, plumbing materials are essential when building water systems. However, due to the presence of plasticizers, hardeners, and other products, plumbing components will support microbial colonization and growth of biofilms (Figs. 3 and 4). As a consequence, water fittings and components should comply with

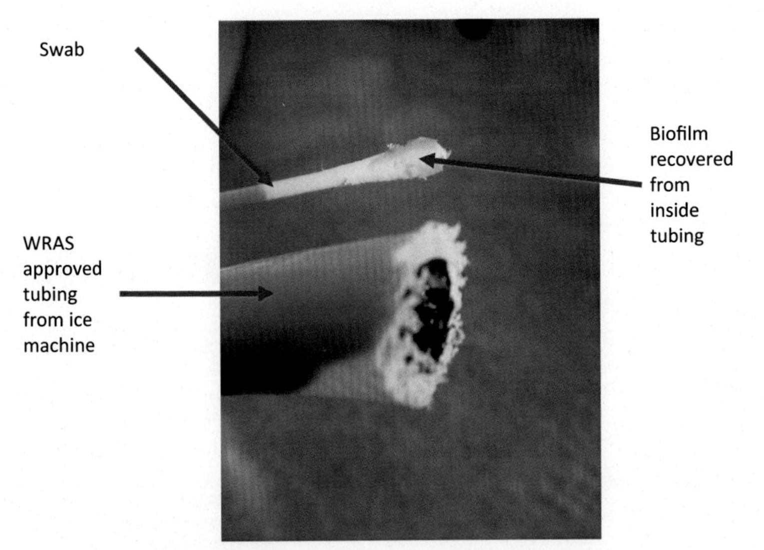

Fig. 3 Evidence of microbial contamination and biofilm growth on WRAS-approved tubing from an ice machine implicated in an outbreak.
Photo courtesy of Peter Hoffman.

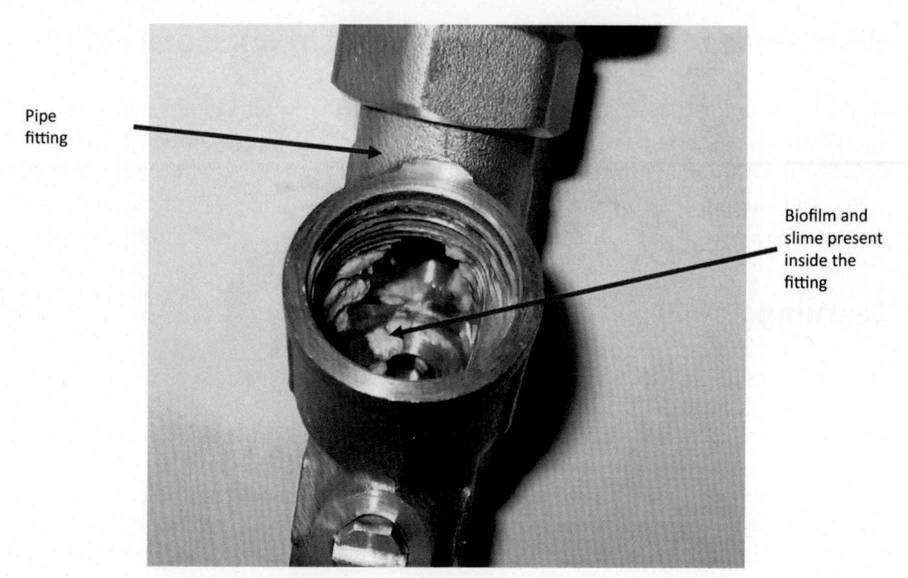

Fig. 4 Presence of biofilm on metallic valves.
Image courtesy of Daniel Pritchard.

the Water Regulation Advisory Scheme WRAS) by WRAS (1999), and these should be specified for the build. However, despite being approved, a number of WRAS-approved materials will still succumb to microbial growth after installation. This reinforces the requirements of a planned preventative maintenance and servicing regimen to ensure that microbial contamination is monitored.

In the long term, biofilms will form on surfaces in healthcare water systems. The microorganisms will sequester nutrients, multiply, and sloughing will occur to contaminate other parts of the system. Nutrients may come from a variety of sources including dirty cistern tanks, uncapped pipes, chemicals leaching out of pipework, soaps, and debris in hand wash stations (DHSC Safe water in healthcare premises (HTM 04-01), 2017).

Case study: Microbial contamination of flexible hoses

Flexible hoses have long been known to present a risk for the growth of waterborne pathogens such as *Legionella* spp. (Chand et al., 2017; Charron et al., 2015; Walker and Moore, 2015; Yui et al., 2021) In this case study, several patients with infections in a liver unit (containing critically ill patients and those undergoing organ transplants) were linked back to the presence of these waterborne pathogens being present within the ice machine (Fig. 3). Nontuberculous Mycobacteria (NTM) were recovered from the flexible tubing suppling the mains water feed and were labeled as food-grade plastic. The ice machine was located in the staff tearoom, and the staff used the ice for patient care.

What areas of the water system present microbial risks?

Basically, if not managed appropriately, all parts of the water system and associated equipment can present a risk to vulnerable users. The following sections and chapters describe the different parameters of the water system from the supply, cold water, hot water, and auxiliary equipment to the distal outlets and will identify the inherent risks that or where design or maintenance issues increase those risks of microbial growth.

Learning points

- In the United Kingdom, water supplied to the healthcare building must be wholesome, i.e., safe for vulnerable users.
- There may be occasional failures in the water quality supplied as a result of accidental ingress of sewage or animal and/human feces or a failure in water treatment.
- Climate change continues to impact the temperature of incoming water beyond 20°C; therefore, those responsible for water quality in healthcare buildings may have to address microbial control.
- Microbiological issues are far more than likely to be found within the building water system rather than in the water supplied from the water company.
- Most parts of an in-premise water system will support biofilm growth and the multiplication of waterborne pathogens.
- Users need to understand how to identify inherent risks in water systems and to think, is this water or this equipment safe for my patients?

References

Anaissie, E.J., Stratton, S.L., Dignani, M.C., Summerbell, R.C., Rex, J.H., Monson, T.P., et al., 2002. Pathogenic Aspergillus species recovered from a hospital water system: a 3-year prospective study. Clin. Infect. Dis. 34, 780–789. https://doi.org/10.1086/338958.

BSI, 2000. BS EN 805:2000—Water Supply. Requirements for Systems and Components Outside Buildings.

BSI, 2015. BS 8558:2015 Guide to the Design, Installation, Testing and Maintenance of Services Supplying Water for Domestic Use Within Buildings and Their Curtilages. Complementary Guidance to BS EN 806.

BSI, 2020. BS 8680—Water Quality. Water Safety Plans. Code of practice. https://shop.bsigroup.com/ProductDetail?pid=000000000030364472. (Accessed 5 January 2021).

Caskey, S., Stirling, J., Moore, J.E., Rendall, J.C., 2018. Occurrence of *Pseudomonas aeruginosa* in waters: implications for patients with cystic fibrosis (CF). Lett. Appl. Microbiol. 66, 537–541. https://doi.org/10.1111/lam.12876.

Chand, M., Lamagni, T., Kranzer, K., Hedge, J., Moore, G., Parks, S., Collins, S., del Ojo Elias, C., Ahmed, N., Brown, T., Smith, E.G., Hoffman, P., Kirwan, P., Mason, B., Smith-Palmer, A., Veal, P., Lalor, M.K., Bennett, A., Walker, J., Yeap, A., Isidro Carrion Martin, A., Dolan, G., Bhatt, S., Skingsley, A., Charlett, A., Pearce, D., Russell, K., Kendall, S., Klein, A.A., Robins, S., Schelenz, S., Newsholme, W., Thomas, S., Collyns, T., Davies, E., McMenamin, J., Doherty, L., Peto, T.E.A., Crook, D., Zambon, M., Phin, N., 2017. Insidious risk of severe *Mycobacterium chimaera* infection in cardiac surgery patients. Clin. Infect. Dis. 64, 335–342. https://doi.org/10.1093/cid/ciw754.

Charron, D., Bédard, E., Lalancette, C., Laferrière, C., Prévost, M., 2015. Impact of electronic faucets and water quality on the occurrence of *Pseudomonas aeruginosa* in water: a multi-hospital study. Infect. Control Hosp. Epidemiol. 36, 311–319. https://doi.org/10.1017/ice.2014.46.

Collier, S.A., Deng, L., Adam, E.A., Benedict, K.M., Beshearse, E.M., Blackstock, A.J., et al., 2021. Estimate of burden and direct healthcare cost of infectious waterborne disease in the United States. Emerg. Infect. Dis. 27, 140–149. https://doi.org/10.3201/eid2701.190676.

Costerton, J.W., 1995. Overview of microbial biofilms. J. Ind. Microbiol. 15, 137–140.

D'Alessandro, D., Fabiani, M., Cerquetani, F., Orsi, G.B., 2015. Trend of legionella colonization in hospital water supply. Ann. Ig. 27, 460–466. https://doi.org/10.7416/ai.2015.2032.

DEFRA, 1998. Water Supply and Sewerage Licensing: Updating Security and Emergency Measures Directions.

DHSC, 2014. Health Building Note 00-07 Planning for a Resilient Healthcare Estate.

DHSC Safe water in healthcare premises (HTM 04-01), 2017. Safe Water in Healthcare Premises (HTM 04-01). [WWW Document]. GOV.UK. URL: https://www.gov.uk/government/publications/hot-and-cold-water-supply-storage-and-distribution-systems-for-healthcare-premises. (Accessed 5 January 2021 (Accessed 4 April 2019).

DWI, 2015. The Water Supply (Water Quality) Regulations 2016.

Eckmanns, T., Oppert, M., Martin, M., Amorosa, R., Zuschneid, I., Frei, U., Rüden, H., Weist, K., 2008. An outbreak of hospital-acquired *Pseudomonas aeruginosa* infection caused by contaminated bottled water in intensive care units. Clin. Microbiol. Infect. 14, 454–458. https://doi.org/10.1111/j.1469-0691.2008.01949.x.

EU, 2021. Drinking Water Legislation—Environment—European Commission.

Falkinham, J.O., Pruden, A., Edwards, M., 2015. Opportunistic premise plumbing pathogens: increasingly important pathogens in drinking water. Pathogens 4, 373–386. https://doi.org/10.3390/pathogens4020373.

Flemming, H.-C., Wingender, J., 2010. The biofilm matrix. Nat. Rev. Microbiol. 8, 623–633. https://doi.org/10.1038/nrmicro2415.

Gómez-Gómez, B., Volkow-Fernández, P., Cornejo-Juárez, P., 2020. Bloodstream infections caused by waterborne bacteria. Curr. Treat. Options Infect. Dis. 12, 332–348. https://doi.org/10.1007/s40506-020-00234-5.

Hayward, C., Ross, K.E., Brown, M.H., Whiley, H., 2020. Water as a source of antimicrobial resistance and healthcare-associated infections. Pathogens 9. https://doi.org/10.3390/pathogens9080667.

HSE, 1974. Health and Safety at Work etc Act 1974—Legislation Explained.

HSE, 2014. HSG 274 Legionnaires' Disease—Technical Guidance Part 2: The Control of Legionella Bacteria in Hot and Cold Water Systems Technical Guidance. http://www.hse.gov.uk/pubns/books/hsg274.htm. (Accessed 5 January 2021).

Kanwar, A., Domitrovic, T.N., Koganti, S., Fuldauer, P., Cadnum, J.L., Bonomo, R.A., Donskey, C.J., 2017. A cold hard menace: a contaminated ice machine as a potential source for transmission of carbapenem-resistant *Acinetobacter baumannii*. Am. J. Infect. Control 45, 1273–1275. https://doi.org/10.1016/j.ajic.2017.05.007.

Li, T., Abebe, L.S., Cronk, R., Bartram, J., 2017. A systematic review of waterborne infections from nontuberculous mycobacteria in health care facility water systems. Int. J. Hyg. Environ. Health 220, 611–620. https://doi.org/10.1016/j.ijheh.2016.12.002.

Marshall, C., Denton, S., Christie, I., Orr, L., 2020. Effect of water chlorination on development and persistence of biofilm in shower heads. Infect. Control Hosp. Epidemiol. 41, s202–s203. https://doi.org/10.1017/ice.2020.745.

Nakamura, S., Azuma, M., Sato, M., Fujiwara, N., Nishino, S., Wada, T., Yoshida, S., 2019. Pseudo-outbreak of *Mycobacterium chimaera* through aerators of hand-washing machines at a hematopoietic stem cell transplantation center. Infect. Control Hosp. Epidemiol. 40, 1433–1435. https://doi.org/10.1017/ice.2019.268.

NHS England, 2021. HBN 00-07 Resilience Planning for NHS Facilities.

Papciak, D., Tchórzewska-Cieślak, B., Domoń, A., Wojtuś, A., Żywiec, J., Konkol, J., 2019. The impact of the quality of tap water and the properties of installation materials on the formation of biofilms. Water 11, 1903. https://doi.org/10.3390/w11091903.

Pauli, B.J., 2020. The Flint water crisis. WIREs Water 7, e1420. https://doi.org/10.1002/wat2.1420.

Rogers, J., Dowsett, A.B., Dennis, P.J., Lee, J.V., Keevil, C.W., 1994. Influence of plumbing materials on biofilm formation and growth of *Legionella pneumophila* in potable water systems. Appl. Environ. Microbiol. 60, 1842–1851. https://doi.org/10.1128/aem.60.6.1842-1851.1994.

Schreiber, P.W., Kohl, T.A., Kuster, S.P., Niemann, S., Sax, H., 2021. The global outbreak of *Mycobacterium chimaera* infections in cardiac surgery—a systematic review of whole-genome sequencing studies and joint analysis. Clin. Microbiol. Infect. 27, 1613–1620. https://doi.org/10.1016/j.cmi.2021.07.017.

UK Government, 1991. Water Industry Act 1991.

UK Government, 2007. The Natural Mineral Water, Spring Water and Bottled Drinking Water (England) Regulations.

UK Government, 2018. The Private Water Supplies (England) Regulations 2016.

van der Kooij, D., Veenendaal, H.R., Italiaander, R., 2020. Corroding copper and steel exposed to intermittently flowing tap water promote biofilm formation and growth of Legionella pneumophila. Water Res. 183, 115951. https://doi.org/10.1016/j.watres.2020.115951.

Waines, P.L., Moate, R., Moody, A.J., Allen, M., Bradley, G., 2011. The effect of material choice on biofilm formation in a model warm water distribution system. Biofouling 27, 1161–1174. https://doi.org/10.1080/08927014.2011.636807.

Walker, J., Moore, G., 2015. *Pseudomonas aeruginosa* in hospital water systems: biofilms, guidelines, and practicalities. In: Journal of Hospital Infection, Proceedings from the 9th Healthcare Infection Society International Conference 89, pp. 324–327, https://doi.org/10.1016/j.jhin.2014.11.019.

Weinbren, M., Inkster, T., 2021. The hospital-built environment: biofilm, biodiversity and bias. J. Hosp. Infect. https://doi.org/10.1016/j.jhin.2021.02.013.

WHO, 2011. Water safety in buildings. https://www.who.int/publications/i/item/9789241548106.

WHO, 2017. Guidelines for Drinking-Water Quality, 4th edition, incorporating the 1st addendum. https://www.who.int/publications/i/item/9789241549950.

WRAS, 1999. The Water Supply (Water Fittings) Regulations 1999.

Yui, S., Karia, K., Ali, S., Muzslay, M., Wilson, P., 2021. Thermal disinfection at suboptimal temperature of *Pseudomonas aeruginosa* biofilm on copper pipe and shower hose materials. J. Hosp. Infect. 117, 103–110. https://doi.org/10.1016/j.jhin.2021.08.016.

Cold water systems

4

Introduction

Drink water supplies are not sterile at the point of entry, and the cold water supply into the building will contain a range of microorganisms, which occur naturally. Where the cold water temperature is below 20°C, these are likely to be in small numbers and will be dormant or only able to grow very slowly unless the building system provides the conditions that allow it to colonize and grow.

Waterborne microorganisms are transported via the cold water supply to the hospital water system. Once in the building, the challenge is to prevent these microorganisms from multiplying in the cold water system. If the bacterial numbers can be controlled, then patients would be better protected. Unfortunately, many water systems provide idyllic conditions for microbial growth. In some cases, as soon as the bacteria enter the water system, they have access to a range of nutrients and will form biofilms on surfaces from which they will sequester further nutrients. Debris and sediment either from the supply itself or corrosion products will accumulate on the base of the storage tank and provide nutrients that support microbial growth. In addition, where cold water storage tanks are oversized, this will promote slow passage or stagnation of the water. This stagnation lengthens the time that the bacteria spend in the storage tanks and gives them greater opportunities to become established. The temperature of the cold water, e.g., below 20°C, should be sufficient to limit the growth of many microorganisms. However, the water storage tanks are often located in warm plant rooms, and the cold water pipes are often sited alongside the hot water pipes and not insulated. It is partly this transient heat transfer that results in warming of the cold water and in favorable conditions for the growth of waterborne pathogens, which leads to greater infection risks to vulnerable users.

The cold water supply

The water supplied to healthcare buildings is supplied by water utilities or private water supplies and may occasionally be sourced from boreholes, springs, and wells (Fig. 1). However, regardless of its source, it must comply with relevant national drinking water regulations for utility suppliers this is DWI (2021) and EU (2020). For boreholes and springs, these fall under "The Private Water Supplies (England) (Amendment) Regulations 2018." The main difference between the two regulations is that utility companies are regulated by the Drinking Water Inspectorate (DWI) who have vast knowledge of the nation's water and large numbers of microbiologists ensuring each water undertaker complies fully with the regulations. However, with the private water regulations, it is the local authority who is the regulator, and they do not usually have such expertise to hand.

Safe Water in Healthcare. https://doi.org/10.1016/B978-0-323-90492-6.00011-2

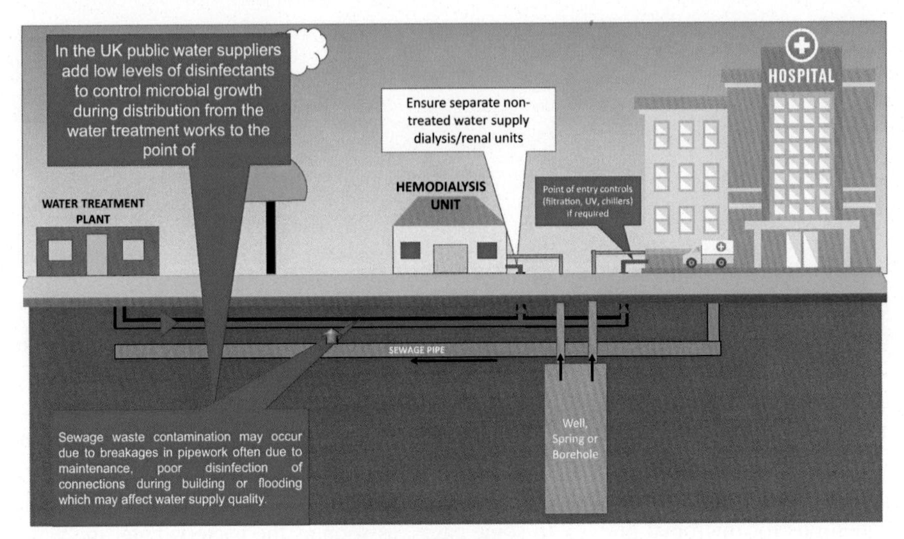

Fig. 1 Supply of mains water to a healthcare building.

Water that is delivered to the hospital is considered "wholesome water," which is fit to use for drinking, cooking, food preparation, or washing without any potential danger to human health (UK Government, 1991).

Microbial hazards in the supply water

While the water supplied at the point of entry is considered not to present any potential danger to the general population, it is not sterile and contains a diverse population of opportunistic microorganisms. Their natural habitat is within biofilms on surfaces from which the water is abstracted. However, once microorganisms enter a hospital water system, they form biofilms on surfaces of the pipework, components, and fittings of the distribution system. The microorganisms may include waterborne pathogens such as *Legionella* spp., *Pseudomonas aeruginosa*, *Stenotrophomonas maltophilia, Cupriavidus pauculus,* and many other opportunistic Gram-negative pathogens as well as nontuberculous mycobacteria (NTMs). Such bacteria are likely to be present in low numbers in the cold water supply and, at these concentrations, are unlikely to cause harm, even to vulnerable patients. However, all uses of water should be risk assessed as in some cases only sterile water should be supplied to some patients, e.g., in a 2012 report on *P. aeruginosa* infections in neonatal intensive care units in Northern Ireland recommended the use of sterile contact with all babies (BSI, 2022; RQIA, 2012).

Water quality within the building

The design, specification, construction, commissioning, maintenance, and management of the cold water system should be undertaken to minimize the risk of microbial growth in the cold water system. However, because of the size and complexity of hospital water systems, where there are additional specialist systems such as reverse osmosis plants and other equipment, the water quality may have deteriorated by the time the water reaches outlets and consequently present a risk to vulnerable patients.

In larger buildings, the cold water is stored and then supplied via a distribution network to all the peripheral outlets, associated equipment, and other systems. One of the first places where cold water deteriorates is where it is stored in tanks otherwise known as cisterns (DHSC, 2017).

Why do you need cold water storage?

Storage tanks maintain a supply of cold water to the hospital if there is a break in the supply, a drop in the supplied water pressure, or if the water demand exceeds the supply capacity (Fig. 2). In tall buildings, there may be cold water tanks in the basement, which may then be boosted to additional tanks in the roof, from which the water is distributed by gravity to the peripheral outlets.

Design and construction of cold water storage vessels

Cold water storage tanks and fittings should be designed and installed to ensure they do not have adverse effects on the water and should be constructed to be accessible for cleaning and maintenance and the design and construction should:

- not encourage microbial growth, which would affect the taste and odor of the water (such as those containing natural rubber, hemp, linseed oil-based jointing compounds, and fiber washers) (WRAS, 1999)

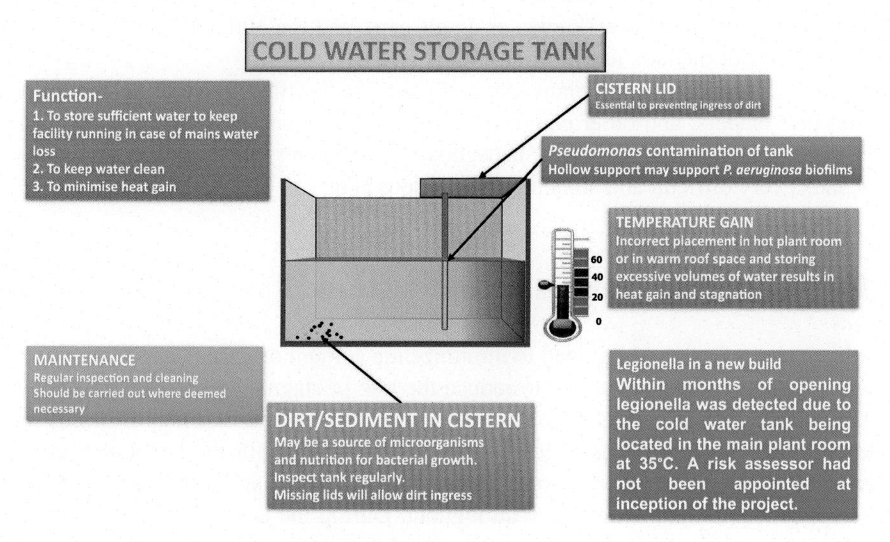

Fig. 2 Schematic of a cold water storage tank.

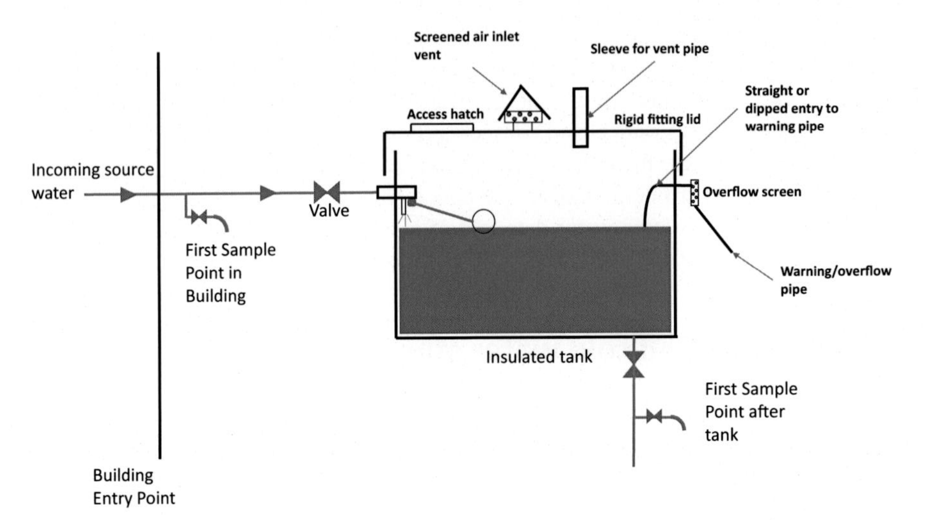

Fig. 3 Schematic of the cold water storage tank demonstrating the tight-fitting lid, inlet, outlet, and overflow screen (schematic is not to scale).

- metals should be resistant to corrosion (e.g., copper or stainless steel)
- comply with national drinking water and water fittings regulations together with relevant standards and guidance (BSI, 2000, 2005, 2015; WRAS, 1999) (BS EN 13280)
- be compatible with any disinfection regime (check with manufacture of materials)
- prevent backflow from the building water system into the supply network, which in the United Kingdom is a legal requirement.

To reduce water stagnation, ensure there is a suitable flow and mixing of water across the tank(s). As such the water flow inlets and outlets should be positioned on opposite sides of the tank with the inlet at a height above the water level that is sufficient to prevent backflow (known as an air gap), and the outlet at a position on the opposite side of the tank and at floor level (Fig. 3).

Access to water tanks in plant rooms is often made difficult due to positioning of pipework, ladders, and other equipment that has subsequently been installed during refurbishments, making effective inspection (any cleaning that might be required) of the tanks very difficult and sometimes hazardous (Fig. 4).

How much cold water should be stored?

Cold water tanks should be sized to minimize the amount of time water resides in the tank and maximize the turnover to reduce the risk of stagnation and deterioration of water quality, which would encourage microbial growth and biofilms. Biofilms that develop in tanks can release bacteria that will colonize both the hot and cold water distribution systems and associated equipment. So managing the tanks correctly is important in maintaining safety throughout. During the design stage, the volumes and sizes of storage tanks should be calculated based on average daily "during use"

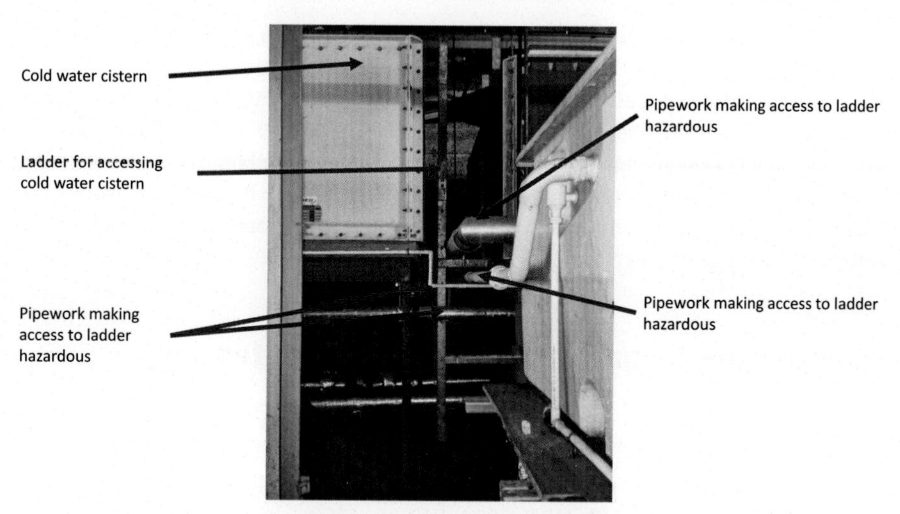

Fig. 4 Demonstration of placement of pipework making access to cold water tank difficult.

consumption in wards and specialist areas with a suggested nominal 12h storage (HSG274 Part 2, HTM 04-01 Part B).

Care should be taken when determining the amount of stored water. In new builds and refurbishments, calculations used are often based on historical data and additional capacity for future developments. For example, in recent years, water use on wards has decreased as many patients are admitted on a day case basis and shower before admission. In-patients who remain for lengthy stays are often too ill to use the ensuite facilities, which pose a real challenge for those managing the water safety to ensure the risk from pathogens is minimized. However, building in capacity for the provision of future developments would lead to excessive water storage, and in such cases, storage capacity should be reduced.

Hospital water use can vary significantly from day to day, with large volumes used on certain days; but not others, this is a real challenge for designers who will always design to the higher demand days, to ensure the hospital has sufficient storage in case of water loss. The problem here is the lower use days can lead to excessive water storage and deterioration of water quality, through stagnation. Designers should consider automatic self-adjusting tank volumes, which can be controlled by the hospitals building management system (BMS), thus always ensuring that the hospital has the recommended nominal 12-h water storage on-site regardless of the water consumption that day.

Within conventional tanks, the volume of water stored should be checked at least annually or more frequently as part of the cold water risk assessment. This is important where there have been changes to the amount of water used in buildings and can be undertaken by reading tank water meters (where fitted) or carrying out a "drop test," where water is prevented from entering the tank (by isolating the mains supply to the tank to prevent water entering the tank) and the volume use checked after a fixed time. This should be done during a period of normal use and may have to be undertaken over a number of different days of the week. It is important to ensure that where multiple tanks or those with more than one compartment are fitted, the cold water flow is shared equally so that one compartment is not used in preference to the other(s), which could lead to stagnation. In reality, it can be difficult to equalize the flow between connected tanks, especially where time release ball-valves are installed, where one tank or compartment could be held off completely by the full flow from the other ball valve. Manufacturers recommend annual synchronization of such valves, which is challenging, but this needs to be as close as possible.

Temperature management of the cold water

Keeping the temperature of the cold water below the growth temperature range of opportunistic waterborne pathogens will help manage risks. Therefore, the cold water should be delivered at the point of use at less than 20°C within 2 min of turning on the tap (BSI, 2000; DHSC, 2017; HSE, 2014; WRAS, 1999).

As such, the aim is to maintain a temperature of less than 20°C throughout the system including those outlets that are furthest from the cold water storage vessel to allow for a maximum increase of 2°C from the temperature at the point of entry. Increases greater than this will indicate poor system design (e.g., long pipe runs next to heat sources, ineffective insulation of pipes, stagnation) or that the system is not operating as intended (lower water use than design). A range of outlets should be monitored to assess heat gain across the cold water system. Maintaining temperatures at less than 20°C is going to be an increasing challenge particularly when year-on-year increases in temperatures are already impacting on cold water supply temperatures, with some healthcare premises reporting incoming supply temperatures of >25°C (Masson-Delmotte et al., 2021).

Where cold water supply temperatures are more than 20°C, then alternative control measures will be required such as flushing, chilling, or the addition of biocides. The temperature in the cold water tank should be checked at least every 6 months (HSE, 2014). Ideally, the water temperature in the tank and at outlets should be monitored in real time and linked to the building management system (BMS) to give an early warning when permitted water temperatures are exceeded (HSE, 2014).

To minimize heat gain, the tank and associated pipework should be insulated from potential sources of heat gain (Fig. 5).

Storage tanks should not be co-located in plant rooms with heat generating equipment (e.g., calorifiers, domestic heating boilers, steam generators, or hot water storage buffer vessels) (Fig. 6) or on top of buildings, where they may be subject to solar heat gain or in inadequately insulated roof spaces where high temperature is likely.

In a BRE study carried out on domestic properties, the co-location of the cold water storage tank above the calorifier was the second highest risk factor for *Legionella* presence (the highest was no lid on the CWST) (personal communication J.V. Lee).

Designers need to take account of the increased risk from rising water supply temperatures when designing new builds to ensure there is sufficient resilience and contingency to protect water systems from heat gain. Reliance on flushing to decrease

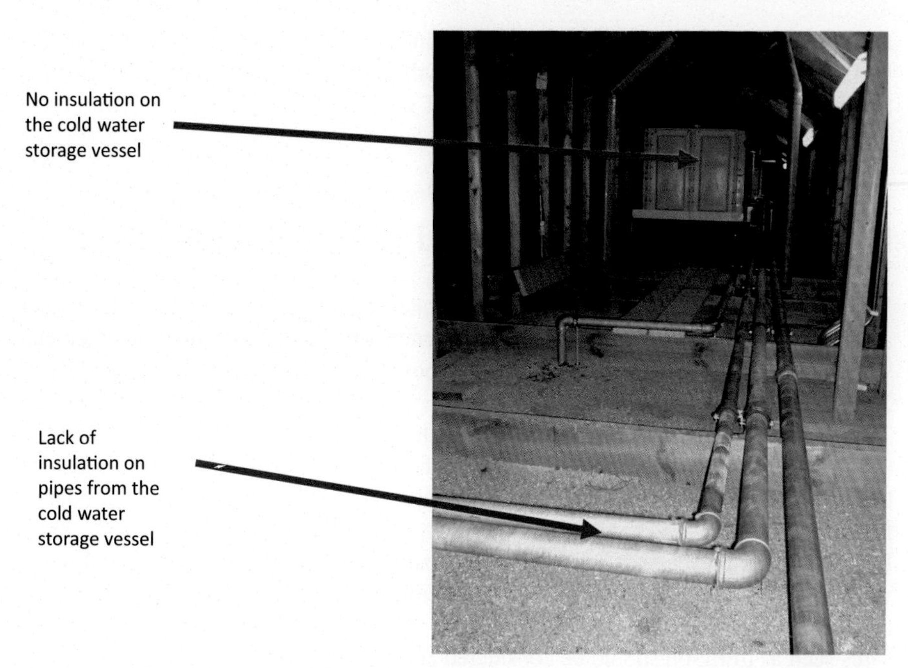

Fig. 5 Cold water vessel and pipework in a roof with no evidence of insulation to minimize heat gain.

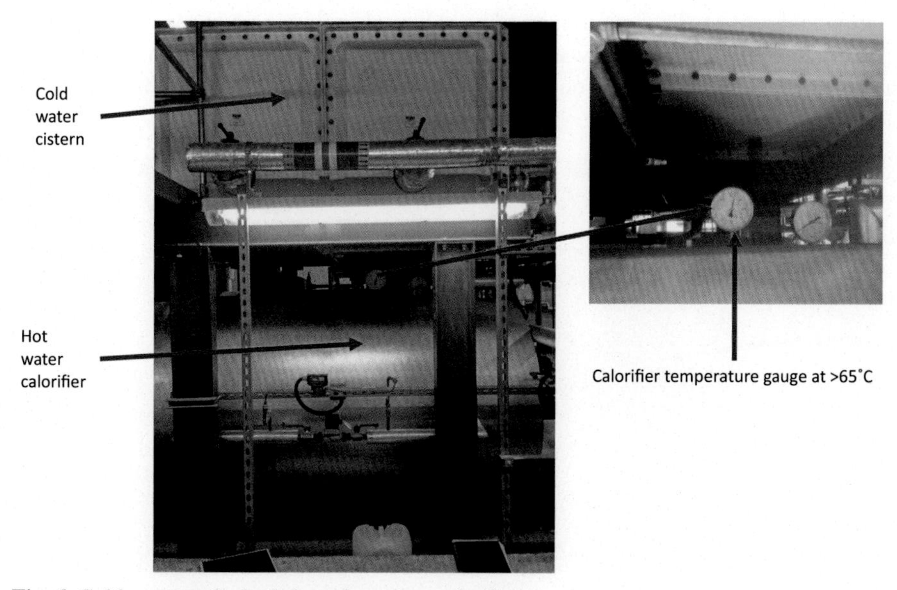

Fig. 6 Cold water tank cited above hot water calorifier.

temperatures due to poor design is not sustainable; water is a precious resource, and it is an offence to wastewater. Poor design will inevitably lead to additional flushing to protect patients, but additional flushing of water to drain is not the answer, better design is.

Microbiological risks associated with cold water tanks

Where the supply water meets the drinking water regulations and systems plus associated equipment properly installed, commissioned, operated, and maintained then cold water systems are less likely to support opportunistic microbial pathogens (DHSC, 2017; HSE, 2014).

Debris and sediment (Fig. 7) either from the water supply itself or corrosion products from the supply pipework or metals in the storage tank (Fig. 8) can settle on the base of the storage tanks and provide nutrients that support microbial growth (Gahrn-Hansen et al., 1995; Peter and Routledge, 2018; Proctor et al., 2020; Sener et al., 2021). In addition, the presence of accumulated deposits on the side wall of the tank at the air water interphase also encourages the formation of biofilm and microbial pathogens (Fig. 8). Where significant debris is present in the incoming water, then control measures such as filtration should be considered at the point of entry (Muzzi et al., 2020).

A lack of suitable covers or lids in cold water tanks is often cited as a reason for debris being present in the tanks, and it is not unusual to find debris, small rodents and birds, including pigeons, in tanks where lids have been damaged or are missing. In addition, contamination can also enter storages tanks when inadequate care is taken during inspection if dirt and debris on the surface of the storage tank lids are disturbed during inspections. The importance of good housekeeping in plant rooms and similar places should be included in water safety plan and should emphasize that dirt should not be allowed to accumulate on or around storage tanks. Workers who have access to

Fig. 7 Sediment and debris (nutrients) on the floor of the water tank and a hollow support for the lid that has been demonstrated to retain stagnant water and high counts of microorganisms.

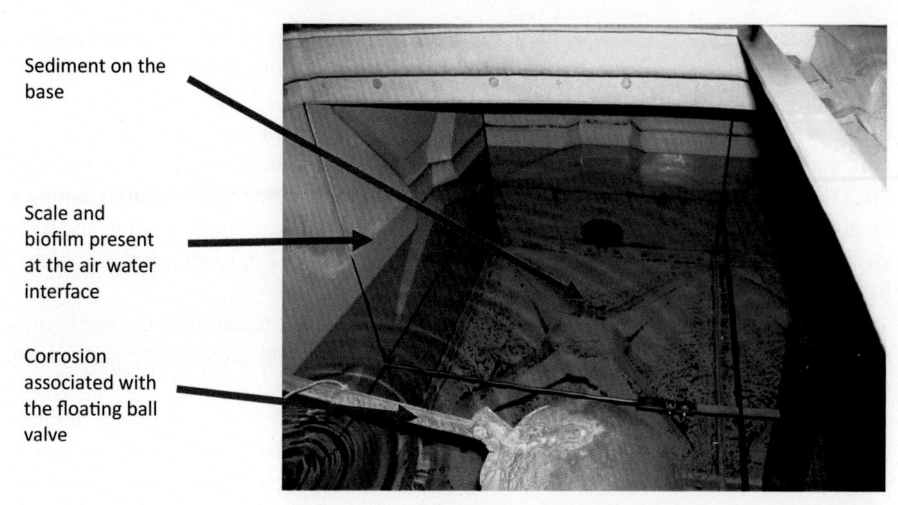

Fig. 8 Sediment (nutrients) deposited on the base of the storage vessel as well as scale at the air water interface and corrosion on the metal (replace with plastic) floating ball valve, all of which can support microbial growth.

these areas should be trained, competent, and understand contamination risks and have a current certificate from an approved contractor training scheme or another national equivalent.

Case study: Cleanliness of storage tanks

Poor maintenance was found to be responsible for a series of gastrointestinal infections due to Yersinia enterocolitica in doctors in a hospital residence (Cafferkey et al., 1993). After reports of particulate matter in the water supplying the hospital theater the tank (which also supplied the hospital residence) was investigated and was found to have a broken cover and contained a dead pigeon. In other hospitals tanks have been found to contain materials such as sponges that were discarded during the building phase and the tanks were never checked during the commissioning stage and may have been present for 2 years (Inkster et al., 2021).

Monitoring of the cold water system

Poor and/or unsafe access to inspect and clean tanks is a common problem and not necessarily limited to old buildings. There should be annual visual checks of the cleanliness of the cold water storage tanks by competent personnel who hold a national water hygiene certificate (to avoid contamination during the inspection process). In large hospital buildings where it is common to have multiple storage tanks, which have several compartments to improve turnover and improve mixing, one tank or compartment can be bypassed for inspection, cleaning disinfection, and repair without interrupting the cold water supplies.

Pipework should be labeled, including with the direction of flow, as poor labeling has been demonstrated as a significant risk of causing contamination of distributed water due to inadvertent cross-connections (WHO, 2011). Where required as part of the risk assessment, water samples should be taken from the first outlet after the tank and not the tank itself to avoid the potential for introducing contamination (BSI, 2021). Only if an investigation is required, following adverse results, should samples be taken from the tank itself by competent personnel, using aseptic technique and with sterile sampling equipment (See Sampling Techniques for Legionella Examinations, https://www.youtube.com/watch?v=-C_ray4Ku_0).

Can provide a trend analysis of total viable counts (a test to estimate the total number of microorganisms in a water sample) and indicates the presence or absence of microorganisms.

Microbiological sampling, using total viable counts to estimate the total number of microorganisms in a water sample, if undertaken regularly as part of a planned maintenance schedule, can provide an indication of the water quality. For example, a gradual decline in water hygiene, or a sudden steep rise, may indicate a contamination event. Taking TVC before and after a disinfection can also be a useful indicator of the success of the disinfection process.

While waterborne pathogens have been recovered from cold water tanks, the presence of hollow lid supports has been shown to retain water that was positive for *Pseudomonas* spp. (DoH, 2013; Ezzeddine et al., 1989; HSE, 2014; Inkster et al., 2021; Pierre et al., 2019). Internal lid support structures should be designed to prevent the retention of water that causes stagnation and harbors microorganisms.

Water hardness deposits on pipework, fittings, and components, resulting in decreased flow and increased surface area, which encourages microbial colonization and growth. Consequently, these deposits are difficult to remove and will harbor waterborne pathogens. In addition, areas with hard water, softening of the cold water supply to the hot water tank may be considered. Such decisions must involve the water safety team to ensure that vulnerable patients are not at risk. For example, softened water cannot be used to prepare baby feeds (DHSC, 2017) (Fig. 9—scale deposits on wall of water tank).

Electronic remote logging linked into building management systems has become more common place and can be retrofitted to monitor key controls such as cold water temperature (tank), water pipes (distribution and outlets), and flow. Alerts can be automated when adverse actions levels have been identified. These monitoring results should be regularly reviewed to identify where the cold water system needs enhanced management measures to control microbial risks.

Case study: Excessive cold water temperatures in new build

As part of a new build project for a teaching hospital, a small stand-alone building was due to be completed first. The contractor had proudly notified the infection control team that the new water systems were so sophisticated that the bacteria had no chance of growing. The hospital water safety group had implemented an enhanced policy of

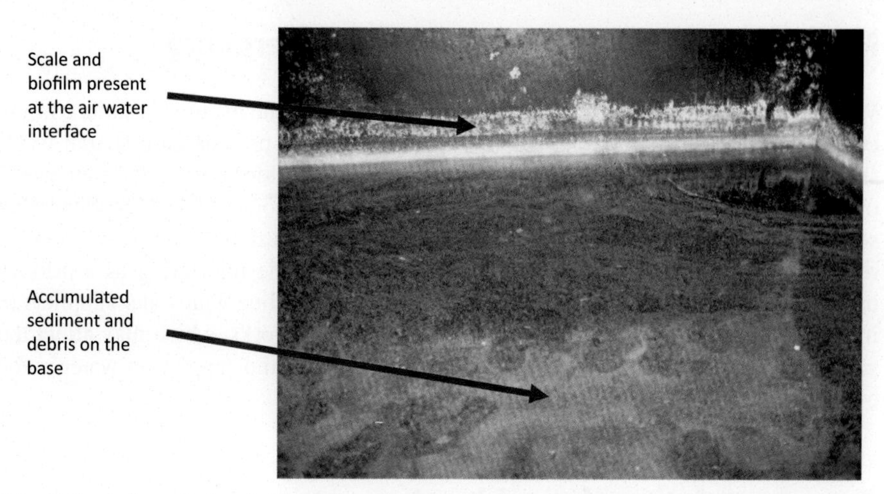

Fig. 9 Scale deposited at the water interface/tideline in hard water areas can encourage microbial growth.

routine water sampling for *Legionella* spp. Within 2 months of opening, *Legionella* spp. was detected in a routine water sample from one of the staff showers. An incident meeting was held, and on further investigation, it was found that the temperature of the cold water supply to the shower was 34°C. The temperature gain was traced back to the cold water tank being located in the plant room where the water temperature in the tank was 36°C.

Case study: Filling sections prior to completion

The design of a major new teaching hospital divided the building into three linked towers each with its own separate water systems. The construction was such that one tower was to be completed 18 months ahead of the other two towers, ready for part occupation. The first tower opened and the lower floors were occupied by women's and children's services. The top floors of the tower remained unoccupied, with the next clinical service due to move in being oncology. Routine Legionella spp. testing detected a contamination of the cold-water system within 8 months of opening. An investigation found numerous underlying factors, for example, a risk assessment performed shortly after the building was opened had highlighted heat gain in the unoccupied areas on the cold water system thought likely to be due to insufficient water turnover. This had not been addressed. The construction team had relied upon regular negative water samples for Legionella spp. rather than finding a solution to the underlying issue of heat gain on the cold water system. Mitigating biofilm in cold water systems is to a certain extent more difficult than hot water systems as the ability to pasteurize the cold water is rarely available. Point-of-use filters had to be installed together with dosing with a biocide, in this instance, chlorine dioxide at 2 ppm (0.5 ppm is the upper limit allowed in drinking water, but at this concentration it will not impact established biofilm).

Maintaining cold water supply in an emergency

Water safety plans need to include contingency plans in the event of a failure in the cold water supply. Public water suppliers have a duty to provide an essential water supply at all times (DEFRA, 1998). This includes giving priority to the domestic needs of the sick, older people, disabled people, hospitals, schools, and other vulnerable sectors of the population. NHS-funded providers should establish a formal working relationship with their water supplier. This should include registering as a priority user (DHSC, 2014) and have a memorandum of understanding with water suppliers to inform the facility when there are planned maintenance works, which may affect the supply and for the development of plans to maintain essential supplies of water in the event of major incidents.

Recommendations for cold water systems

1. The water supply quality, whatever the source, should comply with the relevant drinking water regulations.
2. Any changes planned to the cold water service delivery, which could adversely affect water quality such as change of use, planned closures for refurbishment, recommissioning, or establishment of pop-up wards, should be reviewed to assess patient safety.
3. Water systems should be designed, constructed, commissioned, and managed to ensure they are fit for their intended purpose and there is no risk to patients, staff, or visitors at each point of use.
4. Materials, fittings, and components should not compromise water quality and comply with the relevant national legislation and guidelines.
5. Pipes, fittings, valves, subassemblies, calorifiers, tanks, etc., intended to form part of the cold water service installation should be supplied to site cleaned, free from waterborne pathogens, particulate matter, and other residues.
6. Where temperatures exceed 20°C in the cold water supply, storage, or throughout distribution, then additional control mechanisms are required.
7. Dedicated treatment and supply arrangements will be required for renal and hemodialysis units or for making up infant feeds where concentrations of biocides in the water would be harmful to patients.
8. Where there are taste or odor problems, microbiological monitoring for total viable counts (TVCs) may be considered necessary and where infections involving waterborne microorganisms have been identified the surveillance of the cold water system may be required.
9. There should be ongoing patient surveillance to detect and react to waterborne infections.

Learning points

- By law (Water Act 1991), the cold water supplied to the healthcare building must be wholesome, but there can be failures in the quality of the water supplied usually due to accidental ingress of sewage or animal and/human feces or a failure in water treatment, and contingency plans must be in place.

- Microbiological issues are most likely to manifest within the building cold water system rather than the water supplied from the water company.
- Cold water in storage and distribution systems should be managed to minimize the risk from nutrient ingress, temperature gain, and stagnation. **Keep water cold, keep water clean, keep water moving, and keep records.**
- A negative result for microbial pathogens in a cold water sample does not mean there are no pathogens present in the incoming water. As *Legionella* and other waterborne pathogens such as *P. aeruginosa* and NTMs, etc., are natural inhabitants of water, it should be assumed they are present, and measures put in place to prevent them colonizing and growing within hospital water systems in buildings.

Risk assessing the cold water system

Risk assessment criteria	Yes	No
Do materials comply with the requirements of the Water Supply (Water Fittings) Regulations 1999?		
Do all the nonmetallic materials comply with the requirements of BS 6920?		
Have all pipes, fittings, valves, subassemblies, calorifiers, tanks, etc., been supplied to site cleaned, free from waterborne pathogens, particulate matter and other residues from the manufacturer?		
Have all pipes, fittings, valves, subassemblies, calorifiers, tanks, etc., been assessed as being free from waterborne pathogens, particulate matter and other residues while on site prior to installation?		
Have the tank screen vents actually been checked and overflow pipe? There are several issues in that manufacturers need to make the screens more accessible (currently a complicated process to access the screens) and also the need to provide training so that operatives are aware of where the screens are located and how to get access to them		
Are the plumbing components free from microbial contamination from the manufacturer following wet testing in the production plant?		
Does the cold water storage capacity only enable supply for up to 12 h' normal usage?		
Has the water tank been designed as per current guidance?		
Is the water temperature in the tanks monitored to ensure that heat gain is controlled?		
Are the tanks accessible?		
Liners should not be fitted in the hospital storage tank, but where present they need to be inspected and removed as part of the maintenance plan?		
Have the cold water tanks been inspected at least annually?		
Has sediment accumulation in the tank been assessed to minimize deposits on the base of the tank?		

Risk assessment criteria	Yes	No
Are the tank roof supports of a design to prevent retention of water? Although reports of this are a few years old now it is just worth checking		
Have the water storage tanks been divided into convenient compartments, suitably interconnected and valved to facilitate cleaning, disinfection, repair, modification and inspection, without seriously disturbing the cold-water service?		
Have the cold water service mains been run in separate ceiling spaces, risers or zones to avoid other sources?		
Where hot and cold services are in the same conduit is the hot water piping at a higher level than the cold to minimize heat transfer by convection?		
Have temperatures been monitored at representative to confirm they are below 20°C?		
Are the cold water draw-off points no more than 2°C above the temperature measured in the cold-water header tanks?		
Have the cold water tanks been insulated effectively?		
Is there any evidence of heat gain in the cold water pipes?		
Where temperatures exceed 20°C have additional control mechanisms been implemented, e.g., dosing with a suitable biocide?		
Are there microbiological monitoring records of the tank?		
Are there microbiological monitoring records of the cold water system?		
Have annual checks been carried out to assess new works, e.g., assessing insulation of pipework?		
Has a temperature profile been mapped of the whole cold water system to identify problematic areas?		
Have cold water dead legs been identified to minimize stagnation and placed on the maintenance plan for removal?		
Are records up to date and available?		
Are flushing records available?		
Is a flushing risk assessment available?		
Where present are cold water outlets monitored?		
Is there ongoing patient surveillance to detect and react to waterborne infections?		

Guidance, regulations, and further reading relevant for cold water systems

- CIBSE's Guide G—"Public health engineering" gives further guidance on sizing cold water storage (CIBSE, 2014).
- Estates and Facilities Alert EFA/2013/004—"Cold water storage tanks") (DoH, 2013).

- HTM 04-01 Safe Water in Healthcare (DHSC, 2017).
- HSG274 Legionnaires' disease Part 2: The control of legionella bacteria in hot and cold-water systems (HSE, 2014).
- Cold water services: storage cisterns—general design and installation requirements (WRAS, 2015).

References

BSI, 2000. BS EN 805:2000—Water supply. Requirements for Systems and Components Outside Buildings.

BSI, 2005. BS EN 806-2:2005—Specifications for Installations Inside Buildings Conveying Water for Human Consumption. Design—BSI British Standards.

BSI, 2015. BS 8558:2015 Guide to the Design, Installation, Testing and Maintenance of Services Supplying Water for Domestic Use Within Buildings and Their Curtilages. Complementary guidance to BS EN 806.

BSI, 2021. BS 7592: 2008 Sampling for *Legionella* Bacteria in Water Systems. Code of practice.

BSI, 2022. BS 8580-2:2022—Risk Assessments for *Pseudomonas aeruginosa* and Other Waterborne Pathogens. Code of practice https://standardsdevelopment.bsigroup.com.

Cafferkey, M.T., Sloane, A., McCrae, S., O'Morain, C.A., 1993. *Yersinia frederiksenii* infection and colonization in hospital staff. J. Hosp. Infect. 24, 109–115. https://doi.org/10.1016/0195-6701(93)90072-8.

CIBSE, 2014. Guide G Public Health and Plumbing Engineering.

DEFRA, 1998. Water Supply and Sewerage Licensing: Updating Security and Emergency Measures Directions.

DHSC, 2014. Health Building Note 00-07 Planning for a Resilient Healthcare Estate.

DHSC, 2017. HTM 04-01: SAFE Water in Healthcare Premises.

DoH, 2013. Estates and Facilities Alerts (EFA's) Cold Water Storage Tanks. Health.

DWI, 2021. Legislation, Drinking Water Inspectorate. DWI.

EU, 2020. Directive (EU) 2020/2184 of the European Parliament and of the Council of 16 December 2020 on the Quality of Water Intended for Human Consumption.

Ezzeddine, H., Van Ossel, C., Delmée, M., Wauters, G., 1989. *Legionella* spp. in a hospital hot water system: effect of control measures. J. Hosp. Infect. 13, 121–131. https://doi.org/10.1016/0195-6701(89)90018-2.

Gahrn-Hansen, B., Uldum, S.A., Schmidt, J., Nielsen, B., Birkeland, S.A., Jørgensen, K.A., 1995. Nosocomial *Legionella pneumophila* infection in a nephrology department. Ugeskr. Laeger 157, 590–594.

HSE, 2014. HSG 274 Legionnaires' Disease—Technical Guidance Part 2: The Control of Legionella Bacteria in Hot and Cold Water Systems Technical Guidance. http://www.hse.gov.uk/pubns/books/hsg274.htm. (Accessed 5 January 2021).

Inkster, T., Peters, C., Wafer, T., Holloway, D., Makin, T., 2021. Investigation and control of an outbreak due to a contaminated hospital water system, identified following a rare case of *Cupriavidus pauculus* bacteraemia. J. Hosp. Infect. https://doi.org/10.1016/j.jhin.2021.02.001.

Masson-Delmotte, V., Zhai, P., Pirani, A., Connors, S.L., Péan, C., Berger, S., Zhou, B. (Eds.), 2021. Climate Change 2021: The Physical Science Basis. Contribution of Working Group I to the Sixth Assessment Report of the Intergovernmental Panel on Climate Change.

Muzzi, A., Cutti, S., Bonadeo, E., Lodola, L., Monzillo, V., Corbella, M., Scudeller, L., Novelli, V., Marena, C., 2020. Prevention of nosocomial legionellosis by best water management:

comparison of three decontamination methods. J. Hosp. Infect. 105, 766–772. https://doi.org/10.1016/j.jhin.2020.05.002.

Peter, A., Routledge, E., 2018. Present-day monitoring underestimates the risk of exposure to pathogenic bacteria from cold water storage tanks. PLoS One 13, e0195635. https://doi.org/10.1371/journal.pone.0195635.

Pierre, D., Baron, J.L., Ma, X., Sidari, F.P., Wagener, M.M., Stout, J.E., 2019. Water quality as a predictor of legionella positivity of building water systems. Pathogens 8, 295. https://doi.org/10.3390/pathogens8040295.

Proctor, C.R., Rhoads, W.J., Keane, T., Salehi, M., Hamilton, K., Pieper, K.J., Cwiertny, D.M., Prévost, M., Whelton, A.J., 2020. Considerations for large building water quality after extended stagnation. AWWA Water Sci. 2, e1186. https://doi.org/10.1002/aws2.1186.

RQIA, 2012. Regulation and Quality Improvement Authority—RQIA Inspection Reports | Regulation and Quality Improvement Authority Standards Reports. [WWW Document]. URL: https://rqia.org.uk/reviews/review-reports/2012-2015/rqia-pseudomonas-review/. (Accessed 11 April 2019).

Sener, A., Alkan, S., Önder, T., Karaduman, N., 2021. Ghost in opera: are Legionella bacteria really rare pathogens for hospital plumbing? D. J. Med. Sci. 7, 26–29. https://doi.org/10.5606/fng.btd.2021.25042.

UK Government, 1991. Water Industry Act 1991.

WHO (Ed.), 2011. Water Safety in Buildings.

WRAS, 1999. The Water Supply (Water Fittings) Regulations 1999.

WRAS, 2015. Cold Water Services: Storage Cisterns—General Design & Installation Requirements. https://l8watertrainingservicesltd.co.uk/wp-content/uploads/WRAS-CWST-adive.pdf.

Hot water heating systems

Hot water system

In large healthcare buildings, hot water is usually generated by using a centralized storage hot water system that supplies hot water outlets from a hot water storage vessel, heater, or boiler usually sited in a central position in the property. The water may be heated by a plate heat exchanger, gas boiler, or indirectly by having a heating coil in a calorifier. Temperature is the conventional method to control bacterial multiplication, and to achieve this, hot water in healthcare premises should be ≥55°C throughout the recirculating hot water system. Steam generation on-site may be used as a heat source. Water may be supplied via a cold water storage cistern to open-vented systems. Systems supplied directly from the mains supply are usually unvented, i.e., no vent pipe. In large healthcare buildings, hot water storage vessels (buffer vessels) are generally used to buffer the system during peak use and ensure there is sufficient hot water.

Hot water is then pumped around the building by means of a recirculating pipework (flow) system and back (return) to the heat source for reheating. Hot water is delivered to the highest point in the building by means of pipework located in vertical "risers" running the full height of the building and with horizontally branches off these to supply the individual floors of multistorey buildings. The hot water circulation system is composed of circulating loops, often referred to as the principal loop, subordinate loops, and tertiary loops (HSE, 2014a).

Each floor comprises a looped system with spurs branching off to each outlet to ensure that target temperatures are achieved within 1 min of turning on a tap or shower. Spurs off each loop to feed the outlets should be kept as short as possible to minimize:

- time for hot water to be delivered at the target temperature;
- potential for heat loss; and
- microbial growth.

Achieving the required control temperatures may sound simple, but in practice ensuring target temperatures recirculating through all return branches of the system can be difficult. This will depend on the ability of the system to cope with the amount of water required, ensuring water is pulled through to each outlet and that the water pressures throughout the system are balanced so that the flow is able to maintain the temperatures in each circuit.

The flow rate needed to maintain temperatures in each loop depends on the pipe diameter, the amount of usage, and the inherent temperature loss from the system. In addition, the efficient delivery of hot water at the required temperature ensures that less water is run through the outlets to achieve the target temperature and minimizes the potential for water going down the drain.

Safe Water in Healthcare. https://doi.org/10.1016/B978-0-323-90492-6.00024-0

Monitoring temperature then is a key performance indicator of a hot water system. The presence of scale and corrosion in the pipework will reduce the flow through the pipework and increase the time taken to deliver the target temperatures. Understanding when, where, how, and what the various pitfalls are to achieving representative temperature monitoring are important to maintaining a safe system (see Chapter 25).

Principal loops are the larger bore circulating systems up to the furthest point, and from that point, the water is then on its return back to the water heater (Fig. 1).

Principal loops have branches that feed subordinate loops, which generally consist of smaller bore pipework branching from the principal loop to supply a group of outlets, specialist equipment, or a separate system such as a pool. This then loops back into the principal loop returning to the heat source (Fig. 1).

A further loop, called a tertiary loop, may be fed from and return to the subordinate loop, usually to feed a small number of outlets, e.g., individual patient ensuite facilities or treatment rooms (Fig. 1). It is important that each of these loops is monitored to ensure that they are achieving appropriate temperatures.

Distributed hot water must still comply with the drinking water regulations and is used for a range of purposes, including hand hygiene, personal hygiene of patients and staff, as well as for domestic cleaning activities.

In areas of low or intermittent use, instantaneous hot water heaters where there is no storage may be appropriate as this enables hot water to be delivered when a tap is turned on.

In small localized hot water systems that supply small amounts of hot water to individual outlets or appliances, e.g., dishwashers and washing machines, then temperatures can be relatively easy to manage and control. Large hospital buildings tend to have one large centralized hot water system.

However, experience has shown that temperatures needed to control microbial growth consistently across all parts of large hot water systems are difficult to achieve.

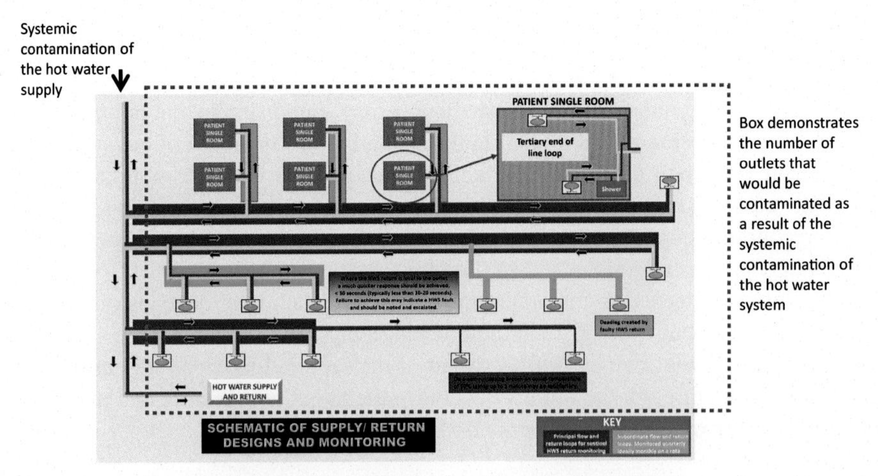

Fig. 1 Diagram to demonstrate principle and subordinate hot water loops.

To achieve the required pressures and circulating temperatures, manual or thermostatic regulation valves in the return loops to all outlets are used to balance the flow to individual pipe branches serving individual or groups of draw-off points, for example, each washroom/toilet and en-suite facility. There should be means of isolation (both up and down stream) of these valves for servicing and access for maintenance without creating a deadleg.

Testing the temperature in the hot water principal, subordinate, and tertiary branches of the circulating systems installed in all departments ensures that the system has been balanced, and that under "no draw-off" conditions 55°C is achieved in the circulating system at all outlets. However, just because the appropriate temperature is achieved on the flow section of the recirculation system, i.e., at the outlets, it is possible for the return valves to malfunction resulting in lower temperatures on the return, which will encourage microbial growth.

Balancing the hot water system flow and return circuits is critical to avoid long lengths of stagnant pipework that is likely to be at a lower temperature.

As in any building, water systems should be designed and installed in accordance with the Water Supply (Water Fittings) Regulations 1999 and relevant parts of BS EN 805, BS EN 806 (Parts 1–5) and BS 8558 (BSI, 2000, 2005, 2015; WRAS, 1999). Advice on the design, installation, and commissioning of water systems in healthcare premises is also provided in HTM 04-01 Part A (DHSC Safe water in healthcare premises (HTM 04-01), 2017, p. 04).

Temperature requirements for hot water calorifier, flow, and returns?

Current UK national guidance in HSG274 Part 2 and HTM 04-01 Parts A and B advise that the water temperature in the calorifier should be at least 60°C (HSE, 2014a). This temperature is required to ensure effective pasteurization (i.e., controlling the microbial load of microorganisms by exposing them to high temperatures) of the water that is to be distributed. However, this does not mean that all microorganisms will be killed, as within the calorifier there is a temperature gradient, hot water flows from the top and returns to near the base, cold water replacing the amount of water used is fed in from close to the base, so the base will be cooler. Sludge also accumulates at the calorifier base, which provides a favorable niche for microbial survival due to the lower temperatures in that location. To minimize the risk of microbial growth at the base, during periods of low use, such as the early hours of the morning, hot water is pumped from the top of the calorifier to the base by an antidestratification pump. Some calorifiers are designed so that the water is recirculated at all times so a pump is not needed. Accordingly, water leaving the calorifier (the "flow") should be at 60°C and arrive at all outlets on the distribution loop(s) at a minimum of 55°C. Hot water returning to the calorifier (the "return") should be a minimum of 50°C. Maintaining hot water at these temperatures in the distribution circuit ensures that any microorganisms, which might have escaped the pasteurization process in the calorifier, do not multiply in the distribution system, and has been shown to be effective in the control of *Legionella* (Bédard et al., 2015; Gavaldà et al., 2019).

Balancing the system

A schematic of a hot water system showing flow and return imparts a reassurance, which is not borne out in practice. In large, complicated water systems as found in a hospital, the hot water pipework may be tens of miles in length. Ensuring that hot water simultaneously flows round every loop of the system can be extremely difficult to achieve. It is therefore important that the return temperature of each loop is monitored and not just the overall return to the calorifier. Cases of Legionnaires disease have occurred because the overall temperature back to the calorifier was compliant even though one loop had a practically closed valve, which meant it did not achieve control temperatures. This was only picked up during the investigation. The process of ensuring hot water circulates round all parts of the circuit is known as "thermal balancing."

Balancing a water system is an essential part of commissioning. Specialist valves are used to regulate the flow of water. Once a system is balanced making changes to one valve can affect the flow of water to other areas. In some circumstances, experience has shown that the balance of the system may be affected by shutting down an isolation valve to be able to work on a particular area but not fully reopening the valve once work is completed.

Case study 1

Routine water sampling for *Legionella* spp. identified contamination at an outlet on a medical ward. As part of the standard operating procedure, water temperatures were measured. This revealed that the hot water temperatures in the area were below the requirement. Temperature mapping of surrounding outlets traced this back to an isolation valve not being fully open. Prior temperature measurements in this area had been satisfactory. However, routine maintenance had recently required closure of the isolation valve, and it is thought likely that the valve had not been fully opened following completion of the work, leading to the suboptimal temperatures.

Balancing a large system is complicated and often rather than balancing each loop, isolating valves on each outlet are used to moderate pressure. When the pressure of the system is such that there is excessive splashing, these valves may be left partially closed. Opening them often results in discolored water as a result of stagnation behind the valve. Thus, temperature profiling the system to check sufficient hot water is flowing through all areas and repeating this regularly is key to ensuring the system is safe.

System design and temperature monitoring

When measuring hot water system temperatures, understanding the design layout of the system is important for interpretation. In an ideal world flow, return and outlet temperatures would be measured, and in new builds this is very achievable as remote monitoring can be installed and linked to the building management system. Remote monitoring is becoming more available and common in hospital due to the ability to identify failures and risks in the water system. Monitoring the water system is discussed in detail in Chapter 25.

Design of a hot water system should ensure that the length of the pipe from the recirculating system to each of the outlets is as short as possible. While measuring water temperatures coming out of an outlet will provide valuable information, it is also important to measure the temperature of the return pipes.

The return temperature should be above 50°C whatever the time of day if the hot water system is functioning correctly.

In Fig. 1, the patient single room box shows a noncirculating branch. In this instance, depending on the flow rate, diameter, and length of the pipe it may take up to 1 min or more for water at the outlet to reach 55°C. It is possible to calculate the time needed to purge a spur to achieve circulating loop temperatures from the pipe length and diameter, and if the system has been well designed, this should only take a few seconds.

As many outlets within hospitals are fitted with thermostatic mixer valves, the outlet temperatures only reflect the temperature of combined hot and cold water and not that of the hot water spur or recirculation loop.

Where thermostatic valves are not present, thermal control can be checked routinely by measuring water temperatures at selected outlets to confirm that hot water reaches 50°C within 30 s and 55°C within 1 min of that outlet being opened.

How is hot water produced?

There are a number of different ways by which water can be heated in a calorifier, but all calorifier designs must comply with current UK regulations (Fig. 2) (HSE, 1998, 2014b; Scottish Water, 2014; WRAS, 1999).

Fig. 2 Hot water calorifiers in the plant room.
Image courtesy of Daniel Pitcher at Water Hygiene Centre.

Typically, the cold water is fed by gravity from the cold water cistern to the base of an indirectly heated calorifier. Calorifiers not only heat the water, they also act as hot water storage vessels. For indirectly heated calorifiers, the steam or high-temperature hot water from the boiler passes through a hollow metallic coil located within the calorifier. Calorifiers are designed so that cold water entering the calorifier mixes with the water that is being heated by the coil before reaching the specified temperature and then the hot water is distributed to the rest of the circulatory system by means of one or more mechanical pumps.

What is stratification?

However, where the cold water enters the calorifier it has an immediate cooling effect as the cold and hot water is blended resulting in water temperatures below 50°C, which have been shown to support microbial growth (Farrell et al., 1990). The cooling effect of the cold supply can also result in thermal stratification. This is a phenomenon caused by the greater density of cold water compared with warmer water, which results in the denser cool water remaining in the bottom of the calorifier vessel. Prolonged periods of stratification can allow harmful bacteria, such as *Legionella*, to grow in the low-temperature zone. The problem can be addressed by means of an antistratification pump (often referred to as a shunt pump), which is fitted to the calorifier vessel and which circulates the water from the top of the calorifier back to the base during the period of least demand, e.g., during the early hours of the morning. The boiler plant (or other calorifier heat source) should be heating while the shunt pump is active to ensure a temperature of at least 60°C is achieved throughout the vessel for at least one continuous hour a day.

Calorifiers that are heated directly at their base tend to be less prone to thermal stratification due to convection currents that are generated at the base that help to circulate the water within the vessel and mix its contents with the cold supply water.

Particulate matter, which enters a calorifier, in the supply water or due to scale and/ corrosion in the returns can accumulate at the base of the calorifier and can exacerbate microbial growth by providing nutrients and protection from thermal inactivation. Therefore, the calorifier should incorporate an easily accessible drain valve to facilitate regular draining and flushing through.

Electrical immersion-type calorifiers

Electrical immersion-type calorifiers tend to hold smaller volumes of water and as a consequence are also usually less prone to thermal stratification. Their smaller storage volume (around 15 L) also helps to reduce overall risk because this feature helps to ensure frequent turnover of the stored water. However, because of their reduced capacity, their use can be limited in healthcare, but they may be considered in certain specialist settings. Similarly, lower-risk hot water generating devices (as described in HSG274

Part 2), such as combination boilers and instantaneous water heaters, which do not store hot water, may have applications in some hospital areas (HSE, 2014a).

Combination water heaters

Combination water heaters, where the heating vessel is fed from an integral cold water storage tank located at the top of the vessel, are not recommended for use in healthcare settings because of the potential for the stored cold water to be heated by the cylinder below and to reach temperatures that encourage the growth of harmful bacteria, such as *Legionella*.

Microbiological problems with calorifiers and hot water systems?

- Water entering the calorifier from the cold water tank may be contaminated by *Legionella* spp., and these can reach dangerously high numbers if conditions in the calorifier allow.
- The capacity of the calorifier is insufficient to meet the demands for hot water. Checking the recovery time following times of peak use can give an indication as to whether the heating capacity is sufficient.
- Temperature stratification—The cold water from the cistern tank can mix with the warm water in the base of the calorifier to create zones where the temperature is not adequate to maintain control of *Legionella* (Farrell et al., 1990). Hence, the importance of the antistratification pump?
- The presence of debris and corrosion deposits in the bottom of the calorifier will also result in areas where microbial multiplication and biofilm development can occur.
- Dead legs are areas where there is little circulation of water. They often arise because when previous modifications have been made to the water system, i.e., an outlet removed, the pipework has not been cut back to the main circulating water pipes. Poorly designed expansion vessels, i.e., not the flow-through variety, or an example of a dead leg. Functional dead legs exist when, for example, an outlet is not used.
- Inadequate insulation—The presence of hot and cold water pipes in the risers and in conduits can lead to a temperature reduction in the hot water pipes especially where insulation has been removed for servicing and/or maintenance and not replaced.
- Failure of hot water to circulate—this can be (1) systemic as in the case of a pump failure or inadequate flow (2) local due to incorrect balancing of the system or local blockage—Faulty manual, blending, or thermostatic regulation valves in the return loops could lead to a reduction in the return hot water.
- Low temperatures in the hot water return entering the calorifier may also reduce the efficiency of the calorifier to maintain high enough temperatures to ensure microbial control is achieved.
- When heated, expanded water can travel along the cold feed to the cistern, which should be prevented as this would warm the water in the cold water tank and encourage microbial growth.
- Spurs to outlets provide a localized area where circulating hot water will not flow, so temperatures will move to the ambient; hence, the requirement for flushing. The longer the spur

from the recirculating system to the outlet, then the greater the heat loss—hence spurs should be kept to a minimum.

- TMVs and TMTs are designed to prevent scalding, and therefore, water downstream of these devices will be at temperatures conducive to bacterial multiplication. When a system is being pasteurized, it is necessary to override the TMV/TMT in order to achieve sufficient temperature to reduce presence of microorganisms.

Case study 2: What to do when a bolus of potentially contaminated material is injected into your hot water system?

Due to a faulty pressure relieving valve, there were simultaneously a number of burst pipes on the hot water system and discolored water appeared at many outlets. The likely source of the discolored material was thought to be debris from the base of the calorifier, an environment where bacteria may survive.

The matter was raised at an incident meeting of the water safety group. The options ranged from doing nothing—the argument being that the hot water temperatures would prevent further multiplication of the bacteria—to doing a system pasteurization. The latter is a demanding and potentially dangerous procedure if not carried out correctly.

Advice was taken from a national expert who viewed the overall risk to be low. There was agreement that the hot water would prevent multiplication of the organisms providing the system was working correctly. However, there will always be areas in the system where the hot water is not circulating and would provide an environment for bacterial multiplication. Such areas would include all spurs emanating from the circulating water system and the strainers before the thermostatic mixing valves, which would trap any debris. Therefore, to mitigate the risk all outlets were sequentially run to flush out any bacteria that may have lodged in the spurs and to have a program of rapidly cleaning all TMV strainers, beginning with the patient high-risk areas first.

Transmission routes for calorifiers and hot water systems?

Where conditions in the base of the calorifier enable microbial growth, then those microorganisms can be transported around the building via the hot water system. This could happen when the contents of the calorifier are turning over (e.g., during periods of high demand for hot water) or when the calorifier has been perturbed by a pressure loss or gain due to expansion. The microorganisms can either be present as free (planktonic) cells in the water column, in fragments of biofilm, or even as biofilm attached to small particles of debris. The microorganisms released from the calorifier can seed other parts of the hot water system where they can multiply to hazardous levels in time. This potential for microbial growth can be encouraged when there is heat loss due to the lack of insulated pipework or in pipework leading to underused outlets where water is static and cooler as a result. When the outlet is subsequently opened, any opportunist pathogens could be released and be transmitted to the environment, individuals, and susceptible patients in the vicinity.

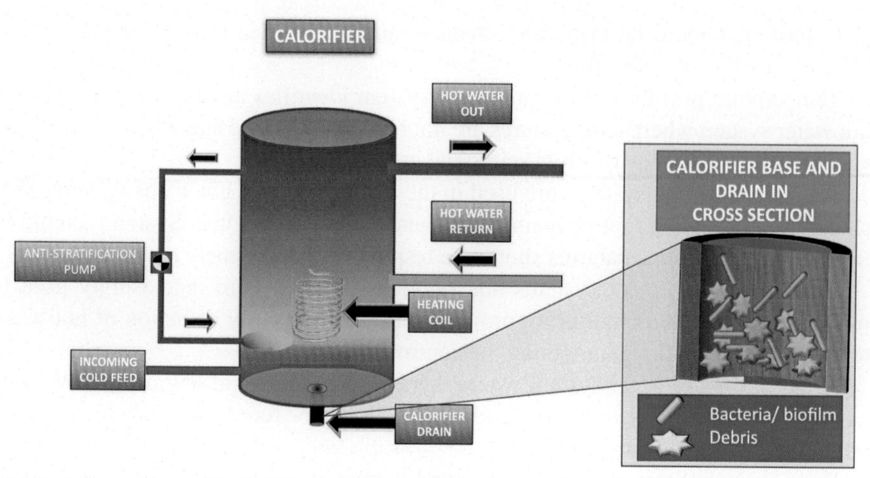

Fig. 3 Schematic of an indirectly heated calorifier.

- Hot water circulation pumps should be used to regulate the balance of the individual pipe branches and different loops but should be alternated regularly and frequently. Ideally, this switching should on an automatic regime to avoid pumps on stand-by, which contain stagnant water (Fig. 3).
- Expansion vessels that are used to regulate water pressure changes in the system contain flexible membranes or bladders made from synthetic rubber, such as EPDM, these provide an ideal surface for microbial colonization and growth. They should be fitted vertically (not horizontally) on the cold side of the system and away from any source of heat. They should have a drain valve and be regularly flushed.
- The presence of inline strainers, used to protect certain plumbing components on the hot water supply, will also provide a breeding ground for the growth of microorganisms including *Legionella* and *Pseudomonas aeruginosa* biofilms, particularly when they have collected sludge and debris attached to them. Inline strainers should be included in routine cleaning, maintenance, and disinfection procedures.
- Hot water pipes should be insulated to prevent heat loss and, ideally, should not occupy the same risers or other conduits as cold water pipes. Where this is not possible (because of the design of the existing distribution system), any hot water pipes should run above the cold water pipes (to minimize heat transfer via convection to the cold pipe).

Monitoring hot water systems

The frequency of inspecting and monitoring the hot water system will depend on the complexity and the susceptibility of those exposed to the water. The risk assessment should define the frequency of inspection and monitoring depending on the type of use and users exposed to this water.

Calorifiers should be subjected to regular manual checks for flow and return temperatures, i.e., there should not be reliance solely on readings taken from gauges, if they are fitted, or measurements taken by a building management system (BMS), if installed.

Calorifiers should be inspected, cleaned, and maintained at least annually or as indicated by the rate of fouling.

Temperature profiling of the hot water system identifies any problems within the hot water system where temperatures are not being maintained and where reduced hot water temperatures may result in microbial growth.

Where monitoring systems are used in the management of hot water systems, they should be subject to a routine maintenance and calibration regime. Systems should be in place to ensure all set alarms should be responded to in a timely manner.

Building management systems and remote monitoring are increasingly used to provide an automated monitoring program, allowing for early detection of hot water temperature failures in maintaining the control regime.

Areas in the hot water system where there is a possibility of low or no flow, such as blind ends, dead legs, and little used outlets, should be removed or where this is not possible, identified for monitoring.

Parts of the hot water system susceptible to heat loss that could support microbial growth should be identified and remedial work to improve insulation carried out.

Identify areas where there is low hot water throughput including, e.g., low-use fittings.

Assess hot water return pipes particularly at points furthest away from the water heater, where circulation has been reduced or failed leading to stagnation and cooling of the hot water.

Temperature monitoring is a key performance indicator of the safety of the system. The water safety group should ensure that:

1. Temperatures are routinely monitored in all the appropriate areas.
2. Areas where target temperatures are not achieved are identified for investigation and remedial work.
3. All monitoring results are kept under regular review.
4. There should be an immediate escalation policy for when temperatures are found to be out of specification.

Sentinel taps, i.e., usually the first and last taps of the hot water recirculating system, should be monitored and can include taps, which represent parts of the recirculating system where monitoring can identify areas needing remedial action to improve control. Where the system consists of several recirculating loops (Fig. 1), the end of each loop should be identified as sentinel points for monthly monitoring. Subordinate loops should be monitored quarterly, ideally at a suitable return leg or from a representative outlet.

Take temperatures at return legs of principal loops (sentinel points) monthly to confirm they are at a minimum of 55°C.

Check temperatures on the surface of pipes at return legs of subordinate loops (quarterly on a monthly rotation), but where this is not practicable, the temperature of water from the last outlet on each loop may be measured and this should be greater than 55°C within 1 min. If the temperature rise is slow, it should be confirmed that the outlet is on a long leg and not that the flow and return has failed in that local area.

Check temperatures at a representative selection of other points (intermediate outlets of single pipe systems and tertiary loops in circulating systems—Fig. 1) to confirm they are at a minimum of 55°C to create a temperature profile of the whole system over a defined time period

In noncirculating systems, take temperatures monthly at sentinel points (nearest, furthest outlet and long branches to outlets) to confirm they are at a minimum of 55°C.

Any parts of the hot water system not represented by sentinels should be identified, and additional outlets selected for less frequent monitoring to create a temperature profile of the whole system over a defined time.

Hot water systems, which supply outlets to high-risk users and incorporate tertiary loops, e.g., showers in healthcare premises, should be identified as areas for additional temperature monitoring.

Timely, appropriate remedial action to poor temperature or other monitoring results should be used as an indicator of the effectiveness and adequacy of the management controls in place.

Where required a range of microbiological monitoring can be used to assess the growth of microorganisms either using total viable counts or trends or for specific microorganisms as identified by the risk assessment or clinical advice.

Where there is a large difference (>10°C) between the flow and the return temperatures to the calorifier, this suggests excessive heat loss and should be investigated.

Where there is only a small difference between the flow and return, this can indicate there may be a bypass and also should be investigated.

Recommendation for hot water heaters

1. Calorifiers should be fitted with a de-stratification pump, where necessary, in order to avoid temperature stratification of the stored water.
2. Where two hot water pumps are installed in parallel, they should be arranged to have individual nonreturn and service valves and be operated such that each the duty and standby pumps are switched regularly, e.g., twice a day. Alternatively, dual pump systems should share the duty to ensure both pumps are operational and control the risk of water stagnation in one or the other.
3. Flushing of the calorifier—Drain should initially be carried out quarterly to minimize the accumulation of sludge and the frequency reviewed depending on the level and rate of fouling.
4. When dismantled for statutory inspection, or every year in the case of indirect calorifiers, calorifiers should be thoroughly cleaned to remove sludge, loose debris, and scale.
5. Calorifiers should have adequately sized access panels to enable inspection and cleaning.
6. Trace heating systems are not recommended as they have been shown not to be reliable.
7. There should be means of isolation, both upstream and downstream. Adequate access for servicing is also essential.

Regulations and guidance for hot water heaters

Hot water services should be designed and installed in accordance with the Water Supply (Water Fittings) Regulations 1999 and relevant parts of BS EN 806-2 and BS 8558 (BSI, 2005, 2015; WRAS, 1999).

In Scotland, the Scottish Water Byelaws 2004 (Scottish Water, 2014) apply and the Provision and Use of Work Equipment Regulations 1998 apply to all calorifiers and, depending on their design, the Pressure Systems Safety Regulations 2000 may be applicable (HSE, 1998; Scottish Water, 2014).

HTM 04-01 Safe Water in Healthcare (DHSC, 2017)

HSG274 Legionnaires' disease Part 2: The control of legionella bacteria in hot and cold water systems (HSE, 2014a)

Risk assessing the hot water system

Risk assessment criteria	Yes	No
Have you identified where and how hot water is stored throughout the hospital?		
Has the calorifier been inspected internally?		
Has the calorifier been drained and cleaned in the last 12 months (Check This Time Period)?		
Have deposits been found in the calorifier?		
Have the temperatures in the hot water tank been assessed for the control of *Legionella* spp. (thermostat settings should modulate as close to 60°C as practicable without going below 60°C) Check calorifier return temperatures (not below 55°C).		
Have antistratification pumps been fitted where needed on calorifiers and hot water buffer vessels and shown to be working in the correct direction, i.e., from the top (hot) down to the base?		
Has insulation been fitted to the appropriate pipes and equipment and has it been retained in place?		
Is the water temperature in all or some part of the system between 20°C and 45°C? (before TMVs)		
Is a building management system (BMS) being used to monitor hot water temperatures of the calorifier flow and return and on each loop and what actions have been taken where adverse readings have been reported?		
Have water temperatures been monitored monthly at return legs of principle loops (sentinel points) to confirm they are at a minimum 55°C?		
Have water temperatures been monitored quarterly at return legs of subordinate loops to confirm they are at a minimum of 55°C within 1 min of running?		

References

Bédard, E., Fey, S., Charron, D., Lalancette, C., Cantin, P., Dolcé, P., Laferrière, C., Déziel, E., Prévost, M., 2015. Temperature diagnostic to identify high risk areas and optimize, *Legionella pneumophila* surveillance in hot water distribution systems. Water Res. 71, 244–256. https://doi.org/10.1016/j.watres.2015.01.006.

BSI, 2000. BS EN 805:2000—Water supply. Requirements for Systems and Components Outside Buildings.

BSI, 2005. BS EN 806-2:2005—Specifications for Installations Inside Buildings Conveying Water for Human Consumption. Design—BSI British Standards.

BSI, 2015. BS 8558:2015 Guide to the Design, Installation, Testing and Maintenance of Services Supplying Water for Domestic Use Within Buildings and Their Curtilages. Complementary Guidance to BS EN 806.

DHSC Safe water in healthcare premises (HTM 04-01), 2017. Safe Water in Healthcare Premises (HTM 04-01). [WWW Document]. GOV.UK. URL: https://www.gov.uk/government/publications/hot-and-cold-water-supply-storage-and-distribution-systems-for-healthcare-premises. (Accessed 5 January 2021 (Accessed 24 August 2020).

Farrell, I.D., Barker, J.E., Miles, E.P., Hutchison, J.G.P., 1990. A field study of the survival of *Legionella pneumophila* in a hospital hot–water system. Epidemiol. Infect. 104, 381–387. https://doi.org/10.1017/S0950268800047397.

Gavaldà, L., Garcia-Nuñez, M., Quero, S., Gutierrez-Milla, C., Sabrià, M., 2019. Role of hot water temperature and water system use on *Legionella control* in a tertiary hospital: an 8-year longitudinal study. Water Res. 149, 460–466. https://doi.org/10.1016/j.watres.2018.11.032.

HSE, 1998. Provision and Use of Work Equipment Regulations 1998 (PUWER)—Work Equipment and Machinery.

HSE, 2014a. HSG 274 Legionnaires' Disease—Technical Guidance Part 2: The Control of Legionella Bacteria in Hot and Cold Water Systems Technical Guidance. http://www.hse.gov.uk/pubns/books/hsg274.htm. (Accessed 5 January 2021).

HSE, 2014b. Safety of pressure systems—L122.

Scottish Water, 2014. Water Byelaws.

WRAS, 1999. The Water Supply (Water Fittings) Regulations 1999.

Systemic contamination

6

Identifying systemic contamination?

Hospital water systems are complex, as is the nature and type of microbial contamination, which occurs in different ways in different parts of the system and can be divided into systemic and peripheral contamination.

Systemic contamination is where microbial contamination occurs in long sections of the water distribution system, e.g., in the risers/droppers, principal and subordinate loops of the hot water system as well as through long sections of the cold water system (Bédard et al., 2019).

Peripheral contamination normally occurs within the last 2 m of the pipework and may affect a number of the asset components, e.g., thermostatic mixer valve, taps, and showerheads). With peripheral contamination, not all the outlets in an area would be expected to be contaminated.

With systemic contamination, one would expect all outlets in the water system or that part of the water system to be equally affected as the water represents the contamination upstream or further back in the system.

The difference between systemic and peripheral contamination is usually reflected in the findings of pre- and postflush water samples, which are taken for microbiological analysis to determine if the contamination is associated with the last 2 m (peripheral) or further back in the system (systemic) (Bédard et al., 2016, 2019).

Preflush represents the sample of water when the outlet is switched on while the postflush samples are taken after the outlet has been switched on for a set period of time and water from further back in the water system has reached the outlet, and the importance of this is explained in greater detail later on in this section.

How does systemic contamination occur?

The main water supply to the building is not entirely free from microorganisms, and potential opportunistic waterborne pathogens are invariably initially introduced into the water system in the water supply. However, in the water system itself, systemic contamination can occur where favorable microbial growth conditions are encountered. These conditions can include a lack of flow (possible stagnation), elevated cold water temperatures >20°C, or reduced hot water temperatures (45°C), which can be exacerbated by a lack of or the presence of ineffective pipework insulation and heat transfer between hot and cold water pipes. Similarly, poor flow within the system can result in water with ineffective concentrations of chemical disinfectants in parts of pulled through the water system, which will lead to microbial proliferation. The use of remote temperature and/or chemical sensors attached to pipework on hot and cold

Safe Water in Healthcare. https://doi.org/10.1016/B978-0-323-90492-6.00014-8

water systems has now been widely adopted in many hospitals and can also provide alerts to aid the identification of problem areas (Chapter 25).

Systemic colonization occurs when these contaminating microorganisms are provided with an opportunity to grow under favorable conditions (inappropriate temperatures and provision of nutrients source) within the water system.

In newly built hospital water systems, contamination can also occur if insufficient care is taken during the construction and commissioning phases (Chapter 2). If pipework and other components are not kept clean and dry prior to installation, contamination of the system at the outset is very likely. Once the system is filled with water, e.g., during pressure testing to check for leaks, contamination can lead to colonization unless meticulous care is taken to prevent it. Managing the water system effectively from the point that water enters the system is, therefore, key to the safe operation of the system when it is brought into normal use.

Commissioning is an extremely important time in the life cycle of water systems. During building, there will inevitably be ingress of dirt and microorganisms and leaching of materials, which will support microbial growth, and all too often water systems are pressure tested without any regard to maintaining water safety due to a commissioning plan not being agreed in advance (e.g., as appeared to happen at the commissioning of new hospitals in Glasgow and Coventry). Temperature control and maintenance if essential for microbial control and failure to achieve the correct water temperatures, e.g., due to stagnation provide an ideal opportunity for planktonic and biofilm formation. During commissioning, flushing rarely produces the equivalent level of water turnover compared with when a building is occupied. In an ideal world, flushing on the cold system would be matched to occupancy levels to ensure that cold water temperatures remain in the safe range, i.e., <20°C. The ritual of chlorinating the system prior to handing over the new building is usually sufficient to mask the presence of planktonic cells and biofilm when water samples are collected. However, the effect of chlorine is likely to be temporary, and as such the contamination is likely to reappear. Although the initial contamination may have been systemic, with regular use, the contamination may only be found at the periphery of the system. Thus, when an outlet is tested, the pre- and postflush counts will suggest local contamination, but if more outlets are tested, many are likely to be positive.

If both the hot and cold water systems have been identified as suffering from systemic contamination, then it is possible that the cold water storage tanks may be the primary source of microbial contamination and should be investigated (Fig. 1). Inadequate routine cleaning and disinfection of tanks could be the root cause, as might oversizing of the tank (lack of turnover, leading to over storage of water, stagnation, and microbial proliferation. Alternatively, locating cold water tanks inappropriately in warm plant rooms or ineffective cleaning and disinfection can be contributory factors, as well as poorly designed or maintained tanks (e.g., lids not being broken or not sealed properly), which allow ingress of debris and nutrients, leading to microbial growth.

Where systemic contamination has only been identified in the hot water system, this could be due to ineffective pasteurization of hot water within the calorifier(s). The temperature of the contents of the entire calorifier vessel should reach a minimum of 60°C, and checks should be made that this is the case, and that the thermal antistratification pumps are working effectively (Chapter 5), and that any accumulated debris in the calorifier vessel is removed during routine inspections.

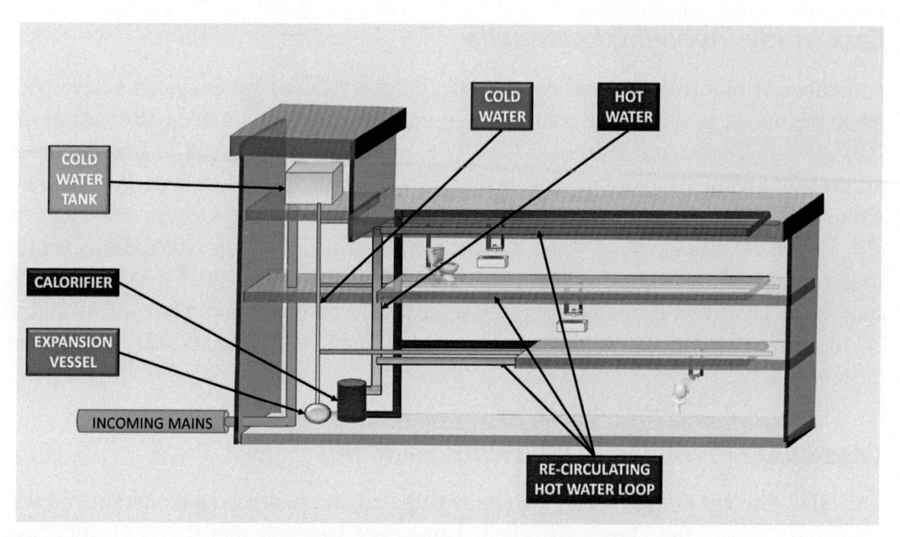

Fig. 1 Examples where systemic contamination can occur in the hot (return pipes) and cold water system (tank and system pipes) as identified by the gradation of color.

What are the impacts of systemic contamination?

Where systemic colonization occurs, there is potential for microbial contamination to be disseminated through the water pipe network to large areas of the hospital, including those where patients who are highly susceptible to infection receive their care (Fig. 2).

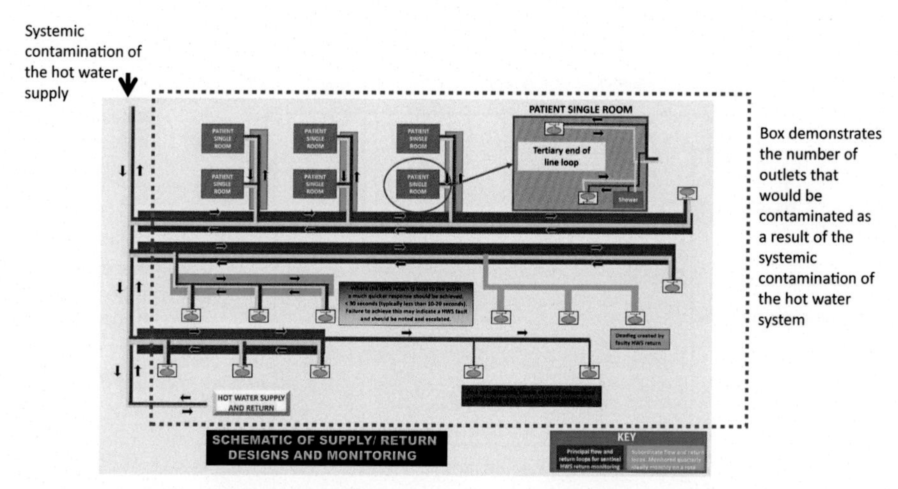

Fig. 2 Example of systemic contamination of the hot water system contamination of all downstream outlets.

Case study involving Legionella

A number of outbreaks caused by systemic colonization of water systems have been reported, and it is likely that many others have gone undocumented (Bédard et al., 2016, 2019; Decker and Palmore, 2013). For example, an outbreak of Legionnaires' disease involving 21 patients that resulted in 5 attributable deaths at the Veterans Affairs Pittsburgh Healthcare System was cited in a review by Decker and Palmore (2013), where 29 out of 44 water samples taken from a variety of locations on the water system (66%) were positive for *Legionella* spp.) (Hicks, 2021). This high percentage of positives, particularly where a particular outbreak strain was identified in environmental samples, is highly indicative of systemic contamination and underlines the significance of its occurrence.

Case study involving Cupriavidus pauculus

The NHS Greater Glasgow and Clyde investigated and managed a contaminated water system across the Queen Elizabeth University Hospital and Royal Hospital for Children with probable linked cases of bloodstream infections due to *Cupriavidus pauculus* between 2016 and 2018. While initial investigations identified positive water samples from a wash hand basin tap outlet, subsequent testing identified widespread contamination of the entire water system (HPS, 2018; Inkster et al., 2021; Weinbren and Inkster, 2021).

While it is acknowledged widely that the majority of contamination incidents with *P. aeruginosa* are at the periphery of the water system, this is not always the case (Garvey et al., 2016). Systemic contamination of the water system with *P. aeuginosa* in this hospital was found to be due to the unusual design of the cold water storage tank. To support the lid, hollow poles (stantions/supports) had been placed in the tank. Being hollow stagnant water was contained within the roof supports (Figs. 3 and 4).

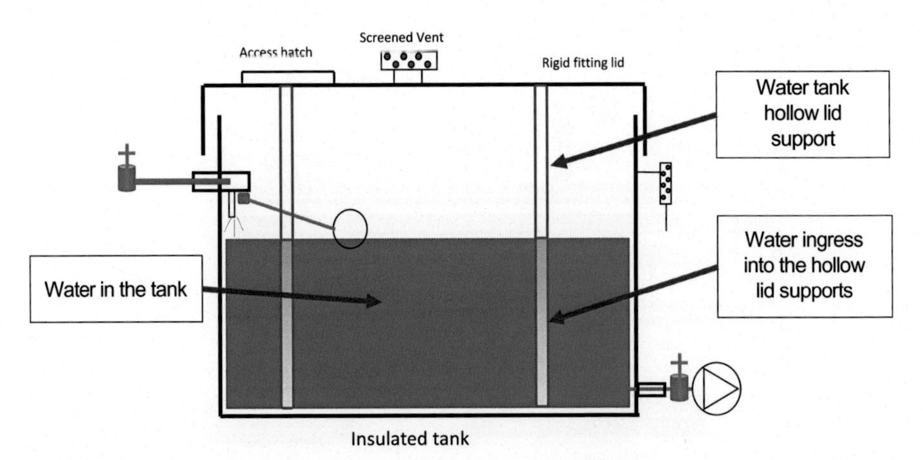

Fig. 3 Schematic of cold water tank demonstrating the hollow lid supports that resulted in systemic contamination.

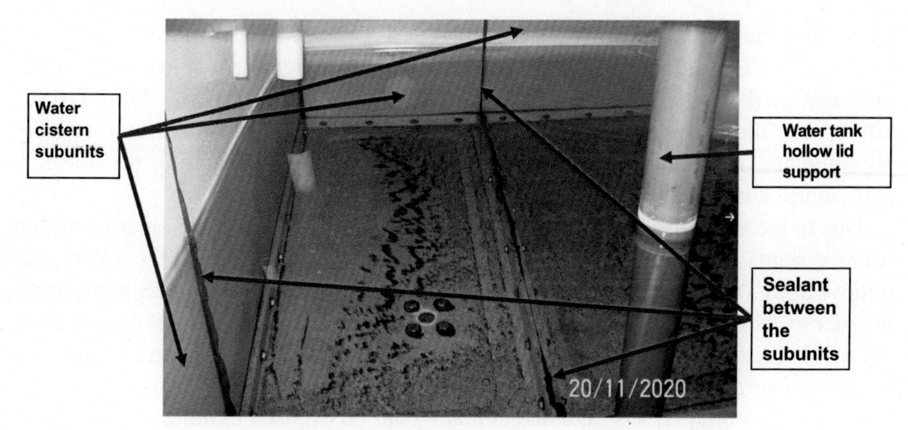

Fig. 4 Water tank subunits, sealant, and hollow legs that resulted in systemic microbial contamination.

This had resulted in biofilm formation with *P. aeruginosa* within the hollow supports and subsequent growth and movement of the organisms into the cold tank.

Case study involving Pseudomonas aeruginosa

At another hospital, and in preparation for the opening of a new Burns unit, soil testing of the bedpan washer disinfector was undertaken (ISO, 2005). The method of testing consisted of making up a soil containing *Enterococcus faecalis*, which was placed in a urine bottle and then put in the bedpan washer disinfector. The urine bottle was then sampled for the presence of *E. faecalis*. No enterococcus was detected, but *P. aeruginosa* was. The test was repeated and yielded the same findings. The origin of the *P. aeruginosa* was the cold water tank. The cold water tank was made up a number of subunits (Fig. 4), which were clamped together after placement of a rubber sealing material shown in red. The material used was not WRAS approved and supported the growth of *P. aeruginosa*. All cold water outlets tested positive for *P. aeruginosa* in keeping with systemic contamination.

Monitoring systemic contamination

Systemic contamination can be identified during routine microbiological monitoring, but only if investigations rule out other types of contamination, such as peripheral contamination that occurs in the last few meters of the water system near to outlets. For example, systemic contamination where large areas of the cold or hot water system are fouled can be identified by sampling a wide range of outlets pre- and postflush (Bédard et al., 2016, 2019; Hicks, 2021). Fundamentally, preflush is the first portion of water taken for sampling when the outlet is opened and represents the microbial content of the water in the immediate vicinity of the outlet. Postflush samples are taken after the preflush sample has been collected and typically after the outlet has been opened (flushed) for 1 or 2 min and represents the microbial content of the water upstream of the outlet and within the system supply pipework (BSI, 2021; DHSC, 2017). It is likely that systemic contamination has taken place where an unusually high number of distal outlets are positive post flush, across an area of the hospital, particularly with the same bacterial strains.

Due to the widespread use of thermostatic mixer valves (TMVs) in many hospital water systems, it may be necessary to either take the samples before the TMV or to remove each TMV prior to taking samples from the hot and cold supplies to the valve in order to avoid contamination of the samples by bacterial biofilms (that may be present within the body of the TMV) and which could give rise to confusing test results.

It is important to remember that widespread systemic contamination has the potential to put more patients (and others that would be exposed to the water) at risk of exposure to harmful microorganisms (DHSC Safe water in healthcare premises (HTM 04-01), 2017).

Where systemic contamination is occurring across multiple or large sections of the water system, and high counts are present in the water tank after treatment, it may be worth monitoring the incoming mains supply to ensure that this is not the cause of the problem (HPS, 2018; Weinbren and Inkster, 2021).

Recommendation for systemic contamination

Monitoring of the water system using pre- and postflush sampling for microbial counts will give an indication as to whether microbial contamination is systemic or localized at the periphery of the water system.

Check that the incoming mains supply meets requirements of the water supply regulations.

Assess the cold water tank(s) for the suitability of its location and ensure that there is an effective programme of inspection and, where necessary, cleaning and disinfec tion to minimum the accumulation of debris and biofilm and ensure that their design is such that ingress of external contamination is minimized, e.g., that the lids are tight fitting.

Monitor heat gain of the cold water pipe network and heat loss in the hot water pipe system.

Extensive use of automated remote sensors and reporting systems should be considered for surveillance of temperature profiling. Sensors should be located throughout the entirety of the hot and cold water systems to ensure they give representative temperature values.

Where systemic contamination has been identified, then cleaning and disinfection of the entire system are likely to be required.

To determine the efficacy of system disinfection, samples for microbiological analysis should be taken between 2 and 7 days after the system is treated as samples taken immediately after a disinfection process might give false negative or artificially low results.

Issues to consider when risk assessing systemic contamination?

During the building of a hospital, a new build extension, or refurbishment, have you been aware of situations where there has been contamination of the clean pipework and joints?

Have you been able to interact with the designers, plumbers, fitters, and builders when building is ongoing to ensure hygienic installation?

Have the temperature monitoring data been reviewed including?

- Cold water tanks
- Cold water network
- Base of the hot water calorifier and primary, secondary, and tertiary loops including return temperatures

Are the cold and hot water pipes thermally insulated?

Have pre- and postflush water samples been taken to determine that systemic contamination has occurred?

Guidance for systemic contamination

HTM 04-01 Safe Water in Healthcare (DHSC Safe water in healthcare premises (HTM 04-01), 2017, p. 04).

HSG274 Legionnaires' disease: Technical guidance (HSE, 2014).

BS 7592 Sampling for Legionella bacteria in water systems—Code of practice (BSI, 2021).

References

Bédard, E., Prévost, M., Déziel, E., 2016. *Pseudomonas aeruginosa* in premise plumbing of large buildings. MicrobiologyOpen 5, 937–956. https://doi.org/10.1002/mbo3.391.

Bédard, E., Paranjape, K., Lalancette, C., Villion, M., Quach, C., Laferrière, C., Faucher, S.P., Prévost, M., 2019. *Legionella pneumophila* levels and sequence-type distribution in hospital hot water samples from faucets to connecting pipes. Water Res. 156, 277–286. https://doi.org/10.1016/j.watres.2019.03.019.

BSI, 2021. BS 7592: 2008 Sampling for *Legionella* Bacteria in Water Systems. Code of practice.

Decker, B.K., Palmore, T.N., 2013. The role of water in healthcare-associated infections. Curr. Opin. Infect. Dis. 26, 345–351. https://doi.org/10.1097/QCO.0b013e3283630adf.

DHSC, 2017. HTM 04-01: Safe Water in Healthcare Premises. https://www.gov.uk/government/publications/hot-and-cold-water-supply-storage-and-distribution-systems-for-healthcare-premises. (Accessed 5 January 2021).

DHSC Safe water in healthcare premises (HTM 04-01), 2017. Safe Water in Healthcare Premises (HTM 04-01). [WWW Document]. GOV.UK. URL: https://www.gov.uk/government/publications/hot-and-cold-water-supply-storage-and-distribution-systems-for-healthcare-premises. (Accessed 5 January 2021 (Accessed 24 August 2020).

Garvey, M.I., Bradley, C.W., Tracey, J., Oppenheim, B., 2016. Continued transmission of *Pseudomonas aeruginosa* from a wash hand basin tap in a critical care unit. J. Hosp. Infect. 94, 8–12. https://doi.org/10.1016/j.jhin.2016.05.004.

Hicks, L.A., 2021. The CDC Investigation of Legionnaires' Disease Among Patients at the VA Pittsburgh Healthcare System. 2013.

HPS, 2018. Summary of Incident and Findings of the NHS Greater Glasgow and Clyde: Queen Elizabeth University Hospital/Royal Hospital for Children Water Contamination Incident and Recommendations for NHS Scotland. [WWW Document]. URL: https://www.gov.scot/binaries/content/documents/govscot/publications/factsheet/2019/02/qe-university-hospital-royal-hospital-children-water-incident/documents/queen-elizabeth-university-hospital-royal-hospital-for-chidren-water-contamination-incident-hps-report/queen-elizabeth-university-hospital-royal-hospital-for-chidren-water-contamination-incident-hps-report/govscot%3Adocument. (Accessed 17 April 2019).

HSE, 2014. HSG 274 Legionnaires' Disease—Technical Guidance Part 2: The Control of Legionella Bacteria in Hot and Cold Water Systems Technical Guidance. http://www.hse.gov.uk/pubns/books/hsg274.htm. (Accessed 5 January 2021).

Inkster, T., Peters, C., Wafer, T., Holloway, D., Makin, T., 2021. Investigation and control of an outbreak due to a contaminated hospital water system, identified following a rare case of *Cupriavidus pauculus* bacteraemia. J. Hosp. Infect. https://doi.org/10.1016/j.jhin.2021.02.001.

ISO, 2005. ISO/TS 15883-5:2005 Washer-disinfectors—Part 5: Test Soils and Methods for Demonstrating Cleaning Efficacy. ISO.

Weinbren, M., Inkster, T., 2021. The hospital-built environment: biofilm, biodiversity and bias. J. Hosp. Infect. https://doi.org/10.1016/j.jhin.2021.02.013.

Peripheral components

7

Introduction

Most microbial problems are going to become apparent with the distal or peripheral components in the last 2 m of the water distribution system as it is here that the most favorable conditions for growth are going to occur, e.g., stagnation when the outlet is not used. In addition, it is from these peripheral components that transmission of waterborne pathogens is going to occur. Such outlets will include taps for hand wash stations, kitchens, baths, and showers that will result in a spray of water leading to splashing and the production of droplets and aerosols (Allegra et al., 2020; Benoit et al., 2021; Chattopadhyay et al., 2017).

Aerosols and droplets are classically defined by their sizes with aerosols being smaller in size and being suspended in the air leading to respiratory inhalation and droplets being large particles leading to surface contamination. Conventional publications tend to define respirable aerosols as less than 5 μm in size (Bennett et al., 2000). The bacteria will not only be surrounded by water, but there will also be salts and proteins that will concentrate as the water in the aerosol evaporates during suspension in the air.

However, the science around the infection route of coronavirus has led to a rethinking of aerosol particles larger than 5 μm, e.g., described as respirable (<10 μm), thoracic (1–20 μm) inhalable (20–100 μm), and ballistic (>100 μm) all of which could play a role in transmission (van Doremalen et al., 2020; Li et al., 2020; Stabile, 2020). We cannot see particles less than 40 μm, and so if you can see particles, they are larger than aerosols.

While showers have traditionally been considered the greatest risk in terms of aerosols production, any outlet where there is sufficient energy produced will create aerosols (Allegra et al., 2016; Bennett et al., 2000; Benoit et al., 2021; Crook et al., 2020).

Droplets are not considered to play a role in respirable or inhalable infections. Indeed, but these large particles present risks as they contain larger numbers of microorganisms and droplets will splatter as they deposit on a surface leading to surface contamination up to 1–2 m from an outlet (Kotay et al., 2019; Tracy et al., 2020). The rate and distance disseminated can be influenced by the flow rate of the water, design of the outlet (the smaller the bore, the higher the velocity and the faster the water flow), type of outlet fitting, design of the hand wash basin (slope of the sides), and position of the drain. Where water flow is directed straight into the drain outlet directly below the outlet, then this will create the greatest amount of splashing with a propensity for contamination of the surrounding area with drain microorganisms.

Droplet dispersal can be important especially where medical activities are carried out adjacent to the outlet and can include:

- Pharmaceutical preparation suites
- Pharmaceutical preparation areas in wards adjacent to hand wash station

Safe Water in Healthcare. https://doi.org/10.1016/B978-0-323-90492-6.00023-9

- Neonatal washing areas
- Areas for the preparation of sterile medical equipment
- Trolleys being used in a ward for a variety of purposes

Wash hand stations can become contaminated through retrograde contamination via:

- Contact of the outlet fitting with contaminated healthcare worker hands
- Washing of patient medical devices
- Being used to dispose of patient wash fluids, floor washing fluids, pharmaceutical products, and food debris.

Why are peripheral components colonized?

As outlets are at the end point of the water supply, i.e., last 2 m and may be used infrequently, then the water will be stagnant more often than not.

When the water outlet is not used, then the water will be not be flowing and the temperature of the hot water will start to reduce, and the cold water temperature will start to increase. When the water temperature reaches 20–45°C, then microbial growth will occur.

There are a range of complicated components, e.g., solenoids, thermostatic mixer valves, electronic taps, and there may also be flexible hoses connecting the distribution pipe to the tap end pipes. In addition, strainers may be present to trap debris that would otherwise be harmful to the plumbing components.

- Thermostatic mixer valves ensure that water hotter than 44°C is not discharged from outlets to prevent scalding. However, this temperature will provide favorable conditions for the growth of microorganisms.
- Solenoids and flexible hoses have rubber EPDM diagrams and liners that encourage the growth of *Legionella* and/or *Pseudomonas aeruginosa* biofilms (Hutchins et al., 2020; Leslie et al., 2021; Yui et al., 2021).
- Flexible hoses are used to connect from the water pipe work to outlet fitting such as taps or shower control units.
- Strainers are used to protect the sensitive plumbing components, but the very material that is entrapped in these strainers provides a niche growth environment for biofilm.
- Water outlets can be very complicated with a range of different components providing a high surface area for biofilm, and where this is at the periphery of the outlet, then there will be sufficient oxygen for microbial growth.
- Wash hand stations are for hand hygiene only and occasionally when there is accidental hand contact with the outlet during hand washing, this may lead to retrograde contamination.
- While wash hand stations should only be for hand hygiene, there is a certain amount of misuse where materials, fluids, and foods are discarded into the basin. Such products present a rich nutrient environment for the bacteria that reside in drains traps leading to retrograde contamination back into the wash hand basin back into the ward and transmitting antimicrobial pathogens to patients.
- The flat area at the back of the wash hand basin is often used as a shelf for the storage of containers.

Case study: Outbreak associated with peripheral components that were contaminated with P. aeruginosa

Outbreak scenario of peripheral contamination: outlet contamination and retrograde contamination of water system components with *P. aeruginosa* in neonatal unit in Belfast in peripheral components.

Background—*P. aeruginosa* is an important nosocomial pathogen that commonly colonizes hospital water supplies, including taps and sinks.

What was the problem? During an investigation of the Northern Ireland incidents, *P. aeruginosa* isolates from water samples, and from biofilms colonizing the thermostatic mixer valve, solenoids valve, isolation valve, tap components including the flow straighteners were found to be indistinguishable by variable number tandem repeat (VNTR) typing from those recovered from the babies (RQIA, 2012; Walker et al., 2014). The infectious microorganisms were associated with components that were located within approximately the last 2.5 m of the pipework suggesting that this outbreak was not due to systemic contamination but due to localized growth. Devices are fitted to, or close to, the tap outlet (e.g., mixing valves, solenoids, or outlet fittings) exacerbated the problem by providing the conditions and nutrients that supported microbial growth (e.g., appropriate temperatures, a high surface area to volume ratio, or a high surface area for oxygenation of water and leaching of nutrients from materials such as ethylene propylene diene monomers (EPDM) found in a number of components.

How was it resolved? The only way to eradicate the presence of the biofilm was to replace the components as decontamination strategies would struggle to completely remove and kill all the biofilm due to the complexity of the various components that

Learning points: Microbiological monitoring and maintenance of peripheral components

Strainers to trap harmful debris

Strainers are used to protect valuable and important components such as TMVs, valves, and taps being damaged by the presence of debris. Strainers generally have an organic rubber outer ring that retains a metal grid that acts as retainer to remove those large particles.

However, debris that is trapped on the grid provides a supportive structure and nutrients for microbial growth that will in addition accumulate other particles. Through time and pressure differentials as the water is switched on and off, bacterial cells and biofilm will slough off and contaminate components downstream including the hands of the healthcare worker and the hand wash basin and drains. It is recommended that strainers should not be used as they easily become contaminated with bacteria (DHSC, 2021).

Fig. 1 represents a strainer that was removed from the inlet of a thermostatic mixer value during investigations of the neonatal outbreak in Northern Ireland (Walker and Moore, 2015; Walker et al., 2014). Microbial biofilm had established, and the microbiological analysis demonstrated that this growth was composed of a *P. aeruginosa* biofilm.

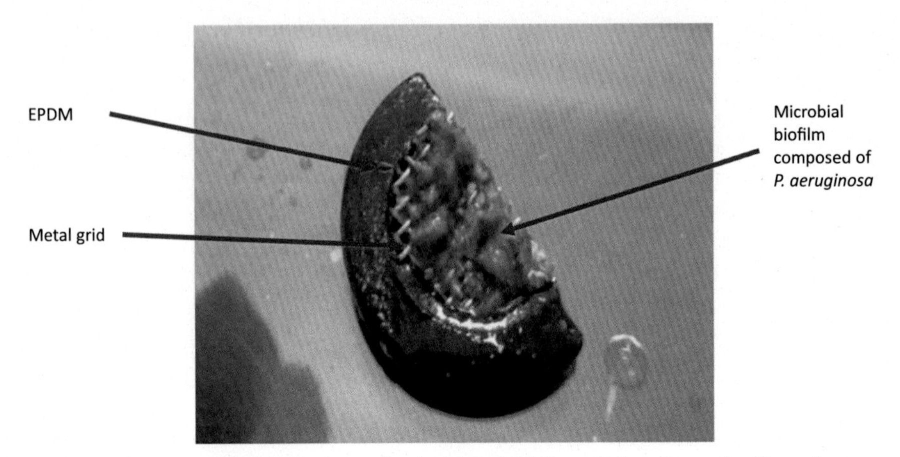

Fig. 1 Biofilm accumulation on the strainer grid from thermostatic mixer valve from the Belfast neonatal outbreak.

Thermostatic mixer valves

What is a thermostatic mixer valve and how does it work?

The TMV is device that blends the hot and cold water before discharge at a stable and set temperature of 42°C (Fig. 2). TMVs should ideally be incorporated directly in the tap fitting to allow mixing at the outlet (known as a thermostatic tap). However, where separate TMV valves are fitted, they should be as close to the outlet as possible to reduce the likelihood of stored blended water. Under such circumstance and within the same hand wash basin, the use of an additional separate cold tap is rarely needed (Fig. 2).

What are the different types of thermostatic mixer valves?

Type 1—a mechanical mixing valve with or without temperature stop (i.e., manually blended)

Type 2—a thermostatic mixing valve: BS EN 1111 and or BS EN (BSI, 2017a, 2017b; Neu et al., 2020)

Type 3—comprising an automatic failsafe, which isolates the hot supply in the event of cold water supply failure. HTM 04-01: Supplement—"Performance specification D 08: thermostatic mixing valves (healthcare premises)" for details relating to the performance requirements, material requirements, and test methods for thermostatic mixing valves (DHSC Safe water in healthcare premises (HTM 04-01), 2017, p. 04)

Why are TMVs used?

They are used to prevent scalding by lowering the temperature (usually 42°C) of the hot water for hand wash basins, bathtubs, and showers to protect the elderly, the young, and vulnerable groups such as patients.

Thermostatic Mixer Valve (TMV)/ Thermostatic Mixer Tap (TMT)

What are they?
TMVs prevent thermal scalding by blending the hot and cold water usually at 42°C.

Where should they be installed?
Iin healthcare is mandatory where wholebody immersion in water may occur (showers, baths) to prevent scalding. Their use should be decided based on a risk assessment see HSE https://www.hse.gov.uk/pubns/hsis6.pdf

What are the differences between thermostatic valve and a thermostatic tap (Exampl 1).

How does a TMV / TMT work?
Within the valve is a thermostatic cartridge which expands or contracts with temperature. This movement is used to blend the hot and cold water supplies to maintain a constant temperature (42°C). Should the cold supply fail the valve will cut off the hot water flow. (Example 2)

Example 1

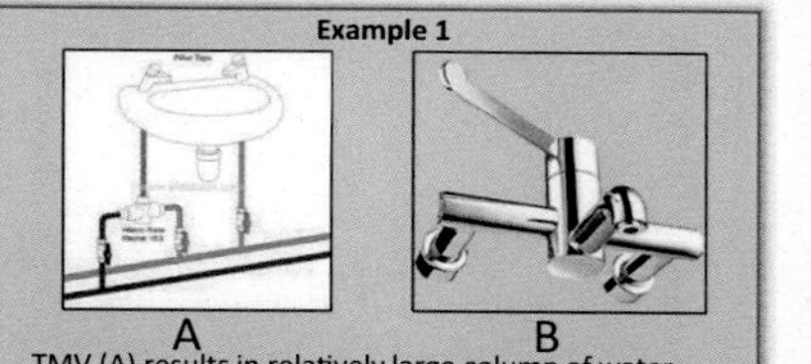

TMV (A) results in relatively large column of water between TMV and outlet at temperatures ideal for bacterial proliferation, whereas in B the mixing valve is located in the outlet tap resulting in a smaller column of water.

Example 2

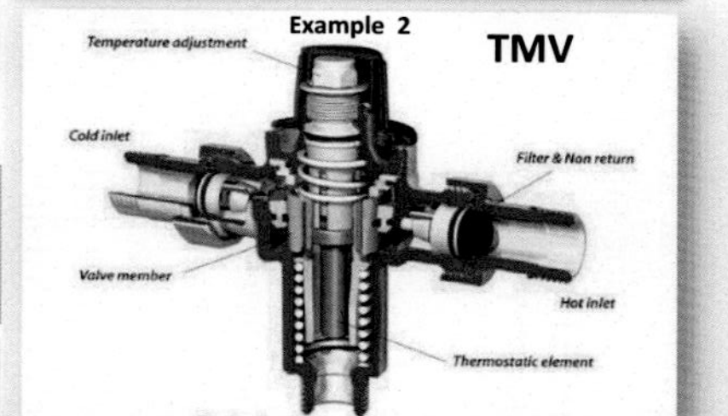

Inside contaminated TMV (Example 3)

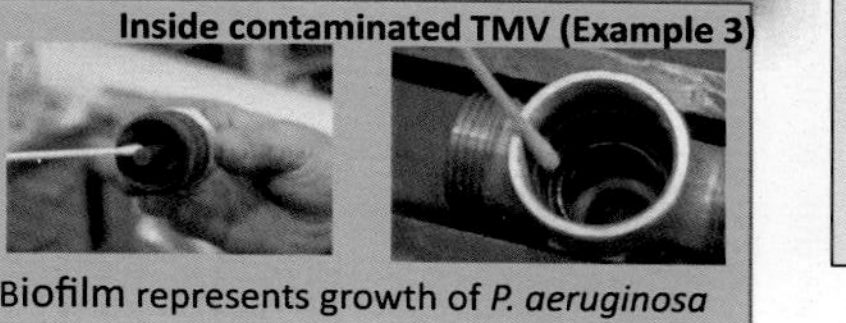

Biofilm represents growth of *P. aeruginosa*

What are the drawbacks to TMVs / TMTs?

1. WRAS approved materials in TMVs prone to biofilm formation and can results in water samples failures (Example 3).
2. Costs for checking function and water testing costs when biofilm (Figure 3) is detected.
3. Used in healthcare setting however on ITU there is little requirement for TMVs as outlets are used by staff who have no problem in reacting to hot water without being scalded.
4. Cross over may occur- when the TMV is faulty cold water which is under greater pressure may pass across into the hot water system which is under lower pressure, reducing the hot water return temperatures.

Issues with maintaining TMV/TMT

1. External companies employed which are highly variable in quality.
2. Some companies may not get access to TMV due to room occupancy may mark job as completed.
3. Need to check that processes are satisfactory by unannounced inspection of how operatives carry out task plus checking they have had basic training including infection control.

Fig. 2 Schematic of a thermostatic mixer valve and issues raised.

Thermostatic valves are used to regulate the water for showers and hair-wash facilities to 41°C, for unassisted baths to 44°C, and for baths for assisted bathing to 46°C and for bidets 38°C.

What are the risks of scalding?

The use and fitting of TMVs should be informed by a comparative assessment of scalding risk versus the legionella infection risk (HSE, 2014). Where a risk assessment identifies the risk of scalding is insignificant, TMVs are not required. Hot water temperatures greater than 44°C are considered a high risk and following full body immersion (bathing and showering particularly for the very young, very elderly, infirm or significantly mentally or physically disabled people, or those with sensory lossy) such temperatures have been associated with serious scalds and have led to a number of fatalities in hospitals and residential care homes (Hartley et al., 2011; Huyer and Corkum, 1997; Schulz et al., 2020). Hot surface (taps, pipes, hoses, or radiators hot water bottles) temperatures greater than 43°C are also considered a high risk. "Prolonged exposure" is a concern within areas that are not continually supervised such as bedrooms and bathrooms and may occur due to individuals falling and being unable to move. Hence, preventative measures need to be taken to mitigate scald and burn risks. Suitable measures may include insulating pipework, installing pipework behind panelling, and ensuring that the correct type of TMV is selected for use. For example, where a high risk of scalding has been identified, then a "Type 3" TMV fitting should be used.

Where should you not install TMVs?

TMVs should not be installed for outlets not intended for hand washing (e.g., sinks in kitchens, dirty utilities, or cleaners' rooms where water at 55°C is required). All such installations require a hot water hazard warning sign as the temperature could equate to the maximum temperature available from the calorifier and will present a scalding risk.

TMVs should not be installed in series with mixing taps (thermostatic or manual), i.e., check that there is not one installed in the run of pipework already and don't install a TMV prior to a thermostatic tap. While this may not sound feasible, this type of installation has been found on a number of occasions and could lead to an unknown cause of microbial contamination.

How should TMVs be maintained?

Where integral, competent and hygienically trained plumbers should inspect, clean, descale, and disinfect any strainers or filters associated with TMVs. To maintain protection against scald risk, TMVs require regular routine (annually or timeline defined by the risk assessment) maintenance carried out by competent persons in accordance with the manufacturer's instructions.

What are the regulations concerning TMVs?

All materials in contact with water should comply with BS 6920 and the current version of the Water Regulations Advisory Scheme's (WRAS) material guidance:

Type 3 TMVs should have undergone third-party testing and certification to the requirements of HTM 04-01: Supplement—"Performance specification D 08: thermostatic mixing valves (healthcare premises)."

Microbial problems associated with TMVs

As the TMV is operated, cold and hot water will be pulled through the outlets. Where the hot water recirculation supply is greater than or equal to 55°C, this will be pulled through to the inlet relatively quickly (Fig. 3). Likewise, the cold should be supplied at less than 20°C. However, as soon as that outlet has been used, the water to and from the TMV and from the cold supply will be stagnant until the outlet is used again (Fig. 4). This will result in:

- a warming of the body of the TMV, which will warm the cold inlet above 20°C
- cold water supply to the TMV will be static and will increase above 20°C

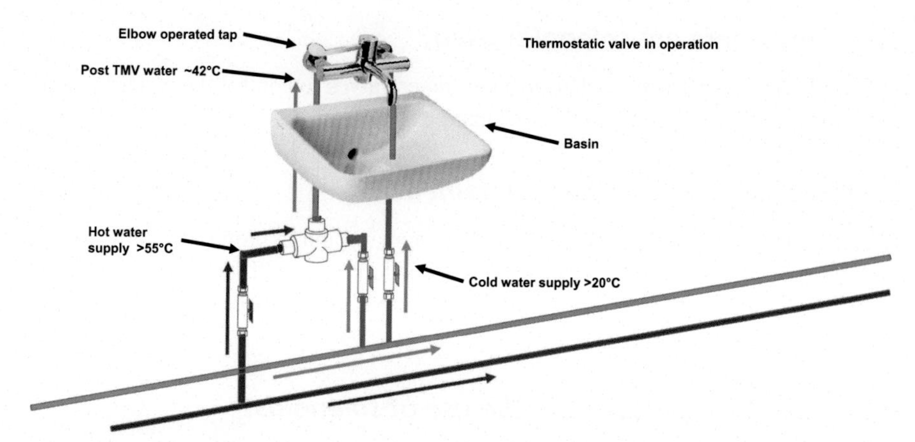

Fig. 3 Schematic of thermostatic valve in operation to supply water to hand wash basin.

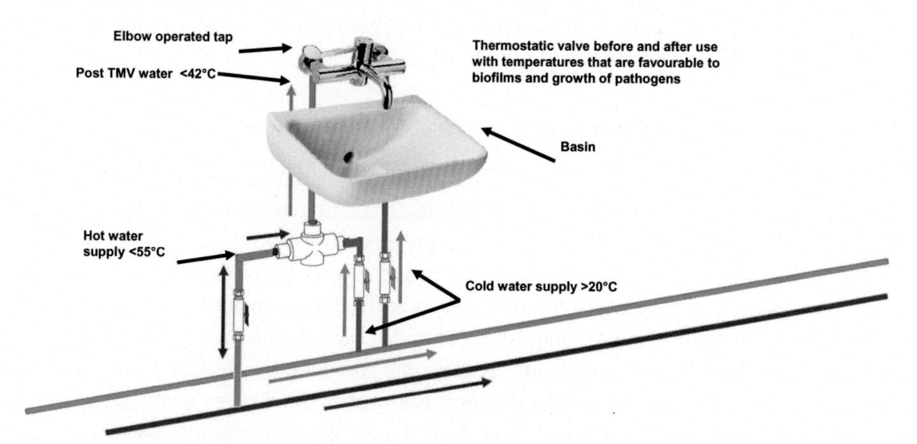

Fig. 4 Thermostatic valve before and after use with temperatures that are favorable to biofilms and growth of pathogens.

- hot water supply to the TMV will be static and will decrease below 55°C
- supply from the TMV to the tap will be static and will decrease below 42°C

All of the above conditions will lead to biofilm formation and provide ideal conditions for the growth of biofilm in the hot water section as it cools, cold section of the TMV as the temperature warms, and at the outlet.

Solenoids

What is a solenoid and how does it work?

A solenoid is a valve that contains a rubber EPDM diaphragm that is activated by a piston when an electronic sensor beam is interrupted at a hand wash basin to allow the water to pass to the faucet.

Why and where are solenoids used?

Solenoids are downstream of thermostatic mixer valves to control the water to an automatic or nontouch taps fitted to wash hand basins.

How should solenoids be maintained?

Because of the presence of the rubber diaphragm, which is prone to biofilm formation, the solenoids should be removed for inspection. This will involve inspecting the strainer components as well as determining if there is any slime or biofilm build-up on the EPDM components.

What are the regulations in the use of solenoids?

Servicing and maintenance periods should be risk assessed as should the need to examine components when an outbreak is ongoing. Where outbreaks occur, a number of components may need to be considered as it is likely that most of those components in the last 2 m could be contaminated with biofilm.

What microbial problems are associated with solenoids?

A solenoid valve includes the rubber (EPDM) diaphragm that is opened and closed using a piston to control the flow of water. While all the materials are WRAS approved, the presence of the diaphragm results in the presence of copious amounts of biofilm. Microorganisms such as *P. aeruginosa* are prone to colonize the rubber diaphragm and, as a result, seed the downstream water as it discharges from the solenoid valve into the outlet and subsequently the water being discharged into the hand wash station. During investigations into water outlets after the incidents in Belfast in which four preterm babies died, the solenoid valve and the rubber diaphragm were found to be colonized with extensive biofilm that had grown on the components. Devices fitted to, or close to, the tap outlet (e.g., mixing valves, solenoids, or outlet fittings) may exacerbate the problem by providing the conditions and nutrients that support microbial growth (e.g., appropriate temperatures, a high surface area-to-volume ratio, or a high surface area for oxygenation of water and leaching of nutrients from materials such as ethylene propylene diene monomers (EPDM)) found in a number of components. The only way to eradicate the presence of the biofilm would be to replace the entire solenoid valve as well as decontamination strategies to completely remove and kill all of the biofilm.

The healthcare team must keep an open mind regarding direct or indirect exposures of patients to contaminated water. Reports have documented transmission of *Aeromonas* spp. to patients via leeches (kept in tanks that were inadequately cleaned) (Sartor et al., 2002). Some *A. hydrophila* isolates are resistant to antibiotics that are used as prophylaxis against infection (Giltner et al., 2013; Wilmer et al., 2013). Appropriate cleaning and decontamination of leech tanks may prevent leech-associated infections.

Flexible hoses

What is a flexible hose?

Flexible hoses, also known as "tails," are often used in the supply of water to connect the hot and/or cold supply to a peripheral outlet (Fig. 5) such as a tap or shower control unit and can include equipment such as wash hand basin, bath, shower, sluice, ice

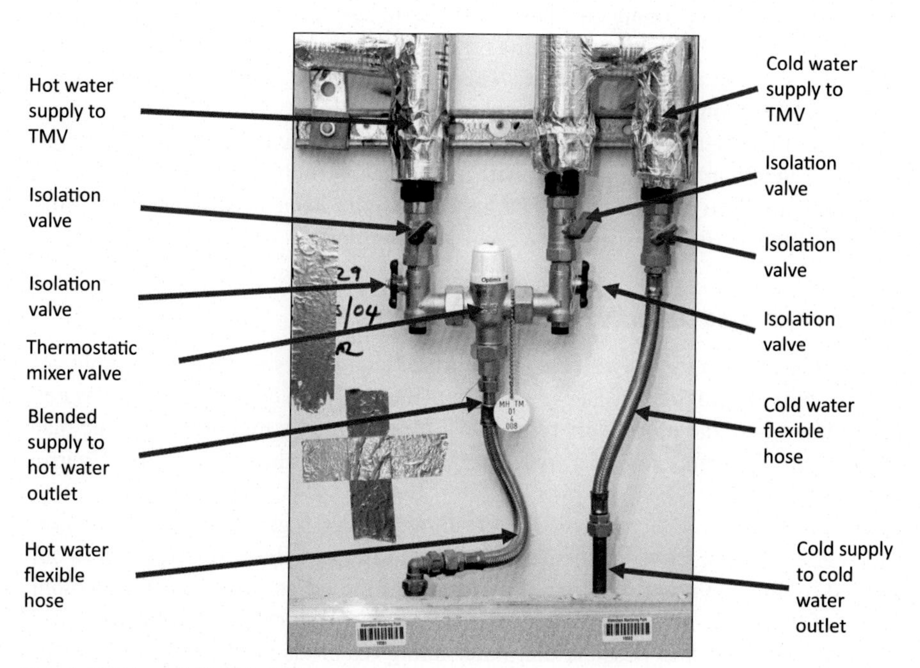

Fig. 5 TMVs and flexible hoses.

making machines, dish/glass washers, high-low baths, drink vending machines, drinking fountains, endoscope washers, clothes washing machines, and hoses for washing down other equipment or areas and any other equipment deemed necessary. They may also be connected to system components such as pressure reducing valves, nonreturn valves, strainers, TMVs, and shower mixers.

What are flexible hoses used for?

Flexible hoses may be used to link between hard pipework and equipment, often for convenience rather than being necessary. The outer casing is typically braided steel or stainless steel with a synthetic rubber inner lining such as EPDM (ethylene propylene diene monomer). Traditionally, these connections would have been manufactured by the plumber on site, and each connection would have been specific for each application. However, due to the skill and the time required to manufacture these individual components, flexible couplings were introduced to save time and money.

What are the microbiological problems with flexible hoses?

The introduction of flexible hoses was no doubt of great benefit to an industry where time is money; however, their introduction has led to the presence of large amounts of EPDM being used in the last 2 m of outlets. While all materials exposed to water will to a certain degree enable the formation of biofilms, EPDM has raised a number of health concerns over a number of years. Flexible hoses vary in length from 250 mm to 1.5 m or longer depending on what is required to be connected and how far they are apart. The high surface-area-to-volume ratio is apparent as the surface is not smooth and contains small microporous areas where bacteria can harbor and build their biofilm networks. Synthetic polymers such as EPDM are composed of a wide range of hardeners, plasticizers, and other carbon compounds that will act as carbon sources for the resident carbon sources that will leech out the rubber providing the biofilm with nutrients (Neu et al., 2020). EPDM and PEX have been shown to result in gross contamination of surfaces by complex multispecies biofilms (Waines et al., 2011).

Outbreaks associated with flexible hoses

Combining those nutrients with the large surface area results in the considerable volumes of biofilm and the presence of microorganisms in high density area such as neonatal units where the hoses were responsible for abnormally high heterotrophic plate counts and in the presence of *P. aeruginosa* (Buffet-Bataillon et al., 2010; Charron et al., 2015).

Health Facilities Scotland (HFS) received reports that high levels of *P. aeruginosa* and *Legionella* spp. bacteria have been found in water samples taken from water outlets fed by flexible hoses, confirmed by testing of the hoses, which revealed colonization of the lining. The lining of the material in these reports was EPDM. However, they indicated that it is possible that other lining materials (and washers within the couplings) could be similarly affected.

What are the alternatives to EPDM?

New lining materials such as PE (polyethylene), PEX (cross-linked polyethylene), LLDPE (linear low-density polyethylene), and PVC C (postchlorinated PVC) are now on the market, and others are likely to follow. However, their long-term performance regarding the growth of microorganisms is still unknown, but it is likely that while these other materials may slow down the growth of the biofilm, it is expected that these microbiological issues will remain. Changes in this situation may be reflected in future guidance.

What are the regulations concerning flexible hoses?

HTM 0401 indicates that:

- organic materials in flexible hoses will increase the risk of microbial colonization and that alternatives should be considered where possible (DHSC Safe water in healthcare premises (HTM 04-01), 2017).
- flexible hoses should be used only for the following applications: to allow for vibration of equipment; to accommodate vertical displacement of high and low baths and sinks; to facilitate essential maintenance and access of bespoke equipment when no alternative is available.
- they should be kept as short as possible.
- flexible EPDM should not be used.

What other types of peripheral components are used?

All of the components within the water system will have a microbiological impact on the water system and downstream components. As the water system reaches the peripheral outlets, there will be an increase in the number of components and often the complexity of these components. There only needs to be one of these components contaminated (whether that is a strainer, TMV, solenoid, or outlet) for it to have a major impact on microbial transmission and the control of that contamination. Where one contaminated component is missed during a service or maintenance schedule, then that contamination will remain. Hence, it is important to understand the complexity and makeup of the water system and the peripheral components that are present.

External influences on the safe use of water in a wash hand station

Taking the hand wash station as an example, there are several external influences that can impact of the safety of that hand wash station (Fig. 6). All professionals need to be aware of their roles in preventing contamination and onward transmission of waterborne pathogens in a healthcare setting.

Architect: the design of a healthcare building is a complex and long drawn out process. Architects like to build in redundancy and/or may follow guidance or opinion that ensures there are too many hand wash stations for the purposes of a ward or do not include an appropriate number of sluices in appropriate places. Too many wash

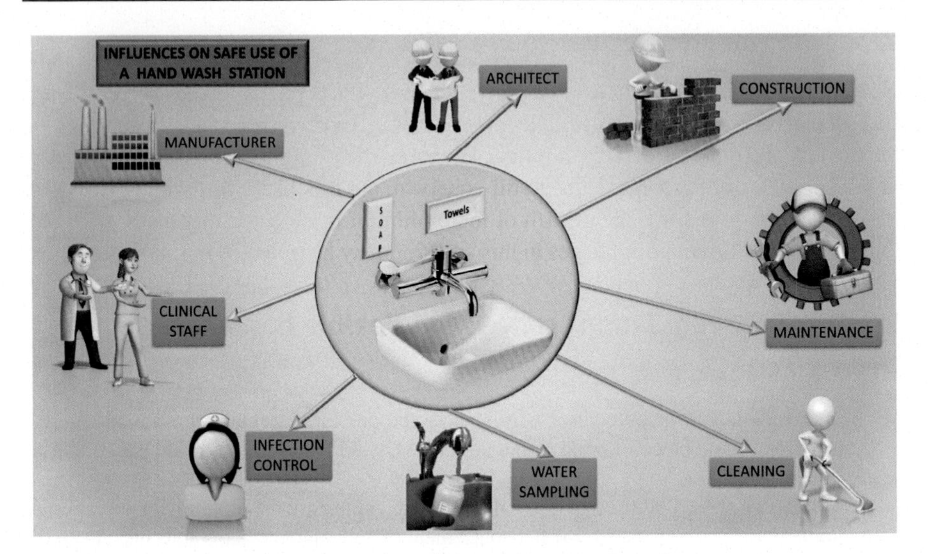

Fig. 6 External influences on the safe use of water in hand wash stations.

had stations result in many not being used and stagnation and biofilm development leading to opportunistic pathogen colonization and dispersal. Have your architects been trained in water system hygiene? Do they understand the guidance in storage and delivery of water? Are they designing with cleanliness in mind and thoughtful of specific components with a reduced propensity to form biofilms or create droplets or aerosols? What competencies (e.g., training) have your architects had to provide you with confidence that they understand safe water in healthcare? Has the architect specified how the hand wash stations should be built? Is there sufficient space between each HWS and other areas to prevent contamination by droplet dispersal or has screen been designed to prevent droplet dispersal?

Manufacturer: wash hand stations have a number of different components, all of which play a part in the delivery of safe water. For the plumbing components, many of these are designed with consideration for microorganisms, but the components still become contaminated or indeed contain organic materials that may leach carbon and nutrients sources that used by the bacteria and biofilms. While some manufacturers try to manufacture the inside of the pipe components to be as smooth as possible, the bacteria themselves are extremely small, and it does not take much for them to be able to find a niche or crevice in such pipework that will lead to the formation of the biofilm. A number of manufacturers have been shown to deliver plumbing components to healthcare building with an already established biofilm containing pathogens such as *P. aeruginosa* as post manufacture, the plumbing parts have been wet tested to ensure functionality, and they unknowingly contaminated what was a pristine component.

Construction: have the plumbers that are building your hospital been hygienically trained? Do they understand why it is important to use clean tools when constructing the new water system, to cap pipes to stop debris and dirt from entering the pipework, to use WRAS-approved materials only, not to use flexible hoses and not to fill each section of water system and then allow it to stagnate as they construct the next section?

Have the plumbers installed the hand wash station as per guidance? Are the elbow levers positioned appropriately allowing staff to operate the levers with their elbows? If the levers are too close to the wall panel at the rear of the hand wash station, then staff will have to use their hands to operate the levers, which will result in contamination.

Maintenance teams: training in hygienic plumbing is essential for plumbers working in hospitals. They need to be able to recognize the importance of hygiene when working on plumbing components in a high dependency unit such as a neonatal unit where multiple parts could be contaminated with P. aeruginosa and understand the dispersal and transmission routes that occur once a system or peripheral component is contaminated. In addition, plumbers need to be trained to understand the importance of decontaminating their tools between a dirty job (where the tools may come into contact with fecal matter such as in a blocked toilet—the most common plumbing job in a healthcare setting) and that those tools need to be clean when working on the plumbing components in a high dependency unit.

Cleaning: cleaning is seen as the culprit in many situations as it is easy to point the finger to blame a nameless and unidentifiable cleaning team. No one should be undertaking their role without training. Yet, if your cleaning domestics are employed by a third party, then how can you ensure that they understand the level of cleanliness required in your hospital? In house training programs fulfill many requirements in enabling the understanding of the importance of their role in reducing waterborne transmission in healthcare. We have all heard of the anecdotal stories of the cleaners diligently cleaning the HWS starting from the bottom of the wash hand basin, i.e., near or around the drain and finishing off with the tap outlet or where they cleaned the wash hand station according to appropriate protocols only for them to move to the next station with the same set of cloths. Everyone has a role to play and when trained and educated, individuals able to act responsibly and competently. Domestic staff should not be disposing of fluids used for other cleaning purposes down the hand wash basin as this will contaminate the basin, outlet, and the drain.

Water sampling: where water samplers are employed by a third party, how confident are you that they are undertaking this sampling in a competent manner? Where are external companies contracted it can often be decided on the lowest tendering price, and this may be reflected in the training and the manner in which samples are taken and also the results? Are samples taken correctly? Have the staff taking the water samples been trained in how to take a sample using the appropriate equipment? Do they understand the difference between pre- and postflush?

Infection control: infection control staff have an overall duty to ensure that all staff understand that a hand wash station is specifically for washing hands and is not to be used for other purposes such as washing patients' medical devices or for disposal of patient fluids or pharmaceutical products. In addition, staff should be aware that water from the HWS should not be used for applications where sterile water should be used. However, the training that the IPC team receives must reflect the relative risks from the exposure to and transmission of waterborne pathogens. Hence, the IPC team must be able to identify where there are problems with a wash hand station including inappropriate tap and/or their positioning, hand wash basin, splashing, and inappropriate use of the hand wash station.

Clinical staff including nurse and doctors: clinical staff should be trained in how to use wash hand station appropriately, how to wash hands effectively and not to abuse the HWS by, using the wash hand basin as shelf, washing patient medical devices, or disposing of patient fluids or pharmaceutical products or disposing of any other items into the wash hand basin. In video evidence only 1 in 25 visits to the wash hand station was for the correct purpose. Clearly, this is not appropriate, and these actions need to be pointed out and staff retrained where appropriately (Figs. 7–9).

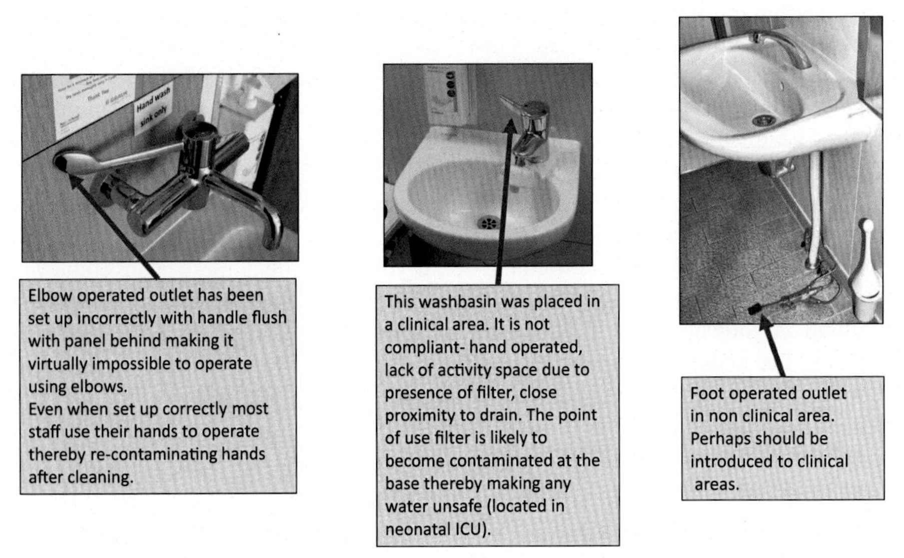

Fig. 7 Manually operated outlets associated with hand wash basins.

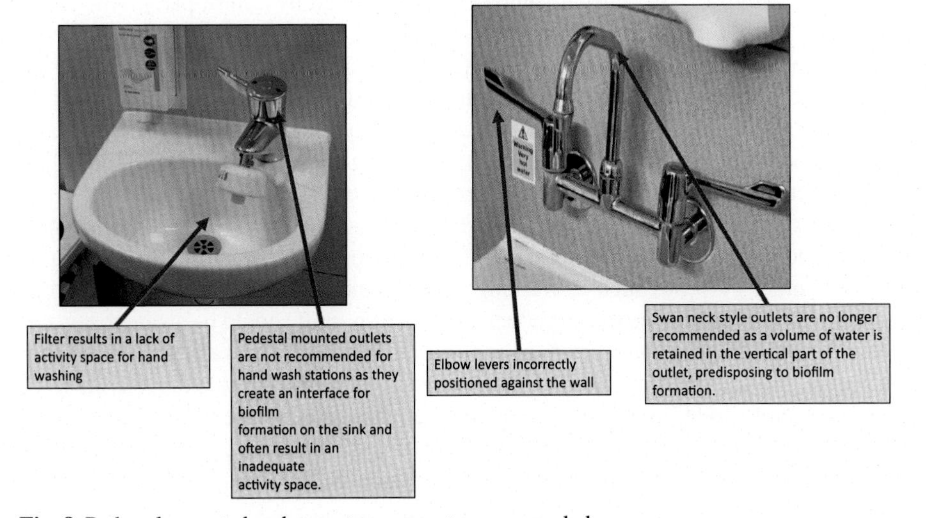

Fig. 8 Pedestal mounted and swan taps are not recommended.

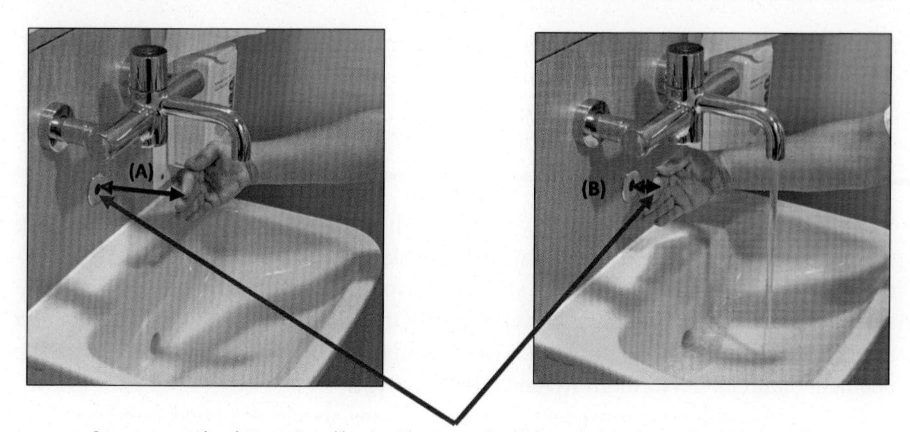

Sensor operated outlets require calibration. The sensor should detect hands when underneath the end of outlet (A). The images show incorrect set up, water only flowing when hands are almost touching sensor (B).

Fig. 9 Wall-mounted sensor-operated taps.

Manual separate hot and cold taps

Manual hot and cold taps are considered the simplest to install, operate, and maintain. Temperature control is the conventional strategy for controlling microorganisms in water systems, and where set up, this should work effectively with such simple manual hot and cold water taps. Within healthcare, manual separate hot and cold taps are commonly used for utility or kitchen sinks.

However, the installation of separate manual for hot and cold taps must be risk assessed from a microbiological and scalding perspective.

For most people, the risk of scalding at this temperature is low as they will be aware that the water is at 55°C. However, the risk assessment should take account of susceptible "at-risk" people including young children, people who are disabled or elderly, and to those with sensory loss for whom the risk is greater (HSE, 2013). As these types of taps should be risk assessed for each location and where they have been cited in a utility or kitchen, then that risk should be low.

Manual operated single-lever taps

Clinically, hand wash stations may have the water outlet controlled using a manual lever that is adjusted for temperature and flow. These are relatively simple units where the tap only controls the flow of water with the TMV further back, e.g., 0.5 m upstream of the tap.

Pedestal taps

Pedestal outlet where the outlet is designed to be set into the hand wash basin (Fig. 8). These are not recommended as that the space between the pedestal and the hand wash basin creates crevices where biofilm can form, and the rubber seal may leach nutrients for microbial growth, and these areas can be very difficult to clean effectively. An additional problem with pedestal fittings is that they can often lead to a lack of activity space required for hand washing (Fig. 8)

Swan neck taps

Swan neck taps were once common on hand wash basins (Fig. 7). However, swan neck taps retain larger volumes of water, which then stagnates, and HFN 30 and HPSC recommend that swan neck tap outlets should be avoided, as they do not empty after use and could be prone to microbial biofouling with microorganisms including *Legionella* and *P. aeruginosa*, the latter of which was associated with a swan neck tap during the neonatal outbreak in Northern Ireland (HPSC, 2015).

From a microbiological aspect, even manual taps should designed to be operated using elbow taps; however, we know that event in a clinical environment that up to 92% of staff have been recorded as using their hand to operate tap levers (Weinbren et al., 2018). Where hands could potentially be contaminated, then this contamination could easily be transferred to the tap and obviously vice versa, resulting in a transmission route to a number of other sites and continued transmission risk from that tap handle.

Thermostatic mixer tap

Thermostatic mixer taps are very common on hand wash basins (Fig. 8). These have the advantage that the thermostatic valve is contained with the body of the tap, and hence, there is a very small volume of water that is retained within the tap. These types of taps are considered to present fewer microbiological issues than outlets that have individual TMV, solenoid, and taps but have still been associated with colonization with *P. aeruginosa*.

Elbow levers

As the hands of healthcare workers are the most common vehicle for the transmission of microorganisms from patient to patient within the healthcare environment, it is clear that hand washing is an important step in preventing that transmission (Pittet et al., 2006). Therefore, hand cleansing should be the primary action to reduce healthcare-associated infection and cross-transmission of antimicrobial-resistant pathogens.

Regardless of the type of tap, manual levers are designed in such a way that they are operated with the elbow rather than using hands, which may already be contaminated. The reasons for using elbow-operated levers are to prevent the healthcare worker from touching the tap control with their hand after touching patients as the hands are likely to be contaminated. However, studies have shown that almost half of elbow levers were set up incorrectly, being flush or within 3.5 cm of the rear panel, making elbow operation extremely difficult (Weinbren et al., 2018) (Fig. 8). Yet even when the lever is set up correctly, it has been found that 92% of staff use their hands to turn on the outlet, potentially contaminating the lever and 68% used their hands to turn the outlet off, potentially re-contaminating their hands (Weinbren et al., 2018). Selection of outlet type according to HBN was most incorrect in the intensive treatment unit but also occurred in the newly built parts of the hospital. Hands were used to turn on the taps in 97% of instances. In 57% of washes, hands were re-contaminated when used to turn the tap off. Only six individuals consistently used their elbows to turn outlets off. Surprisingly, more individuals used their elbows to operate taps whose handles were flush with the inspection panel behind (Weinbren et al., 2017).

Knee or foot-operated water outlets

Other types of manual tap operations are available, and these include knee and foot-operated controls. These make sense in the context of absolutely preventing staff from using their hands contaminating the tap. They are not that common but are occasionally implemented in hospitals.

Sensor-operated tap

Water distribution outlets are available that are operated by a sensor. Again, there is no need for the healthcare worker to touch the tap fittings, hence alleviation of any potential for hand contamination. The sensor requires a power supply and detects the movement of the hands of the healthcare worker. The solenoid is then activated to allow water to flow through the body of the tap and out of the outlet. The sensor can be adjusted such that the reaction time till the flow starts can be altered or even the length of the time that the flow period can be shortened or lengthened and even used to implement automated flushing. Flushing is a major control strategy for removing microbial growth from the stagnant water, and therefore, automating this removes the manual aspect that is so often not carried out effectively. There are a number of potential microbial contamination issues with sensor-operated taps. These include healthcare staff not being able to operate the sensor, becoming frustrated, and moving their hand under the tap outlet to the extent they touch the outlet and hence contaminate the outlet fittings. Sensor taps themselves have been the center of a number of studies identifying the issue of microbial contamination with high heterotrophic plate counts or even the presence of waterborne pathogens including *P. aeruginosa*, which resulted in the neonatal deaths in Northern Ireland (Hargreaves et al., 2001; Walker et al., 2014). The sensor-operated tap would have been preceded by a TMV and a solenoid, and as discussed earlier, once one of these components is contaminated, it is difficult prevent contamination in the last 2 m.

Sensor-operated TMVs are inherently complex. The following images (Figs. 10 and 11) demonstrate that such units are highly complex with multiple sections and different materials. This brief investigation was the result of a water sample from a sensor-operated TMV being identified positive for *P. aeruginosa*. The unit that had yielded the positive sample and a unit that was negative were removed from the hospital and investigated. The unit that had yielded the positive *P. aeruginosa* was found to contain visible biofilm on a number of components (Fig. 11) and *Pseudomonas* spp. was recovered. However, these results indicate that biofilms will readily form on the

Sensor operated thermostatic mixer valves are hidden away behind panelling and little thought is given to them until a water sample becomes positive. In this case a water sample from was positive for *P. aeruginosa* in a healthcare setting. This microbially positive unit was removed and investigate along with a unit where the water samples were deemed to be negative.

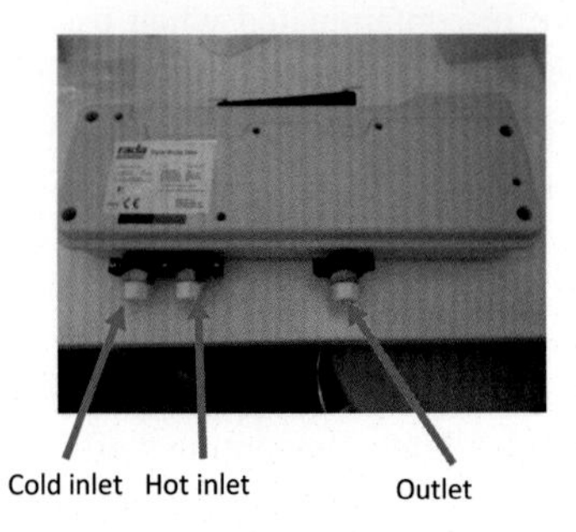

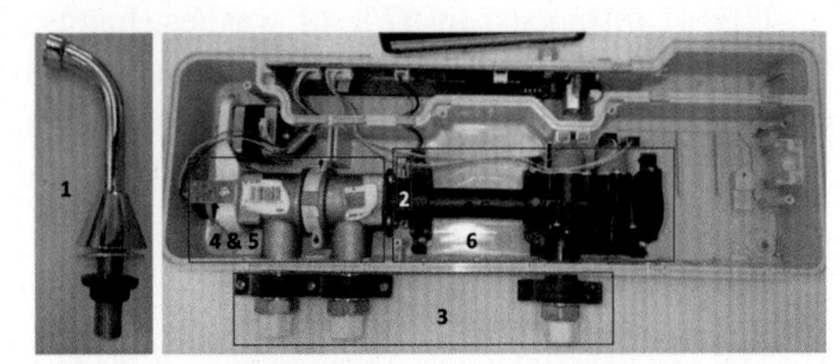

Components tested for *P. aeruginosa*. 1) Water tap outlet; 2) Temperature probes; 3) Hot and cold inlets and outlet (rhs); 4) Plastic mixers; 5) TMV metal casings; 6) Solenoid casing and interior components

Fig. 10 An example of a sensor-operated thermostatic device that was related to the presence of *P. aeruginosa*.

The various components of the systems were dismantled for investigation of biofilm growth and microbial recovery. As can be seen below there were a number of components where biofilm growth was clearly evident. The microbiological investigation revealed that the growth was due to *Pseudomonas* spp. but *P . aeruginosa* was not recovered (as per the original water samples).

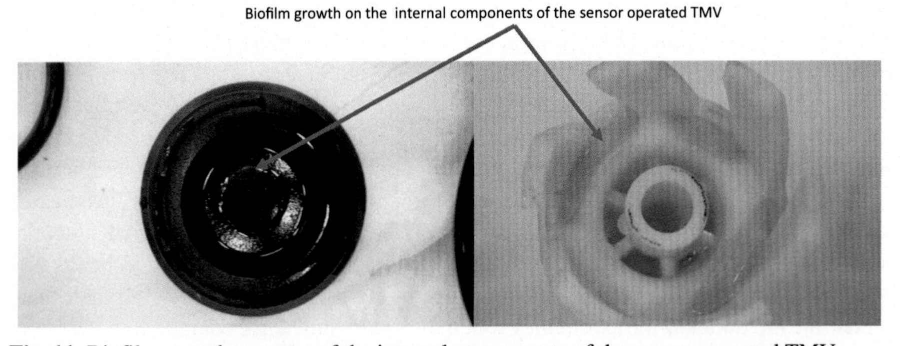

Fig. 11 Biofilm growth on some of the internal components of the sensor-operated TMV.

different components, even if they have been WRAS approved. Areas that were visually observed to be fouled with biofilm included the temperature sensor probes, TMV metal casing interior, plastic water mixers, and the solenoid outer casing cap (which being at the end of the unit formed a dead leg).

Different type of outlets fittings

There are a wide range of outlet fittings used in taps and all for different purposes. They may be fitted to reduce flow, aerate, or even straighten the flow. There has been tendency for these components to be manufactured from plastics. While these plastic outlet fittings will have passed a WRAS test, there is no doubt that their complicated structure, high surface-area-to-volume ratio, potential for contact with healthcare workers hand, contamination from the drain, and exposure to oxygen encourage microbial colonization and biofilm proliferation (Fig. 9).

Aerators

The aerator is a small attachment that either fits onto the end of the tap or can be inserted inside of the existing spout to reduce the required flow of water from taps, which will in turn reduce your water usage. They are also called flow regulators and work by simply mixing air into the flow to reduce the volume of water passing through the tap. These components have a high surface-area-to-volume ratio and are highly complex (Figs. 8 and 12).

Flow straighteners

Flow straighteners use an open flow fitting that provides a stream of water from the outlet that is straighter (Fig. 9). This device does not have the same mesh surface and fine orifices of a pressure compensating aerator or flow regulator and so the flow regulated and therefore if a flow straightener is used, remembering these devices have no flow control capability and therefore do not comply with WELS, then that fitting may deliver an unregulated water stream of approximately 15–20 Lpm. This impacts greatly on pipe sizing calculations; it also creates splashing issues, further increasing bacterial growth risks and slip hazards around the basin environment. Additionally, many basins are not capable of draining at a rate greater than 6–8 Lpm (some even less), and in hospitals no overflow is used, potentially this is a major safety, health, and hygiene hazard (Fig. 13).

There was alarm during the outbreak in the neonatal outbreaks in Northern Ireland when the outlet fittings were found to be extensively fouled only months after a refurbishment. These fitting and other components were found to be extensively colonized by biofilm and particularly *P. aeruginosa*. The consequence is that everyone, including healthcare workers, who operated the taps, would have had their hands showered in a soup of *P. aeruginosa* (Fig. 14).

Aerators or flow regulators are used to reduced the volume of water used by mixing air with the water. However, they are inherently complex and present a high surfaces area to volume ratio that enable bacteria to colonies the surfaces. This image presents the large number of complex components that form an aerator.

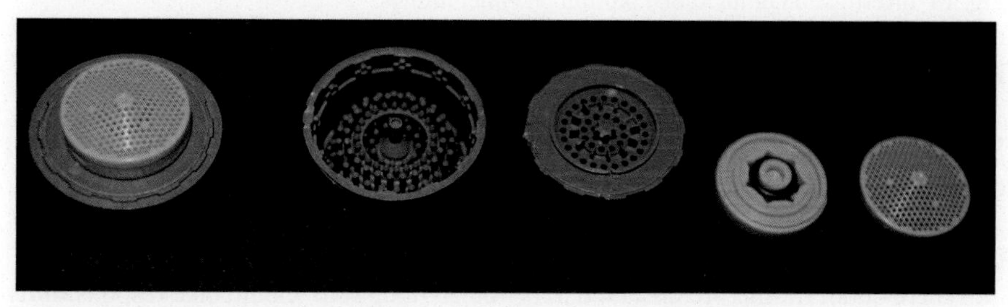

Fig. 12 Demonstration of the complexity and high surfaces area of an aerator. Photo courtesy of Susan Paton, PHE.

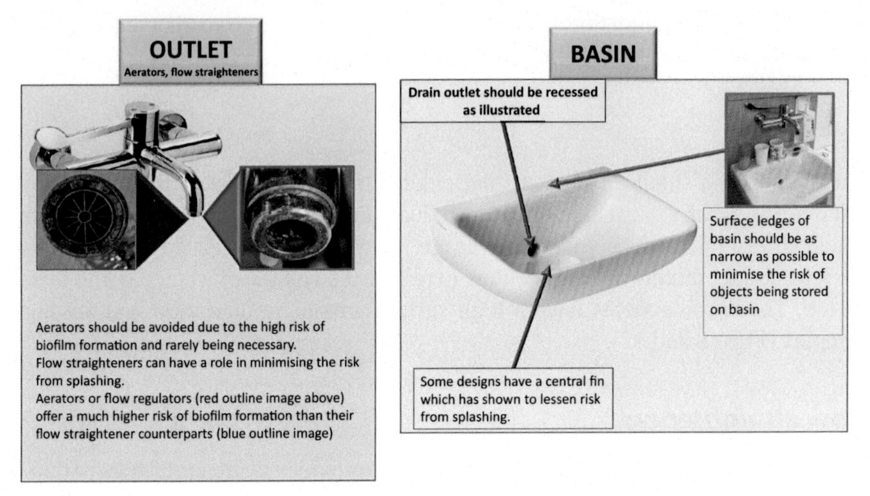

Fig. 13 Microbiological issues associated with hand wash basin and outlet fittings.

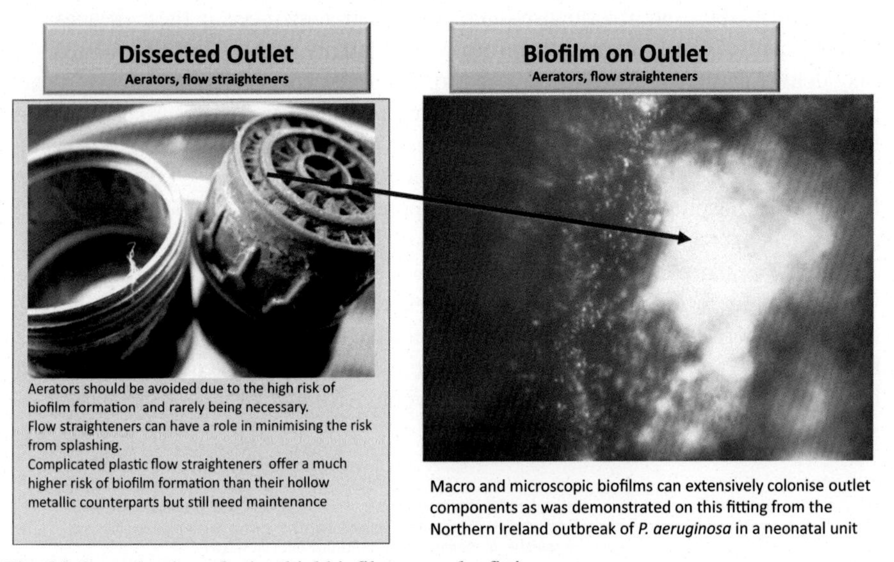

Fig. 14 Investigation of microbial biofilm on outlet fittings.

Recommendations for peripheral components

Schematic diagrams should be available to illustrate the layout of all peripheral components.

Ensure that there is a regime for the inspection, maintenance, cleaning, disinfection, or replacement of all peripheral components and fittings where required.

Where outlet fittings are observed to be fouled, then staff should take responsibility to alert the WSG and estates department.

There should be ease of access to all peripheral components to enable periodic inspection.

Isolation valves should be present to facilitate maintenance.

A monitoring regimen should be in place to assess the microbiological status of the water from each outlet on a scheduled written maintenance plan.

TMVs should be incorporated directly in the tap fitting, as mixing at the point of outlet is preferable.

Reduce the use of flexible hoses and replace with fixed copper piping.

Remove swan neck taps where located as no longer recommended.

Remove flexible hoses and replace with solid fittings.

Ensure that wash hand basins are only used for hand washing.

Where not integral to components, in-line strainers should be fitted within the water pipework system to protect vulnerable peripheral components such as valves and fittings.

Training should ensure that staff are empowered and risk assess appropriate disposal of patient fluids or cleaning fluids and that they are not disposed of down hand wash basins.

Implement training regimens for all staff involved with the use of hand wash stations to ensure that staff are aware of the risks to vulnerable patients.

Risk assessment considerations for peripheral components

A policy should be in place to decide on applicable standards, procurement decisions, and approval processed, responsibility for the finance and upkeep of components (e.g., TMVs). This policy, completed by competent person(s), should enable the WSG to risk assess the impact of component contamination and the dissemination of waterborne pathogens on susceptible patient cohorts.

Are schematic diagrams available to accurately illustrate the layout of the different peripheral components relevant to the growth of microorganisms in the water distribution system?

Have manufacturers of peripheral plumbing components provided evidence that where components have been wet tested in the factory that these components have been disinfected and do not retain any microbial contamination?

The WSG should establish controls to ensure that only microbiologically safe components can be purchased for installation and work with the procurement team so that they understand the risks and are able to ask manufacturers for evidence of testing prior to purchasing.

Has a comparative assessment of Legionella infection risk versus the scalding risk been carried out to determine if TMVs are required? Where a risk assessment identifies the risk of scalding is insignificant, then TMVs are not required. In addition, TMVs should not be installed where outlets are not for hand washing (e.g., sinks in kitchens, dirty utilities, or cleaners' rooms),

Are all TMVs and peripheral components part of an asset register?

Where present have TMVs and other appropriate peripheral components including integral and inline strainers and filters been inspected, cleaned, descaled, and disinfected on an annual basis or as agreed through the risk assessment?

Have all outlets been assessed for suitability from a functionality as well as an infection control perspective including placement of manual operating levers and/or sensors.

To minimize the risk of scalding, TMVs require regular routine maintenance carried out by competent persons in accordance with the manufacturer's instructions (www.hse.gov.uk/healthservices/scalding-burning.htm).

When selecting new taps, has HBN 00-09 been reviewed to ensure that aerators, strainers, and flow restrictors are not used at the point of discharge.

Has the use of flexible hoses been reviewed to ensure that they are only used to allow for vibration of equipment, for vertical movement of high and low baths and sinks, and finally, only where their use allows maintenance and access to bespoke equipment when no alternative is available?

Regulations for outlet fittings

HTM 04-01 Safe Water in Healthcare (DHSC, 2017, pp. 04-01)

The Water Supply (Water Fittings) Regulations 1999 (DWI, 2015)

HSG274 Legionnaires' disease Part 2: The control of legionella bacteria in hot and cold water systems (HSE, 2014)

References

Allegra, S., Leclerc, L., Massard, P.A., Girardot, F., Riffard, S., Pourchez, J., 2016. Characterization of aerosols containing *Legionella* generated upon nebulization. Sci. Rep. 6, 33998. https://doi.org/10.1038/srep33998.

Allegra, S., Riffard, S., Leclerc, L., Girardot, F., Stauffert, M., Forest, V., Pourchez, J., 2020. A valuable experimental setup to model exposure to Legionella's aerosols generated by shower-like systems. Water Res. 172, 115496. https://doi.org/10.1016/j.watres.2020.115496.

Bennett, A.M., Fulford, M.R., Walker, J.T., Bradshaw, D.J., Martin, M.V., Marsh, P.D., 2000. Microbial aerosols in general dental practice. Br. Dent. J. 189, 664–667. https://doi.org/10.1038/sj.bdj.4800859.

Benoit, M.-È., Prévost, M., Succar, A., Charron, D., Déziel, E., Robert, E., Bédard, E., 2021. Faucet aerator design influences aerosol size distribution and microbial contamination level. Sci. Total Environ. 775, 145690. https://doi.org/10.1016/j.scitotenv.2021.145690.

BSI, 2017a. Sanitary Tapware. Thermostatic Mixing Valves (PN 10). General Technical Specification. BS EN 1111:2017.

BSI, 2017b. Sanitary Tapware. Low Pressure Thermostatic Mixing Valves. General Technical Specification BS EN 1287:2017—TC.

Buffet-Bataillon, S., Bonnaure-Mallet, M., de la Pintière, A., Defawe, G., Gautier-Lerestif, A.-L., Fauveau, S., Minet, J., 2010. Heterotrophic bacterial growth on hoses in a neonatal water distribution system. J. Microbiol. Biotechnol. https://doi.org/10.4014/JMB.0906.06049.

Charron, D., Bédard, E., Lalancette, C., Laferrière, C., Prévost, M., 2015. Impact of electronic faucets and water quality on the occurrence of *Pseudomonas aeruginosa* in water: a multi-hospital study. Infect. Control Hosp. Epidemiol. 36, 311–319. https://doi.org/10.1017/ice.2014.46.

Chattopadhyay, S., Perkins, S.D., Shaw, M., Nichols, T.L., 2017. Evaluation of exposure to *Brevundimonas diminuta* and *Pseudomonas aeruginosa* during showering. J. Aerosol Sci. 114, 77–93. https://doi.org/10.1016/j.jaerosci.2017.08.008.

Crook, B., Willerton, L., Smith, D., Wilson, L., Poran, V., Helps, J., McDermott, P., 2020. Legionella risk in evaporative cooling systems and underlying causes of associated breaches in health and safety compliance. Int. J. Hyg. Environ. Health 224, 1–7.

DHSC, 2017. HTM 04-01: Safe Water in Healthcare Premises.

DHSC, 2021. Health Building Note 00-09: Infection Control in the Built Environment 47.

DHSC Safe water in healthcare premises (HTM 04-01), 2017. Safe Water in Healthcare Premises (HTM 04-01). [WWW Document]. GOV.UK. URL: https://www.gov.uk/government/publications/hot-and-cold-water-supply-storage-and-distribution-systems-for-healthcare-premises. (Accessed 5 January 2021 (Accessed 4 April 2019).

DWI, 2015. The Water Supply (Water Quality) Regulations 2016.

Giltner, C.L., et al., 2013. Ciprofloxacin-resistant *Aeromonas hydrophila* cellulitis following leech therapy. J. Clin. Microbiol. 51 (4), 1324–1326.

Hargreaves, J., Shireley, L., Hansen, S., Bren, V., Fillipi, G., Lacher, C., Esslinger, V., Watne, T., 2001. Bacterial contamination associated with electronic faucets: a new risk for healthcare facilities. Infect. Control Hosp. Epidemiol. 22, 202–205. https://doi.org/10.1086/501889.

Hartley, D., McCarthy, A., Greenwood, J.E., 2011. Water temperature from hot water outlets in a major public hospital: how hot is our water? Eplasty 11, e49.

HPSC, 2015. Guidelines for the Prevention and Control of Infection from Water Systems in Healthcare Facilities.

HSE, 2013. Legionnaires' Disease. The Control of Legionella Bacteria in Water Systems. Approved Code of Practice Legionnaires' Disease. ACOP. https://www.hse.gov.uk/pubns/books/l8.htm. (Accessed 5 January 2021).

HSE, 2014. HSG 274 Legionnaires' Disease—Technical Guidance Part 2: The Control of Legionella Bacteria in Hot and Cold Water Systems Technical Guidance. http://www.hse.gov.uk/pubns/books/hsg274.htm. (Accessed 5 January 2021).

Hutchins, C.F., Moore, G., Webb, J., Walker, J.T., 2020. Investigating alternative materials to EPDM for automatic taps in the context of *Pseudomonas aeruginosa* and biofilm control. J. Hosp. Infect. 106, 429–435. https://doi.org/10.1016/j.jhin.2020.09.013.

Huyer, D., Corkum, 1997. Reducing the incidence of tap-water scalds: strategies for physicians. CMAJ 156, 841–844.

Kotay, S.M., Donlan, R.M., Ganim, C., Barry, K., Christensen, B.E., Mathers, A.J., 2019. Droplet- rather than aerosol-mediated dispersion Is the primary mechanism of bacterial transmission from contaminated hand-washing sink traps. Appl. Environ. Microbiol. 85. https://doi.org/10.1128/AEM.01997-18. e01997-18.

Leslie, E., Hinds, J., Hai, F.I., 2021. Causes, factors, and control measures of opportunistic premise plumbing pathogens—a critical review. Appl. Sci. 11, 4474. https://doi.org/10.3390/app11104474.

Li, Y., Qian, H., Hang, J., Chen, X., Hong, L., Liang, P., Li, J., Xiao, S., Wei, J., Liu, L., Kang, M., 2020. Evidence for probable aerosol transmission of SARS-CoV-2 in a poorly ventilated restaurant. medRxiv 2020.04.16.20067728. https://doi.org/10.1101/2020.04.16.20067728.

Neu, L., Cossu, L., Hammes, F., 2020. Towards a probiotic approach for building plumbing—nutrient-based selection during initial biofilm formation on flexible polymeric materials. bioRxiv. https://doi.org/10.1101/2020.04.10.033217.

Pittet, D., Allegranzi, B., Sax, H., Dharan, S., Pessoa-Silva, C.L., Donaldson, L., Boyce, J.M., WHO Global Patient Safety Challenge, World Alliance for Patient Safety, 2006. Evidence-based model for hand transmission during patient care and the role of improved practices. Lancet Infect. Dis. 6, 641–652. https://doi.org/10.1016/S1473-3099(06)70600-4.

RQIA, 2012. Independent Review of Incidents of Pseudomonas aeruginosa Infection in Neonatal Units in Northern Ireland. https://www.rqia.org.uk/RQIA/files/ee/ee76f222-a576-459f-900c-411ab857fc3f.pdf.

Sartor, C., Limouzin-Perotti, F., Legré, R., Casanova, D., Bongrand, M.C., Sambuc, R., Drancourt, M., 2002. Nosocomial Infections with *Aeromonas hydrophila* from Leeches. Clin. Infect. Dis. 35 (1), E1–E5.

Schulz, A., Grigutsch, D., Alischahi, A., Perbix, W., Daniels, M., Fuchs, P.C., Schiefer, J.L., 2020. Comparison of the characteristics of hot tap water scalds and other scalds in Germany. Burns 46, 702–710. https://doi.org/10.1016/j.burns.2019.10.001.

Stabile, L., 2020. Estimation of Airborne Viral Emission: Quanta Emission Rate of SARS-CoV-2 for Infection risk Assessment 13.

Tracy, M., Ryan, L., Samarasekara, H., Leroi, M., Polkinghorne, A., Branley, J., 2020. Removal of sinks and bathing changes to control multidrug-resistant gram-negative bacteria in a neonatal intensive care unit: a retrospective investigation. J. Hosp. Infect. 104, 508–510. https://doi.org/10.1016/j.jhin.2020.01.014.

van Doremalen, N., Bushmaker, T., Morris, D.H., Holbrook, M.G., Gamble, A., Williamson, B.N., Tamin, A., Harcourt, J.L., Thornburg, N.J., Gerber, S.I., Lloyd-Smith, J.O., de Wit, E., Munster, V.J., 2020. Aerosol and surface stability of SARS-CoV-2 as compared with SARS-CoV-1. N. Engl. J. Med. https://doi.org/10.1056/NEJMc2004973.

Waines, P.L., Moate, R., Moody, A.J., Allen, M., Bradley, G., 2011. The effect of material choice on biofilm formation in a model warm water distribution system. Biofouling 27, 1161–1174. https://doi.org/10.1080/08927014.2011.636807.

Walker, J., Moore, G., 2015. *Pseudomonas aeruginosa* in hospital water systems: biofilms, guidelines, and practicalities. In: Journal of Hospital Infection, Proceedings from the 9th Healthcare Infection Society International Conference 89, pp. 324–327, https://doi.org/10.1016/j.jhin.2014.11.019.

Walker, J.T., Jhutty, A., Parks, S., Willis, C., Copley, V., Turton, J.F., Hoffman, P.N., Bennett, A.M., 2014. Investigation of healthcare-acquired infections associated with *Pseudomonas aeruginosa* biofilms in taps in neonatal units in Northern Ireland. J. Hosp. Infect. 86, 16–23. https://doi.org/10.1016/j.jhin.2013.10.003.

Weinbren, M., Bree, L., Sleigh, S., Griffiths, M., 2017. Giving the tap the elbow? An observational study. J. Hosp. Infect. 96, 328–330. https://doi.org/10.1016/j.jhin.2017.05.009.

Weinbren, M.J., Scott, D., Bower, W., Milanova, D., 2018. Observation study of water outlet design from a cross-infection/user perspective: time for a radical re-think? J. Hosp. Infect. https://doi.org/10.1016/j.jhin.2018.11.007.

Wilmer, A., Slater, K., Yip, J., Carr, N., Grant, J., 2013. The role of leech water sampling in choice of prophylactic antibiotics in medical leech therapy. Microsurgery 33 (4), 301–304.

Yui, S., Karia, K., Ali, S., Muzslay, M., Wilson, P., 2021. Thermal disinfection at suboptimal temperature of Pseudomonas aeruginosa biofilm on copper pipe and shower hose materials. J. Hosp. Infect. 117, 103–110. https://doi.org/10.1016/j.jhin.2021.08.016.

Wash hand stations

8

How do wash hand stations become contaminated?

Wash hand stations include all the equipment required for hand hygiene and are only to be used for that purpose (Hillier, 2020). However, the hand wash basin and associated equipment become contaminated with microorganisms in a number of different ways including:

- The water supply can be a source of contamination where peripheral components could have been contaminated through the growth of microorganisms and biofilm (Franco et al., 2020; Kinsey et al., 2017).
- Tap (elbow) levers, handles, soap, and hand towel dispensers can become contaminated through direct contact with healthcare workers' hands (Fig. 1) (Ding et al., 2021; Weinbren et al., 2017).
- Wash hand basins have been used as receptables for the disposal of a range of items including patient medicines, patient waste, food sources, and it has to be said, general rubbish (Feng et al., 2020).
- The rear ledge of the sink is often used as a shelf on which to place a number of containers underneath which moisture, dirt, and nutrients will collect and biofilm will grow (Fig. 1).
- The drains of wash hand basins are a recurrent contamination and recontamination source as microorganisms flow from the basin into the drains and microorganisms from the drains access the basin (either through blockages and water overflowing or through microbial motility) and become dispersed into the surrounding environment (Fig. 1) (De Geyter et al., 2021; Hajar et al., 2019).
- Taps mounted into wash hand basins or on the walls all have contact points, crevices, and niches (in the tap handles and levers and where they attach in the wall), and during use, debris, dirt, and bacteria can accumulate.
- Wash hand stations can present a risk if they are not available for use. This often happens where items or equipment are placed in the wash hand basin or where equipment is placed near the wash hand station that prevents access for hand hygiene (Fig. 2). Such practices should be pointed out as the lack of use will lead to stagnation and biofilm growth including pathogens.

What are the transmission routes and microbial risks?

Wash hand stations are an environment where high standards of infection control are required. They should only be used for hand washing to break the chain of infection. However, once contaminated, the wash hand station acts as an unlikely transmission vehicle of waterborne as well patient-derived strains of microorganisms (Grabowski et al., 2018). There are a number of different ways in which hand wash stations can act as vehicle to transmit pathogens including:

- Direct contact of microbial pathogens in the water flow discharging from the outlet into the hands of healthcare workers as well as hand contact with contaminated levers, handles, soap, and hand towel dispensers (Garvey et al., 2018).

Safe Water in Healthcare. https://doi.org/10.1016/B978-0-323-90492-6.00005-7

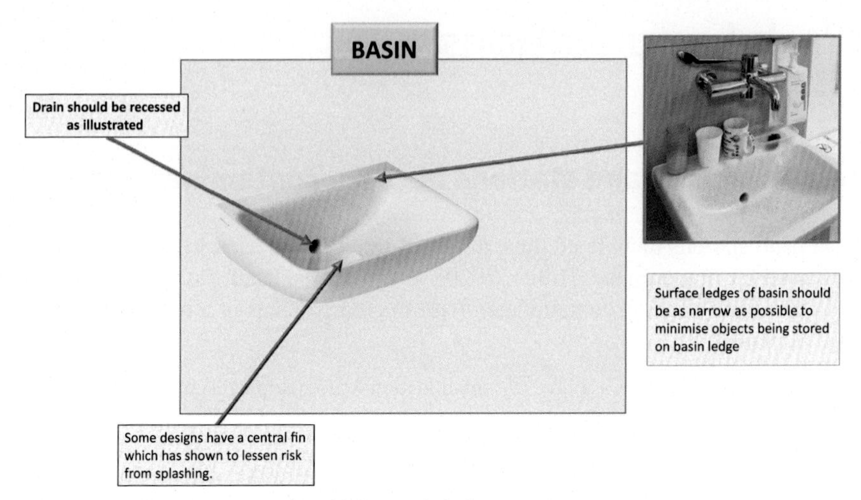

Fig. 1 Design of wash hand basin with integral drain.

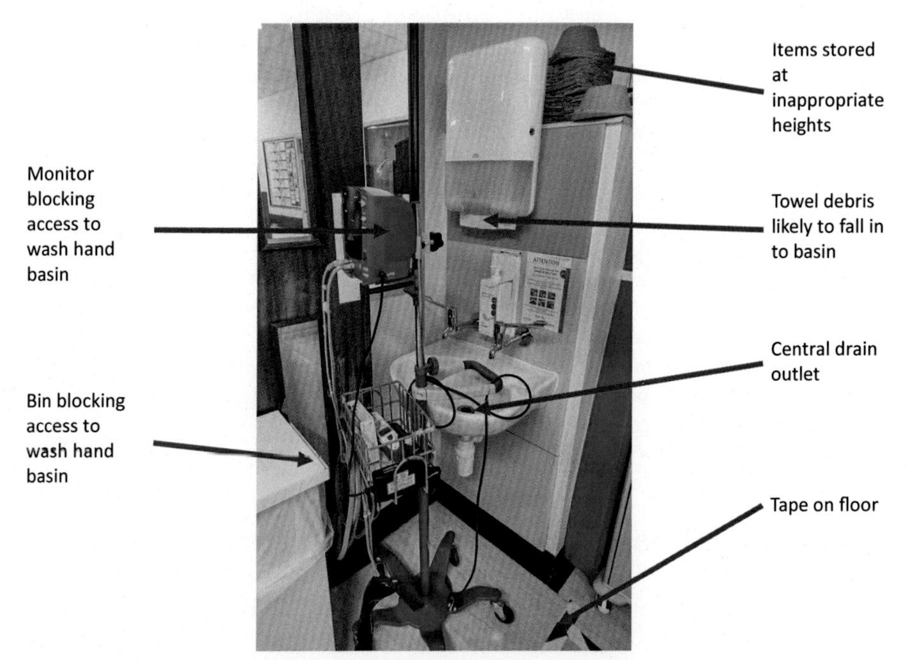

Fig. 2 Poor practices resulting in increased risks with the use of the wash hand station for hand hygiene.

- Dispersal of droplets during hand hygiene contaminating the tap levers and surrounding basin and gel/soap dispensers including sterile pharmaceutical preparation areas or sterile products stored nearby on shelves areas within a 2 m distance (Hajar et al., 2019; Kotay et al., 2019; Leitner et al., 2015).
- Physical contamination of clean hands can occur when there is accidental contact with contaminated outlet fittings (Weinbren, 2018).

- Soap dispensers, which are refillable, should not be used as the bulk soap containers are prone to contamination as bacteria are able to degrade compounds in the soap liquid mixture as nutrients (Schaffner et al., 2018; Zapka et al., 2011). So even if disposable sealed soap dispensing systems are used, if an infection breaks out, then do not rule out that the soap dispenser and/or nozzles could be contaminated (Boyle et al., 2020). These soap contamination issues may be exacerbated where there is low usage of the hand wash station.
- Bars of soap have been associated with outbreaks and should not be used on wash hand stations as the soap bars and associated surfaces become dirty and contaminated with microorganisms (Zeiny, 2009).
- Retrograde contamination, i.e., when the water outlets or hand wash station areas come into contact with microbes as a result of external activities can occur through the following mechanisms (Inkster et al., 2021):
  - From the drains where biofilms permanently reside (Moloney et al., 2020).
  - Where the contaminated hands of the healthcare workers accidentally come into contact with the outlet fitting resulting in contamination with patient-acquired strains (antibiotic resistant). These antibiotic resistance strains are able to survive in the biofilms in and around the wash hand basin, and there is the potential for continued and repeated dissemination over long periods of time.
  - Contaminated patient prosthesis being washed in the wash hand basin.
  - Disposal of items into the sink that supply nutrients and infectious bacteria.

It is very easy to clutter a wash hand station, and items such as soap dishes, soap dispensers, paper towels, pillows, and medical equipment should not be placed on or in the wash hand basin (Fig. 1). The wash hand basin should be fitted to the wall such that there are no crevices where biofilm and dirt can build up.

A wash hand station is a wash hand station, and a storage room is a storage room. But sometimes there is an overlap (Fig. 7) where shelves are placed adjacent to the wash hand station.

Activity space

Wash hand stations are designed only for hand hygiene, and there needs to be sufficient space present between the outlet and the bottom of the basin (Fig. 3). This space is critical and is known as the activity space to allow for the natural movement of hands when performing handing washing. Design and dimensions of the activity space must be taken into account for the instalment of point of use filters to control microbial contamination from the water source. The fitting of filters will reduce the activity space and may compromise effective and efficient washing of the hands.

Splashing and dispersal from wash hand basins

Studies have demonstrated that splashing and dispersal of droplet can easily occur from poorly set up wash hand stations (Gestrich et al., 2018). Either where an inappropriate basin design has been used or water flow is too high for the basin resulting in droplets and splashes being evident (Fig. 4). On one site visit to a unit, the water

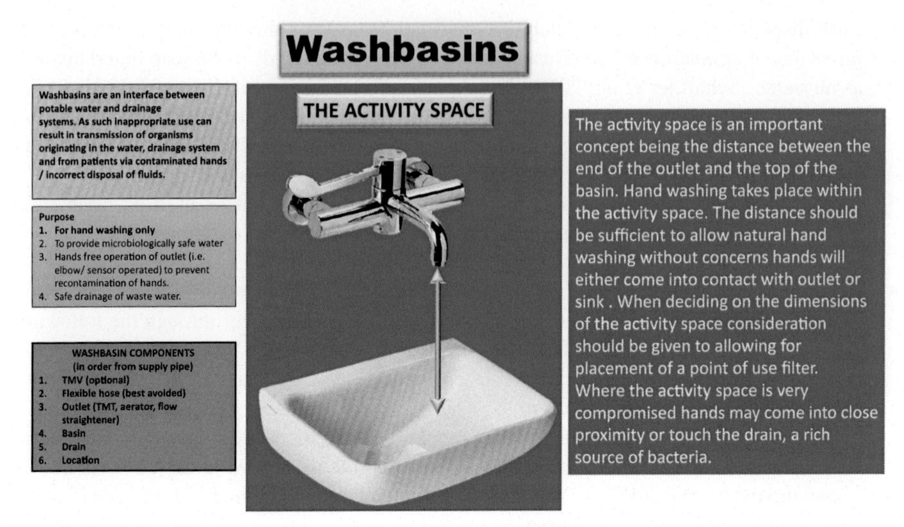

Fig. 3 Wash hand basins and the activity space.

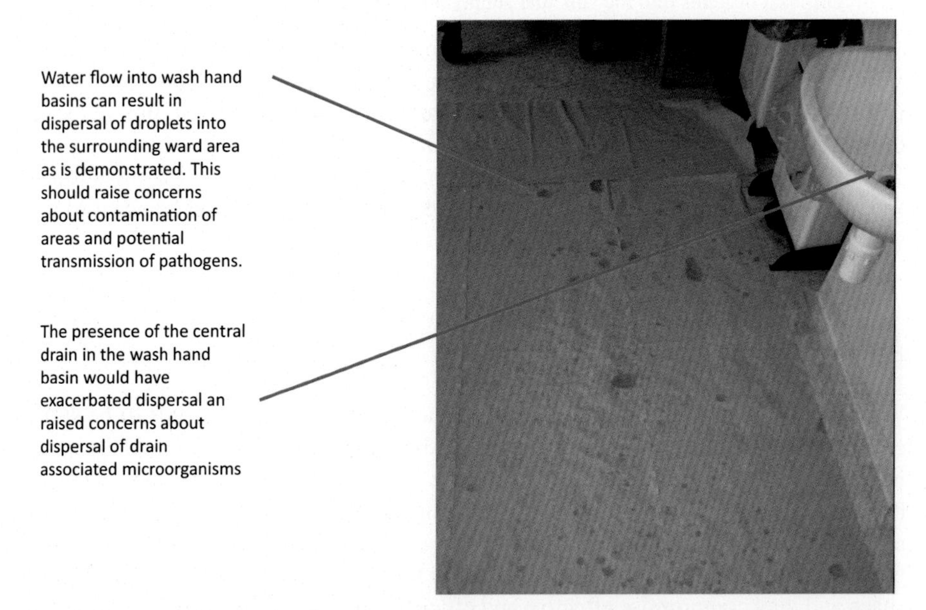

Fig. 4 Dispersal of water droplets from the wash hand basin where the water flow is too high.

volume was so high that point-of-use filters had been fitted specifically to reduce the water flow and hence the amount of splashing. Obviously, there were issues with the water flow that needed to be dealt with as a primary concern not just in terms of dispersal of *Pseudomonas* spp. over 1 m from the handwash basin but also the damage that was occurring to the floor underneath the wash hand basin that would have led to the growth of molds and environmental contamination issues (Fig. 5).

Fig. 5 Demonstration of floor damage due to water splashing from wash hand basin leading to the growth of molds and infection control and prevention issues.

Often there are preparation areas for sterile pharmaceutical medicines or even baby bathing areas next to wash hand stations. Where a wash hand station is contaminated with waterborne pathogens, e.g., *P. aeruginosa* or antibiotic strains from the drain, then clearly it does not take much for droplet dispersal during wash handing to occur and contaminate the surrounding area. We have learnt much during COVID-19 and that 2 m rule of distance has become enshrined in our behavior. Hence, there should be a 2 m clear zone surrounding a wash hand station to prevent the dispersal and contamination due to waterborne pathogens (Fig. 8).

Where this is not feasible, then the placement of a splash screen may assist in reducing or minimizing the dispersal of droplet to the left or right of the wash hand station (Figs. 6 and 7).

What does a safer wash hand station look like?

The design and setting up of a hand wash station are obviously not as easy as it seems, and the different configurations are discussed in HBN 00-10 (DH, 2013). Schematic diagrams of how the basin and taps should be placed are all very well, but documents such as HBN 00-10 give very little information as to how the wash hand station should be set up.

Fig. 8 describes some of aspects of a safer hand wash station environment for vulnerable patients.

Wash hand stations should only be installed where necessary. This is particularly important as oversupply of wash hand stations leads to a lack of use and increases the microbial risk.

There are a number of issues demonstrated;

1. This is a storeroom so does not require a hand wash station. Lack of use will predispose to biofilm formation.
3. The angle of the tap handle has been set up incorrectly such that it cannot be turned on with an elbow.
2. The placement of the hand wash station is such that any splashing is likely to contaminate clean equipment.

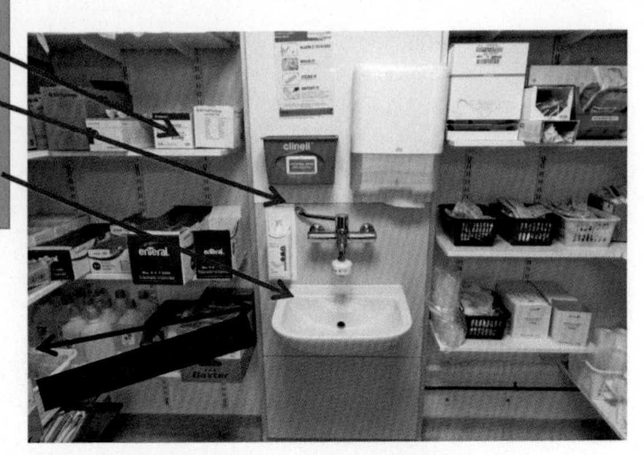

Fig. 6 Microbial dispersal from wash hand basins and how to minimize risk.

Fig. 7 Installation of a screen fitted to a hand wash station to prevent splashing.

There is obviously abuse of wash hand stations leading to these becoming microbially contaminated whether that is through accidental hand contact or blockage of the drains or by disposal of inappropriate material into the sink.

Access to wash hand stations could also be restricted at any time for a number of reasons including poor staff behavior including patient equipment blocking the route or equipment placed in the basin and in other cases, just redundancy, where there are too many (Fig. 2).

Yet, equipment issues preventing the use of wash hand stations happen with an increasing frequency due to the pressures of clinical care. Where hand wash stations present a risk, then they should be taken out of service and not returned to use until the water has been microbiologically tested and satisfactory results obtained.

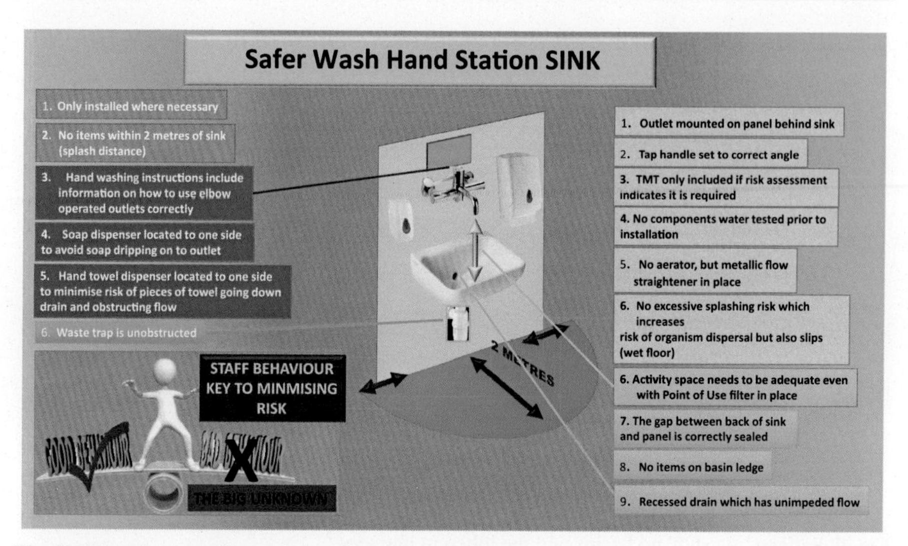

Fig. 8 What does a safer wash hand station look like?

Instructions on hand washing should be visible, including how the elbow levers should be positioned and how the levers and the hand wash station should be used.

Leak-free soap dispensers with single-use sachets should be located to the side of the wash basin. This would prevent spillage and soap deposition onto the wash basin ledges and taps that could result in microbial contamination.

Hand towel dispensers should be located to the side to prevent pieces of towel dropping into and blocking the drain.

There are issues with the placement of the elbow-operated levers, the volume of activity space required for hand washing, the placement of the drain hole, and positioning of the soaps, lotion, gel, and hand towel dispensers, and so many of these issues can lead to microbial contamination. The positioning of these items is all integral to setting up a safer wash hand station. Examples have also been cited by infection control staff where the dispensers have been microbiologically sampled and shown to be positive for *P. aeruginosa* (Boyle et al., 2020). These hand contact points would lead to continual recontamination of healthcare workers' hands.

Bars of soap and soap dishes should not be used at clinical wash hand stations or in shower rooms. Bars of soap and the soap dish can rapidly become dirty and contaminated with microorganism (Boyle et al., 2020). The soap dish itself will retain water, which creates a moist-laden reservoir in which biofilm will grow.

Patient wash hand station

The design features of clinical wash hand stations to reduce risk of exposure to opportunistic waterborne pathogens and drainage organisms surprisingly have not been translated to the design of patient washbasins, which are often en suite.

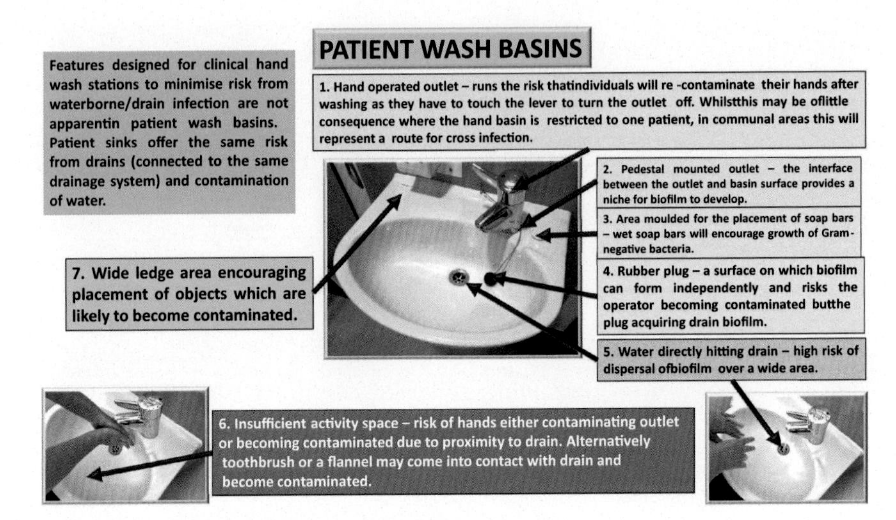

Fig. 9 Patients' wash hand basins and the inherent dangers.

Patient washbasin tap outlet components tend to be typical of most domestic installations and are pedestal mounted and manually hand-operated (Fig. 9). Like most domestic washbasins, they often empty directly into a central drain and have insufficient activity space, leading to hands either touching the outlet or coming into contact with the drain. In most instances, a plug is provided and there is a wide rear deck, or ledge, on the basin, which unfortunately allows patients to place items, such as soap and wash cloths, which will encourage debris and the growth of biofilms. The design of many taps fitted to patient washbasins means that, in most cases, there is insufficient space for the placement of point-of-use filters when they are needed. As such where it is possible to fit a filter (e.g., by using an adapter), the activity space is reduced even further, which increases the likelihood of touching either the drain area or the filter casing.

In one incident, there was an extended outbreak of CPE (Turner et al., 2020). Following an initial outbreak where the drains were identified as the source of contamination, a further outbreak occurred. In this latter outbreak, the acrylic sink tap handles in a patient's toilet, drain traps near the nurses' station, and sink in the offending patient bay were positive for CPE organisms (Fig. 10). The fact that a single bay was exclusively affected was indicative that the outbreak was associate with that location. Providing sensor-activated outlets could mitigate the risk of transferring organisms via hand contact. The drains would also have to be treated and staff assessments in using wash hand stations refreshed, i.e., for wash handing only.

Hand-operated outlets are of particular concern in communal bathrooms as these may act as a vehicle for transmission of organisms; Turner et al., report a CPE outbreak linked to a contaminated sink tap in a communal bay (Turner et al., 2020). Drains are increasingly recognized as a reservoir for the transmission of antibiotic-resistant organisms (and also sensitive organisms, but these are rarely detected and reported as they blend in with the endemic flora). Washbasin drain contamination can result from the use of the basin for

Fig. 10 Acrylic manual tap handle on a patient sink (not recommended) is difficult to clean and has implicated in an extended outbreak of CPE.
From Turner, C., Mosby, D., Partridge, D., Mason, C., Parsons, H., 2020. A patient sink tap facilitating carbapenemase-producing enterobacterales transmission. J. Hosp. Infect. 104, 511–512. https://doi.org/10.1016/j.jhin.2019.12.020.

disposal of contaminated secretions. Alternatively, equipment (used for unblocking drains) can cause contamination, or it can be via retrograde contamination from the main drainage system as water backs up in the system due to a blockage.

Lack of activity space may be through poor design or the installation of inappropriate fittings and can result in patients conducting activities in too close a proximity to the drain. Thus, in another CPE outbreak, acquisition of the infecting organism was linked to the patient being infected when brushing their teeth as they were in close proximity to the drain (Jung et al., 2020). Such incidents could be exacerbated by the position of the tap over the drain creating splashing when the water is running. Provision of a plug can create further risks because the plug (handled by the operator) will inevitably encounter biofilm around the plughole and become contaminated. Furthermore, filling these basins, which only hold small volumes of water, is likely to result in contamination with drain-dwelling organisms, particularly if the drainage rate is poor and the U-bend is overwhelmed.

Traditionally, the risk to patients from waterborne organisms, such as *P. aeruginosa,* has been restricted mainly to augmented care units. Of these units, the highest risks arise where susceptible patients (e.g., haematology/oncology) are able to use water services themselves. For these patients, their contact with water is direct and the risk of acquiring an infection might be expected to be higher, which again goes against the rationale for lower grade of design of patient sinks. Some patients receiving augmented care occupy single room accommodation and perversely are often so unwell they do not get up to use en suite water services. This lack of use of the en suite can result in stagnation of water in the supply pipework, which, if not flushed proactively as part of a control scheme, can promote biofilm formation and increase risks of infection when the services are used eventually.

Clinical wash hand stations in patient rooms (e.g., en suite) are designed for one purpose only, and that is hand decontamination. In high-risk areas at least the patient wash basin should be for only one purpose, which is also hand decontamination. But even for this sole purpose due to poor design, this is a risk.

Recommendation for wash hand stations

Staff must be trained on the appropriate use of hand washing and use of hand wash stations.

Consideration should be given to the number and placement of the wash hand station as oversupply leads to underuse and potential microbiological issues.

Wash hand basins must only be used for washing hands.

Taps should be operated without the use of hands

Wall-mounted single-lever action or sensor-operated taps with the appropriate TMVs and self-draining spouts should be used.

Elbow levers should be correctly fitted to enable operation with elbows.

Can point-of-use filters be attached to the outlet and still leave sufficient activity space?

Leak-free soap dispensers should not be refillable.

Bars of soap should not be available on wash hand basins.

Towel dispensers should be off set to prevent the towels from falling into the basin and possibly leading to blockages.

Do not place patient waste, medicine products, food, or general rubbish into the wash hand basin.

Wash hand basins should only have drains on the rear wall of the basin.

Plugs should not be available for wash hand basins.

Fitters should only follow manufacturer's instructions in fitting the outlet drain fittings and not use sealant if not recommended by the manufacturer.

Where strainers are used in the drains, these should be serviced and maintained to prevent biofilm build-up.

Do not block access to or place other equipment in the wash hand basin, which would render them not usable for their intended purpose.

Ensure that the water is draining freely from the wash hand basin and the water is not backing up to prevent water collecting in the basin as this will lead to splashing and contamination of surrounding areas.

Assessing the risks associated with wash hand basins

Risk assessment criteria	Yes	No
Are all wash hand stations included in the asset register?		
Are up to date as-fitted drawings and schematic diagrams available?		
Have the wash hand stations been designed and fitted appropriately?		
Is there knowledge of building occupancy and use including vulnerability of patient's groups in the areas where wash hand basins are situated?		
Has the number of wash hand stations been assessed to determine usage?		
Has a *Legionella* risk assessment been carried out?		

Continued

Risk assessment criteria	Yes	No
Has a *Pseudomonas aeruginosa* risk assessment been carried out due to the risk of water retention and biofilm growth on this outlet?		
Are outlet fittings appropriate?		
Are outlet fittings reviewed regularly to assess biofilm and scale build-up?		
Is there sufficient activity space in the basin once filters are fitted?		
Do the tap outlets have the appropriate fittings that can fit filters?		
Once a filter has been fitted has the flow been assessed as sufficient for hand washing?		
Have outlets in healthcare facilities where susceptible patients been assessed for regular use either manually or preferably using automated methods?		
Has the water draw-off to achieve required temperature control been assessed for both hot and cold water flow through?		
Has a risk assessment been carried out to assess the risk of scalding in clinical wash hand basins?		
Do all basins flow freely to prevent back up of water in the basin?		
Has the rear basin drain area been assessed for presence of biofilm?		
Has each basin been assessed for its potential to create splashing and aerosols during normal and abnormal use?		
Does the water from the outlet cause splash back from the surface?		
Are items stored adjacent to the basins could be contaminated during abnormal splashing?		
Are items stored adjacent to the basins could prevent access for hand washing?		
Are the soap dispensers leak-free and towel dispensers off set?		
Are screens present to prevent splashing of items stored nearby?		
Are servicing and maintenance carried out as per the protocols in place for components in the last 2 m of the water supply such as TMVs, strainers, outlet fittings?		
Are policies in place for the cleaning of wash hand basins? Are policies and procedures in place for the safe disposal of clinical effluent (i.e., not in wash hand basins)?		
Are the cleaning services assessed to ensure that the appropriate cleaning of the wash hand basins is being undertaken?		
Are there signs of damage to the laminate, walls, or flooring adjacent to the sink that would be hard to clean and harbor mold, fungi, or bacteria?		
Have patient basins been risk assessed?		

Regulations for wash hand stations

Health Building Note 00-09: Infection control in the built environment (DHSC, 2021b).

Health Technical Memorandum 04-01: Safe water in healthcare premises HTM07-04: Water management (Water Management and Water Efficiency (HTM 07-04), 2013). HTM00-02: Sanitary spaces (DHSC, 2021a).

References

Boyle, M.A., Kearney, A.D., Sawant, B., Humphreys, H., 2020. Assessing the impact of handwashing soaps on the population dynamics of carbapenemase-producing and non-carbapenemase-producing Enterobacterales. J. Hosp. Infect. 105, 678–681. https://doi.org/10.1016/j.jhin.2020.04.037.

De Geyter, D., Vanstokstraeten, R., Crombé, F., Tommassen, J., Wybo, I., Piérard, D., 2021. Sink drains as reservoirs of VIM-2 metallo-β-lactamase-producing *Pseudomonas aeruginosa* in a Belgian intensive care unit: relation to patients investigated by whole-genome sequencing. J. Hosp. Infect. 115, 75–82. https://doi.org/10.1016/j.jhin.2021.05.010.

DH, 2013. Health Building Note 00-10 Part C: Sanitary Assemblies.

DHSC, 2021a. Health Building Note 00-02: Designing Sanitary Spaces. https://www.england.nhs.uk/publication/designing-sanitary-spaces-likebathrooms-hbn-00-02/ like bathrooms.

DHSC, 2021b. Health Building Note 00-09: Infection Control in the Built Environment 47.

Ding, Z., Qian, H., Xu, B., Huang, Y., Miao, T., Yen, H.-L., Xiao, S., Cui, L., Wu, X., Shao, W., Song, Y., Sha, L., Zhou, L., Xu, Y., Zhu, B., Li, Y., 2021. Toilets dominate environmental detection of severe acute respiratory syndrome coronavirus 2 in a hospital. Sci. Total Environ. 753, 141710. https://doi.org/10.1016/j.scitotenv.2020.141710.

Feng, Y., Wei, L., Zhu, S., Qiao, F., Zhang, X., Kang, Y., Cai, L., Kang, M., McNally, A., Zong, Z., 2020. Handwashing sinks as the source of transmission of ST16 carbapenem-resistant *Klebsiella pneumoniae*, an international high-risk clone, in an intensive care unit. J. Hosp. Infect. 104, 492–496. https://doi.org/10.1016/j.jhin.2019.10.006.

Franco, L.C., Tanner, W., Ganim, C., Davy, T., Edwards, J., Donlan, R., 2020. A microbiological survey of handwashing sinks in the hospital built environment reveals differences in patient room and healthcare personnel sinks. Sci. Rep. 10, 8234. https://doi.org/10.1038/s41598-020-65052-7.

Garvey, M.I., Bradley, C.W., Holden, E., 2018. Waterborne *Pseudomonas aeruginosa* transmission in a hematology unit? Am. J. Infect. Control 46, 383–386. https://doi.org/10.1016/j.ajic.2017.10.013.

Gestrich, S.A., Jencson, A.L., Cadnum, J.L., Livingston, S.H., Wilson, B.M., Donskey, C.J., 2018. A multicenter investigation to characterize the risk for pathogen transmission from healthcare facility sinks. Infect. Control Hosp. Epidemiol. 39, 1467–1469. https://doi.org/10.1017/ice.2018.191.

Grabowski, M., Lobo, J.M., Gunnell, B., Enfield, K., Carpenter, R., Barnes, L., Mathers, A.J., 2018. Characterizations of handwashing sink activities in a single hospital medical intensive care unit. J. Hosp. Infect. 100, e115–e122. https://doi.org/10.1016/j.jhin.2018.04.025.

Hajar, Z., Mana, T.S.C., Cadnum, J.L., Donskey, C.J., 2019. Dispersal of gram-negative bacilli from contaminated sink drains to cover gowns and hands during hand washing. Infect. Control Hosp. Epidemiol. 40, 460–462. https://doi.org/10.1017/ice.2019.25.

Hillier, M., 2020. Using effective hand hygiene practice to prevent and control infection. Nurs. Stand. 35. https://doi.org/10.7748/ns.2020.e11552.

Inkster, T., Peters, C., Wafer, T., Holloway, D., Makin, T., 2021. Investigation and control of an outbreak due to a contaminated hospital water system, identified following a rare case of *Cupriavidus pauculus* bacteraemia. J. Hosp. Infect. https://doi.org/10.1016/j.jhin.2021.02.001.

Jung, J., Choi, H.-S., Lee, J.-Y., Ryu, S.H., Kim, S.-K., Hong, M.J., Kwak, S.H., Kim, H.J., Lee, M.-S., Sung, H., Kim, M.-N., Kim, S.-H., 2020. Outbreak of carbapenemase-producing *Enterobacteriaceae* associated with a contaminated water dispenser and sink drains in the cardiology units of a Korean hospital. J. Hosp. Infect. 104, 476–483. https://doi.org/10.1016/j.jhin.2019.11.015.

Kinsey, C.B., Koirala, S., Solomon, B., Rosenberg, J., Robinson, B.F., Neri, A., Halpin, A.L., Arduino, M.J., Moulton-Meissner, H., Noble-Wang, J., Chea, N., Gould, C.V., 2017. *Pseudomonas aeruginosa* outbreak in a neonatal intensive care unit attributed to hospital tap water. Infect. Control Hosp. Epidemiol. 38, 801–808. https://doi.org/10.1017/ice.2017.87.

Kotay, S.M., Donlan, R.M., Ganim, C., Barry, K., Christensen, B.E., Mathers, A.J., 2019. Droplet-rather than aerosol-mediated dispersion Is the primary mechanism of bacterial transmission from contaminated hand-washing sink traps. Appl. Environ. Microbiol. 85. https://doi.org/10.1128/AEM.01997-18. e01997-18.

Leitner, E., Zarfel, G., Luxner, J., Herzog, K., Pekard-Amenitsch, S., Hoenigl, M., Valentin, T., Feierl, G., Grisold, A.J., Högenauer, C., Sill, H., Krause, R., Zollner-Schwetz, I., 2015. Contaminated handwashing sinks as the source of a clonal outbreak of KPC-2-producing *Klebsiella oxytoca* on a hematology ward. Antimicrob. Agents Chemother. 59, 714–716. https://doi.org/10.1128/AAC.04306-14.

Moloney, E.M., Deasy, E.C., Swan, J.S., Brennan, G.I., O'Donnell, M.J., Coleman, D.C., 2020. Whole-genome sequencing identifies highly related *Pseudomonas aeruginosa* strains in multiple washbasin U-bends at several locations in one hospital: evidence for trafficking of potential pathogens via wastewater pipes. J. Hosp. Infect. 104, 484–491. https://doi.org/10.1016/j.jhin.2019.11.005.

Schaffner, D.W., Jensen, D., Gerba, C.P., Shumaker, D., Arbogast, J.W., 2018. Influence of soap characteristics and food service facility type on the degree of bacterial contamination of open, refillable bulk soaps. J. Food Prot. 81, 218–225. https://doi.org/10.4315/0362-028X.JFP-17-251.

Turner, C., Mosby, D., Partridge, D., Mason, C., Parsons, H., 2020. A patient sink tap facilitating carbapenemase-producing enterobacterales transmission. J. Hosp. Infect. 104, 511–512. https://doi.org/10.1016/j.jhin.2019.12.020.

Water Management and Water Efficiency (HTM 07-04), 2013. GOV.UK.

Weinbren, M.J., 2018. The handwash station: friend or fiend? J. Hosp. Infect. 100, 159–164. https://doi.org/10.1016/j.jhin.2018.03.023.

Weinbren, M., Bree, L., Sleigh, S., Griffiths, M., 2017. Giving the tap the elbow? An observational study. J. Hosp. Infect. 96, 328–330. https://doi.org/10.1016/j.jhin.2017.05.009.

Zapka, C.A., Campbell, E.J., Maxwell, S.L., Gerba, C.P., Dolan, M.J., Arbogast, J.W., Macinga, D.R., 2011. Bacterial hand contamination and transfer after use of contaminated bulk-soap-refillable dispensers. Appl. Environ. Microbiol. 77, 2898–2904. https://doi.org/10.1128/AEM.02632-10.

Zeiny, S.M.H., 2009. Isolation of some microorganisms from bar soaps and liquid soaps in hospital environments. Iraqi J. Pharm. Sci. 18.

Sinks in ward kitchens or domestic areas

What is a sink and what is it used for?

A number of hospital wards have utility rooms and small kitchen rooms that may contain deep stainless steel sinks for the cleaning of drinking and eating utensils prior to decontamination in a dish washer. These types of sinks tend to be stainless steel with a large drain hole and a hot and cold mixer tap placed into the stainless steel sink (Fig. 1).

The surrounding area may be used for the preparation of food materials.

In addition, due to the depth of the stainless steel sinks, they may be used to fill patient water jugs. This can be a time-consuming activity where 20–30 jugs may have to be filled.

What are the microbial contamination sources of stainless steel sinks?

Water supplied to the sink unit taps could be a source of waterborne pathogens.

The drain area of the sink, which will be supplied with a wide range of rich nutrients, due the food content being washed down the drain will harbor a wide range of pathogens, which will be both waterborne and potentially clinically relevant and may include antibiotic-resistant strains.

What are the transmission routes?

Due to depth of the stainless steel sink unit as soon as the tap is switched on the force of the water will be sufficient to result in aerosols and droplets of the supplied water (e.g., *Legionella* spp. and *Pseudomonas aeruginosa*) to be disseminated across the room and surrounding surfaces, which may include food preparation areas (Decraene et al., 2018).

Many sink areas are used without thought to the repercussions on the potential onward transmission of waterborne to patients (Fig. 1).

Swan neck taps—due to their design, swan neck taps retain larger volumes of water that does not discharge after use and are prone to microbial biofouling with microorganisms including *Legionella* and *P. aeruginosa*, and guidance recommends that this type of tap should be avoided (HFN 30 and HPSC (HPSC, 2015)).

Safe Water in Healthcare. https://doi.org/10.1016/B978-0-323-90492-6.00007-0

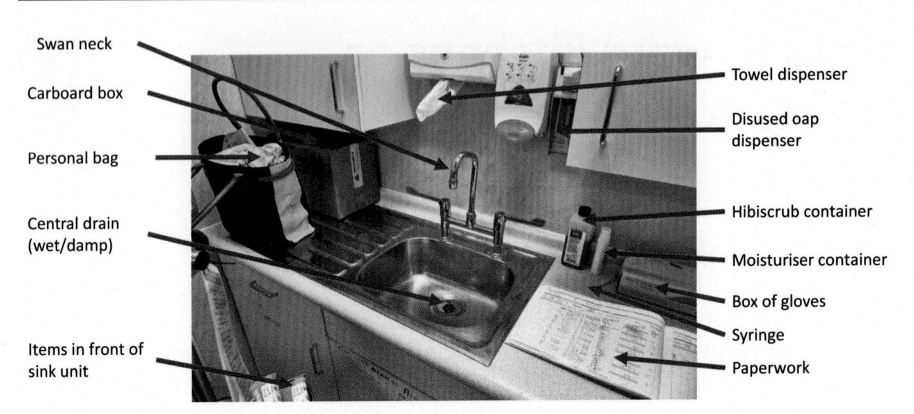

Fig. 1 Inappropriate placement and use of sink area.

For this mixer tap, there is no indication whether the hot water is supplied directly from the hot flow or through a TMV? If directly from the hot flow should there be an indicator that water will be hot, i.e., present a scalding risk?

Elbow operated taps—in this instance, due to the depth of the sink from the front to the back, the taps are most likely to be out of reach of the elbows of the majority of staff who will be placed in the position of then having to operate the taps with their hands.

Carboard box—the storage of cardboard boxes on sink draining board is not appropriate. Water dispersal will lead to the cardboard becoming damp and/or water being retained on the bottom of the box leading to mold, fungal, and bacterial growth (Inkster and Weinbren, 2021).

Personal bag—the placement of a personal bag on the sink draining board is inappropriate. Potential for splashing onto bag and contamination from and to handles and contents from the sink and hands in this vicinity.

Central drain in sink—drains have become an increasing area of concern and the outbreak of antibiotic-resistant carbapenemase producing *Enterobacteriaceae* (CPE) at the Manchester Heart Hospital demonstrated the direct route from the sink to the infections in patients (Decraene et al., 2018). The area round the drain may still be wet/damp, which will encourage growth of the biofilm microorganisms from the drain to the sink surface. As the outlet is used, the water from the swan neck tap will cascade onto the drain surface and will disperse droplets, containing antibiotic microorganisms from the sink onto the surrounding areas, surfaces, stored items, pharmaceutical or medical preparation areas (Aranega-Bou et al., 2018, 2021; Kotay et al., 2019).

Items on floor in front of sink unit—the positioning of the trolley/items on the floor may lead to poor access to the sink hindering the sink being used for its intended purpose. In addition, these items will become contaminated from splashes containing microorganisms from the drain.

Towel dispenser—the positioning of this towel dispenser will result in parts of tissues falling into the sink and into the drain potentially leading to blockages.

Disused soap dispenser—this dispenser is no longer used and creates unnecessary clutter around the sink station.

Hibiscrub and moisturizer container—unnecessary clutter and objects on the work surface. Being placed close to the sink unit, contaminated splash water from the sink will collect under these containers leading to biofilm growth and contamination.

Box of gloves—when the sink is used, splash droplets containing antibiotic resistance microorganisms will contaminate the surface of this box and when open, the gloves contained therein. Water will collect under the cardboard glove box leading to the growth of biofilms.

Syringe—the presence of this syringe indicates that medical procedures are occurring within the vicinity of this sink. This is alarming and suggests an increased water/infection risk with this sink unit. The dispersal of contaminated splashes from the sink could potentially contaminate this syringe leading to patient infections.

Paperwork—the positioning of the paperwork so close to the sink will lead to potential contamination. The rim of the sink unit will permanently retain moisture and biofilm, and the paperwork butts right up to the side of the sink rim. Splashes from the sink when the water is flowing will lead to droplets being dispersed onto the paperwork, which may contaminate the hands of the healthcare worker when making notes, adding labels, using a pen and the syringe, etc.

The drain itself will be a source of a wide range of waterborne and clinically relevant microorganisms, e.g., CPE that will contaminate the surrounding sink surfaces (Decraene et al., 2018). Contaminated sink drains lead to dissemination of microbial contamination through:

- the splashing of the tap water into the sink, which will distribute these drain-associated microorganisms into the surrounding environment and onto clinical equipment.
- contamination of the bottom of water jugs and other utensils that are placed into the sink for filling with water and then distributed into the wards.

Outbreaks associated with stainless steel utility sinks

A hospital-wide CPE outbreak

Outbreak 1

In an outbreak report at a large hospital had an endemic CPE strain with a low frequency of isolation with approximately 1 new case every 2 months (Fig. 1). In July, the number of cases increased dramatically (Weinbren, 2020). A case-control study identified salads as the risk factor for CPE acquisition. Testing of salad ingredients on arrival on site failed to identify the CPE. However (Fig. 1), it was noted that the salads were prepared near a sink. Water from the outlet directly hit the drain, which could be shown to be dispersing the CPE strain into the environment and contaminating the salads. This only partially explained the outbreak as the hospital main kitchen whether salads were prepared was in a separate building on a separate drainage system. It was unclear how the endemic hospital strain would have found its way into the kitchen in a separate building. On further investigations, it was discovered that a drain in the kitchen had become blocked, and this was cleared by using a rodding coil, which had been previously used to clear blockages in the main sewer of the hospital. Microbiological sampling of the rodding coil yielded the outbreak strain (Personal communication with Martin Exner) (Fig. 2).

Outbreak 2

An outbreak of *Klebsiella pneumoniae* carbapenemase (KPC)-producing *Escherichia coli* and wider CRE trends were investigated at the Central Manchester University Hospital NHS Foundation Trust (CMFT) (United Kingdom) over 8 years (Decraene et al., 2018).

Control interventions were undertaken along with cohorting, rectal screening (n = 184,539 screens), environmental sampling, enhanced cleaning, and ward closure and plumbing replacement

Genomic analysis (n = 268 isolates) identified the spread of a KPC-producing *Escherichia coli* outbreak clone (strain A, sequence type 216 [ST216]; n = 125) among patients and in the environment, particularly on two cardiac wards (wards 3 and 4), despite control measures. ST216 strain A had caused an antecedent outbreak and shared its KPC plasmids with other *Escherichia coli* lineages and *Enterobacteriaceae* species.

The presence of these strains was identified in kitchen sink drains and contamination of eating and drinking utensils that were then taken back into the ward. Following closure of wards 3 and 4 and replacement of plumbing components, CRE acquisition incidence declined suggesting an environmental contribution.

However, ward 3/ward 4 wastewater sites were rapidly recolonized with CRE and patient CRE acquisitions recurred over time, albeit at lower rates. Patient relocation and extensive plumbing replacement were associated with control of

Fig. 2 Illustration of contamination of food products from a kitchen domestic sink and reported isolates per month.

the clonal KPC producing *Escherichia coli* outbreak. However, environmental contamination with CRE and patient CRE acquisitions recurred rapidly following this extensive intervention. The large numbers of cases and the persistence of blaKPC in *Escherichia coli*, including pathogenic lineages, and the identification of the drains as an environmental reservoir should be to all healthcare facilities (Figs. 3–5).

This sink unit was implicated in a series of outbreaks of *Klebsiella pneumoniae* carbapenemase (KPC)-producing *Escherichia coli* at the Manchester Heart Centre. The KPC-producing *E. coli* clone was isolated from the kitchen utility sink on multiple occasions following numerous remedial interventions (replacing drainage traps, sections of pipework and treatment with 1000ppm hypochlorite solution). Patient jugs had been placed in the sink to be filled and had been contaminated from drain.

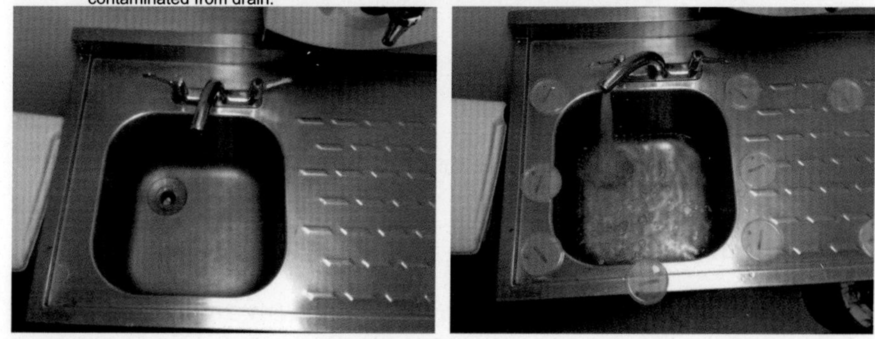

Fig. 3 Ward kitchen sink where patients jugs were filled and were contaminated from the drain recess.
Images courtesy of Dr Ryan George.

Patient jugs had been placed in the sink to be filled. As can be observed in this image the jugs were placed directly over the drain which may have resulted in contamination of the water jugs from microorganisms in the drain.

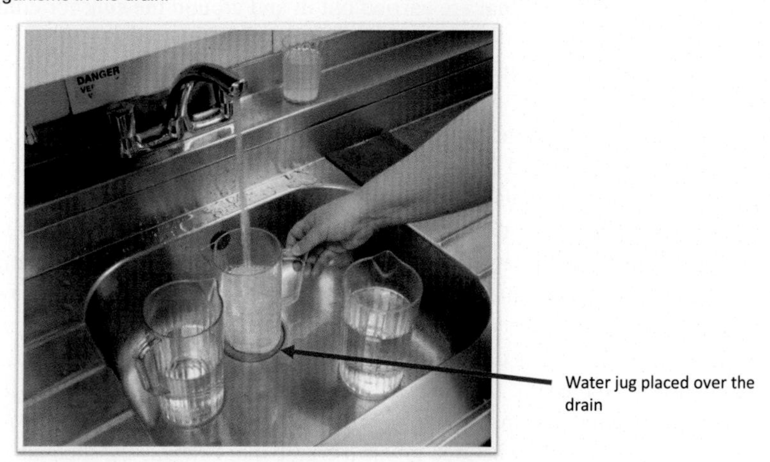

Fig. 4 Demonstration of water jugs being filled in domestic kitchen sink.

Fig. 5 Use of a splash barrier to reduce contamination of surrounding surfaces.

Guidance on sinks

Health Building Note 00-10 Part C: Sanitary Assemblies (DH, 2013)

Safe water in healthcare premises (HTM 04-01) (Department of Health, 2016)

BS EN 13310:2003 Kitchen sinks. Functional requirements and test methods (BSI, 2003)

Recommendation on sinks

Risk-assess all activities that are carried out in and around the sink for contamination events.

Ensure risk assessments are in place for the cleaning and decontamination of the sink unit.

Sinks used in clinical areas (such as dirty utilities) should not have tap holes (i.e., wall-mounted taps should be used).

Sinks used in general use/domestic services (e.g., in cleaners' rooms and kitchens) should have tap holes for use with pillar taps (separate hot and cold water).

Overflows to sinks not recommended, as they constitute a constant infection control risk much more significant than the possible risk of damage due to water overflowing.

Servicing and maintenance of the sink taps and drain should be carried out to ensure that these vital areas do not present a contamination risk.

Do not design the sink such that the tap water flows directly into the drain.

Do not place utensils that are going out into the patient environment on the bottom of the sink, e.g., you may be able to use a cradle that is keeps the utensils off the sink floor that can be cleaned and decontaminated.

Do not prepare food materials in the vicinity of the sink as splashing and contamination will occur.

For filling jugs and utensils with water, then use a stand-alone filling station designed for this purpose.

Risk assessments for sinks

Items for consideration	Yes	No
Has a risk assessment been carried out for the usage of each sink?		
Have staff using the sink been given appropriate training by competent personnel?		
Is this a domestic area?		
Are patients treated within this area?		
Is this a clinical area containing susceptible patients?		
What is the susceptibility of patients treated in this area where the sinks are located?		
Are schematic diagrams available for each sink water components?		
Are the sinks and components identified on an asset register?		
Do the sinks have a TMV to the hot flow?		
Is the hot water directly from the hot supply?		
Should there be a hot water warning label?		
Has the TMV been serviced/maintained?		
Is the swan neck appropriate?		
Can staff use the elbow taps appropriately?		
Has the outlet cleanliness been assessed from hardness/microbial growth?		
Does the flow of water result in dispersal of droplets?		
Are items that could become contaminated placed on the work surfaces within the splash zone?		
Has a *Legionella* risk assessment been carried out due to the risk of aerosol production?		
Has a *Pseudomonas aeruginosa* risk assessment been carried out due to the risk of water retention and biofilm growth on this outlet?		
Is there unnecessary storage of items with the splash zone?		
Is there appropriate storage space for equipment so that there is access to the sink?		
Are items placed in the sink or within the splash zone that come into contact with patients?		

Questions on sinks

Water flows in stainless steel sinks will create splashing resulting in the distribution of droplets into the surrounding environment.

Kitchen utensils can be placed into the stainless steel sinks as there will not be risk of contamination.

Stainless steel sinks can be contaminated from a range of microorganisms from the drain.

Directing the tap such that the flow of water flows directly into the drain will result in microbial contamination of the surrounding area.

All uses of stainless steel sinks should be risk assessed.

References

Aranega-Bou, P., George, R.P., Verlander, N.Q., Paton, S., Bennett, A., Moore, G., Aiken, Z., Akinremi, O., Ali, A., Cawthorne, J., Cleary, P., Crook, D.W., Decraene, V., Dodgson, A., Doumith, M., Ellington, M., Eyre, D.W., George, R.P., Grimshaw, J., Guiver, M., Hill, R., Hopkins, K., Jones, R., Lenney, C., Mathers, A.J., McEwan, A., Moore, G., Neilson, M., Neilson, S., Peto, T.E.A., Phan, H.T.T., Regan, M., Seale, A.C., Stoesser, N., Turner-Gardner, J., Watts, V., Walker, J., Sarah Walker, A., Wyllie, D., Welfare, W., Woodford, N., 2018. Carbapenem-resistant *Enterobacteriaceae* dispersal from sinks is linked to drain position and drainage ratesin a laboratory model system. J. Hosp. Infect. https://doi.org/10.1016/j.jhin.2018.12.007.

Aranega-Bou, P., Cornbill, C., Verlander, N.Q., Moore, G., 2021. A splash-reducing clinical handwash basin reduces droplet-mediated dispersal from a sink contaminated with gram-negative bacteria in a laboratory model system. J. Hosp. Infect. 114, 171–174. https://doi.org/10.1016/j.jhin.2021.04.017.

BSI, 2003. BS EN 13310:2003. Kitchen Sinks. Functional Requirements and Test Methods. https://shop.bsigroup.com/en/ProductDetail/?pid=000000000030092766.

Decraene, V., Phan, H.T.T., George, R., Wyllie, D.H., Akinremi, O., Aiken, Z., Cleary, P., Dodgson, A., Pankhurst, L., Crook, D.W., Lenney, C., Walker, A.S., Woodford, N., Sebra, R., Fath-Ordoubadi, F., Mathers, A.J., Seale, A.C., Guiver, M., McEwan, A., Watts, V., Welfare, W., Stoesser, N., Cawthorne, J., Group, the T.I, 2018. A large, refractory nosocomial outbreak of *Klebsiella pneumoniae* carbapenemase-producing *Escherichia coli* demonstrates carbapenemase gene outbreaks involving sink sites require novel approaches to infection control. Antimicrob. Agents Chemother. 62. https://doi.org/10.1128/AAC.01689-18.

Department of Health, 2016. Health Technical Memorandum 04-01: Safe Water in Healthcare Premises—Part B: Operational Management. Part B 98.

DH, 2013. Health Building Note 00-10 Part C: Sanitary Assemblies.

HPSC, 2015. Guidelines for the prevention and control of infection from water systems in healthcare facilities. https://www.hpsc.ie/abouthpsc/scientificcommittees/sub-committeesofhpscsac/waterguidelinessub-committee/Water%20systems%20in%20healthcare%20facilities%20FINAL%20amended%20May%202017.pdf.

Inkster, T., Weinbren, M., 2021. Water springing to life the fungal desert. J. Hosp. Infect. 111, 65–68. https://doi.org/10.1016/j.jhin.2021.02.015.

Kotay, S.M., Donlan, R.M., Ganim, C., Barry, K., Christensen, B.E., Mathers, A.J., 2019. Droplet-rather than aerosol-mediated dispersion Is the primary mechanism of bacterial transmission from contaminated hand-washing sink traps. Appl. Environ. Microbiol. 85. https://doi.org/10.1128/AEM.01997-18. e01997-18.

Weinbren, M.J., 2020. Dissemination of antibiotic resistance and other healthcare waterborne pathogens. The price of poor design, construction, usage and maintenance of modern water/ sanitation services. J. Hosp. Infect. 105, 406–411. https://doi.org/10.1016/j.jhin.2020.03.034.

Showers

10

Showers and accessories

Showers are used in a wide range of environments in hospitals including patients' bathrooms where a separate shower may be present. However, shower hoses and shower heads also appear in many different areas of the hospitals including burns unit where they are used for debridement, in assisted baths, birthing pools, sluices.

What are the microbiological risks with showers, hoses, and showerheads?

Like many of the peripheral outlets in a hospital water system, the water in showers can harbor a wide range of organisms including *Legionella* spp., *Pseudomonas aeruginosa,* and nontuberculous mycobacteria that can lead to a range of infections associated with bathing (Kline et al., 2004; Koide et al., 2007). The issue with showers is that relative to other devices attached to the potable water system, the shower head produces significant inhalable aerosol, which places them at higher risk for causing *Legionella* infections. Shower facilities vary in design and include wall-mounted units, where the shower head may be fixed or be on a flexible hose and where the patient may stand, be seated, or placed on a shower trolley; where the shower may be an attachment on an assisted bath and where the shower head and hose may be adapted to fit most outlets to enable a shower to be installed.

There are various issues and problems associated with showers (Fig. 1). For example, if the shower hose is too long, there is a risk that the showerhead will touch the floor. This may lead to bacterial contamination of the showerhead with drain-associated bacteria.

Showers are often designed such that the patient is forced to stand directly on top of the drain, which is the most contaminated area in the bathroom. In the Fig. 1 hair, which provides a nidus for bacterial proliferation, can be seen collecting on top of the drain sieve and the drain will concentrate hair and debris.

Showers in staff areas are often not used. Fig. 1 demonstrates the shower being used as a storage area. In such cases when the shower is returned to patient use, how do we know that this shower is safe for vulnerable patients to use? In addition, care has to be taken to ensure that the shower hose length will not allow the shower to be contaminated with the drain and that soap dishes can easily become contaminated and so should be removed (Fig. 2). In patient areas, it is perversely the most

Safe Water in Healthcare. https://doi.org/10.1016/B978-0-323-90492-6.00016-1

Fig. 1 Issues and problems associated with using showers rooms as storage areas.

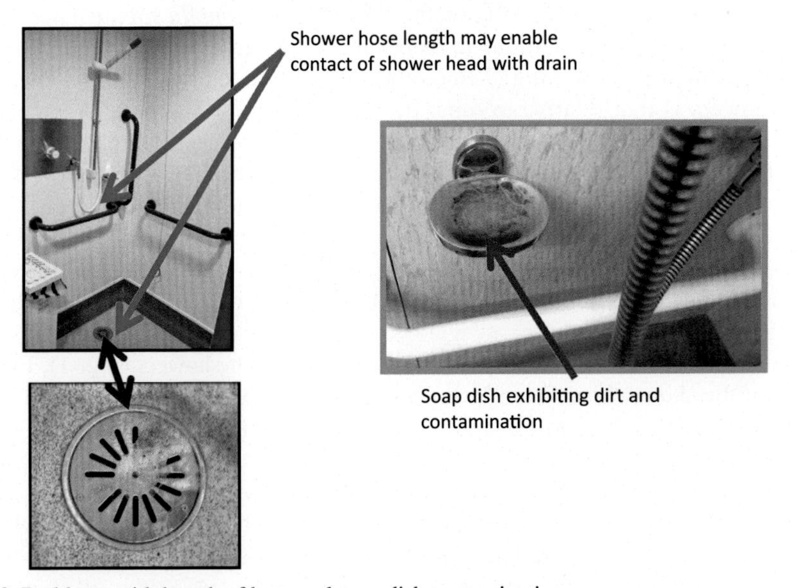

Fig. 2 Problems with length of hose and soap dish contamination.

susceptible patients who put themselves at increased risk as they may be too unwell to use the facility sometimes for weeks.

When damp, bars of soap provide an environment for bacterial proliferation (Fig. 2). The placement of soap trays should be avoided as this inadvertently encourages the use of bars of soap. Shower gel is preferable, and containers must not be refilled as this may lead to contamination of the dispenser.

Shower controller unit

Showers also have particular issues related to potential scalding, and type 3 TMV valves should be fitted to ensure that the maximum water temperature delivered is 41°C. However, like other TMVs used in outlets, there will be residual stagnant water that will encourage the growth of microorganisms and biofilms. The TMV as previously discussed combines the hot water (50–55°C) and cold <20°C to produce water at 41°C. This temperature of 41°C will be the temperature of the water as it leaves the TMV and passes through the shower hose and the shower head. For the time that the water flows, the temperature will be 41°C. However, as soon as the controller is turned off, the water flow will cease, in effect turning the TMV and shower into a dead leg. The hot water (50–55°C) to the TMV will start to decrease in temperature to a temperature that will start to encourage the growth of any residual bacteria. As the hot cools, the cold water, which is now stationary and is physically close to the hot water, will start to rise in temperature, which will again encourage any resident microorganisms to proliferate and form biofilms.

Thermostatic mixing valves should be accessible in such a way as to prevent damage to the shower installation, the supply pipework, or the removable panels themselves in wet areas. Access for safe inspection and removal should be given high priority.

Where patients have heavily contaminated areas that need washing, i.e., burn raw areas, pressure sores, or leg ulcers, the water will mobilize surface microorganisms leading to widespread dispersal in the surrounding environment including hand contact areas.

Hence, shower controls that operate water flow also need to be considered from a hygiene perspective for when they become contaminated the surfaces will be a source of cross infection to the next patient. This will happen with the various types of manual controls that are available, and some control units with manual lever are easier to clean than others. Compare the manual controls with the sensor-operated system shown (Fig. 3).

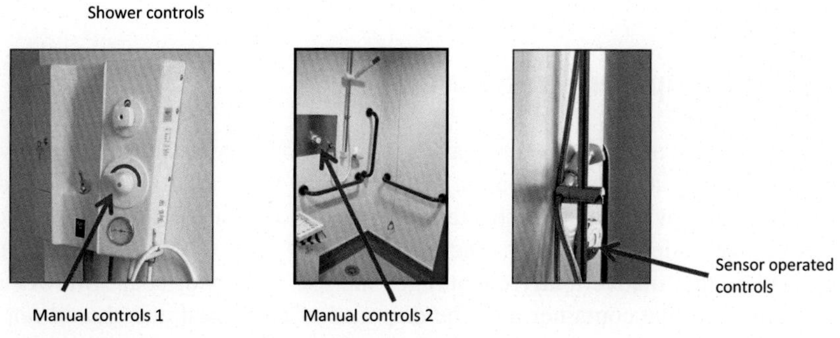

Fig. 3 Manual and sensor-operated shower controls.

Shower hoses

The temperature of 41°C of the flowing water within the shower hose will enable microbial growth. When the controller is switched off, the flow of water will stop and the temperature of the stagnant water in the hose will start to decrease. Shower hoses have traditionally been manufactured from EPDM, which has been shown to generate extensive biofilms (Waines et al., 2011; WRAS, 1999). Hence, the inner walls of the shower hose, which may be as long as 2 m in length, can harbor substantial compositions of biofilm (Moat et al., 2016). Where flexible hoses and moveable shower outlets are provided, the outlet should not be capable of being accidentally immersed into a drain, WC, or other potential source of contamination.

Shower heads

Showers with fixed heads are preferred for prevention of backflow and minimization of contamination. Adjustable shower heads are not recommended in healthcare. The heads should be selected based on ease of cleaning, descaling, and disinfection. The flow of some showerheads can be adjusted by selecting different sets of nozzles (fine spray, pulsating flow, etc.); as this will exacerbate possible stagnation problems, but this type should not be installed in healthcare premises.

It is not just the creation of aerosols that creates a risk with showers. In some cases, their lack of use results in this peripheral outlet becoming a dead leg i.e., an outlet through which water flows, but the outlet is unused/rarely used.

The shower head itself should also be risk assessed as this is a device from which substantial amounts of aerosols will be emitted. The shower heads will have a number of rubber washers and where the hose is particularly long, it may be left hanging where the shower head may come into contact with the floor and drain. Shower heads themselves become a niche where biofilm will proliferate to the extent that slime may even be visible. If the biofilm is visible, then the inside of the shower head will be particularly colonized. While not all biofilms contain pathogens, the ability of the showerhead to produce aerosols ensures that shower as an entity should be treated with the utmost respect when it comes to patient risk.

Ancillary equipment in showers

Bars of soap would once upon a time have been commonplace in showers. However, damp bars of soap in holders create a mass of soap, water, and accumulated dirt that provides an environment for bacterial proliferation and potential cross infections from one patient to another. The installation of soap trays or dishes should be avoided as this inadvertently encourages the use of bars of soap. Shower gel is preferable, and the container must be disposed of once used and should not be refilled.

Shower rooms often become a room for storage of small pieces of equipment, and over time as the shower is used less and less, then more and more equipment get stored in the shower room. Eventually the situation is reached where the shower room is no longer used at all, such that the dead leg becomes a dead end. At this point when the shower is brought back into use, there is an increased microbiological risk from the discharged water. Always risk-assess water outlets that have been left stagnant and unused and where necessary ensure that such systems are disinfected.

Deluge showers

Deluge or emergency showers are intended for use in an emergency where staff member or patients have suffered external chemical contamination and need to be doused with water. These showers should not be installed on the end of lines. These showers should also be dismantled, cleaned, descaled, and disinfected regularly e.g., quarterly in accordance with the risk assessment, and should be flushed in accordance with the recommendations in HSG274 Part 3.

Shower trolleys

Shower trolleys are used to support the patient while they are showered. A disposable plastic liner is often placed on top of the shower trolley to minimize trolley contamination, but inevitably some water will get through and the trolley mattresses obviously become covered in water and any washings from the patient. For cleanliness, the water has to drain away from the shower trolley mattress. However, one of the attributes of water is its ability to mobilize, transport, and disperse organisms over large distances.

In burns units, showers are used to debride wounds for the removal of dead (necrotic) or infected skin tissue or to remove foreign material from tissue to help a wound heal. Typically, debridement is used for old wounds that aren't healing properly and for chronic wounds that are infected and getting worse or wounds that are likely to become contaminated.

Therefore, it is likely that significant discharge of tissue will occur, which will flow onto mattress. The liner mattress has a drainage hole to allow the discharged effluent from the wound to drop from the liner onto the floor (Fig. 4). As the trolley may be approximately 4 feet off the floor, then this will result in significant widespread environmental contamination as the effluent splatters on the floor and is dispersed 2–3 m across the wet room. In the case of a burn's patient or patients with other heavily contaminated and infected wounds, then the contamination will be extensive. Over time there is extensive discoloration as the debris and microorganisms stain the floor (Fig. 5). A shower sluice can be installed above which the shower trolley can be positioned so that all the discharged wastewater is collected though there is likely to be a lot of splatter when alignment is not quite correct (Fig. 5). To minimize environmental contamination, any splashes from falling water hitting the receptacle must be

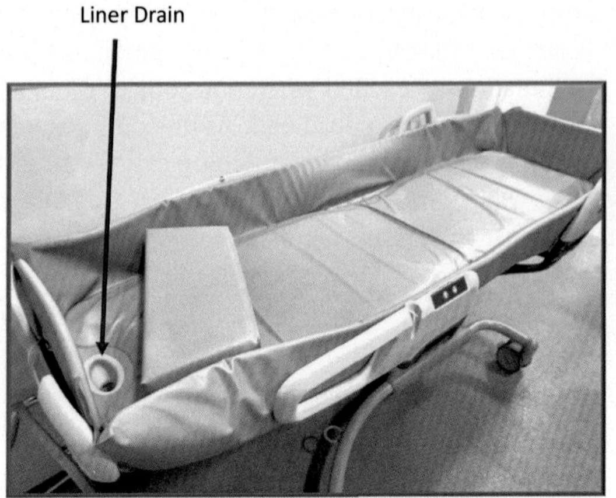

Fig. 4 Shower trolley and identification of the drainage outlet.

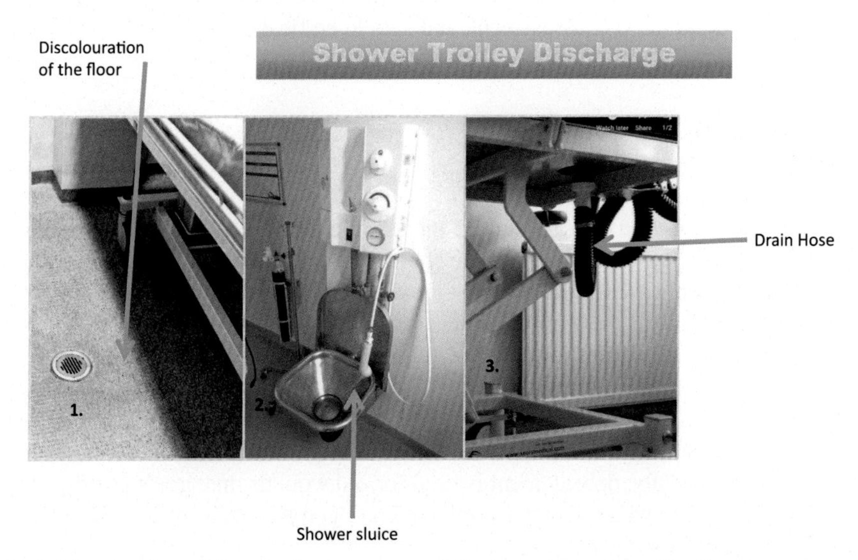

Fig. 5 Discoloration of floor and management of discharge from the trolley liner/mattress.

contained, and again there is likely to be dispersal of droplets as the discharge dispenses into the shower sluice from the trolley above. While it may be clear, the shower head is fitted with a point-of-use filter, but this is touching the receptacle risking post filter contamination and the length of the flexible hose means that when the hose is dropped that it will end up on the floor with resulting endogenous contamination (from the room onto the shower head). Other alternatives include fitting a flexible drainage tube to the drain of the trolley. While this may help reduce environmental contamination by directing the water directly into a drain hose, there is a risk that the drain outlet and hose will be extensively colonized and will be a source for cross-contamination and infection to other patients.

Trolley liners/mattresses are exposed to the copious amounts of discharge and debridement of tissue from wounds when the patient is washed. Where the liner surface is insufficiently cleaned and disinfected, then biofilm will form on the surface and in crevices.

The liners will often be manhandled when cleaning and disinfecting and moving around the trolley, and therefore, breaches in integrity of liner through tears or creases that have worn. These breaches expose the foam and enclosed PVC liners, and allow water ingress with associated contamination that will not be alleviated by disinfection as the disinfectant agent will not penetrate into the liner breach. Breaches or holes in the PVC may be obvious (Fig. 6).

When a patient is placed on the liner, the pressure of their weight will cause contaminated water to egress.

As a consequence, liners with breaches in their integrity will be heavier due to the ingress of water. In some burns unit, they weigh the mattresses to check for water ingress. As there would be a record of the weights of the integral liner and mattress then clearly if and when it had gained weight, then this would be indicative of water ingress and time to replace the mattress.

Patient risk and risk to patients

Patients should be risk assessed for their appropriateness for showering. For example, if a patient has a line in place, then a risk assessment may indicate that this patient should not be showered as the points of entry of the line may become contaminated. If the patient must be showered, then make sure the lines are sufficiently covered to prevent the ingress of water and microorganisms including *P. aeruginosa* and NTM. Fig. 7 demonstrates a case study of patient who became ill from a *Mycobacterium chelonae* infection following showering.

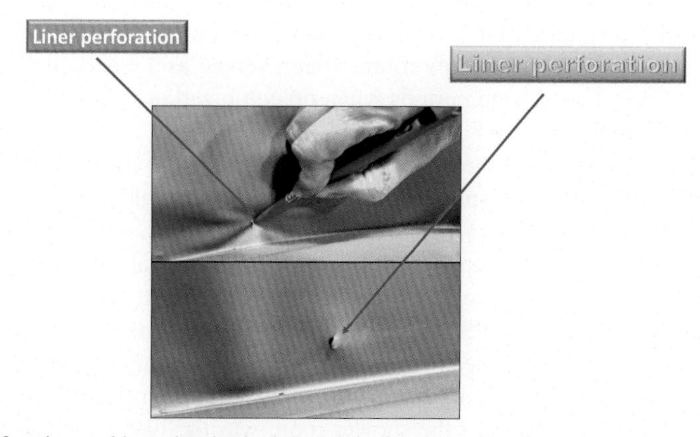

Fig. 6 Perforation and breaches in the burns unit shower trolley liner.

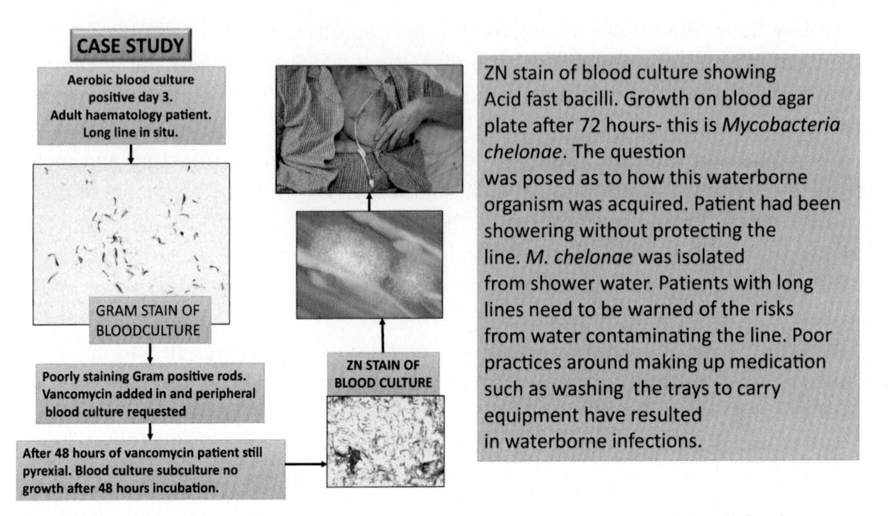

Fig. 7 Case study of patient who became ill from a *Mycobacterium chelonae* infection following showering.

Microbial outbreaks associated with showers rooms

Tissot et al., investigated a 2-year outbreak of *P. aeruginosa* in 23 patients who were hospitalized in a burns unit (Tissot et al., 2016). Patients were showered 1–3 times per week using showers that had microbial retention filters attached. Environmental sampling and typing recovered the unique isolate from floor traps, sink traps, shower trolleys, and the shower mattress of the hydrotherapy rooms. During audits it was observed that inappropriate disinfectants were being used for surface cleaning including the mattresses, contact times for other disinfectants were also inadequate and old repairs were found to be positive for *P. aeruginosa*. Disinfection protocols were revised, wet surfaces were dried after disinfection, damaged mattresses were replaced, and all sink traps in all rooms were treated with bleach. The authors considered that the contamination of burns patients during overlapping periods most likely occurred in the hydrotherapy room, which served as a reservoir allowing the persistence of the clone during periods when no colonized or infected patients were hospitalized in the unit.

Embil et al. reported an outbreak involving an index case and five secondary cases in a multiinstitution tertiary hospital burns unit caused by a single strain of methicillin-resistant *Staphylococcus aureus* (MRSA) (Embil et al., 2001). All strains were identical by pulsed-field gel electrophoresis. During environmental sampling the grip of the handheld shower and the corroded area of the stretcher frame for showering in the hydrotherapy room of the PSBU were culture positive for the outbreak strain and were considered as the route means of transmission. Showering was stopped and replaced by bedside sterile burn wound compresses, which terminated the outbreak. The investigators concluded that the shower room was the likely environmental site of the potential source of the nosocomial MRSA and that other institutions should be notified of the issues with the shower room.

Hanrahan et al. investigated seven cases of nosocomial legionellosis in a small community hospital in Upstate New York. All seven were cases of *Legionella pneumophila* serogroup 1 (Hanrahan et al., 1987). The epidemiological investigation suggested that significant risk factors for acquiring legionellosis were related to longer hospital stays and the proximity of patients' rooms to ward showers. The environmental investigation demonstrated that the ward showers and the hospital hot water system were contaminated with *Legionella pneumophila* serogroup 1. Monoclonal antibody subtyping performed on isolates obtained during the outbreak investigation confirmed that the hot water system and patient isolates had an identical pattern of reactivity. The outbreak demonstrates that legionellosis can be a significant cause of nosocomial pneumonia in a community hospital and that transmission can occur from contaminated potable hot water sources, potentially via shower aerosols.

The growth of *Legionella* spp. is of particular concern in shower units due to dispersal of aerosols and potential for immunocompromised patients to be infected with Legionnaires' disease (Tobin et al., 1980).

Microbiological monitoring

Showers should listed be on the WSP asset register with a risk assessment, maintenance, and microbiological monitoring plan in place (DHSC, 2017).

Particular concern should be related to the presence of *Legionella* strains in the water being delivered to the showers and the potential contamination of water from the shower. Where the legionella count is 1000–10,000 cfu/L, then the shower should be taken out of use. The water system, shower hose, and head associated with the shower should be risk assessed including disinfection/replacement and retesting before being reintroduced for use. Where the shower cannot be taken out of use, then consider fitting point-of-use microbial retention filters.

In addition, staff should be aware of other microorganisms contaminating the surrounding wet room and associated equipment and therefore a strategy of swabbing and/or sampling should be considered.

To ensure the sample is representative of the water flowing around the system and not just of the area downstream of the fitting, samples should be taken from separate hot and cold outlets rather than through mixer taps or outlets downstream of TMVs or showers.

Requirements and recommendations for showers

Showers should be placed on the asset register.

Risk-asses the requirement for the showers and consider removal.

Assess whether the showers are actually required.

Assess the exposure to the susceptibility of the patients.

Ensure that the showers and accessories are serviced and maintained as per manufacturer's specifications and as per WSP.

Ensure that type 3 TMVs are fitted in showers.

Showers with fixed heads are preferred for prevention of backflow.

The outlet or shower head should not be capable of being immersed in or being in contact with the drain, WC, or other source of contamination.

Showerheads that can be adjusted by selecting different sets of nozzles (fine spray, pulsating flow, etc.) should not be installed in healthcare premise.

Shower outlets and strainers should be cleaned and be free from contamination and limescale.

Access for safe inspection and removal of thermostatic mixing valves should be given high priority to prevent damage to the shower installation, the supply pipework, or the removable panels themselves in wet areas.

In in-patient accommodation where ensuite facilities are provided, it is recommended that the hot water circulation be extended to draw-off points in series; for example, the supply to a basin, bath, and/or shower should be run as one circuit.

Damaged equipment should be replaced including those that are corroded, torn, or punctured.

Risk assessing showers and their equipment

Risk assessment criteria	Yes	No
Is each shower unit on the asset register?		
Are schematic diagrams available for each shower water system?		
Has a *Legionella* risk assessment been carried out?		
Has a *Pseudomonas aeruginosa* risk assessment been carried out, particularly where debridement takes place?		
Has a frequency of use survey been carried out for showers to ascertain usage?		
Are the temperatures of the hot water supplies to the showers monitored frequently?		
Has a scalding risk assessment been carried out?		
Has the retention of the shower unit been risk assessed?		
Is the TMV integral to the body of the shower?		
Has the shower unit TMV been serviced and maintained including descaling and decontaminating?		
Are the shower hoses and heads part of a maintenance programme e.g., quarterly clean, descale, disinfection, or replacement?		
Are the shower heads in a fixed position (adjustable shower heads are not recommended)?		

Risk assessment criteria	Yes	No
Are the shower hoses of a suitable length to prevent the shower head from contacting the floor?		
Are soap trays/dishes present in the shower room?		
Are the shower facilities including the shower unit, walls, floors and drain clean with an absence of mold and slime?		
Is each shower room free of clutter and inappropriately stored items?		
Are the shower drain strainers serviced to remove biofilm and hair on a regular basis?		
Have shower trolleys been assessed for damage and punctures?		
Is the shower trolley drainage area clean and free from slime and mold?		
Are the trolley liners appropriately cleaned and disinfected?		
Have patients been risk assessed for their appropriateness of showering, e.g., presence of lines, venous catheters, surgical wounds, or susceptibility to pneumonia?		
Where showers have been removed, has all redundant pipework been removed?		
Have deluge showers been risk assessed?		

Guidance

Health Building Note (HBN) 00-02 on designing bathrooms, shower rooms, changing areas and toilets in healthcare settings (Designing Sanitary Spaces Like Bathrooms (HBN 00-02), 2013).

Safe water in healthcare premises (HTM 04-01) (Department of Health, 2016).

Managing the risks from hot water and surfaces in health and social care (HSE, 2012).

Health Building Note 00-09: Infection control in the built environment (Health Building Note 00-09: Infection control in the built environment, 2013).

Legionnaires' disease Part 2: The control of legionella bacteria in hot and cold water systems (HSE, 2014).

Examining food, water, and environmental samples from healthcare environments Microbiological guidelines (PHE, 2020).

References

Department of Health, 2016. Health Technical Memorandum 04-01: Safe Water in Healthcare Premises—Part B: Operational Management. Part B 98.

Designing Sanitary Spaces Like Bathrooms (HBN 00-02), 2013. GOV.UK.

DHSC, 2017. Health Technical Memorandum (HTM) 04-01: Safe Water in Healthcare Premises. https://www.gov.uk/government/publications/hot-and-cold-water-supply-storage-and-distribution-systems-for-healthcare-premises. (Accessed 5 January 2021).

Embil, J.M., McLeod, J.A., Al-Barrak, A.M., Thompson, G.M., Aoki, F.Y., Witwicki, E.J., Stranc, M.F., Kabani, A.M., Nicoll, D.R., Nicolle, L.E., 2001. An outbreak of methicillin resistant *Staphylococcus aureus* on a burn unit: potential role of contaminated hydrotherapy equipment. Burns 27, 681–688. https://doi.org/10.1016/s0305-4179(01)00045-6.

Hanrahan, J.P., Morse, D.L., Scharf, V.B., Debbie, J.G., Schmid, G.P., McKinney, R.M., Shayegani, M., 1987. A community hospital outbreak of Legionellosis; transmission by potable water. Am. J. Epidemiol. 125, 639–649. https://doi.org/10.1093/oxfordjournals.aje.a114577.

Health Building Note 00-09: Infection control in the built environment, 2013. 47.

HSE, 2012. Managing the Risks From Hot Water and Surfaces in Health and Social Care. https://www.hse.gov.uk/pubns/hsis6.pdf.

HSE, 2014. HSG 274 Legionnaires' Disease—Technical Guidance Part 2: The Control of Legionella Bacteria in Hot and Cold Water Systems Technical Guidance. http://www.hse.gov.uk/pubns/books/hsg274.htm. (Accessed 5 January 2021).

Kline, S., Cameron, S., Streifel, A., Yakrus, M.A., Kairis, F., Peacock, K., Besser, J., Cooksey, R.C., 2004. An outbreak of bacteremias associated with *Mycobacterium mucogenicum* in a hospital water supply. Infect. Control Hosp. Epidemiol. 25, 1042–1049. https://doi.org/10.1086/502341.

Koide, M., Owan, T., Nakasone, C., Yamamoto, N., Haranaga, S., Higa, F., Tateyama, M., Yamane, N., Fujita, J., 2007. Prospective monitoring study: isolating *Legionella pneumophila* in a hospital water system located in the obstetrics and gynecology ward after eradication of *Legionella anisa* and reconstruction of shower units. Jpn. J. Infect. Dis. 60, 5–9.

Moat, J., Rizoulis, A., Fox, G., Upton, M., 2016. Domestic shower hose biofilms contain fungal species capable of causing opportunistic infection. J. Water Health 14, 727–737. https://doi.org/10.2166/wh.2016.297.

PHE, 2020. Examining Food, Water and Environmental Samples from Healthcare Environments—Microbiological Guidelines. https://assets.publishing.service.gov.uk/government/uploads/system/uploads/attachment_data/file/865369/Hospital_F_W_E_Microbiology_Guidelines_Issue_3_February_2020__1_.pdf.

Tissot, F., Blanc, D.S., Basset, P., Zanetti, G., Berger, M.M., Que, Y.-A., Eggimann, P., Senn, L., 2016. New genotyping method discovers sustained nosocomial *Pseudomonas aeruginosa* outbreak in an intensive care burn unit. J. Hosp. Infect. 94, 2–7. https://doi.org/10.1016/j.jhin.2016.05.011.

Tobin, J.O., Dunnill, M.S., French, M., Morris, P.J., Beare, J., Fisher-Hoch, S., Mitchell, R.G., Muers, M.F., 1980. Legionnaires' disease in a transplant unit: isolation of the causative agent from shower baths. Lancet 316, 118–121. https://doi.org/10.1016/S0140-6736(80)90005-7.

Waines, P.L., Moate, R., Moody, A.J., Allen, M., Bradley, G., 2011. The effect of material choice on biofilm formation in a model warm water distribution system. Biofouling 27, 1161–1174. https://doi.org/10.1080/08927014.2011.636807.

WRAS, 1999. The Water Supply (Water Fittings) Regulations 1999.

Assisted baths

11

Assisted baths for patient care

Assisted baths are often used to provide hygiene solutions for patients who require assisted bathing. High-low baths are engineered to enable the bath to be lowered and raised for the safety of the patient for entering/leaving the bath. To enable this to happen, the bath is adapted with flexible hoses to enable the bathtub to be extended while still being connected to the water supply. As a consequence, long lengths of flexible hosing are used to achieve the functionality of the bath to assist in the raising and lowering of bath. However, as discussed, flexible hoses with their EPDM liners are prone to extensive biofilm formation.

There have been a number adverse incident reports published by the FDA relating to the presence of *Legionella* spp. and *Pseudomonas aeruginosa* (FDA, 2013a,b)

The shower hoses and shower heads of high-low baths should also be treated with caution.

Baths by their very nature are infrequently used, and the water in the pipes and hoses will be stagnant at times. Due to the infrequent use of assisted baths, the room they located in is often used as a storage space, which further limits the use of bath (Fig. 1).

While it may be possible to clean the surfaces of the baths and the tubing, which has become fouled with biofilm, it is not feasible for the tubing to be treated and cleaned effectively. Where they are installed, risk assessments need to be implemented to ensure regular maintenance, servicing, and microbial testing.

Microbiological problems associated with assisted baths

Assisted baths will be potentially contaminated in a number of different ways and include:

Microbial contamination of the water supply (Bédard et al., 2016; Hoebe et al., 1999).

Biofilm formation of plumbing components and outlets supplying the assisted bath (Capelletti and Moraes, 2016).

Contamination of the bath tub water with the skin flora of the bather as well as any pathogens that may be associated with the patient including *P. aeruginosa* and methicillin-resistant *Staphylococcus aureus* (MRSA), and as such assisted baths are frequently contaminated by Gram-negative and Gram-positive microorganisms (Baier et al., 2020). In addition, enterococci and *Escherichia coli* may also be present (Thöni et al., 2010).

Safe Water in Healthcare. https://doi.org/10.1016/B978-0-323-90492-6.00008-2

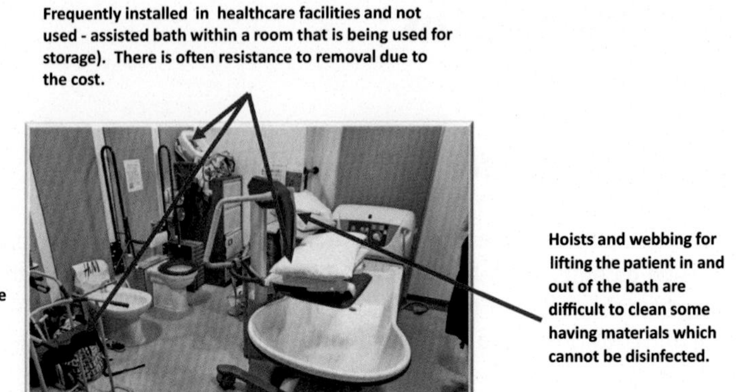

Fig. 1 Assisted bath as a storage unit.

There have also been reports of where new high-low baths have been tested prior to installation in the hospital and found to be positive for *P. aeruginosa*. In such circumstances, it may be conceivable that the contamination occurred following wet testing by the manufacturer as has happened with a number of plumbing components.

Assisted baths present other problems for patients. Staff need patient aids such as hoists, chairs, and supports to handle the patients safely. Straps (webbing) may be used to ensure that the patient is safe while in the bathing zone. However, the plastic areas and webbing may be difficult to clean, and the webbing has been known to become contaminated with pathogens resulting in outbreaks. Therefore, the staff have to consider that any area that is going to retain moisture will support the growth of microorganisms (Fig. 2).

Bath accessories and bath toys

Nosocomial outbreaks of *P. aeruginosa* in pediatric hospitals frequently involve neonates and immunosuppressed patients and can cause significant morbidity and mortality. Bathroom accessories such as toys for children can play a large part in their treatment and recovery while in hospital. This would include bath toys that bring together a number of factors including flexible plasticized rubbers, potable water, microbial contamination, and consistent biofilm formation. (Buttery et al., 1998; Neu et al., 2018). While the outside of the bath toys can be cleaned and even disinfected, the insides can be contaminated with biofilms that will time and again contaminate the surrounding areas and anyone who is playing with the toy. However, when that contamination includes multiresistant *P. aeruginosa* that is responsible for eight clinical illnesses including bacteremia and infections of the skin, central venous catheter site, and urinary tract infections, then it is clear that such toys can create major infection risks. The study by Buttery et al. was able to demonstrate that the environmental

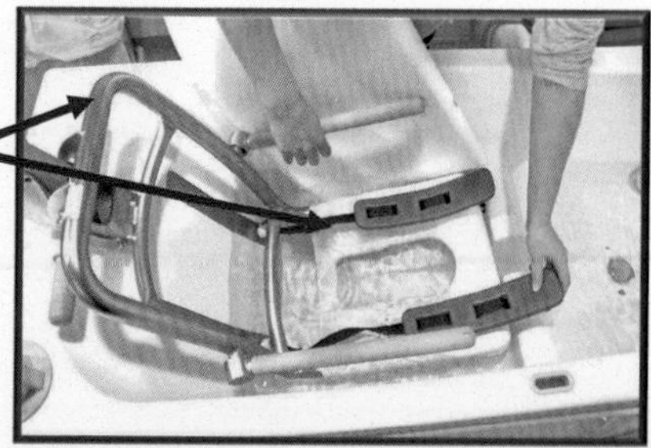

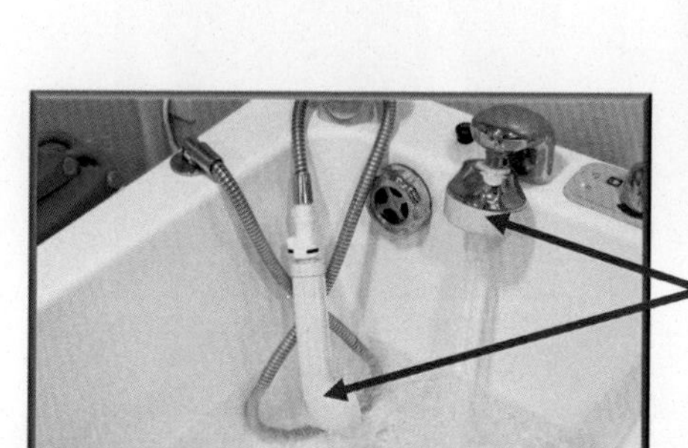

Fig. 2 Issues associated with accessories, e.g., showers, hoses, and hoists used with assisted baths.

ward yielded isolates of multiresistant *P. aeruginosa* from a toy box containing water-retaining bath toys, as well as from three of these toys. Laboratory analysis demonstrated identical isolates from patients, toys, and toy box water. Clearly, infection control should caution against the use of water-retaining bath toys in wards treating immunocompromised children (Fig. 3).

Monitoring of assisted baths

A microbiological monitoring plan should be in place (DHSC Safe water in healthcare premises (HTM 04-01), 2017). Testing should be carried out on the drinking water quality parameters of the water entering the assisted bath (PHE, 2020).

To ensure the water sample is representative of the water flowing around the system and not just of the area downstream of the fitting, samples should be taken from separate hot and cold outlets rather than through mixer taps or outlets downstream of TMVs or baths.

A visual inspection should be carried out of accessories and associated equipment to ensure that surfaces are not deteriorating, e.g., accumulation and entrapment of debris.

Recommendations for assisted baths

Assisted baths should be manufactured to a CE Standard (EEC93/42) and comply with quality assurance ISO 13485 and ISO 9001 qualifications. This will mean that "manufacturers of baths designed to compensate for a disability and not intended for use by

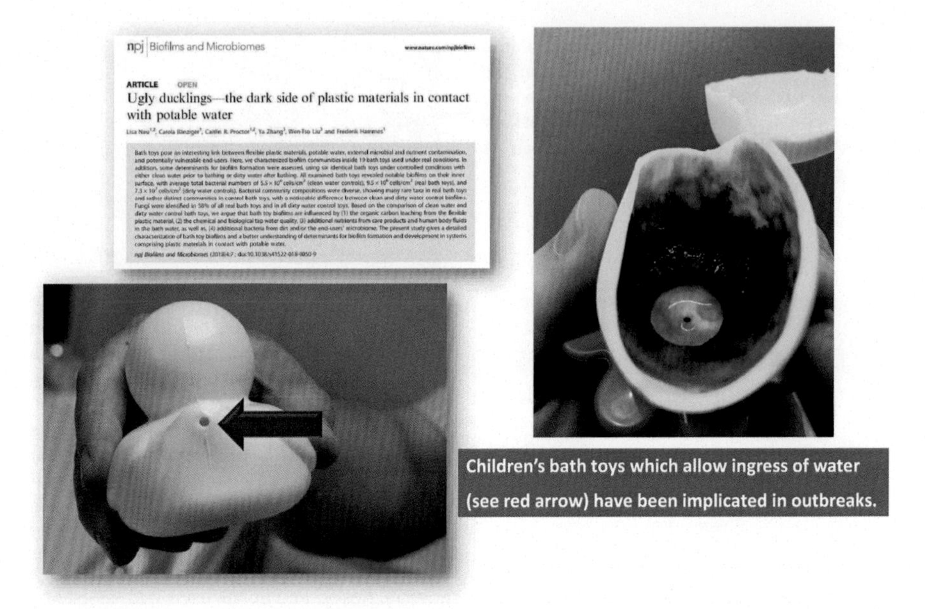

npj | Biofilms and Microbiomes www.nature.com/npjbiofilms

ARTICLE OPEN

Ugly ducklings—the dark side of plastic materials in contact with potable water

Lisa Neu[1,2], Carola Bänziger[1], Caitlin R. Proctor[1,2], Ya Zhang[3], Wen-Tso Liu[3] and Frederik Hammes[1]

Bath toys pose an interesting link between flexible plastic materials, potable water, external microbial and nutrient contamination, and potentially vulnerable end-users. Here, we characterized biofilm communities inside 19 bath toys used under real conditions. In addition, some determinants for biofilm formation were assessed, using six identical bath toys under controlled conditions with either clean water prior to bathing or dirty water after bathing. All examined bath toys revealed notable biofilms on their inner surface, with average total bacterial numbers of 5.5×10^6 cells/cm^2 (clean water controls), 9.5×10^6 cells/cm^2 (real bath toys), and 7.3×10^7 cells/cm^2 (dirty water controls). Bacterial community compositions were diverse, showing many rare taxa in real bath toys and rather distinct communities in control bath toys, with a noticeable difference between clean and dirty water control biofilms. Fungi were identified in 58% of all real bath toys and in all dirty water control toys. Based on the comparison of clean water and dirty water control bath toys, we argue that bath toy biofilms are influenced by (1) the organic carbon leaching from the flexible plastic material, (2) the chemical and biological tap water quality, (3) additional nutrients from care products and human body fluids in the bath water, as well as, (4) additional bacteria from dirt and/or the end-users' microbiome. The present study gives a detailed characterization of bath toy biofilms and a better understanding of determinants for biofilm formation and development in systems comprising plastic materials in contact with potable water.

npj Biofilms and Microbiomes (2018)4:7 ; doi:10.1038/s41522-018-0050-9

Fig. 3 The hidden biofilm inside bathroom accessories and toys.

able bodied people" need to be registered with the MHRA (Medicines and Healthcare products Regulatory Agency) as manufacturers of Class 1 Medical Devices."

The EU regulation covering Class 1 Medical Devices is Council Directive 93/42 EEC. Your bath should be clearly marked with a CE 93/42 notice affixed to it. If it is not, you should seek further clarification from your supplier and satisfy yourself that they or the manufacturer is registered with the MHRA.

Ensure that all nonmetallic material is WRAS approved.

Measures to optimize water consumption to avoid risk of stagnation should be considered at design stage to use showers rather than baths wherever practicable, but both showers and baths should be risk assessed to prevent water stagnation and microbial proliferation.

Type 3 thermostatic mixing valve should be used for the supply of blended water to assisted baths.

Due to the issues with biofilm growth with flexible hoses, their use in healthcare should be restricted to accommodate for the vertical displacement of high and low baths, i.e., assisted baths.

Where en suite facilities are provided, it is recommended that the hot water circulation be extended to draw-off points in series; for example, the supply to a bath and/or shower should be run as one circuit.

Bath fill temperatures of more than 44°C should only be considered in exceptional circumstances where there are particular difficulties in achieving an adequate bathing temperature. If a temperature of more than 44°C is to be used, then a safe means of preventing access to the hot water should be devised to protect vulnerable patients from scalding.

For assisted baths, the maximum recommended set delivery temperature of the water is 46°C to allow for the cold mass of the bath. Note that prior to patient immersion, water MUST be checked with a thermometer.

Where dual-function delivery devices, i.e., bath/shower diverter are used for the filling of the assisted bath and use of a shower, type 3 valves should be used to deliver the temperature appropriate to each outlet, e.g., bath max 44°C or 46°C, shower 41°C. (Refer also to the commissioning procedure section in HTM 04-01: Supplement—"Performance specification D 08: thermostatic mixing valves (healthcare premises).")

The WSG should consider the maintenance and cleaning of wash hand basins and associated taps, specialist (assisted) baths, and other water outlets associated in the locality.

Risk assessing assisted baths

Risk assessment criteria	**Yes**	**No**
Is each unit on the asset register?		
Are schematic diagrams available for each bath water system?		
Has a *Legionella* risk assessment been carried out?		
Has a *Pseudomonas aeruginosa* risk assessment been carried out?		
Has a survey been carried out for each bath to ascertain frequency of usage?		
Has the retention of the bath unit been risk assessed?		
Are the temperatures of the hot water supplies to the baths monitored frequently?		
Has a scalding risk assessment been carried out or each bath?		
Has a type 3 TMV been fitted to the bath?		
Is there potential for a hot metal burn on the tap body?		
Is the TMV integral to the body of the outlet?		
Are there flexible hoses to the hot or cold outlets?		
Has the bath unit TMV been serviced and maintained including descaling and decontaminating?		
Has the maximum recommended set delivery temperature of 46°C been assessed for assisted bathing?		
Is the temperature of the water in the bath assessed with a valeted thermometer prior to the immersion of the patient?		
Where present are shower hoses short enough to prevent the shower head and hose from being placed in the bath water to prevent backflow?		
Do soap trays/dishes present in the bathing room?		

Risk assessment criteria	Yes	No
Are the bath facilities including walls, floors, and drain clean with an absence of mold and slime?		
Is each bathroom free of clutter and inappropriately stored items?		
Are the bath drain strainers serviced to remove biofilm and hair on a regular basis?		
Is the bath hoist including webbing clean and free from slime, mold, and rust?		
Are all surfaces of the hoist cleanable with no broken or cracked surfaces?		
Are all accessories cleanable and capable of being disinfected?		
Do any of the accessories retain water?		
Where accessories are not cleanable, are they single use?		

Guidance for assisted baths

HTM 04-01 Safe Water in Healthcare Part A.

HTM 04-01: Supplement—"Performance specification D 08: thermostatic mixing valves (healthcare premises)."

MHRA (www.mhra.gov.uk).

References

Baier, C., Ebadi, E., Mett, T.R., Stoll, M., Küther, G., Vogt, P.M., Bange, F.-C., 2020. Epidemiologic and molecular investigation of a MRSA outbreak caused by a contaminated bathtub for carbon dioxide hydrotherapy and review of the literature [WWW Document]. Can. J. Infect. Dis. Med. Microbiol. https://doi.org/10.1155/2020/1613903.

Bédard, E., Prévost, M., Déziel, E., 2016. *Pseudomonas aeruginosa* in premise plumbing of large buildings. Microbiologyopen 5, 937–956. https://doi.org/10.1002/mbo3.391.

Buttery, J.P., Alabaster, S.J., Heine, R.G., Scott, S.M., Crutchfield, R.A., Bigham, A., Tabrizi, S.N., Garland, S.M., 1998. Multiresistant *Pseudomonas aeruginosa* outbreak in a pediatric oncology ward related to bath toys. Pediatr. Infect. Dis. J. 17, 509–513. https://doi.org/10.1097/00006454-199806000-00015.

Capelletti, R.V., Moraes, Â.M., 2016. Waterborne microorganisms and biofilms related to hospital infections: strategies for prevention and control in healthcare facilities. J. Water Health 14, 52–67. https://doi.org/10.2166/wh.2015.037.

DHSC Safe water in healthcare premises (HTM 04-01), 2017. Safe Water in Healthcare Premises (HTM 04-01). [WWW Document]. GOV.UK. URL: https://www.gov.uk/government/publications/hot-and-cold-water-supply-storage-and-distribution-systems-for-healthcare-premises. (Accessed 5 January 2021) (Accessed 4 April 2019).

FDA, 2013a. MAUDE Adverse Event Report: *Legionella* spp. Detected in ARJO Bath.

FDA, 2013b. MAUDE Adverse Event Report: *Pseudomonas* Detected in ARJO Bath.

Hoebe, C.J., Cluitmans, J.J., Wagenvoort, J.H., van Leeuwen, W.J., Bilkert-Mooiman, M.A., 1999. Cold tap water as a source of fatal nosocomial pneumonia due to *Legionella pneumophila* in a rehabilitation center. Ned. Tijdschr. Geneeskd. 143, 1041–1045.

Neu, L., Bänziger, C., Proctor, C.R., Zhang, Y., Liu, W.-T., Hammes, F., 2018. Ugly ducklings—the dark side of plastic materials in contact with potable water. NPJ Biofilms Microbiomes 4, 1–11. https://doi.org/10.1038/s41522-018-0050-9.

PHE, 2020. Examining Food, Water and Environmental Samples from Healthcare Environments—Microbiological Guidelines. https://assets.publishing.service.gov.uk/government/uploads/system/uploads/attachment_data/file/865369/Hospital_F_W_E_Microbiology_Guidelines_Issue_3_February_2020__1_.pdf.

Thöni, A., Mussner, K., Ploner, F., 2010. Water birthing: retrospective review of 2625 water births. Contamination of birth pool water and risk of microbial cross-infection. Minerva Ginecol. 62, 203–211.

Birthing pools

12

Birthing pools

Warm water immersion during labor including birth for relaxation and pain relief has been promoted since the 1970s and has become increasingly popular and accepted across many countries and particularly in midwifery-led care settings in hospitals. The use of water during labor and birth appeals to women and their carers, particularly those women seeking a relaxed intervention-free, "normal" experience.

In 1993, the use of water immersion during labor gained acceptance as a care option in the United Kingdom when the "Changing Childbirth" publication recommended that a pool facility should be an option available to women in all UK maternity units (Department of Health, 1993). Professional recognition of the use of water during labor and birth has been done by the RCM and United Kingdom Central Council for Nursing, Midwifery and Health Visiting (Royal College of Midwives, 2000; UKCC, 1994).

On very rare occasions, there have been serious adverse effects for newborns including respiratory distress and drowning from birthing pool water aspiration, pneumonia, sepsis, seizures, and perinatal asphyxia (lack of oxygen) (Fehervary et al., 2004; Kassim et al., 2005; Pinette et al., 2004; Sidebottom et al., 2020).

While these incidents have been very rare, those using such pools have to be aware of the risks associated with water birth to reduce the risks and make a fully informed decision (Garland, 2010; Harper, 2014). Other risks associated with birthing pool include manual handling (HSE, 2018).

What are microbiological risks from using birthing pools?

The potential microbiological risks arise from:

1. Waterborne organisms entering the pool from the outlet.
2. Organisms from previous occupants of pool—blood, urine, feces, birth canal fluids often present in pool.
3. Potential for contamination from drain organisms.

The majority of cases of microbiological neonatal infections related to water births appear to be related to home births (Barton et al., 2017; Collins et al., 2016; Fritschel et al., 2015; Nagai et al., 2003; Phin et al., 2014; Soileau et al., 2013). PHE recommends that heated birthing pools (incorporating a recirculation pump and heater), filled in advance of labor, should not be used for labor or birth in the home setting (ADHR, 2016; Phin et al., 2014). This recommendation does not apply to hospital birthing pools or to pools in the home that are filled immediately prior to birth (Phin et al., 2014).

Safe Water in Healthcare. https://doi.org/10.1016/B978-0-323-90492-6.00002-1

Specific microbial infections have occurred due to *Legionella, Pseudomonas aeruginosa, Haemophilus parainfluenza*, Group B *Streptococcus,* and adenovirus (Franzin et al., 2004; Thöni et al., 2010). Apart from the *Legionella* and *P. aeruginosa* cases, it is not clear that the other infections would not have occurred if the delivery had been water-free.

By limiting water births to low-risk deliveries, the extra maturity of the neonatal immune system compared with preterm deliveries significantly reduces the risk from Gram-negative infections such as *P. aeruginosa* (NICE, 2017).

The microbiological risks relating to water births are very low provided water systems are maintained correctly and the pools and all equipment in contact with water are adequately disinfected between patients. Birthing pool manufacturers have different cleaning and disinfectant protocols, which should be followed in line with the WSG risk assessments.

While neonatal cases of Legionnaires' disease have been recorded, these have either occurred in home births where the pools had recirculating heated water (due to their size and time taken to fill, pools are filled in advance providing a period for bacterial multiplication) (Collins et al., 2016; Phin et al., 2014) or from inadequately maintained hospital water systems (Franzin et al., 2004).

The large volumes of water required to fill birthing pools will have a dilution effect. Perversely taking small volumes of water from an outlet on the neonatal intensive care unit may expose babies to an extremely high bacterial load of *P. aeruginosa* if located at the periphery of the system. Running large volumes of water through an equally contaminated outlet will result in significant dilution.

Although not currently reported, there could be a microbiological risk from poorly designed or operating birthing pool drains. Companies must comply with the water supply regulations to ensure that back flow into the water does not occur (Department of Health, 2016; DWI, 2009; WRAS, 1999).

In recent years, the role of drains in the dispersal of highly antibiotic-resistant organisms has become increasingly recognized. The threat from antibiotic-resistant organisms is at least twofold: firstly, the risk of infection, and secondly, a risk from colonization as these organisms are adept at disseminating antibiotic resistance through highly promiscuous plasmids. Gram-negative organisms entering the pool via the drain are unlikely to give rise to infection in the neonate or mother but could lead to colonization. In a normal delivery, the latter is likely to go undetected so a link between poor design and inadequate drainage will not be established. It is therefore key that birthing pools are correctly designed including the drainage system and staff are trained staff to report issues relating to slow drainage or blockages.

Those responsible for birthing pools need to be aware of the potential microbiological risks to the mother, the infant as well as staff being exposed to the water from microorganisms such as *Legionellae* and *P. aeruginosa*.

Where birthing pools are only be used periodically, then risk assessments should determine if flushing is required to prevent biofilm formation through water stagnation.

Hence, water used within the birthing center for the birthing pool, hand wash stations, and any en-suite facilities should be risk assessed to reduce the potential for microbiological contamination.

As the water in the birthing pool is at body temperature, bacteria can multiply quickly and there are concerns for the microbiological contamination of the mother during giving birth as well the newborn after water delivery due the fecal and skin flora of the mother and environmental bacteria from the water system.

How do birthing pools become contaminated?

1. Microbial contamination may occur from the water distribution system where the water entering the birthing pool could be contaminated with a range of waterborne pathogens including *Legionella* spp. and *P. aeruginosa*. This could be due to systemic contamination where long section of the water is contaminated (Chapter 6) or through biofilm contamination of different components such as thermostatic mixer valves or valves at the periphery of the water system, e.g., last 2 m (Chapter 7). The type of microorganisms would be typical waterborne pathogens including *Legionella* spp., *P. aeruginosa* that could pose a risk to the mother and newborn baby
2. On rare occasions during the birthing process, the pool could also be contaminated with fecal matter (Franzin et al., 2001, 2004; Kaushik et al., 2015; Nagai et al., 2003; Thoeni et al., 2008).

There are concerns that the "protective" reflex that prevents the newborn from aspirating (breathing in) is sometimes overridden especially in stressed or compromised infants at birth resulting in a lung infection known as aspiration pneumonia (Johnson, 1996; Thach, 2007). Data by Vanderlaan and Hall (2020) did not support concerns of water aspiration, but did identify other potential risks including infection risk, optimal management of compromised water-born infants, and the potential association between immersion practice and hyponatremia (Vanderlaan and Hall, 2020). While risks cannot be eliminated, there is limited evidence that water birth causes harm to neonates compared with land birth (Taylor et al., 2016).

As a result, water births have been deemed safe for infants as the risk of bacterial colonization of infants does not seem to differ between normal births and water births (Thoeni et al., 2005).

To prevent infections and cross-contamination in birth pools and equipment, it is advisable to make as much of the accessories as possible either single use, disposable, or to ensure that they can be disinfectant and cleaned. Having infection control policies in place for all birth settings is necessary to prevent serious infections from occurring.

Case study 1: Black particle appeared in the birthing pool water

Midwives reported discolored water coming from an outlet when filling a birthing pool. No discoloration of the water was noted in nearby hand wash basins, showers, or toilets. Unfortunately, the water had been discarded before samples could be taken. A review of water temperatures and usage did not detect any abnormalities. The water system supplied other areas of the hospital and no problems with discoloration of water were reported. Two weeks later, the same occurred. The midwives seem to notice

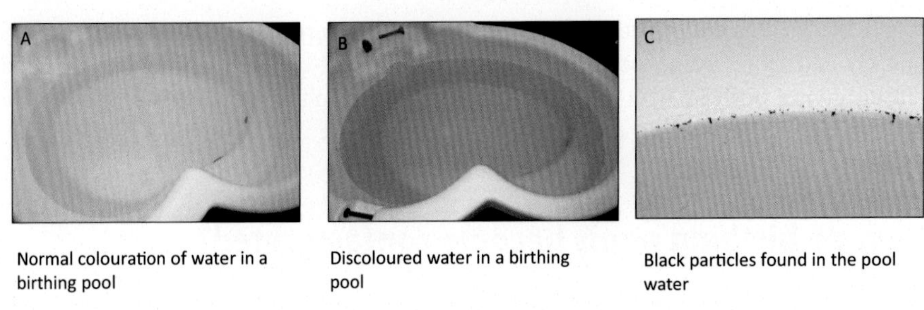

Fig. 1 Contrast between normal and the contaminated birthing pool water.

that on both occasions, the particles appeared when filling both birthing pools simultaneously. A standing order had been left to collect water should this happen again. Fig. 1, image A demonstrates the water quality expected when a birthing pool is filled. Image B shows the difference in the appearance of the color of the water with the contamination. Water samples were taken for microbial contamination (*Legionella* spp., *P. aeruginosa,* and enteric organisms), and the results were negative. The situation then recurred several weeks later. Black particles seen in image C were sent to the public analyst for testing. This showed the particles to be made up of rubber. Quarter turn valves are large, and significant numbers may be present in water systems. They were considered to be a potential source of the rubber particles, and so initially, three were chosen for removal and examination. Fig. 2, image A shows a new quarter turn valve, in comparison, images B and C show corroded and damaged valves removed from the system. The metal valve was highly corroded (had the appearance of barnacles on

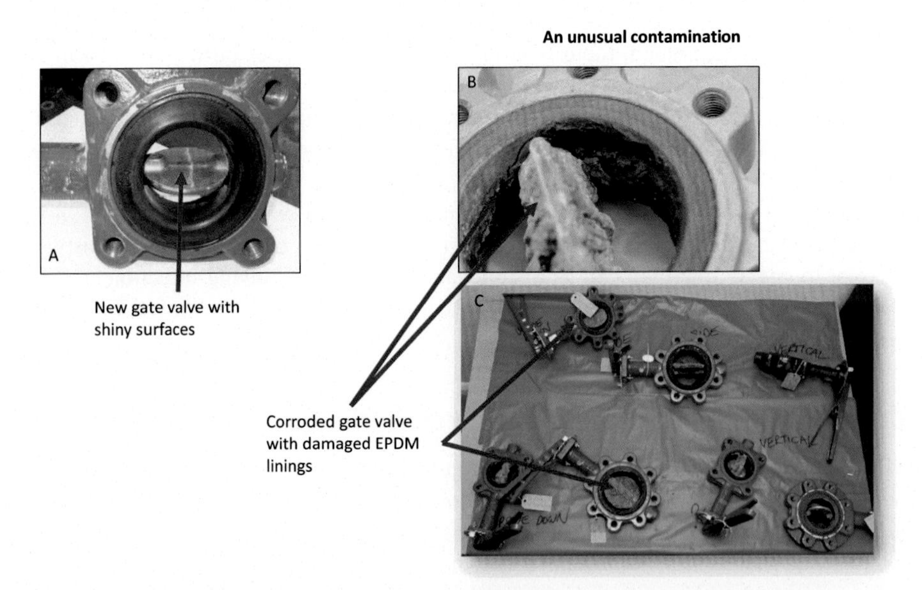

Fig. 2 Corrosion of the quarter turn gate valve and degradation of the EPDM lining that was at first considered to be a microbial/biofilm problem.

the surface), and the EPDM rubber had broken off in places (Image C). These black rubber EPDM pieces presumably found their way into the birthing pool. What was left of the EPDM rubber on the valve had become spongy and when pressure was exerted on the rubber, water was expressed. Water collected from the birthing pool containing the discolored water looked clear in small sample pots. The reason why the water discoloration was only detected in the birthing pools is potentially explained by two reasons. Firstly, the water discoloration could only be detected in large volumes of water so other water receptacles such as sinks and toilets held insufficient water to make the discoloration apparent. Secondly, it is thought that the rubber running along the pipework may have temporary blockages in some of the hot water returns. These were potentially freed by the drop in pressure when birthing pools were run simultaneously, which pull large volumes of water.

Case study 2: Birthing pool drain may have resulted in an outbreak

The rate of neonatal referral from a new birthing pool unit was perceived to be above the expected rate, leading to formation of an incident team to investigate possible causes. One feature picked up on by the team was the unusual design of the drain. The birthing pool had no plug. The drain from the pool was connected to a flexible hose, which fed into a "dry trap." There was a further length of hose connected to a valve, which was used to control emptying of the pool. In essence, the drain had become an extension of the pool when filled with water. A coarse sieve was across the drain, but its size was such that debris from the birth could lodge in the dry trap interfering with its function. There was a slight slope to the floor, but it was not clear whether this met current guidance (Fig. 3). Fluorescein dye was placed in the pipe just above the valve. On opening and closing the valve, fluorescein could be observed under UV light tracking back up the drain toward the bath water. Berrouane et al. (2000) described an outbreak of multidrug-resistant *P. aeruginosa* in a whirlpool attributing the outbreak due to an unusual design of drain (Berrouane et al., 2000). The drain closed 2.54 cm below the sieve. It was from this small area of drain, which remained in contact with the pool water when filled, that the outbreak strain was cultured.

Case study 3: Inflatable pools used in healthcare facilities

As indicated earlier in this section, the majority of cases of birthing pool neonatal infections and fatalities, e.g., due to Legionnaires' disease have occurred in the home setting (Barton et al., 2017; Collins et al., 2016; Fritschel et al., 2015; Nagai et al., 2003; Phin et al., 2014; Soileau et al., 2013). As a consequence in the United Kingdom, PHE recommended not to use heated birthing pools (incorporating a recirculation pump and heater) that are filled and heated in advance, for labor or birth in the home setting (ADHR, 2016; Phin et al., 2014). While this PHE guidance does not apply to hospital birthing pools or to pools in the home that are filled immediately prior to birth (Phin et al., 2014), a number of trusts have considered using professional inflatable birthing

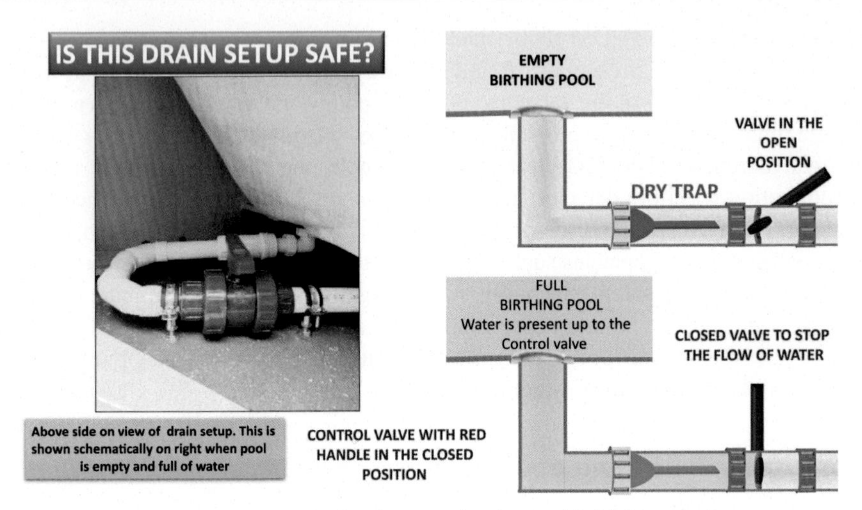

Fig. 3 Concerns raised over potential back contamination in this birthing pool.

pools in the healthcare setting. There are a number of risks associated with this approach including:

- Filling or topping up of the pool with a hose, which will present a risk of the hose in the hose slot from falling into the water. This will result in a lack of back flow protection from fluids that will pose a Category 5 risk to the mains water supply, i.e., presence of pathogenic organisms from any other source or fecal material or other human waste (WRAS, 1999).
- Flexible hoses or piping used to fill inflatable pools will retain stagnant water, which will increase the risk of biofilms and the growth of pathogens, which will present a potential hazard in the water birth pools.
- Liners should be used to prevent microbial contamination of the inside of the pool surfaces and base and should be single use.
- There needs to be an appropriate and evidence-based and tested protocol for the cleaning and decontamination for the surfaces of the pool including the handles after each of us—request the testing reports, which should have been carried out by an accredited laboratory.
- Accessories should be single use.

Microbiological case reports associated with birthing pools

Incident related to a case of pneumonia due to *Legionella pneumophila* serogroup 1 in a 7-day-old neonate following a water birth (Franzin et al., 2001)

Franzin et al. described a case of pneumonia due to *L. pneumophila* serogroup 1 in a 7-day-old neonate following a water birth. Environmental investigations were performed on the water supply of the hospital. *L. pneumophila* serogroup 1 was isolated by culture from central hot water tanks and from hot water outlets (tap and shower head of delivery room's pool and sink tap in the delivery room and the patient's room) at 300–2000 cfu/L. *L. pneumophila* was not isolated from the patient's home. The authors went on to explain that the fact that the hospital water, and particularly the pool water for water birthing, was contaminated by the same *L. pneumophila* serogroup 1 that was responsible for the child's infection strongly suggests that he was infected after prolonged delivery in contaminated water, perhaps by means of aspiration. They also suggested that during water birth, infection control policies (pool maintenance and decontamination for Legionella species) are highly recommended to prevent Legionella transmission.

Incidents describing microbial contamination of birthing pool water (Thoeni et al., 2008).

Thoeni et al. performed microbiological analyses on water samples taken from birthing pools before and after water births and compared the rate of neonatal infection arising after water births with that arising after conventional delivery (Thoeni et al., 2008). The first samples (sample A) were taken after the tub was filled with tap water; the second samples (sample B) were taken after the delivery. Of the A-samples, 29% were positive for *L. pneumophila*, 22% for *P. aeruginosa*, 18% with enterococci, 32% with coliforms, and 8% with *E. coli.* Where water samples (B-samples) were taken after the birth, 82% were positive for coliforms, 64% *E. coli*, and 8%–12% were positive for *P. aeruginosa, Staphylococcus aureus* and *Candida* spp. were also present. The authors suggested that the microbiological contamination of the water in the birthing pool after it was filled may have partly originated from the insufficient cleaning and disinfections of the birthing pool. This contamination was controlled by changing the cleaning protocols. There was no significant risk to neonates following water births compared with conventional births.

Regulations and recommendations for consideration for birthing pools

1. When choosing a healthcare birthing pool;
   - Avoid overflow drains in water birth pools as they constitute a constant infection control risk.
   - Avoid Internal water inlets on the inside of the pool just above the water line as when the water level rises, there is a high risk of back flow enabling bacteria to enter the system creating a risk of cross infection.
   - Do not install handheld showers and bath/shower mixers in water birthing pools as they are a Fluid Category 5 risk to the mains water supply, i.e., potential for back flow of a fluid, which represents a serious health hazard because of the concentration of pathogenic organisms from any other source or fecal material or other human waste (WRAS, 1999). Do not use flexible piping or branch pipes that will hold stagnant water, which will present a potential hazard in the water birth pools.
   - Do not implement a heating system to the water birth pool.
2. Manufacturers' instructions regarding installation, routine maintenance, and disinfection must always be followed.
3. The healthcare water safety plan and water safety group should ensure that the quality of the water delivered to the birthing unit is safe. From a legionella perspective, ensuring control mechanisms are in place, are meeting the defined parameters, and monitored is paramount.
4. Remove any debris from the pool, using the sieve, before emptying the pool (to prevent debris blocking the pool outlet).

5. Keep birthing pools clean using a detergent followed by disinfection as agreed with the microbiology department and in accordance with the disinfectant and pool manufacturer's guidelines. Ensure the disinfectant has sufficient contact time with the bath before running to drain.
6. Ensure the tap is cleaned first, so as not to transfer microorganisms from the "dirty" pool area to the cleaner tap region. Rinse well with warm water
7. Single-use disposable strainers should be used to remove debris in the water during labor and prior to drainage.
8. Single-use plugs, disposable scrubbing brushes should be used for removal of stubborn debris and to clean inside plug holes.
9. Single-use liners with integrated plugs have also been considered by some manufacturers.
10. Any other associated equipment, e.g., pool thermometer, mirror should be either disposable or cleaned and disinfected appropriately after use.
11. The outside of the pool, window ledges, sink, and its tap are cleaned with an Actichlor Plus solution.
12. All staff should follow Trust guidelines on infection prevention by ensuring that they effectively "decontaminate their hands" before and after each procedure to prevent contamination of the water.
13. Regular flushing is required to avoid stagnation of water if the pools and associated water services in the area are not used regularly, e.g., run the taps for 10 min once a week.
14. Following cleaning, an appropriate disinfectant should be poured into the drain.
15. Periodically monitor the microbiological quality of the water.

Risk assessment criteria for birthing pools

Risk assessment criteria for birthing pool	Yes	No
Is the birthing pool and associated equipment on the asset register?		
Has a *Legionella* risk assessment been carried out?		
Has a *Pseudomonas aeruginosa* risk assessment been carried out?		
Are infection control policies in place?		
Is the pool and associated equipment effectively cleaned, disinfected and monitored?		
Are the taps mounted in the pool?		
Have those responsible for cleaning the pool been competently trained?		
Is there an overflow drain in the pool?		
Has the drain outlet been designed to ensure water is not retained in the pipework?		
Have those who are responsible for the birthing pool been trained appropriately in the risks posed by water?		
Have any hand wash facilities and ensuite facilities been risk assessed?		
Has the TMV been serviced and maintained as per the written protocols?		

Risk assessment criteria for birthing pool	Yes	No
Is the water temperature assessed with a validated thermometer prior to entry by the mother?		
Are accessary items, e.g., strainers, plugs, thermometer, and scrubbing brushes single use?		
Are policies in place to record adverse issues with the water quality?		
Are flexible hoses used on the tap fittings?		
Is the birthing pool water periodically sampled for microbiological analysis to assess the quality of the water to ensure decontamination is appropriate?		
Where showers are associated with the bath, is it possible for the shower head and hose to enter the water?		
Have manufacturer's instructions for the pool been provided?		
Following birth, is debris always removed from the pool?		
Has the microbiology department been consulted on the cleaning and disinfection procedures?		
Are the outside areas of the pool including any horizontal areas cleaned as per the protocols?		
Do all staff wash their hand before and after each procedure?		
Is the drain disinfected after each use?		

Guidance and regulations relevant for birthing pools

HTM 04-01 Safe Water in Healthcare.

Health Building Note 09-02: Maternity care facilities.

HSE—RR1132—Manual handling risks to midwives associated with birthing pools literature review and incident analysis.

References

ADHR, 2016. Guidelines for Water Immersion and Water Birth. [WWW Document]. URL: https://www.azdhs.gov/documents/licensing/special/midwives/training/guidelines-for-water-immersion-water-birth.pdf.

Barton, M., McKelvie, B., Campigotto, A., Mullowney, T., 2017. Legionellosis following water birth in a hot tub in a Canadian neonate. CMAJ 189, E1311–E1313. https://doi.org/10.1503/cmaj.170711.

Berrouane, Y.F., McNutt, L.A., Buschelman, B.J., Rhomberg, P.R., Sanford, M.D., Hollis, R.J., Pfaller, M.A., Herwaldt, L.A., 2000. Outbreak of severe *Pseudomonas aeruginosa* infections caused by a contaminated drain in a whirlpool bathtub. Clin. Infect. Dis. 31, 1331–1337. https://doi.org/10.1086/317501.

Collins, S.L., Afshar, B., Walker, J.T., Aird, H., Naik, F., Parry-Ford, F., Phin, N., Harrison, T.G., Chalker, V.J., Sorrell, S., Cresswell, T., 2016. Heated birthing pools as a source of Legionnaires' disease. Epidemiol. Infect. 144, 796–802. https://doi.org/10.1017/S0950268815001983.

Department of Health, 1993. Changing Childbirth. Part 1. HMSO, London.

Department of Health, 2016. Health Technical Memorandum 04-01: Safe Water in Healthcare Premises—Part B: Operational Management. Part B 98.

DWI, 2009. Drinking Water Safety: Guidance to Health and Water Professionals.

Fehervary, P., Lauinger-Lörsch, E., Hof, H., Melchert, F., Bauer, L., Zieger, W., 2004. Water birth: microbiological colonisation of the newborn, neonatal and maternal infection rate in comparison to conventional bed deliveries. Arch. Gynecol. Obstet. 270, 6–9. https://doi.org/10.1007/s00404-002-0467-4.

Franzin, L., Scolfaro, C., Cabodi, D., Valera, M., Tovo, P.A., 2001. *Legionella pneumophila* pneumonia in a newborn after water birth: a new mode of transmission. Clin. Infect. Dis. 33, e103–e104. https://doi.org/10.1086/323023.

Franzin, L., Cabodi, D., Scolfaro, C., Gioannini, P., 2004. Microbiological investigations on a nosocomial case of *Legionella pneumophila* pneumonia associated with water birth and review of neonatal cases. Infez. Med. 12, 69–75.

Fritschel, E., Sanyal, K., Threadgill, H., Cervantes, D., 2015. Fatal legionellosis after water birth, Texas, USA, 2014. Emerg. Infect. Dis. 21, 130–132. https://doi.org/10.3201/eid2101.140846.

Garland, D., 2010. Revisiting Waterbirth: An Attitude to Care, first ed. Palgrave Macmillan, Houndmills, Basingstoke, Hampshire; New York, NY.

Harper, B., 2014. Birth, bath, and beyond: the science and safety of water immersion during labor and birth. J. Perinat. Educ. 23, 124–134. https://doi.org/10.1891/1058-1243.23.3.124.

HSE, 2018. RR1132—Manual Handling Risks to Midwives Associated With Birthing Pools Literature Review and Incident Analysis. https://www.hse.gov.uk/research/rrhtm/rr1132.htm.

Johnson, P., 1996. Birth under water—to breathe or not to breathe. Br. J. Obstet. Gynaecol. 103, 202–208. https://doi.org/10.1111/j.1471-0528.1996.tb09706.x.

Kassim, Z., Sellars, M., Greenough, A., 2005. Underwater birth and neonatal respiratory distress. BMJ 330, 1071–1072.

Kaushik, M., Bober, B., Eisenfeld, L., Hussain, N., 2015. Case report of *Haemophilus parainfluenzae* sepsis in a newborn infant following water birth and a review of literature. AJP Rep. 5, e188–e192. https://doi.org/10.1055/s-0035-1556068.

Nagai, T., Sobajima, H., Iwasa, M., Tsuzuki, T., Kura, F., Amemura-Maekawa, J., Watanabe, H., 2003. Neonatal sudden death due to *Legionella pneumonia* associated with water birth in a domestic spa bath. J. Clin. Microbiol. 41, 2227–2229. https://doi.org/10.1128/JCM.41.5.2227-2229.2003.

NICE, 2017. Intrapartum Care for Healthy Women and Babies.

Phin, N., Cresswell, T., Parry-Ford, F., Incident Control Team, 2014. Case of Legionnaires disease in a neonate following a home birth in a heated birthing pool, England, June 2014. Euro Surveill. 19. https://doi.org/10.2807/1560-7917.es2014.19.29.20857.

Pinette, M.G., Wax, J., Wilson, E., 2004. The risks of underwater birth. Am. J. Obstet. Gynecol. 190, 1211–1215. https://doi.org/10.1016/j.ajog.2003.12.007.

Royal College of Midwives, 2000. The Use of Water in Labour and Birth. Position paper no. 1a. RCM www.rcm.org.uk/data/info_centre/data/position_papers.htm.

Sidebottom, A.C., Vacquier, M., Simon, K., Wunderlich, W., Fontaine, P., Dahlgren-Roemmich, D., Steinbring, S., Hyer, B., Saul, L., 2020. Maternal and neonatal outcomes in hospital-based deliveries with water immersion. Obstet. Gynecol. 136, 707–715. https://doi.org/10.1097/AOG.0000000000003956.

Soileau, S.L., Schneider, E., Erdman, D.D., Lu, X., Ryan, W.D., McAdams, R.M., 2013. Case report: severe disseminated adenovirus infection in a neonate following water birth delivery. J. Med. Virol. 85, 667–669. https://doi.org/10.1002/jmv.23517.

Taylor, H., Kleine, I., Bewley, S., Loucaides, E., Sutcliffe, A., 2016. Neonatal outcomes of waterbirth: a systematic review and meta-analysis. Arch. Dis. Child. Fetal Neonatal Ed. 101, F357–F365. https://doi.org/10.1136/archdischild-2015-309600.

Thach, B.T., 2007. Maturation of cough and other reflexes that protect the fetal and neonatal airway. Pulm. Pharmacol. Ther. 20, 365–370. https://doi.org/10.1016/j.pupt.2006.11.011.

Thoeni, A., Zech, N., Moroder, L., Ploner, F., 2005. Review of 1600 water births. Does water birth increase the risk of neonatal infection? J. Matern. Fetal Neonatal Med. 17, 357–361. https://doi.org/10.1080/14767050500140388.

Thoeni, A., Ploner, F., Zech, N., 2008. Water Contamination and the Rate of Infections for Water Births 5–10.

Thöni, A., Mussner, K., Ploner, F., 2010. Water birthing: retrospective review of 2625 water births. Contamination of birth pool water and risk of microbial cross-infection. Minerva Ginecol. 62, 203–211.

UKCC, 1994. United Kingdom Central Council for Nursing, Midwifery and Health Visiting. Position Statement on Waterbirths. Annexe 1 to Registrar's Letter 16/1994. UKCC, London.

Vanderlaan, J., Hall, P., 2020. Systematic review of case reports of poor neonatal outcomes with water immersion during labor and birth. J. Perinat. Neonatal Nurs. 34, 311–323. https://doi.org/10.1097/JPN.0000000000000515.

WRAS, 1999. The Water Supply (Water Fittings) Regulations 1999.

Hydrotherapy pools

13

What are hydrotherapy/aquatic physiotherapy pools?

Hydrotherapy/aquatic physiotherapy pools are warm water pools, also known as therapeutic pools, designed to provide medical and physical care in water under the control of a competent Aquatic Physiotherapist (AP) (BSI, 2018).

What is aquatic physiotherapy?

"A physiotherapy programme utilises the properties of water, designed by a suitably qualified physiotherapist. The programme should be specific for an individual to maximise function, which can be physical, physiological or psychological. Treatments should be carried out by appropriately trained personnel, ideally in a purpose built, and heated pool" (ATACP, 2015, 2020, 2022).

Aquatic therapy is used to treat a range of musculoskeletal, neurological, and medical conditions, to improve mobility and general health as well as having a positive effect on patients with sensory and mental health conditions. Warm water helps muscles relax, eases pain, and makes exercise easier as the water helps to support body weight relieving pressure on joints resulting in increased patient movement.

Where the water temperature is below 32°C, there is a risk of patients becoming cold. Patients with limited function, increased neurological muscle tone, or pain may find that their tone or pain increases. It is contraindicated for hydrotherapy pool water to be above 35.5°C due to increased cardiac demand (ATACP, 2015).

Design

It is important that hydrotherapy pools are designed to be accessible and safe for all patients and staff (ATACP, 2020; HSE, 2018a; PWTAG, 2017). Hydrotherapy pools require a larger footprint than traditional swimming pools, both for reception and pool side for patients' trolleys, wheelchairs, and other mobility aids and equipment. There should also be sufficient space for storage before entry into the pool area as well as poolside. The design should also take into account of the need for extra space in the showers, changing rooms for patients who may require showering in a wheelchair or on a trolley.

New pool designs should take account of all mobility and condition profiles of the patient groups. Expert advice will be required to prepare the tender specification to take account of the following considerations:

- Placement of pool on the ground floor for ease of access by patients with poor mobility.
- Consideration given to how patients would be safely evacuated in the event of a fire or chemical incident (e.g., chlorine gas release).

Safe Water in Healthcare. https://doi.org/10.1016/B978-0-323-90492-6.00022-7

- Patients should flow;
  - through reception where there should be sufficient storage for equipment
  - to changing rooms where wheelchair transfer would be undertaken with access to a poolside toilet, shower, and the pool showers on exit.
- The design of the pool and areas should minimize the risk of contamination from outside by having a demarcation between dirty side and clean areas with space for wheelchairs and trolleys. For existing pools without sufficient space, the provision of overshoes and wheel covers for wheelchairs/pushchairs can reduce contamination.
- There should be sufficient space around the pool and in the changing and shower rooms to maneuver trolleys and wheelchairs, which should not obstruct the poolside or visibility of patients in the pool (Fig. 1).
- There should be:
  - an emergency pull cord/button for use in the pool.
  - separate changing/locker rooms and showers for staff.
  - space away from the poolside for disposing of clinical waste.
  - Some patient groups benefit from sensory lights and music, but there should be no direct lights over the pool and appropriate materials used to improve acoustics (especially deck level pools).
  - Drinking water should be available
- Location and design of the plant room should include sufficient space for routine maintenance and replacement of plant such as sand filters as required.
- Chemicals should be delivered and stored safely away from public areas and liquid chemicals separated by type to avoid explosions or chlorine gas release (Fig. 2) (HSE, 2018a).

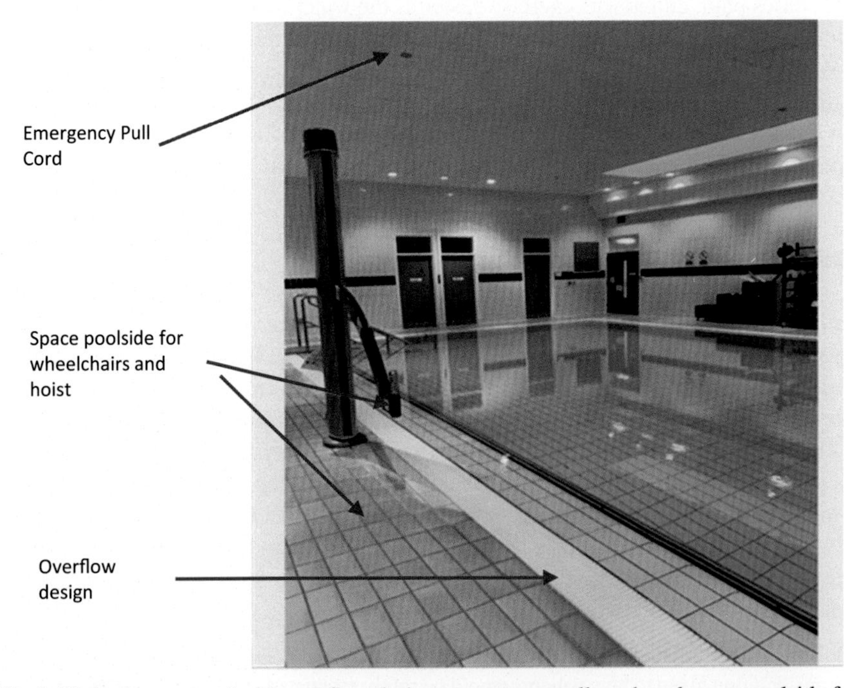

Fig. 1 Hydrotherapy pool with overflow design, emergency pull cord, and space poolside for wheelchairs and trolleys.

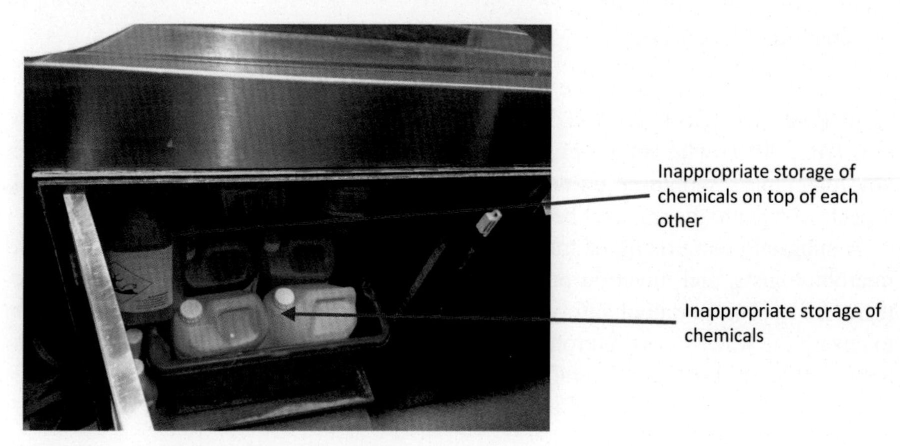

Fig. 2 Unacceptable storage of chemicals.

Liquid chemicals should be stored in bunds capable of containing at least 110% of the total volume. Chemical risk assessments are required under health and safety legislation (Control of Substances Hazardous to Health Regulations (COSHH), and the design should also comply with HSG 179, PWTAG, and ATACP guidance as well as HSG 274 and the CDM regulations (HSE, 2014, 2018a, 2020)).

Commissioning

As with any new build or major refurbishment, commissioning is important. The designer should prepare a commissioning brief that details filling, supervision, and monitoring. The approved commissioning plan requires input from independent consultants with expertise in pool design, hydrodynamics, equipment, treatment, and operational experience as well as clinical input.

The commissioning process should ensure the pool is filled from a potable source of water and disinfected as close to handover as possible and that there is a sampling plan. The pool and pool plant should be appropriately commissioned to ensure they perform as detailed in the design specification, and the disinfection systems are validated to achieve water quality targets for maximum design bather load.

Governance

General

Design and management of hydrotherapy pools and associated equipment should be overseen by the water safety group (DHSC, 2016). Risk assessments are required to ensure the water does not pose a risk. There should be oversight of water management, routine monitoring, and microbiological testing. There needs to be communication between those responsible for water management including pool plant operators, aquatic physiotherapists, assistants and technical staff, and those responsible for clinical care.

Hydrotherapy pools

UK national guidance from the Pool Water Safety Advisory Group (PWTAG) and ATACP advises that there should be a competent designated aquatic physiotherapist (PWTAG, 2019) who should ensure the pool remains safe. The designated aquatic physiotherapists should be trained in both pool water management and the clinical aspects of aquatic therapy and be represented on the water safety group.

A subgroup comprising of trained pool plant operators, aquatic physiotherapists, microbiologists, and infection prevention and control specialists will help to ensure there are appropriate pool safety operating procedures and emergency action plans to ensure the water meets microbiological and chemical water quality targets. This is important where there are several pools to ensure long-term resilience and appropriate actions are taken in the event of untoward events.

Bather loads

Pools and their water treatment systems are designed to ensure the pool remains safe for the designed bather load. This should take account of the maximum numbers of bathers at a given time, the volume and surface area of the pool, the water turnover time, and the risk from bathers contaminating the water. The immune status of the users should also be taken into account as failures in effective disinfection could have more serious consequences than for the general population.

It is essential that there is effective disinfection during increased bather load as there is a risk that undesirable levels of combined chlorine products may appear. Exceeding the bather load may compromise disinfection and bather load calculations are available from ATACP (ATACP, 2020).

How does a hydrotherapy pool function?

The recommended configuration for hydrotherapy pools is for deck level overflow, where water from the pool is displaced into the overflow channels to the balance tank and returns to the pool via the water treatment system (Fig. 3). Deck level pools enable recovery of patients as well as for the removal of floating debris from the pool surface.

Pool water circulation is maintained as water is pumped from the drain outlets at the lowest point of the pool to remove dirt and debris from the pool floor. The drains should be opposite the filling point to ensure cross flow and reduce areas of low flow.

The pool flow rate should be calculated to ensure there is no risk of entrapment of hair and limbs and safety grilles are covering the outlets. The pool water circulates through strainers, located before the pumps, to remove large particulate matter and protect the pumps from damage. A flocculation step before the filters improves the ability of the filters to remove microorganisms, particularly Cryptosporidia (resistant to normal pool chlorine concentrations). Water is then pumped through the filter (usually sand), through ultraviolet light treatment, to the heater, followed by pH adjustment, and finally, biocide addition before returning to the pool (Fig. 3).

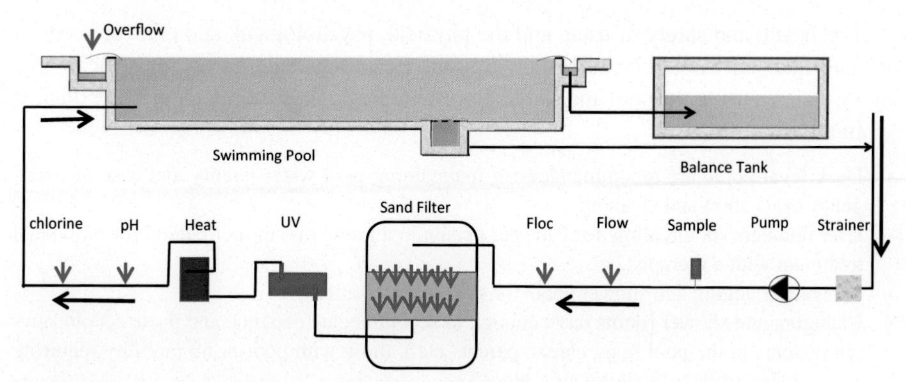

Fig. 3 Schematic of the water flow through a hydrotherapy pool and treatment plant. *Modified* from original provided by Devin Consulting UK.

Backwashing

Backwashing of the filter bed ensures replacement of pool water by freshwater and restores the filters efficiency. The flow of water is reversed through the filter bed to remove debris to improve flow rates. The backwashed water is then run to waste and the pool volume made up with fresh water. Backwashing frequency is based on the pressure gradients recommended by the manufacturer but should be carried out at least once a week.

During backwashing, there is no effective filtration, and it can take several hours to resettle and regain efficiency. Ideally there should be a window in the filter so that the top of the filter bed can be observed. Backwashing should always be carried out after the last session of the day to allow the filter bed to resettle. Alternatively, ensure an afternoon session is timetabled (free from patient use) for backwashing, pool cleaning, and cleaning and disinfection of equipment and surrounding environment.

Day-to-day management

The pool safety operating procedures including normal operating procedures and emergency action plan should be up to date (EAP). These should be based on risk assessments by competent assessors familiar with pool operation and relevant hazards (chemical, biological, and physical), hazardous events, and clinical aspects. There should be an asset register of all systems, equipment, major components of the water system, which require a risk assessment and maintenance and schematics (a legal requirement in the United Kingdom).

Only chlorine-based disinfectants are approved to treat hydrotherapy pools as there has been evidence of rashes caused by bromine products through aquatic therapists spending many hours in the pool.

The health and safety of users and the physical, psychological, and clinical needs of patients and staff should be taken into account. Ensure patients, carers, those accompanying patients using the pool and staff, shower, before pool use to maintain water quality.

In addition:

- Deck level pools are recommended for maintaining pool water quality and ease of emergency evacuation and cleaning.
- User numbers should allow for $2\,m^2$ per person in a group or $4\,m^2$ per person for individual treatment with a therapist.
- Separate changing and shower areas for staff and for patients.
- Changing and shower rooms large enough to accommodate patients, and those accompanying patients in the pool (e.g., carers, parents etc.), those with poor or no mobility requiring wheelchairs, trolleys, walking aids, etc.
- Sufficient room around the pool itself for wheelchairs, hoists, etc.
- Drainable—check with the manufacturer that this procedure will not damage the pool fabric/shell.
- Rest area with drinking water available.
- Appropriate changing room areas and clinical waste bins.
- Appropriate clinical waste disposal and hand washing facilities.

Processes should be in place to ensure that operating plans and risk assessments are updated after there have been significant changes relating to pool management, treatment, and water quality.

An overview of the hazards applicable to hydrotherapy pools

Potential microbial hazards in pools including hydrotherapy pools used for aquatic physiotherapy can be designated as fecal and nonfecal pathogens (Fig. 4).

While there are few microbial outbreaks associated with hydrotherapy pools, those responsible must be aware of the microbiological risks to users and staff (Chapuis et al., 2004, BS 8580-2) including physical hazards (drowning, slips trips, and falls) and chemicals hazards (BSI, 2020).

Patient screening

Patients must be screened for contraindications or precautions prior to treatment in the pool to prevent the pool water from being contaminated and to minimize the risk of hospital-acquired infections. Policies should be in place to manage patients safely in the pool both including;

- diarrhea in the past 2 days with a contraindication to entering the pool due to the risk of Cryptosporidium.
- Infections are only a contraindication if there is system infection, i.e., a raised temperature (which needs to be 24 h clear) or if the wound cannot be waterproofed.
- Applying precautions for patients with stomas, catheters, fixators

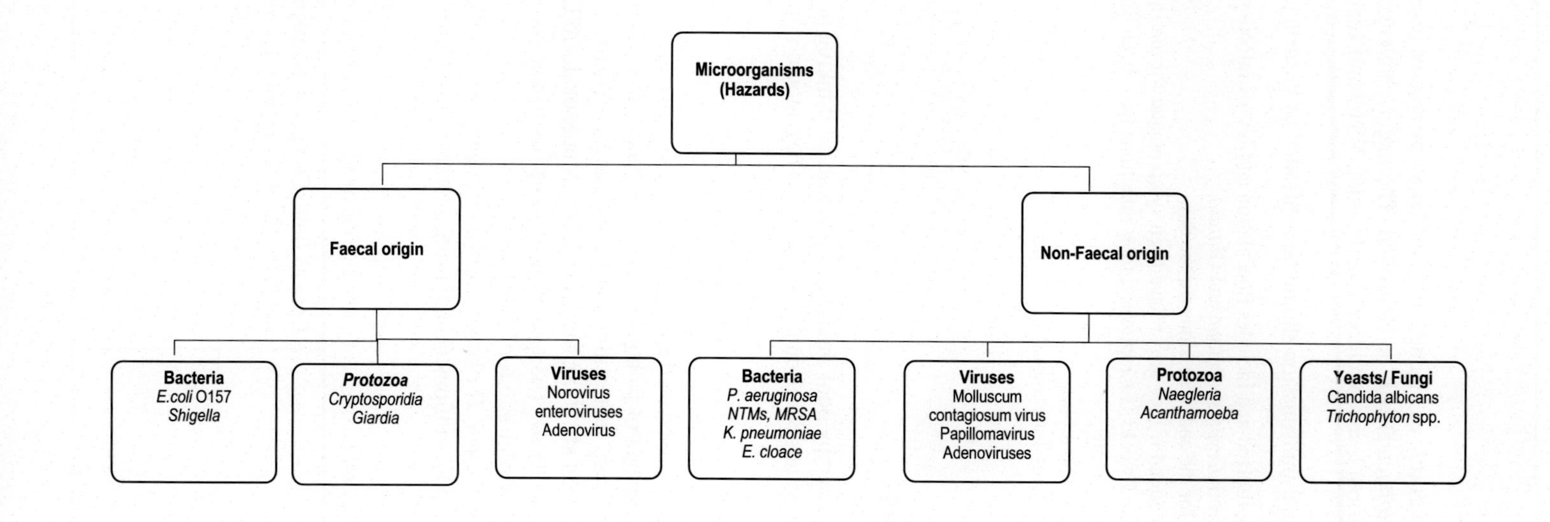

Fig. 4 Examples of microbial pathogens relevant to hydrotherapy pools. *Modified* from World Health Organization 2006.

Fecal pathogens

If the treatment system is working as planned, then the target pathogens (e.g., *E. coli* O157, *Salmonella, Campylobacter*) will be controlled. The major challenge is from parasites resistant to the normal levels of chlorine in the pool (WHO and Team, 2006) (Table 1).

The risk of gastrointestinal transmission should be controlled by ensuring:

- the pool water quality is optimized and includes flocculation and UV disinfection to minimize the risk of infection from *Cryptosporidium* and *Giardia.*
- backwashing is scheduled for the end of sessions for the day,
- patients are advised not to attend if they are or have been symptomatic for diarrhea and/or vomiting in the last 48 h (or 2 weeks if *Cryptosporidium* infections have been diagnosed), and this is confirmed when booking
- patients and staff shower before pool use
- there are appropriate procedures for removal of formed and unformed stools, vomit, and blood in the pool

Nonfecal pathogens

Water supplied to the hydrotherapy pool should comply with all drinking quality parameters and meet the agreed target residual disinfectant. They should not be filled through a hosepipe.

Microorganisms should not pose a risk in appropriately treated and managed pools. However, patients and staff may introduce microorganisms from the skin and feces into a pool. Where the pool and associated equipment are not well managed, there will be a risk of infection from a range of pathogens including multidrug-resistant strains of *Pseudomonas aeruginosa* and methicillin-resistant *S. aureus* (MRSA) (Aspinall and Graham, 1989; Berrouane et al., 2000; Meldrum, 2001; Moore et al., 2002; Penny, 1991; Weber and McManus, 2004). The exception is Cryptosporidia, which is not removed by routinely used concentrations of chlorine.

Biofilm including *Legionella* and *P. aeruginosa* will remain a risk for the life cycle of the pool and related pool equipment, requiring greater contact times for the

Table 1 Contact time required for control of a range of microorganisms using chlorine.

Microorganisms	Contact time (Ct) in minutes required to kill 99.9% of the population with 1 mg/L chlorine at pH 7.5 at 25°C
E. coli O157:H&	~1
Hepatitis A	~16
Giardia	~45
Cryptosporidium	~ 15,300

same concentration of disinfectant (BSI, 2022; DHSC, 2016; HSE, 2014). *P. aeruginosa* biofilms will also colonize floatation aids, pool covers, swim aids, pool toys, and damp inflatables. The water safety group should review the operating procedures, audit water treatment and cleaning practices, inspect overflow channels, grids, and equipment to ensure they are cleaned and disinfected regularly.

Hydrotherapy pools and associated showers and hand wash basins in and around the pool may also be used intermittently. This will increase the risk of legionellosis and *P. aeruginosa* infections as a result of stagnant water. Measures should be implemented to prevent the water and outlets from becoming stagnant either by installing automatic flushing devices (with remote monitoring) or having an auditable flushing program.

There is increasing evidence that pool users should not wear contact lenses in pools as Acanthamoebae, which occur naturally in water associated with biofilms, can attach to the lens and cause keratitis (Seal, 2003).

Main challenges for hydrotherapy pools

There are challenges to ensuring healthcare hydrotherapy pools remain safe for patients and staff. Patients can be of all ages including young children; some with little or no mobility so the design for the pool and pool environment needs to consider the need for space for storage of wheelchairs and equipment both in reception and in the changing area to prevent contamination being introduced.

Ensure that equipment can be cleaned and disinfected and that this is documented.

Therefore, both design and operational issues need to be considered to ensure the pool is safe for all users including:

- microbiological risk assessments are required to ensure patients are not exposed to opportunistic pathogens in the pool.
- appropriately managed pools should not smell of chlorine; however, disinfectant chemicals breakdown faster in warm water, which increases the concentrations of harmful chloramines (biocide breakdown products).
- ventilation should maintain humidity (60%) in the pool hall at 25–28°C and should remove airborne chloramines to prevent overexposure of users.
- Pools may be used intermittently with bather loads ranging from two (one to one), to larger groups of patients. Some patients, e.g., dependent adults may need multiple supporters in the pool.
- There should be guidance for patients who may have gastrointestinal symptoms, infected wounds, indwelling catheters, external fixators, stomas, or urinary/fecal incontinence as to when they can be readmitted to the pool including:
- Patients who are more susceptible to infection should be treated when the pool water quality is at its least contaminated, i.e., first thing in the day.
- Leasing the pool to diverse groups of patients including baby swim groups, may pose additional challenges for the pool environment (facilities for prams and nappy changing and disposal, etc.) and pool water quality. For example, one NHS Trust hydrotherapy audited by the author allowed baby swim classes at the weekend, which compromised the water quality and resulted in the cancellation of patient treatment while remedial action was implemented.

Emergency action plans

A multidisciplinary group should identify what emergency action plans are needed and that they contain communication details for service providers, infection prevention control, and estates teams. Emergency action plans for predictable untoward events should include procedures following monitoring results indicating a risk to health, dosing pump failures, fecal fouling or vomiting, cardiac arrest, slips, trips, and falls, and cases of infection associated with the pool.

Outbreak 1

Schlech et al., described an outbreak of *P. aeruginosa* folliculitis related to a new hydrotherapy pool. Those infected included:

5/45 (45%) physiotherapists, 6/29 (21%) outpatients, 4/12 (25%) inpatients (including one wound infection). The cause was a failure of an automatic chlorinator, which was not functioning so they had resorted to erratic manual dosing usually twice/day. The filter system was also not working properly, with organic debris in pool, seals were failing causing accumulation of debris around peeling sealant.

Rashes developed in 45% of the physiotherapists who had used the pool, 21% outpatients and 33% of inpatients who had used the facility; surgical wound infections also developed in 25% inpatients.

The pool was hyperchlorinated at a free residual chlorine (3–4 ppm) and structural changes including acid wash and resealing the pool, repairs to the automatic chlorinator and filter.

Hydrotherapy tanks/tubs/pools used for burns, etc.

Burns patients may be immersed in water for immediate cooling of burns and for removal of adherent dressings or wound debridement. Limiting the damage caused by heat is the immediate requirement and as sterile water may not be available at the scene, patients may acquire infections during this process. Patients with open wounds, especially burns, are particularly susceptible to infection and treatment should be carried out with sterile water where possible. Preventing cross-contamination of burn wounds from equipment and water can be challenging and may involve polymicrobial infections and antibiotic-resistant organisms, including MRSA and *P. aeruginosa* (Baier et al., 2020; Ribeiro et al., 2010). Tap water supplied into the hospital is not sterile and contains a range of waterborne pathogens. However, if the water system is not well managed, pathogens can multiply and pose a risk of infection to susceptible patents in high-risk intensive care units including haematology oncology, burns, and transplant units. Processes need to be in place to ensure that equipment is safe and fit for purpose and that risk assessments are up to date and should include:

- water used for filling, cleaning, and decontamination does not pose a risk to patients
- the quality of water used for cleaning of surfaces such as trolleys used for transport and treatment of patients is safe
- risk assessment of splashing contaminating equipment including tanks, whirlpool baths, tubs, sinks, and toilets relative to patients and any equipment used by patients or for patient treatment is minimized,
- supporting programs in place to ensure there is appropriate audit, monitoring, and training for all staff involved also needs

Outbreak 2

Tredget et al. investigated the infections associated with burns patients who were involved in full body immersion for removal of adherent dressing and wound debridement (Tredget et al., 1992). The outbreak involved a near fatal episode of septic shock due to a multidrug-resistant *P. aeruginosa* that was not present on admission. Another patient who died of sepsis became infected with the same microorganism. Swab cultures of sinks and drains in each patient's room and of hydrotherapy equipment were obtained, and samples were taken from the water supplied to the hydrotherapy pool. *P. aeruginosa* was detected in the water supply, the hydrotherapy tank, and the transportation equipment used for hydrotherapy treatment. The outbreak occurred despite weekly surveillance cultures of the hydrotherapy equipment and the use of standardized protocols for its disinfection between uses. Routine environmental surveys of the burn unit over a 4-year period revealed that there were eight deaths associated with *P. aeruginosa* among burn patients. There were 28 *Pseudomonas*-positive patient isolates with 15 positive samples from a small hydrotherapy tub and 10 from a large Hubbard tank (a particular style of hydrotherapy pool). The discontinuation of hydrotherapy resulted in a cessation of the outbreak and eradication of the outbreak strain. The authors indicated that eradication or control of *P. aeruginosa* in whirlpools and other hydrotherapy equipment requires regular, intense monitoring, and treatment of both water supply and equipment.

Whirlpool baths and assisted baths

Assisted baths (sit baths) are commonly used for bathing patients with mobility issues and for therapeutic purposes. They usually have shower attachments and may have jets and bubble induction systems as in a whirlpool bath. Water in assisted baths and whirlpool bath, unlike a shot tub or a spa pool, is not filtered or chemically treated but should be drained and cleaned between each bather. However, these can retain water within the pipework and jet systems can become contaminated with *P. aeruginosa* and legionellae biofilms (Hollyoak et al., 1995). Considerations for new assisted baths should include:

- Has the bath been filled with water prior to delivery and if so you should request a decontamination and microbial testing certificate from the manufacturer to that the bath is free from waterborne pathogens.

- Checking the active ingredient in any cleaning and disinfection products, including oxidizing biocides such as hypochlorite to ensure the cleaning and disinfection regime are effective.
- Assessing whether the system is compatible with hypochlorite shock dosing.

Factors to consider in a risk assessment

- Is there an approved cleaning and disinfection protocol?
- Is there a maintenance contract by a manufacturer-approved contractor?
- Are shower heads and hoses included in the cleaning and disinfection regime?
- Are the jets/bubble systems accessible for cleaning and disinfection?
- Is the bath used regularly, and if not, is there a flushing regime in place?

Spas, hot tubs, footbaths, etc.

Whirlpool baths, spa pools, and hot tubs used in healthcare can generate aerosol risks from *P. aeruginosa* and *Legionella* (Dennis and Lee, 1988). Process needs to be in place to reduce the risk to users and those who may be in the vicinity (HSE, 2014). National guidance in the United Kingdom for the safe management of spa pools and hot tubs should be taken into account (HSE, 2013, 2018b).

Risk assessment considerations include:

- Can the pool be fully drained?
- Whether corrugated pipes, which increase the surface-area-to-volume ratio, for colonization and biofilm formation are used?
- Are all parts of the system including water jets accessible for cleaning and disinfection?
- Are the jets removable for cleaning and disinfection?
- Has the manufacturer's cleaning and disinfection regime been independently assessed?
- Written assurance for new systems should be sought as to the microbiological safety of such systems upon delivery as these will generally have been leak tested by the manufacturer and may pose a risk from residual stagnant water.

Outbreak 3

Baier et al. investigated an MRSA outbreak, which affected four adult patients from different wards related to the use of a hydrotherapy bathtub (Baier et al., 2020). The rehabilitation unit had two hydrotherapy tubs. After each usage, the bathtubs were emptied, cleaned, and then disinfected by dedicated hydrotherapy staff using proprietary wipes. All patients acquired nosocomial MRSA on day 3 or later of the hospital stay and did not have a prior history of MRSA. Environmental specimens were collected at different locations tub surface at several positions, sink, showerhead, water tap, patient lifter, neck roll, disinfectant container, and disinfectant plastic packaging. Analysis of movements and treatment data showed that the first three patients were repeatedly treated in the same bathtub immediately prior to their first MRSA findings and on one occasion, all three patients were treated at the same day in this tub. All of the patients' MRSA isolates showed an identical antimicrobial susceptibility phenotype. Environmental sampling showed growth of MRSA at three different locations (bathtub sink, showerhead, and exterior surface of disinfectant container). The isolates from the showerhead and the disinfectant container showed the same phenotypic resistance profile as the patients' isolate. In addition, growth of methicillin-susceptible Staphylococcus aureus (bathtub surface and plug) and Enterobacter species (disinfectant plastic packaging) was found in three additional environmental samples. The outbreak reported here and other outbreaks link MRSA cross-transmission in the hospital settings to hydrotherapy and other bathing facilities. It was demonstrated that MRSA contaminated the surfaces in bathing facilities, which may serve as an environmental reservoir and underscores the need for clear concepts regarding surface cleaning and disinfection to guarantee high adherence levels.

Microbiological monitoring of hydrotherapy pools

The following recommendations are a guide to microbiological sampling for hydrotherapy pools (Table 2).

Further guidance on sampling and interpretation of microbiological, including what action to take when results are unsatisfactory, is available from UKHSA (PHE, 2020) (Table 3).

Table 2 Guidance for microbiological sampling of hydrotherapy pools.

Microbial test	Results (satisfactory)
Aerobic colony count (total viable count) at 37°C for 24h	NOT >10CFU
Coliforms	Absent in 100mL (<10 per 100mL if not consecutive samples or *E. coli* or colony count <10CFU)
Escherichia coli	Absent in 100mL (<10 per 100mL if not consecutive samples or *E. coli* or colony count <10CFU)
Pseudomonas aeruginosa	Absent in 100mL (<10 per 100mL if not consecutive samples and no *E. coli* or colony count <10CFU)

Table 3 Range of microorganisms that would be deemed as being unacceptable in terms of water quality.

	Microorganism	Rout of infection	Sensitivity to chlorine
Bacteria	*E. coli* O157	Gastrointestinal route, very low dose, can cause hemolytic uremic syndrome in children	Very sensitive to chlorine
Viruses	Norovirus (also known as Norwalk-like agent SRSV), Enteroviruses including Hepatitis A	Gastrointestinal	
Parasites/protozoa	Cryptosporidium, Giardia	Gastrointestinal	Most persistent
Staphylococcus aureus	By person to person	Wound infections, bacteremia sepsis	

Criteria for risk assessing birthing pools

Risk assessment criteria for hydrotherapy pools	Yes	No
Is there a WSP to BS 8680 in place, which takes account of all pools, baths, and associated equipment used for patient hygiene and treatment?		
Is there a hydrotherapy pool policy in place?		
Is the hydrotherapy pool and associated equipment on the asset register?		
Has a *Legionella* risk assessment been carried out to BS 8580-1 and reviewed by the WSG?		
Has a *Pseudomonas aeruginosa* risk assessment been carried out to BS 8580-2?		
Are there pool safety operating procedures and NOP?		
Is there a remedial action plan been drawn up based on the risk assessment findings?		
Has the pool safety operating procedure including training, competency, and implementation of risk assessments been completed, reviewed, and approved by the WSG, and is it up to date?		
Are there EAP in use to take account of all predictable untoward events including in the accidental fecal release/vomiting/blood in the pool?		
Is the hydrotherapy pool included on the agenda for the water safety group regular review?		
Are there internal and external audit reports available?		

Risk assessment criteria for hydrotherapy pools	Yes	No
Is an accurate daily log maintained for both poolside and pool plant monitoring as appropriate (including chemical disinfection, water quality monitoring, water and air temperatures, humidity levels, backwashing, and microbiological testing)?		
Do trained personnel understand the potential microbial, chemical, and physical hazards and the risks to health associated with them?		
Do the relevant staff (aquatic physiotherapist, trained pool operators/ engineers, and microbiologists) have an effective working relationship to ensure smooth running, daily maintenance, and an accurate daily log that should include items such as chemical disinfection and microbiological testing?		
Are clear responsibilities for all involved with the pool to ensure smooth running with the stated procedures for all potential hazards that could occur?		
Are the cleaning schedules/regimes should be set out with those responsible identified for all areas and equipment?		
Is microbiological sampling for hydrotherapy pools carried out weekly?		
Has the disinfectant treatment regime been validated to ensure it is effective prior to patient using the pool?		
Have those who sample the water been trained, e.g., to prevent sample contamination and the appropriate location to take samples?		
Is the microbiological analysis carried out in a laboratory accredited for testing pool waters for aerobic colony count, coliforms, *Escherichia coli*, and *Pseudomonas aeruginosa?*		
Is there a record of all microbiology reports?		
Are all drains fitted with Category Ensure that the drains can be treated effectively with disinfectants?		
Has the bather load been calculated and adhered to?		
Have physiotherapists completed the ATACP Foundation Aquatic Therapy course or its equivalent?		

Guidance and regulations relevant for birthing pools

HTM 04-01 Safe Water in Healthcare.

Approved Code of the Control of Legionella Bacteria in Water Systems: Practice L8.

PWTAG: Code of Practice The management and treatment of swimming pool water. Pool Water Treatment Advisory Group (PWTAG, 2019). https://www.pwtag.org/code-of-practice/.

Terminology

The terminology for systems and equipment used for hydrotherapy treatment can be confusing as the term "“pool” is also used to refer to swimming pools, whirlpool baths, footbaths, spa pools, hot tubs, and birthing pools. Pools, tanks, and tubs are also used interchangeably for equipment used for immersion of the whole body or individual limbs for cooling and treatment of burns, treatment of diseases of the skin, soft tissues, and wound debridement. Assisted baths are also covered in this section, which are intended mainly for patients with limited mobility often with hoist access and used for personal hygiene and mental health therapy.

References

Aspinall, S.T., Graham, R., 1989. Two sources of contamination of a hydrotherapy pool by environmental organisms. J. Hosp. Infect. 14, 285–292. https://doi.org/10.1016/0195-6701(89)90068-6.

ATACP, 2015. Guidance on Good Practice in Aquatic Physiotherapy.

ATACP, 2020. Hydrotherapy Pools—Expert Clinical Considerations in Planning and Design. Aquatic Therapy Association of Chartered Physiotherapists.

ATACP, 2022. Aquatic Therapy Association of Chartered Physiotherapists. Aquatic Therapy Association of Chartered Physiotherapists.

Baier, C., Ebadi, E., Mett, T.R., Stoll, M., Küther, G., Vogt, P.M., Bange, F.-C., 2020. Epidemiologic and molecular investigation of a MRSA outbreak caused by a contaminated bathtub for carbon dioxide hydrotherapy and review of the literature [WWW Document]. Can. J. Infec. Dis. Med. Microbiol. https://doi.org/10.1155/2020/1613903.

Berrouane, Y.F., McNutt, L.A., Buschelman, B.J., Rhomberg, P.R., Sanford, M.D., Hollis, R.J., Pfaller, M.A., Herwaldt, L.A., 2000. Outbreak of severe *Pseudomonas aeruginosa* infections caused by a contaminated drain in a whirlpool bathtub. Clin. Infect. Dis. 31, 1331–1337. https://doi.org/10.1086/317501.

BSI, 2018. BS EN 15288-2:2018 Swimming Pools for Public Use Safety Requirements for Operation. https://www.en-standard.eu.

BSI, 2020. BS 8680—Water Quality. Water Safety Plans. Code of practice https://shop.bsigroup.com/ProductDetail?pid=000000000030364472. (Accessed 5 January 2021).

BSI, 2022. BS 8580-2:2022—Risk Assessments for *Pseudomonas aeruginosa* and Other Waterborne Pathogens. Code of practice https://standardsdevelopment.bsigroup.com.

Chapuis, C., Gardes, S., Tasseau, F., 2004. Management of infectious risk associated with therapeutic pools. Ann. Readapt. Med. Phys. 47, 233–238. https://doi.org/10.1016/j.annrmp.2004.02.009.

Dennis, P.J., Lee, J.V., 1988. Differences in aerosol survival between pathogenic and non-pathogenic strains of *Legionella pneumophila* serogroup 1. J. Appl. Bacteriol. 65, 135–141. https://doi.org/10.1111/j.1365-2672.1988.tb01501.x.

DHSC, 2016. HTM 04-01: Safe Water in Healthcare Premises.

Hollyoak, V., Allison, D., Summers, J., 1995. *Pseudomonas aeruginosa* wound infection associated with a nursing home’s whirlpool bath. Commun. Dis. Rep. CDR Rev. 5, R100–R102.

HSE, 2013. Legionnaires’ Disease—Technical Guidance—Part 3: The Control of Legionella Bacteria in Other Risk Systems.

HSE, 2014. HSG 274 Legionnaires’ Disease—Technical Guidance Part 2: The Control of Legionella Bacteria in Hot and Cold Water Systems Technical Guidance. http://www.hse.gov.uk/pubns/books/hsg274.htm. (Accessed 5 January 2021).

HSE, 2018a. Health and Safety in Swimming Pools—HSG179.

HSE, 2018b. Control of Legionella and Other Infectious Agents in Spa-pool Systems—HSG282—HSE.

HSE, 2020. Control of Substances Hazardous to Health (COSHH). [WWW Document]. URL: https://www.hse.gov.uk/coshh/. (Accessed 1 June 2021).

Meldrum, R., 2001. Survey of *Staphylococcus aureus* contamination in a hospital's spa and hydrotherapy pools. Commun. Dis. Public Health 4, 205–208.

Moore, J.E., Heaney, N., Millar, B.C., Crowe, M., Elborn, J.S., 2002. Incidence of *Pseudomonas aeruginosa* in recreational and hydrotherapy pools. Commun. Dis. Public Health 5, 23–26.

Penny, P.T., 1991. Hydrotherapy pools of the future—the avoidance of health problems. J. Hosp. Infect. 2000 (18), 535–542. https://doi.org/10.1016/0195-6701(91)90068-J.

PHE, 2020. Examining Food, Water and Environmental Samples from Healthcare Environments—Microbiological Guidelines. https://assets.publishing.service.gov.uk/government/uploads/system/uploads/attachment_data/file/865369/Hospital_F_W_E_Microbiology_Guidelines_Issue_3_February_2020__1_.pdf.

PWTAG, 2017. Swimming Pool Water—Essential Guide for the Pool Industry. PWTAG—The Pool Water Treatment Advisory Group.

PWTAG, 2019. Code of Practice: The Management and Treatment of Swimming Pool Water. Pool Water Treatment Advisory Group.

Ribeiro, N.F.F., Heath, C.H., Kierath, J., Rea, S., Duncan-Smith, M., Wood, F.M., 2010. Burn wounds infected by contaminated water: case reports, review of the literature and recommendations for treatment. Burns 36, 9–22. https://doi.org/10.1016/j.burns.2009.03.002.

Seal, D.V., 2003. Acanthamoeba keratitis update—incidence, molecular epidemiology and new drugs for treatment. Eye 17, 893–905. https://doi.org/10.1038/sj.eye.6700563.

Tredget, E.E., Shankowsky, H.A., Joffe, A.M., Inkson, T.I., Volpel, K., Paranchych, W., Kibsey, P.C., Alton, J.D., Burke, J.F., 1992. Epidemiology of infections with *Pseudomonas aeruginosa* in burn patients: the role of hydrotherapy. Clin. Infect. Dis. 15, 941–949. https://doi.org/10.1093/clind/15.6.941.

Weber, J., McManus, A., 2004. Infection control in burn patients. Burns 30, A16–A24. https://doi.org/10.1016/j.burns.2004.08.003.

WHO, Team, S. and H, 2006. Guidelines for safe recreational water environments. Volume 2, Swimming pools and similar environments.

Sources of patient drinking water 14

Sources of drinking water

Many people are of the impression that water from sources other than the tap is of a superior microbial quality and prefer to drink water from alternative sources including chilled and nonchilled drinking water dispensers, water from bottles including those refilled by friends and relatives. However, bottled water has caused *Pseudomonas aeruginosa* outbreaks when given as an alternative to distributed water, and therefore, there should not be an assumption that these alternative sources of drinking water are safer for patients. Water coolers are not fit and forget equipment as they need daily cleaning, regular servicing, and disposal of consumables.

For some vulnerable patients, water is used to dilute thickeners used in drinks, and staff need to ensure that the water is safe for that patient.

Only sterile water may be appropriate for some patients who may be, e.g., immunocompromised, and this needs to be taken into consideration when that patient is relocated or moved to other departments and ensure that provision of safe water to that patient is maintained.

Drinking water needs to be safe for patients. Therefore, the WSG needs to risk-assess the source and quality of water for the needs of different patients and never assume that the source is actually safe.

Chilled drinking water dispensers

Many people prefer to drink water that has been chilled, and in the hospital environment, the situation is no different, with patients, visitors, and staff expressing a preference for chilled drinking water and some patients benefitting from the soothing effect that it can provide. However, chilled water coolers and other dinks dispensers are prone to microbial contamination (Lévesque et al., 1994; Liguori et al., 2010), which can be accounted for by the ways in which that they are designed, installed (e.g., carbon filters and control strategies including ultraviolet light and point-of-use microbial retention filters), used, and maintained. HTM 04-01 Part A gives some practical advice on installation, and Part B highlights potential issues that can occur if water coolers are switched off for periods between uses.

A range of waterborne pathogens can contaminate and colonize drinking water dispensers, including *Legionella* spp., *P. aeruginosa,* and NTM. In addition, surveys have shown that poor hygiene can result in fecal contamination of dispenser nozzles as indicated by the presence of *Escherichia coli* and coliforms. It is of great importance, therefore, that care is exercised by WSGs when considering whether or not chilled water dispensers should be installed in certain parts of the hospital, and precisely

Safe Water in Healthcare. https://doi.org/10.1016/B978-0-323-90492-6.00019-7

what cleaning, disinfection, maintenance, and microbiological monitoring programs are required to ensure that the water dispensed from them is safe for users. The situation can often be exacerbated by difficulties in knowing exactly how many dispensers have been installed and where they are located because it is not uncommon for these devices to be donated to hospitals by patients' relatives and charitable groups.

In some hospital areas (e.g., wards that provide augmented care), risks to patients from water from contaminated dispensers can be high, and WSGs should consider whether these devices are appropriate in those circumstances; chilled water for patient use, if it is required, can be provided by other means, e.g., by using microbiologically filtered water that has been dispensed into sterile containers and then chilled in a fridge. WSGs should consider the guidance given in HTM 04-01 Part C, which advises that chilled water dispensers should not be installed in augmented care units.

When deciding on provision of cooled water dispensers, the WSG should first produce an accurate inventory of the chilled water dispensers across the estate. It should then consider the suitability of manufacturers' cleaning and maintenance procedures for these devices when used in the hospital setting and decide whether additional precautions are required, e.g., if more frequent cleaning and disinfection of filters and nozzles are required and whether microbiological monitoring for pathogens, such as *P. aeruginosa,* might be beneficial, e.g., in wards where some patients may be more susceptible to infections caused by opportunistic waterborne pathogens. Once decided, the WSG should ensure that all control measures are recorded, and any positive microbiological test results are acted upon in line with the advice given in HTM 04-01 Part B.

Chilled water drinking fountains are used to provide cold water (4°C) for patients and hence microbiological contamination is not common. Drinking fountains normally include a reservoir to assist in the cooling cycle, and if machines are turned off, water quality can deteriorate. Provision of bottle dispensers should be approved only by the WSG. Where carbon filters and UV are fitted, these should be maintained as per the manufacturer's instructions. Carbon filters are used to remove disinfectant such as chlorine, and therefore, microbial growth will occur downstream of the filter and will result in biofilm formation on the tubing, e.g., flexible polyvinylchloride (PVC) tubing (Guyot et al., 2013). Additional cleaning to ensure adequate hygiene of nozzles, etc., should be put in place as recommended by the manufacturer and the WSG.

A range of waterborne pathogens will contaminate and colonize drinking water fountains including *Legionella* spp., *P. aeruginosa, Mycobacterium gordonae, Stenotrophomonas maltophilia,* carbapenemase-producing *Enterobacteriaecae,* and NTM (Costa et al., 2015; Guyot et al., 2013; Jung et al., 2020; Lalande et al., 2001).

Outbreak scenarios

Outbreak scenario 1

In an ear, nose, and throat (ENT) department, a cluster of cases of *P. aeruginosa* occurred and of all the water sources that were sampled only the drinking water fountain were found to be positive and found to be identical to the clinical isolates

(Costa et al., 2015). The drinking water fountain had been plumbed into the mains supply in the ward and had a carbon filter, a 0.45 μm cartridge filter, and a cooling system. The drinking fountain had also been serviced according to the manufacturer's recommendations including an annual deep clean by the inhouse plumbing department. However, the water from the outlet was contaminated with high concentrations of *P. aeruginosa* indicating that even with maintenance and sampling (6 months previous), the *P. aeruginosa* biofilm that had led to this outbreak in the ENT unit had become established within the tubing in the drinking water fountain.

Outbreak scenario 2

In another hospital outbreak, patients in a chest medicine department were found to be infected with *M. gordonae* (Lalande et al., 2001). The results of this outbreak were interesting as samples from the patients' rooms, nurses' offices and from the drinking water fountains were all positive, but more microorganisms were recovered from the drinking fountain than the other sampling points. Analysis demonstrated that the strains from the patients and the drinking fountain had identical patterns. Strains from the other samples differed from the outbreak strain but could have led to infections at other times. In this case, the drinking water fountain was connected to the ward water supply using a rubber hose. While the ward water supply was found to be positive (albeit at a low concentration), the water sample taken from the end of the rubber hose was heavily contaminated indicating the presence of an extensive biofilm. The rubber hose provided an environment niche and nutrients that encouraged the growth of the *M. gordonae* biofilm.

Monitoring of drinking water sources

The water inlet and outlets should be monitored to assess whether microbial growth is occurring with the units. For example, a 2 log increase in TVC would give an indication of whether there is excess microbial proliferation compared with the incoming water.

Where outbreaks have occurred, then sampling should be undertaken to investigate for the presence of the microorganisms of interest by taking appropriate water and biofilm samples.

Guidance for chilled drinking water fountains

HTM 01-04 Parts A and B Health Technical Memorandum 04-01: Safe water in healthcare premises.

BS 8680: 2020 Water quality—water safety plans—Code of practice identifies cross-contamination of plumbed water dispensers with microorganisms, fecal pathogens, in particular.

Requirements and recommendations for chilled drinking water fountains

- When separate drinking water systems have been provided, the policy is normally to distribute directly from the mains without storage (nor softening) and strategies need to be considered to prevent stagnation of the water to negate microbial and biofilm growth.
- Cold water supply pipework to dispensers should be copper and fitted with a local isolation valve and drain valve.
- Pipework should be as short as possible from take-off point (mains water tee).
- Flexible pipe connector if required should be kept as short as possible.
- Do not install chilled drinking water systems in augmented care.
- Do not use softened water.
- Do not install drinking fountains at the end of the line (potential dead leg).
- Do not use flexible EPDM hoses to connect the ward supply to the drinking water fountain.
- Do fit a double-check valve to prevent backflow.
- Biofilms will become established after the carbon filter—ensure that the tubing is changed regularly
- Ensure that there is responsibility for hygienic cleaning of outlets and surfaces.
- Ensure there is good communication with wards/units so they understand the microbiological risks including that where fitted carbon filters will provide high nutrient supplies for bacteria.
- There are systems in place with procurement to alert estates if water dispensers require plumbing into the CWS and are requested.
- Ensure the WSP identifies the processes and design criteria for alternative drinking water source approval.
- Contractors are aware of restriction in plumbing in such devices.
- There is appropriate staff and patient advice given on admission to high-risk areas on consumption of water.
- IPC staff are trained to understand the risks from colonization and poor hygiene associated with water dispensers.
- There are systems for cleaning, maintenance, and audit of water dispensers.

Bottled drinking water dispensers

Bottled water dispensers have become common in hospitals for the supply of drinking water. However, many of these dispensers may be located in areas that are warm and may be prone to microbial contamination (Tischner et al., 2021). These bottled water dispensers are primarily for staff and visitors and should not be used for providing water to vulnerable patients who require safe sources of water. There need to be policies in place to ensure that these units are located in appropriate place, are serviced, maintained, and cleaned according to manufacturer's instructions.

Recommendations for bottled water dispensers

Provision of bottle dispensers should be approved only by the WSG.

Ensure that there are systems in place for cleaning (potential for biofilm on inlets and outlets), maintenance, and audit of such devices.

A policy should be in place to decide on applicable standards, procurement decisions and approval process, responsibility for the finance and upkeep of cooler (e.g., retention of records). This policy should enable the WSG to make decisions on either agreeing to or rejecting the request on the basis of a risk assessment completed by a competent person(s).

The WSG should establish controls to ensure that only authorized drinking water coolers can be purchased or rented for installations and working with the procurement team so that they understand the risks.

An application process should be in place with approval criteria that enables requests for water dispensers to be dealt with systematically and could include. For example:

- What susceptible/vulnerable patients will be exposed to the water?
- Has the finance for the purchase of the unit and the ongoing service contract been agreed as per the Trusts WSG protocols for water coolers?
- A detailed installation survey should be carried out to establish the precise location to ensure there is a suitable water quality source (within 3 m of the proposed site and upstream of frequently used outlets), electrical supply, drains, and there are no localized heat sources.
- Installers should be suitably qualified and competent and always work under the supervision of the Trust's Authorized Person (Water)] to ensure that the requirements of the Trust's Water Safety Plan, risk assessment, and any further instructions from the WSG are implemented.
- Has the tubing material used for connections been assessed to reduce the like hood of biofilm growth?
- Servicing and maintenance of carbon filters in water dispensers should be risk assessed as they are a high nutrient source for bacteria.
- Has frequency of use been assessed to avoid stagnation and deterioration of the water?
- Have the staff who will undertake the daily hygiene regime been identified, trained, and are they deemed competent (Figs. 1 and 2)?
- Is the water dispenser easy to clean and maintenance and has the person responsible for the daily cleaning of the device been identified and appropriately briefed about their responsibilities?

Fig. 1 Lack of cleaning of dispensing tray in a bottle water dispenser.

Fig. 2 Lack of cleaning of the surfaces of a bottled water dispenser.

Risk assessing water coolers and other drinking sources for patients

Risk assessment criteria	Yes	No
Have patients been assessed for the type of water that they should consume, e.g., drinking water or sterile water?		
Has the drinking water been assessed as being safe for patients?		
Where a drinking water unit has been fitted, have the inlet and outlet been microbiologically assessed?		
Are the drinking water units identified on the asset register?		
Is the pipework to the drinking dispenser as short as possible?		
Has a flexible hose been fitted to the dispenser?		
Is the drinking dispenser supplied with softened water?		
If present, has the carbon filter been changed at the appropriate service points?		
Has the drinking water unit been fitted at the end of the line (not recommended)?		
Has a double-check valve been fitted to prevent backflow?		
Has hygienic cleaning been identified as a regular maintenance issue?		
Have appropriate staff been identified for undertaking the cleaning of the unit?		

Guidelines for bottled water dispensers

HTM 04-1 Safe water in healthcare Part B.

References

Costa, D., Bousseau, A., Thevenot, S., Dufour, X., Laland, C., Burucoa, C., Castel, O., 2015. Nosocomial outbreak of *Pseudomonas aeruginosa* associated with a drinking water fountain. J. Hosp. Infect. 91, 271–274. https://doi.org/10.1016/j.jhin.2015.07.010.

Guyot, A., Turton, J.F., Garner, D., 2013. Outbreak of *Stenotrophomonas maltophilia* on an intensive care unit. J. Hosp. Infect. 85, 303–307. https://doi.org/10.1016/j.jhin.2013.09.007.

Jung, J., Choi, H.-S., Lee, J.-Y., Ryu, S.H., Kim, S.-K., Hong, M.J., Kwak, S.H., Kim, H.J., Lee, M.-S., Sung, H., Kim, M.-N., Kim, S.-H., 2020. Outbreak of carbapenemase-producing *Enterobacteriaceae* associated with a contaminated water dispenser and sink drains in the cardiology units of a Korean hospital. J. Hosp. Infect. 104, 476–483. https://doi.org/10.1016/j.jhin.2019.11.015.

Lalande, V., Barbut, F., Varnerot, A., Febvre, M., Nesa, D., Wadel, S., Vincent, V., Petit, J.C., 2001. Pseudo-outbreak of *Mycobacterium gordonae* associated with water from refrigerated fountains. J. Hosp. Infect. 48, 76–79. https://doi.org/10.1053/jhin.2000.0929.

Lévesque, B., Simard, P., Gauvin, D., Gingras, S., Dewailly, E., Letarte, R., 1994. Comparison of the microbiological quality of water coolers and that of municipal water systems. Appl. Environ. Microbiol. 60, 1174–1178. https://doi.org/10.1128/AEM.60.4.1174-1178.1994.

Liguori, G., Cavallotti, I., Arnese, A., Amiranda, C., Anastasi, D., Angelillo, I.F., 2010. Microbiological quality of drinking water from dispensers in Italy. BMC Microbiol. 10, 19. https://doi.org/10.1186/1471-2180-10-19.

Tischner, Z., Sebők, R., Kredics, L., Allaga, H., Vargha, M., Sebestyén, Á., Dobolyi, C., Kriszt, B., Magyar, D., 2021. Mycological investigation of bottled water dispensers in healthcare facilities. Pathogens 10, 871. https://doi.org/10.3390/pathogens10070871.

Drink vending machines

15

Drink vending machines

Vending machines offer a 24-h meal option for hot beverages for both patients and their families, as well as any visitors the patient may get while they are in hospital and have virtually replaced what was the traditional tea lady (Hunter and Burge, 1986; Schillinger and Du Vall Knorr, 2004).

In hot-drink vending machines, drink powders, powdered milk/whitener, and sugar are stored in plastic canisters. Vending machines that provide beverages will need to be connected to the local water supply, are used intermittently, and have an automatic rinse cycle. As each drink is chosen, the relevant amount of each powder is dispensed into the mixing apparatus where it is mixed with hot water to form the drink, which is then dispensed into a cup through a short length of food-grade plastic tubing. The mixing apparatus consists of a mixing bowl, steam hood, and a whipper unit to mix the powder and water. The powder used for the drinks has a sufficient lack of water, and the acidity of the drinks has been found to between a pH of 1 and 3, which may also restrict microbial growth. In addition, vending machines have automatic flushing, with hot water, and should be serviced and maintained by trained personal.

The vending machine is connected to the local water supply through a flexible hose, and any chlorine present in the supplied water is removed via a carbon filter installed between the water supply and the vending machine. As such the quality of water from vending machines depends on the quality of the water supplied as well on the design and maintenance of the vending machine.

Microbiological issues associated with drink vending machines

Vending machines are subject to the same regulations that govern the food industry. The law requires food to be "fit for human consumption" and of the "nature, quality and substance demanded." All food (and drinks) must be prepared, handled, and transported in premises and equipment that meet relevant food safety standards (EC, 2004).

As drink vending machines are supplied with water and the drink powder can supply nutrients, then there are a number of areas where contamination of the water and drinks can occur, for example:

- Water supply within the local area where the vending machine is situated.
- In the flexible hose used to connect the local supply to the vending machine.

Safe Water in Healthcare. https://doi.org/10.1016/B978-0-323-90492-6.00018-5

- Due to the presence of a carbon filter, the residual chlorine in the water supplied to the vending machine will be removed, which will enable control microbial growth within the vending machine.
- Stagnant water within the vending machine.
- Once mixed with hot water, any powder residue left within the mixing bowl, dispense tubing, or around the dispenser nozzle will have a raised water activity suitable for microbial growth.
- The drain may have deposits of previous drinks and will be moist and provide an environmental niche for microbial growth.

While no reports have been found of microbiocidal issues with vending machines in hospitals, it is clear that these vending machines do provide an environment in which microbial growth will occur. Indeed there appear to be limited recent publications on vending machines (Hall, 2008; Hall et al., 2007). Most of the literatures relate to studies in the 1980s. Studies have shown that a range of microorganisms can be detected in both the flavored drinks and in the potable water dispensed from vending machines. Hunter and Burge (1986) were able to demonstrate that 44% of drinking water samples and 6% of flavored drinks contained coliforms (Hunter and Burge, 1986). They recommended that vending machines should not be used to dispense potable water. Robertson (1987) sampled the dispenser nozzle, drip trays, and splash areas of both hot and cold machines and found 13% of the nozzles were contaminated with *E. coli* (Robertson, 1987).

Hall et al. evaluated microbial growth over a 19-week period using total viable counts (Hall et al., 2007). Over the first 8 weeks, there was a significant increase in microbial counts from a range of components within the vending machine and a significant increase in microbial counts in the drinks dispensed by the end of the trial.

Bacterial poisoning from vending machines is very rare and has only been reported once where four employees at a Minneapolis manufacturing plant reported nausea and vomiting after drinking hot chocolate from a vending machine based in the plant cafeteria (Nelms et al., 1997). During the investigation, *Bacillus cereus* (5.23 log CFU/g) were found in the vended drink, and although they had no evidence, they suspected that the microbial contamination was occurring in the mixing cup as no *B. cereus* was detected in the milk chocolate powder. During a wider testing of vending machines in the city dispensing hot chocolate, 17% of licensed machines were found to be contaminated, with two machines having *B. cereus* concentrations capable of causing illness.

The Automatic Vending Association of Britain has produced detailed guidelines for the microbial quality of drinks vending machines (AVA, 1997).

Testing of water from vending machines should show compliance with the water regulations, and testing should ensure the quality of water passing through the machine and cleaning adequacy. It states that (10,000 CFU/mL) in water from a vending machine are not uncommon. However, if levels above (10,000 CFU/mL) are found, it would be good practice to sample from the stopcock and the end of the inlet hose to identify where growth might be occurring. However, it is not surprising that the microbial counts in vending machines will be higher than in the water to the mains supplied to the building. These increased counts themselves may not necessarily pose a risk; however, it would be prudent to assess such products for vulnerable patients in healthcare (Barrell et al., 2000).

The AVA guidance does imply that samples must be taken but implies they are taken by their members. This does not appear to be common when installing servicing or cleaning of such machines is undertaken, and unless such samples are taken by trained operatives, this is likely to cause more issues than it would resolve. The only samples that are usually undertaken by your vending supplier are related to water chemistry, e.g., for hardness and disinfectant to determine what filter would best serve your machine.

Risk assessments have to take into consideration:

- the frequency of use of the vending machine, e.g., low use
- is the unit on long run?
- on a long run, does it have a long flexible hose?
- or is the machine close to high-risk patients?

All such vending machines should be included along with all other water use devices, at the water safety group meetings.

It would be prudent in hospitals that vending machines are not sited in ward areas with immunocompromised patients at particular risk such as intensive care, neonatal, AIDS, and transplant units.

The list for augmented wards will be designated by the WSG, which also has to consider transportation of patients to other areas of the hospital where they should not be exposed to drinks from vending machines.

Beverage vending machines could be a potential risk to immunocompromised patients, and the WSG has to ensure that this information is provided to installers, those who maintain such machines as well as clinical and nursing staff responsible for such patients.

Monitoring of vending machines

Vending machines should listed be on the WSP asset register with a risk assessment, maintenance, and monitoring plan in place (DHSC Safe water in healthcare premises (HTM 04-01), 2017). Testing should be carried out for drinking water quality parameters of the water entering the machine and of the water within the machine to verify that the quality of water is not being adversely affected by the machine and/or handling practices (PHE, 2020).

Where an outbreak is suspected, then microbiological monitoring would include looking for specific microorganisms based on clinical surveillance.

Recommendations for vending machines

Do not install vending machines at the end of the line (potential dead leg).

Do not have flexible hoses that are longer than necessary.

Do ensure that immunocompromised patients are not exposed to the hot drinks from vending machines.

Risk assessing vending machines

Risk assessment criteria for vending machines	Yes	No
Have patients been assessed for the type of water that they should consume, e.g., drinks from vending machines?		
Has the water inlet been microbiologically assessed?		
Has the outlet water been microbiologically assessed?		
Have the vending units been identified on the asset register?		
Is the pipework as short as possible?		
Has a flexible hose been fitted to dispenser?		
If present, has the carbon filter been changed at the appropriate service points?		
Has a double-check valve been fitted to prevent backflow?		
Has hygienic cleaning, e.g., flexible, hoses, dispenser nozzle, mixing bowl, and drain, been identified as part of a regular maintenance schedule?		
Have appropriate staff been identified for undertaking the cleaning of the unit?		
Is servicing and maintenance part of the service contract?		
How frequently is the vending machine used?		
Is the vending machine situation in an area where high-risk patients could be exposed?		
Have staff been educated in the risks of providing vended drinks to high-risk patients?		

Regulations for vending machines

Reference should also be made to the Food Safety (Temperature Control) Regulations 1995 and Food Safety (General Food Hygiene) Regulations 1995 (MAFF, 1995).

The Automatic Vending Association of Britain (AV) "Guide to good hygiene practice in the vending industry should be followed regarding hygiene and water quality and hygienic operation of vending machines" (AVA, 1997).

References

AVA, 1997. Automatic Vending Association of Britain. Quality of Water Supplied From Vending and Dispensing Machines. AVAB, Banstead, p. 1997.

Barrell, R., Hunter, P., Nichols, G., 2000. Microbiological standards for water and their relationship to health risk. Commun. Dis. Public Health 3, 6.

DHSC Safe water in healthcare premises (HTM 04-01), 2017. Safe water in healthcare premises (HTM 04-01). [WWW Document]. GOV.UK. URL: https://www.gov.uk/government/publications/hot-and-cold-water-supply-storage-and-distribution-systems-for-healthcare-premises. (Accessed 5 January 2021 (Accessed 4 April 2019).

EC, 2004. Regulation (EC) No 852/2004 of the European Parliament and of the Council of 29 April 2004 on the Hygiene of Foodstuffs. https://webarchive.nationalarchives.gov.uk/eu-exit/https://eur-lex.europa.eu/legal-content/EN/TXT/?uri=CELEX:02004R0852-20090420.

Hall, A., 2008. Microbiological Risks Associated With Hot-drinks Vending (PhD thesis). University of Wales.

Hall, A., Short, K., Saltmarsh, M., Fielding, L., Peters, A., 2007. Development of a microbial population within a hot-drinks vending machine and the microbial load of vended hot chocolate drink. J. Food Sci. 72, M263–M266. https://doi.org/10.1111/j.1750-3841.2007.00473.x.

Hunter, P.R., Burge, S.H., 1986. Bacteriological quality of drinks from vending machines. J. Hyg. (Lond.) 97, 497–500. https://doi.org/10.1017/s0022172400063683.

MAFF, 1995. The Food Safety (General Food Hygiene) Regulations 1995.

Nelms, P., Larson, O., Barnes-Josiah, D., 1997. Time to *B. cereus* about hot chocolate. Public Health Rep. 112, 240–244.

PHE, 2020. Examining Food, Water and Environmental Samples From Healthcare Environments—Microbiological Guidelines. https://assets.publishing.service.gov.uk/government/uploads/system/uploads/attachment_data/file/865369/Hospital_F_W_E_Microbiology_Guidelines_Issue_3_February_2020__1_.pdf.

Robertson, P., 1987. The modern drinks vending machine—a link in the food poisoning chain? Environ. Health 94, 281–285.

Schillinger, J., Du Vall Knorr, S., 2004. Drinking-water quality and issues associated with water vending machines in the city of Los Angeles. J. Environ. Health 66, 25–31. 43; quiz 45–46.

Expansion vessels

16

What are expansion vessels?

An expansion vessel (or tank) is a small container that is used to protect the hot water systems from excessive pressure. The vessel contains a rubber diaphragm that contains air. As the hot water heats up, it expands in volume, and as a consequence, the water pushes against the bladder, which is full of air. The bladder then acts like a spring, or shock absorber, to absorb the excessive pressure as the water expands and contracts, helping to keep the system stable. Basically, the expansion vessel is used as an overflow tank for the hot water system (Fig. 1).

Expansion vessels are typically placed in a vertical in orientation and should preferably be of a design in which water passes through the vessel entering at low level and exiting at high level (Figs. 1 and 2).

Expansion vessels should be appropriately sized as they need to be installed and operated in a manner that prevents the accumulation of debris, pockets of stagnating water, and increases in temperature. The size of the expansion should be designed to stimulate flow within the vessel.

Microbiological sources and problems associated with expansion vessels?

EPDM bladders—The internal bladders are often made of synthetic rubber such as EPDM, similar to what has been used in other plumbing components, which supports the growth of microbial biofilms including *Legionella* (Bleys et al., 2018; Hengesbach et al., 1993) (Fig. 2). As a consequence, the Health and Safety Executive has identified these vessels as a potential legionella risk as they can suffer from low water flow or stagnation problems, especially where system pressures and temperatures remain steady.

Design—Expansion vessels bladders that do not enable flow through may have a greater propensity to enable the formation of biofilm formation. Those vessels that have a "flow through" design (as opposed to the water passing in and out) may provide less opportunity for water to stagnate and become contaminated as the water flowing through the device may be more likely to remove of the biofilm from the bladder surface.

Stagnation—However, expansion vessels in systems operating at steady temperature and pressure may have long periods without exchanging any significant amount of water, and therefore, stagnation of water may also lead to a risk of promoting microbial growth.

Heat gain—In addition, expansion vessels installed in plant rooms maybe exposed to increased temperatures, which introduces a potential problem of microbial colonization, including *Legionella*, as plantroom temperatures usually exceed that of the incoming water.

Safe Water in Healthcare. https://doi.org/10.1016/B978-0-323-90492-6.00028-8

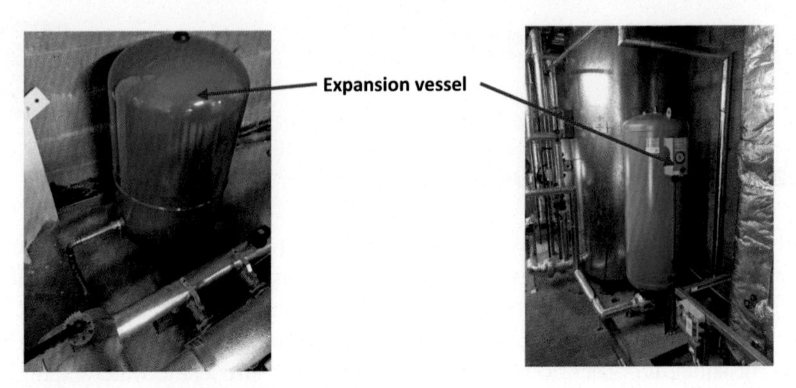

Fig. 1 Expansion vessels located in a plant room.
Photos courtesy of Daniel Pritchard.

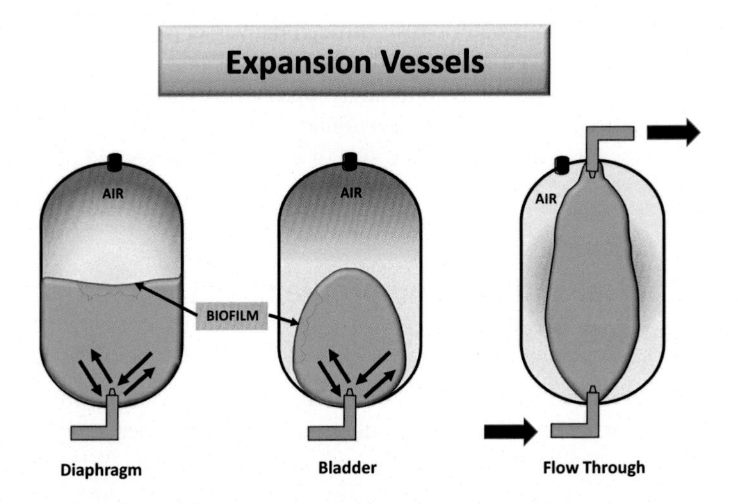

Fig. 2 Schematic of expansion vessels with biofilm growth on the bladder.

Microbial contamination of expansion vessels and seeding the system?

Water enters into the expansion vessel as the hot water expands bringing with it any bacteria that may be present. The vessel will then empty, and this transient presence of water will leave pockets of water and the walls of the vessel moist. Where no bladder is present, then this will result in the presence of water in contact with the metal components, which will eventually rust. resulting in debris and extensive biofilm formation. Where bladders are present, then the microorganisms will form biofilms on the rubberized material.

Microbial biofilms forming on the bladders of expansion vessels will be released due to the pressure surges as the hot water expands and contracts. The microorganisms that are sloughed off from the biofilm on the bladders will be pumped through the pipework to seed the rest of the system and will reattached to pipework when the conditions in the pipework allow at peripheral outlets (e.g., hand wash basins or showers), enabling the release of the microbial pathogens via droplets or aerosols.

Microbiological monitoring of expansion vessels

While there will not be routine sampling of expansions vessel, there may be a requirement to assess the microbiological quality of the water in the expansion vessel in circumstances where microbial contamination of the water supply nearby has been identified and/or an outbreak has occurred, and the source needs to be identified. When high microbial numbers have been detected and there is an expansion vessel nearby, then sampling and monitoring should be considered.

Regulations relevant to expansion vessels

Department of Health, 2016. Health Technical Memorandum 04-01: Safe water in healthcare premises—Part B: Operational management. Part B 98.(Department of Health, 2016)

HSE, 2014. HSG 274 Legionnaires' disease—Technical guidance Part 2: The control of legionella bacteria in hot and cold water systems Technical Guidance http://www.hse.gov.uk/pubns/books/hsg274.htm (HSE, 2014).

BS 6920-1:2014 Suitability of nonmetallic materials and products for use in contact with water intended for human consumption with regard to their effect on the quality of the water. Specification.

Recommendations for expansion vessels

Where practicable, all pressurization/expansion vessels shall be of the flow-through type.

Ideally flow through designs should be used, if not, then isolation valves and drains need to be fitted.

Expansion vessel should be flushed monthly—6 monthly.

Where expansion vessels are of the single entry type, they must be fitted with appropriate flow-through valves or drain valves to facilitate flushing of the unit.

Vessels where the diaphragm or bladder is designed to be replaceable may be considered as they will facilitate routine checking and/or replacement when contaminated.

Assess whether if the bladder materials in the expansion vessels are approved against BS 6920.

Risk assessing expansion vessels

Risk assessment criteria for vending machines	**Yes**	**No**
Have you identified the expansion vessels in your water system		
Are the expansion vessels part of the Legionella risk assessment?		
Are the expansion vessels on the asset register?		
Are the expansion vessels placed as close to the incoming water as possible?		
Are the expansion vessels in cool areas to minimize heat gain and reduce microbial growth in the bladder?		
Are the expansion vessels mounted vertically to minimize trapping debris?		
Are the bladder materials in the expansion vessel approved against the relevant regulations?		
Have isolation valves and drain valves been fitted to aid flushing and sampling?		
Have the expansion vessels been sized appropriately to minimize volume retention?		
Have the expansion vessels been designed to stimulate flow?		
Are flow-through vessels fitted where appropriately with water passing through the vessel at low level and exiting at a high level?		
Is the operation of expansion vessels assessed regularly to ensure that they are working correctly?		

References

Bleys, B., Gerin, O., Dinne, K., 2018. The Risk of *Legionella* Development in Sanitary Installations., p. 8.

Department of Health, 2016. Health Technical Memorandum 04-01: Safe Water in Healthcare Premises—Part B: Operational Management. Part B 98.

Hengesbach, B., Schulze-Röbbecke, R., Schoenen, D., 1993. *Legionella* in membrane expansion vessels. Zentralbl. Hyg. Umweltmed. 193, 563–566.

HSE, 2014. HSG 274 Legionnaires' disease—Technical Guidance Part 2: The Control of Legionella Bacteria in Hot and Cold Water Systems Technical Guidance. http://www.hse.gov.uk/pubns/books/hsg274.htm. (Accessed 5 January 2021).

Ice making machines

17

Background

Ice is used for a number of therapeutic practices in healthcare. There are a number of methods of producing ice in healthcare, and risk assessments have to be implemented to ensure that the ice does not become contaminated with waterborne pathogens.

Ice production in healthcare

- Ice machines are only one method of acquiring ice in healthcare, and a number of significant risks are associated with the technology. Ice producing units should be of a design suitable for purpose and which minimizes the potential for cross-contamination with a chute system and pan area used to funnel and deliver ice. The ice machine should be attached to a potable water supply and waste outlet drain and have a storage compartment for ice. Older machines may require ice removal by scoop.
- Purchasing sealed frozen ice packs for cooling tissues and specimens avoids direct contact with water. Packs need to be sanitized between patients and refrozen in a clean process in a dedicated freezer (Fig. 1).
- Placement of ice cubes in clean food-grade plastic bags—carefully placing ice into a new clean plastic bag using aseptic technique to avoid contamination of the outside and sealing can be used to prevent contact with ice or melt water.
- Ice chips, bullets, or nuggets may also be utilized for patients who may be nearing the end of life to ensure that their mouths are kept relatively moist and also promote hydration (Pinho Reis et al., 2018). End-of-life patients will often report a positive impact on their overall well-being, especially if they are suffering from nausea or swallowing difficulties.

Why is ice used for patients?

Ice is used for several reasons in healthcare including:

- patient treatment and comfort—sucking ice for nil by mouth patients who are fasting prior to a general anaesthetic
- for preserving organs, tissues, or pathology specimens
- comfort cooling
- inducing hypothermia
- musculoskeletal conditions including trauma.

Chemotherapy patients, and head and neck cancer patients, who undergo radiotherapy treatment, often complain of dry mouth and mucositis. Pain and inflammation in patients with head and neck cancer can last for a number of weeks during the acute stages of treatment, but also have chronic lasting effects on saliva production. Sucking on ice reportedly helps to alleviate some of these painful symptoms, by improving

Safe Water in Healthcare. https://doi.org/10.1016/B978-0-323-90492-6.00017-3

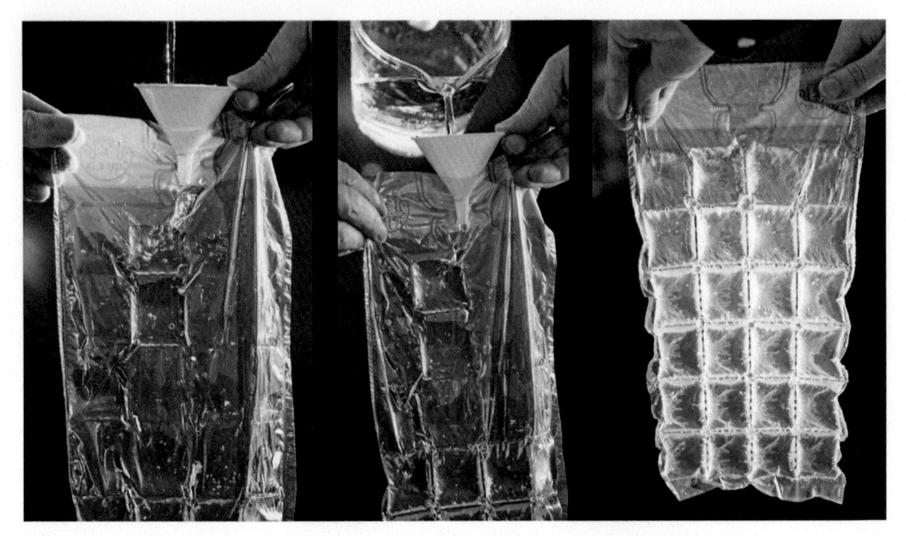

Fig. 1 Ice bags used for production of noncontaminated ice using sterile water.

patient's comfort due to its cooling and soothing effects (Ameen et al., 2019; Nikoletti et al., 2005). However, such patients are severely immunosuppressed and are at the highest risk of negative outcomes from microbially contaminated ice.

Ice is used for:

- Measuring and limiting fluid intake, which is particularly useful in patients suffering from acute nausea and vomiting.
- Limiting the deterioration of organs and tissues for transplant or prior to pathology investigations.
- Rehabilitation of patients who have dysphagia (difficulty in swallowing) under the direction of speech and language therapists, with people who may be struggling to initiate the swallow process or appear unable to sense residual food in the mouth after swallowing. This is a common issue seen after conditions such as acute stroke, head injury, and certain progressive neurological conditions.
- For therapeutic techniques with patients to help stimulate the sensory pathways (chemoreceptors) involved in the swallowing mechanism (Pisegna and Langmore, 2018). This may allow the patients to slowly begin to eat and drink other consistencies as the sensation within their throat improves.
- Other therapeutic purposes include reducing hyperthermia, inducing hypothermia, for example, to reduce injury, and local application for patient treatment for musculoskeletal conditions (Schlesinger et al., 2002). Ice can also be used to speed up cooling by placing ice packs in front of fans.
- Chilled drinks or consumed by sucking is defined as food and must be made, stored and handled so that it is not contaminated and complies with the requirements of the Food Safety (General Food Hygiene) Regulations 1995 (as amended) (MAFF, 1995).

Design and components of ice machines

- Ice machines should be connected to a potable water supply without the need for flexible connectors and with appropriate backflow protection to prevent the potable supply becoming

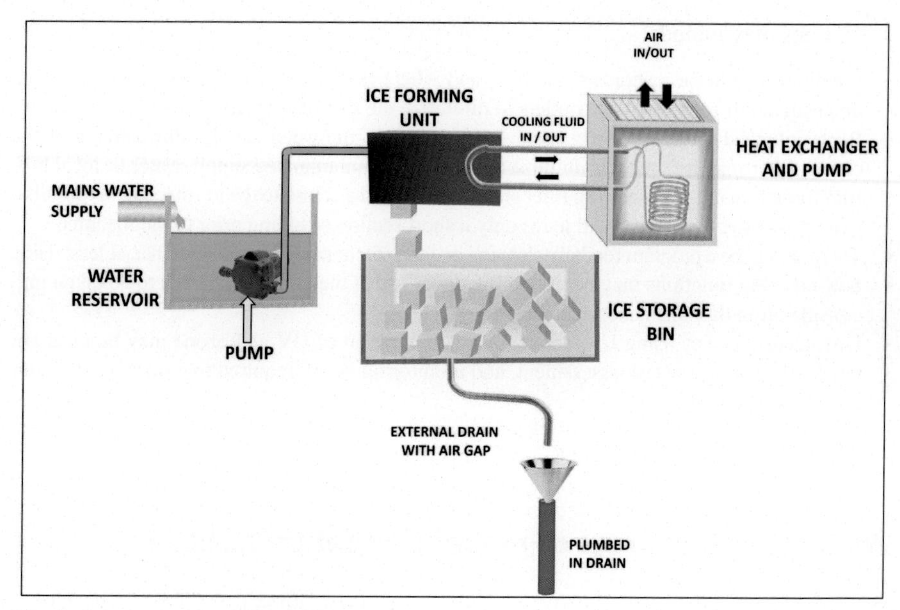

Fig. 2 Schematic of an ice machine.

contaminated (Fig. 2). Where flexible hoses are required (to allow for movement, e.g.), these should not be lined with EPDM. Alternative materials should be WRAS-approved products, but even these hoses will still result in biofilm formation and, as a consequence, should be replaced as part of regular maintenance.

- Internal tubing and pipework are usually of food-grade silicone, which can become colonized with biofilms (consortia of microorganisms on a surface) and should therefore be changed at intervals as advised by the manufacturer.
- The turnover of water used by an ice machine is likely to be low, leading to water stagnation and further predisposing to biofilm formation. Pipelines should be as short as possible, and where possible ice machines should be fitted downstream of regularly used outlets to minimize the risk of stagnation.
- Carbon filters are placed in the pipework to remove disinfectants, such as chlorine, to improve the taste of the ice. Removing the disinfectant results in the growth and proliferation of microbial biofilms in the tubing and dispensers downstream of the carbon cartridge. Unless these are changed at the intervals indicated by the manufacturer, filters can also become colonized with microbial biofilm and lead to downstream contamination.
- Ice machines are prone to heat gains and so should be fitted according to the design ambient operating temperatures, water flow, and input temperatures. They should also be sited such that there is an adequate airflow (especially air-cooled models) of approximately 150 mm around the sides and rear of the unit, depending on the design of the machine. If the pipework is placed behind the ice machine, local heat production from the heat exchanger and pump may warm water in the hose between plumbed in outlet and ice machine, which would encourage microbial growth. Where the ice machine is located at a lower height than the local waste outlet, then a built-in drain pump will be required.
- Scoops should be smooth and food-safe and should not be kept in the ice machine. There should be a program for daily cleaning and disinfection, e.g., in a dishwasher, at least daily.

Options may include:

- Pipelines should be a short as possible, and where possible ice machines should be fitted downstream of regularly used outlets to minimize the risk of stagnation.
- Bacteriological-grade filters (to remove bacteria from the water supply) downstream of the carbon filter may be used to minimize the risk from contaminated supply water or local biofilm formation in the hoses. The filter should be located as close to the ice machine as possible to be most effective as that will leave only a short section of tubing prior to the machine.
- There should be a program for daily cleaning and disinfection, e.g., in a dishwasher, at least daily.
- Self-cleaning functions may be part of the operation of the unit, but it needs to be taken into consideration that this is not a disinfection cycle.
- Disinfection cycles using UV, ozone, or a combination of UV and ozone may be used but need to be part of the risk assessment, and monitoring is still required to ensure that the unit is microbiology safe.
- Manufacturers may recommend cleaning products suitable for use in contact with food; however, these products should be part of the risk assessment.

What are the sources of microbial contamination?

The sources of microbial contamination in ice machines include:

1. Antegrade contamination such as:
   - The incoming potable water supply.
   - Microbially contaminated/colonized filters or tubing.
2. Retrograde contamination:
   - Poor hand hygiene leading to cross-contamination from hands when collecting ice, scoops, and other receptacles used to collect ice, ice returned to the machine, and poor hygiene practices when cleaning (Mills et al., 2019).
   - Ice returned to the machine.
   - Lack of cleaning, decontamination, and servicing.

What are the microbial transmission routes?

- Ingestion—swallowing contaminated ice or melt water can lead to gastrointestinal infections (Ravn et al., 1991).
- Aspiration—the inhalation of melting water from ice will predispose patients to pneumonia with a range of organisms that may be entrapped within the ice, such as *Legionella* spp. or *P. aeruginosa* (Blatt et al., 1993; Graman et al., 1997).
- Contact—An outbreak of wound infection in cardiac surgery patients caused by *Enterobacter cloacae* arising from cardioplegia ice[a] (Breathnach et al., 2006). The ice chute and pan area can be contaminated with biofilms (Kanwar et al., 2018)

[a] Not all ice is for consumption by the patient. In one ice-related outbreak, a surgeon used semi-frozen Hartmann's solution (used for replacing fluids and electrolytes in those who have low blood volume or low blood pressure) to achieve cardioplegia, i.e., the intentional and temporary cessation of cardiac activity, primarily for cardiac surgery. The freezer used for this was swabbed and yielded *Enterobacter cloacae*, indistinguishable from the clinical isolates, and it was hypothesized that this organism contaminated the freezer, and that the contamination was passed on to the ice/slush solution, thus infecting the patients (Breathnach et al., 2006).

Contaminated ice can lead to secondary decontamination of food, drinks, clinical specimens, and medical solutions, which require cold temperatures for either transport or holding.

What are the microbiological risks with ice machines?

Potable water

While microbial contaminants are unlikely to grow in ice, they can survive, and regrowth will occur once the ice melts, and the resulting water warms, which might then be in contact with or consumed by the patient.

The water supply to the ice machine has been shown to be associated with a range of waterborne infections including *Legionella* spp., Nontuberculous mycobacteria (NTM), and *P. aeruginosa* (Kanwar et al., 2017; Millar and Moore, 2020). Although rare, there have also been incidents of the water supply being contaminated with faecal microorganisms, including Norwalk virus (causing diarrhea and vomiting), and parasites such as *Giardia lamblia* and *Cryptosporidium parvum* (causing diarrhea).

Another likely source of microbial hazards likely to cause infection is from those handling the ice.

Environmental contamination

"Environmental contamination of ice machines, either through patient handling or healthcare worker handling, has been associated with a risk of infection from a range of pathogens such as *Acinetobacter* species, coagulase negative staphylococci, *Salmonella enteritidis*, and *Cryptosporidium parvum*. The above list is not exhaustive, and ice machines should be considered a possible source in outbreak settings."

Outbreaks and incidents associated with ice machines

Ice machines have been linked to several outbreaks of hospital-acquired infections including from *L. pneumophila*, (Bencini et al., 2005; Graman et al., 1997; Schuetz et al., 2009; Stout et al., 1985), *Mycobacterium chelonae*, (Iroh Tam et al., 2014; Laussucq et al., 1988; Millar and Moore, 2020), *Enterobacter cloacae* (Breathnach et al., 2006), and *P. aeruginosa* (Newsom, 1968).

Sampling water and ice from ice machines has also identified a significant risk of transmission from a wide range of opportunistic pathogens including *P. aeruginosa*, *Acinetobacter* spp., *Burkholderia cepacia*, *Klebsiella oxytoca* (Panwalker and Fuhse, 1986; Wilson et al., 1997; Yorioka et al., 2016).

Yorioka et al. (2016) demonstrated that carbon filters located within ice machines can be a source of environmental contamination of ice; all of the water samples they tested from ice machines contained *P. aeruginosa* but were negative following removal of the carbon filters.

Carbon filters are used to remove disinfectants, such as chlorine, to improve the taste of the water. However, removing the disinfectant results in the growth and proliferation of microbial biofilms in the tubing and dispenser components and drains downstream of the carbon cartridge.

The risk of Legionnaire's disease by aspiration is underestimated, particularly in patients with swallowing difficulties. There is strong evidence of aspiration as a mode of infection, including from sucking ice made in ice machines (Blatt et al., 1993; Graman et al., 1997; Newsom, 1968; Prasad et al., 2020; Yu and Stout, 2010).

Microbiological monitoring

Ice machines should be listed on the water safety plan asset register with a risk assessment, maintenance, and monitoring plan in place (DHSC Safe water in healthcare premises (HTM 04-01), 2017). Testing on a monthly basis for drinking water quality parameters of the water entering the machine and of the ice will verify that the quality of ice is not being adversely affected by the machine and/or handling practices (PHE, 2020).

Microbial targets and action alerts for ice machines (Table 1) (PHE, 2020)

Where an outbreak is suspected, then microbiological monitoring would include the above, plus the specific outbreak micro-organism.

Where ice is bought prepackaged, this should comply with the standards in the European Package Ice Association guidance http://www.europeice.com/ Manufactured ice is defined to a microbiological standard according to HACCP approach as follows (Table 2) (EPIA, 2021):

Risk assessing the requirement for ice

Due to the difficulties in ensuring the microbial quality of ice made by ice machines, these should only be used for low-risk patient or visitor areas and only for the purposes agreed by the WSG.

Table 1 Microbial targets and action alerts for ice machines.

Microorganism	Microbial targets
Escherichia coli	0/100 mL
Enterococci	0/100 mL
TVC at 22°C and 37°C	No more than 1 log above the incoming supply to the machine
Pseudomonas aeruginosa	May be included if the risk assessment indicates a need based on the susceptibility of the user group, type of intended use and previous *P. aeruginosa* monitoring results from the water system feeding the machine

Table 2 Microbial targets and action alerts for prepackaged ice that is purchased.

Microorganism	Microbial targets
Escherichia coli coliforms	Negative in 250 mL
Enterococci	0/100 mL
TVC at 22°C	≤100/mL
TVC at 22°C and 37°C	≤20/mL
Pseudomonas aeruginosa	May be included if the risk assessment indicates a need based on the susceptibility of the user group, type of intended use and previous *P. aeruginosa* monitoring results from the water system feeding the machine

Note: Account should be taken of national guidance (HTM 04-01 part C) which indicates where ice machines should not be used (see recommendations below) (DHSC Safe water in healthcare premises (HTM 04-01), 2017).

To prevent infection caused by contaminated ice, the quality of ice required for each use should be predetermined to take account of the susceptibility of the patients and its intended use.

Where the consumption of ice is deemed clinically necessary for cooling drinks, or for patients to suck who are at high risk of acquiring waterborne infection, a risk assessment may determine that ice made with filter sterilized or boiled cooled water in food-grade ice bags or presterilized trays, using aseptic techniques and kept in allocated freezers may be appropriate.

Purchasing and installation

The water safety group should ensure that there are procedures in place in the water safety plan (WSP) to ensure that the water safety group is informed of any requests for the purchase of ice machines and that those purchased are specifically designed for use in the healthcare environment (Fig. 3).

The type of ice machine required will depend on the quantity and intended purpose of ice required, ranging from table-top machines to commercial machines with inbuilt cleaning and defrost cycles. Ice machines can be either air or water-cooled. They will need a potable water supply (which meets the design criteria for flow and temperature) and a wastewater drain (air gap will be required). A drain pump may be necessary depending on the intended location. Tubing should be WRAS approved and be suitable for use in food premises (DHSC Safe water in healthcare premises (HTM 04-01), 2017; WRAS, 1999).

Ensure that ice machines are installed by competent plumbers with a Watersafe or similar qualification. They should then be commissioned and tested following manufacturer's recommendations and procedures, which are specified in the Water Safety Plan.

At the time of purchase, the Water Safety Group should ensure that the specified equipment is added to the asset register, and they should review the Water Safety Plan to ensure that lines of responsibility are clear, and that there is an agreed schedule

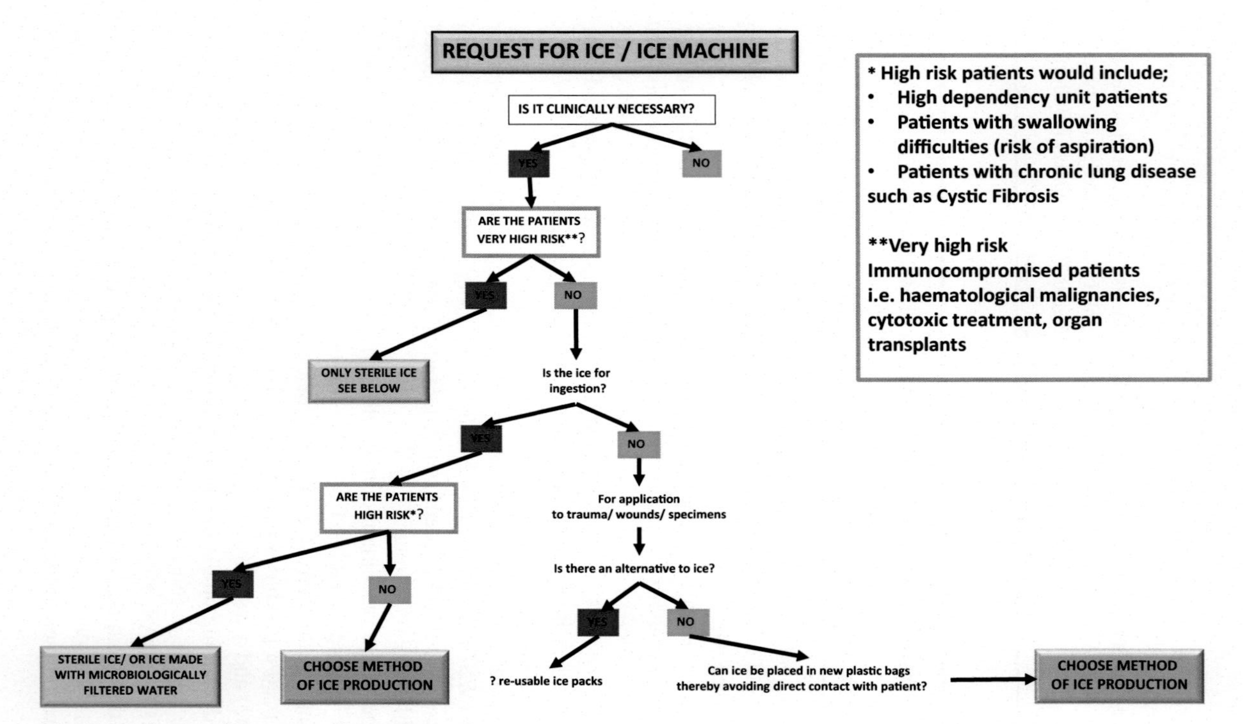

Fig. 3 Reviewing the requests for ice for patients.

for regular cleaning, disinfection, and maintenance, including replacement of supply tubing as specified by the manufacturer, to minimize the risk of microbiologically unsafe ice.

In one outbreak investigation, which identified gross contamination of the surfaces within an ice machine in a ward kitchen (S Surman-Lee personal communication), there was no management or maintenance being carried out as the ward staff thought the Estates Team were responsible for maintenance and vice versa, so no maintenance, cleaning, etc., were being carried out at all.

When ice is purchased from a commercial source, due diligence must be taken when specifying the quality of water needed to produce the ice, depending on its intended use and susceptibility of the patient to infection and possible routes for transmission of infection.

In summary:

- Where the use of ice is required, a risk assessment is needed prior to its introduction, which should involve discussions with the clinical and microbiological teams. The quality of ice required for each use should be predetermined and needs to take into account the intended use of the ice and the susceptibility of the patients.
- The whole process must be risk assessed including necessary training, maintenance, and cleaning to understand the inherent risks associated with ice machines (Fig. 4).

Ice machines have been associated with a wide range of infections including investigations of an outbreak involving nontuberculous mycobacteria in a liver unit (critically ill and transplant) (Figs. 5–7). The ice machine was in the staff tearoom, and the staff were supplying the patients ice contaminated with nontuberculous mycobacteria that subsequently resulted in an outbreak.

When the unit was dismantled, and the insides of the pipe connections were swabbed, considerable amounts of biofilm were evident. The growth of microorganisms was extensive and demonstrates the intrusive investigations that are required when an outbreak occurs. The tubing supplying the main water feed to the ice machine was labeled as food-grade plastic.

Microbially contaminated ice machines have also been suspected as the potential source for transmission of carbapenem-resistant *Acinetobacter baumannii* (Kanwar et al., 2017). It is feasible that the ice machine became contaminated by the hands of personnel that were contaminated due to contact with a colonized patient. The inside lumen of the ice machine outlet spout and drain, which were then contaminated with *A. baumannii* biofilm, would then persistently colonize the ice and lead to subsequent transmission of infection (Kanwar et al., 2017).

As the ice is manufactured, the microorganisms from the biofilms will slough off the pipe walls, will be encompassed in the ice being formed, and will survive the freezing process. If this ice is contaminated with *Legionella* species and given orally to patients, this could then be aspirated by the patient into their lungs and cause a severe pneumonia (as opposed to the more common transmission routes of aerosols) (Bencini et al., 2005). Where ice is used on wounds, if the water from the ice cubes comes into direct contact with the wound, then the ice could contaminate the wounds with contamination microorganisms such as *P. aeruginosa*.

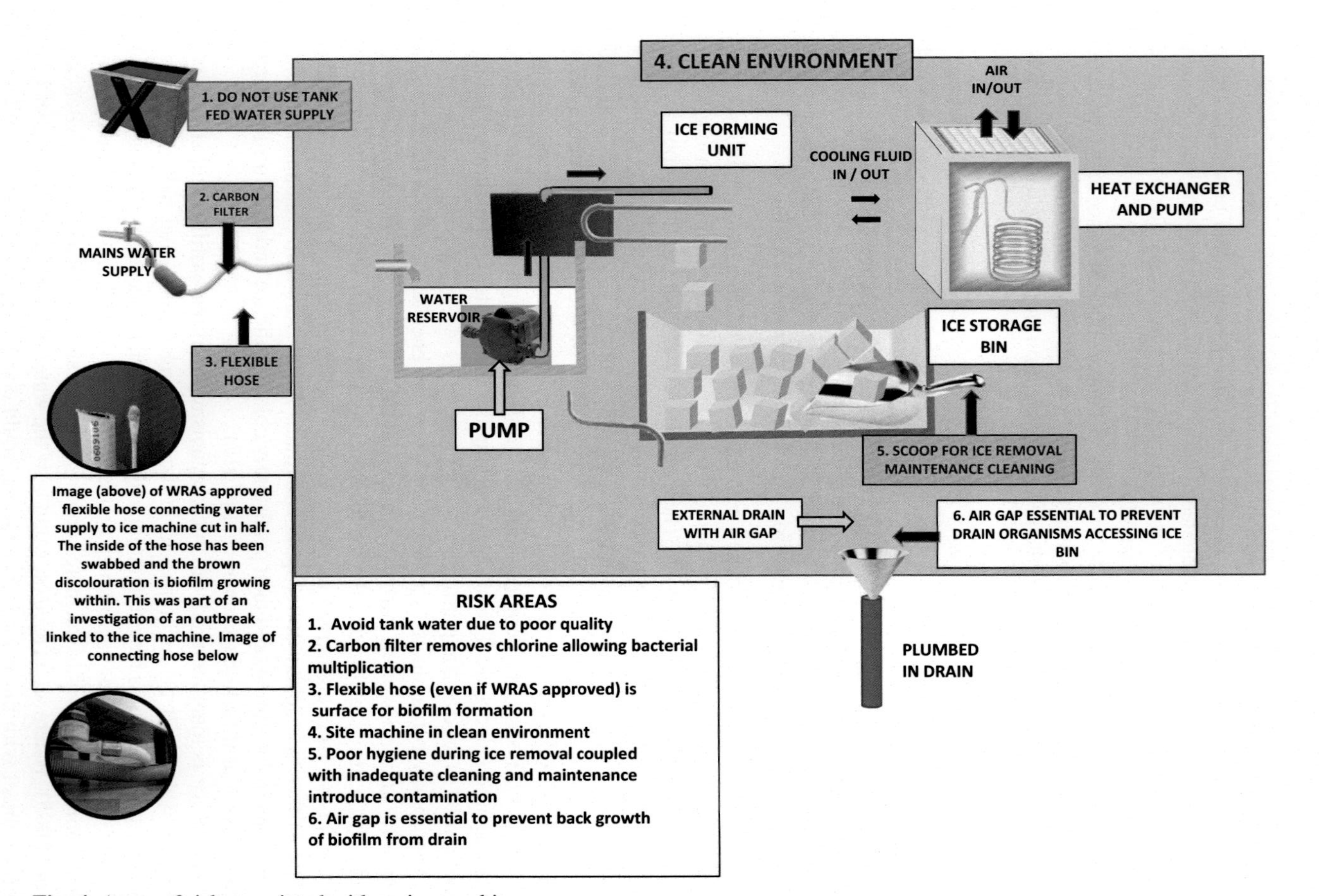

Fig. 4 Areas of risk associated with an ice machine.

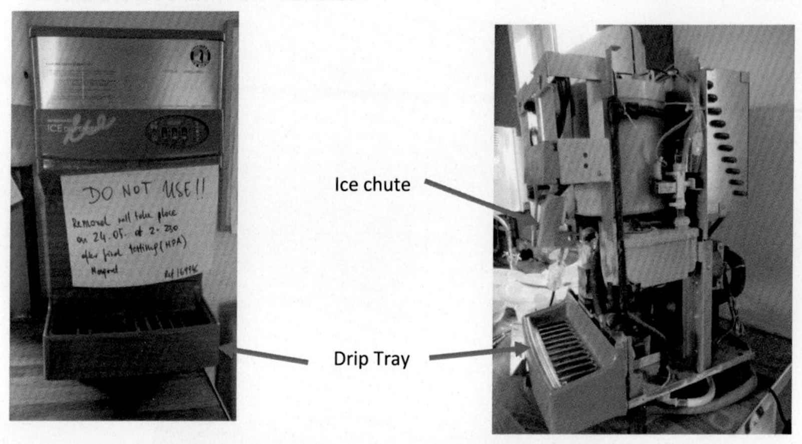

Fig. 5 An ice machine taken out of service due to microbial infections. Images courtesy of Peter Hoffman, PHE.

Fig. 6 Water supply pipe from the contaminated ice machine demonstrating extensive biofilm. Images courtesy of Peter Hoffman, PHE.

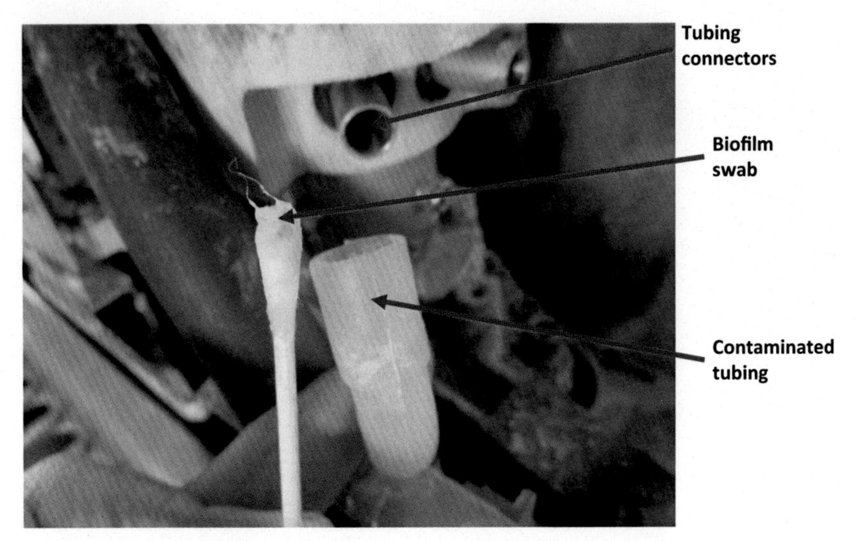

Fig. 7 Presence of extensive biofilm from one of the tubing on the inside of the ice machine indicating microbial fouling.

Recommended practice for ice machines

Ice machines should only be purchased and installed with prior approval of the WSG.

Ice machines should be placed on the water asset register.

Water supply to the ice machine should be ideally direct from a mains supply and not via a water storage tank.

Ensure the ice machine is located in a clean area, e.g., free from contamination by droplets from sinks.

Risk assessment should be carried out to determine if the location and supply are suitable and be agreed with the water safety group in the water safety plan.

Maintain ice-making machines in accordance with the manufacturer's recommendations.

Cleaning, servicing, and maintenance protocols should be available from the manufacturer in conjunction with the infection control team.

Water safety group should review reports on cleaning and maintenance of any ice machines and any failures or breakdowns.

Ice required for treatment purposes should be made using water obtained through a microbiological bacterial retention filter (to prevent ingress of microorganisms) or boiled water in sterile ice trays (ensure trays are adequately cleaned and disinfected) or ice bags.

Ensure staff are trained on how to use the ice machines, collect ice, and dispense hygienically.

Do not allow patients to use the ice machines directly.

Infection control teams should have a policy and protocol to follow in the event of suspected infections associated with the use of ice.

In line microbiological filters should be considered to minimize the risk of contaminated water ingress and should never be by-passed.

Other microbial control systems could be considered, e.g., alternative disinfection system including UV or other systems recommended by the manufacturer.

Do clean and disinfect spouts, scoops, trays, and drains on a planned preventative maintenance schedule.

Do not allow the water supply to the ice-making machine to gain heat as this will encourage microbial growth.

Do not place ice machines in augmented care.

Do not allow ice to stagnate in an ice-making machine's storage tray/bin.

Tubing, especially plastic tubing, should be regularly replaced and the system cleaned and disinfected following the manufacturer's instructions.

Example of a cleaning and maintenance regimen

A regime for regular cleaning and maintenance needs to be in place and audited. Staff undertaking cleaning should be competent and trained.

An example of the general steps for cleaning and maintenance of ice machines

The following may be considered when there are no specific instructions from the manufacturer; however, the manufacturer should still be consulted on the procedures and protocols followed. The authors take no responsibility for outcomes.

1. Disconnect unit from power supply.
2. Remove and discard ice from bin or storage chest.
3. Allow unit to warm to room temperature.
4. Disassemble removable parts of machine that make contact with water to make ice. Check for cracked, scratched, and scuffed surfaces as these increase the risk of biofilm formation
5. Thoroughly clean machine and parts with water and detergent (this should be as indicated by the manufacturer and suitable for use in contact with food.
6. Dry external surfaces of removable parts before reassembling.
7. Check for any needed repair.
8. Replace feeder lines, as appropriate (e.g., at a frequency indicated by the manufacturer or when damaged, old, or difficult to clean).
9. Ensure presence of an air space in tubing leading from water inlet into water distribution system of machine.
10. Inspect for rodent or insect infestations under the unit and treat, as needed.
11. Check door gaskets (open compartment models) for evidence of leakage or dripping into the storage chest.
12. Clean the ice-storage chest or bin with fresh water and detergent; rinse with fresh tap water.
13. Disinfect machine by circulating a 50–100 parts per million (ppm) solution of sodium hypochlorite (i.e., 4–8 mL sodium hypochlorite/gallon of water) through the ice-making and storage systems for 2 h (100 ppm solution), or 4 h (50 ppm solution). This may negate a guarantee as will adversely affect seals, valves, etc., also need COSHH assessment and neutralization with sodium thiosulphate
14. Drain sodium hypochlorite solutions and flush with fresh tap water.
15. Allow all surfaces of equipment to dry before returning to service.

Risk assessment criteria for ice machines

Risk assessment criteria for ice machines	Yes	No
Has a *Legionella* risk assessment been carried out?		
Has a *Pseudomonas aeruginosa* risk assessment been carried out?		
Have the design, location, and installation been overseen and approved by the WSG		
Are infection control policies in place?		
Has the water supply been taken from a wholesome supply via a double-check valve to prevent backflow?		
Has the ice machine been installed upstream of a regularly used outlet with the minimum of intervening pipe-run of less than 3 m?		
Has pipework (fixed or flexible) been kept as short as possible?		
Is the cold water monitored to ensure that it does not become stagnant and does not increase above 20°C?		
Has the unit been situated in an area where it will not be exposed to warm exhausts?		
Is maintenance been undertaken according to manufacturer's instructions?		
Are actions being taken to ensure that the ice does not stagnate?		
Are approved cleaning and hygiene protocols being undertaken?		
Is there a protocol to ensure the nozzles and ice trays are being cleaned?		
Is the ice dispensed by a nontouched nozzle?		
Is the ice machine installed in a nonaugmented care area?		
Is ice used for immunocompromised patients safe for use, free of pathogens?		
Has any microbiological monitoring of the ice machine and ice been carried out?		
Has water for ice for treatment purposes been passed through a bacterial retention filter?		
Are all staff using the ice machine trained in the microbial risks?		

Guidelines for ice machines

HBN 00-09—"Infection control in the built environment" (DHSC, 2021).
HTM 04-01 (DHSC, 2016).

References

Ameen, B.I.H., Alzubaidee, A.F., Majid, S., 2019. Comparative evaluation of the efficacy of ice cubes versus sodium bicarbonate mouthwash both as prophylactic measure and as treatment of oral mucositis induced by systemic anticancer therapies. Tabari Biomed. Stud. Res. J. https://doi.org/10.18502/tbsrj.v1i4.2244.

Bencini, M., Yzerman, E.P.F., Koornstra, R.H.T., Nolte, C.C.M., den Boer, J.W., Bruin, J.P., 2005. A case of Legionnaires' disease caused by aspiration of ice water. Arch. Environ. Occup. Health 60, 302–306. https://doi.org/10.3200/AEOH.60.6.302-306.

Blatt, S.P., Parkinson, M.D., Pace, E., Hoffman, P., Dolan, D., Lauderdale, P., Zajac, R.A., Melcher, G.P., 1993. Nosocomial Legionnaires' disease: aspiration as a primary mode of disease acquisition. Am. J. Med. 95, 16–22. https://doi.org/10.1016/0002-9343(93)90227-g.

Breathnach, A.S., Riley, P.A., Shad, S., Jownally, S.M., Law, R., Chin, P.C., Kaufmann, M.E., Smith, E.J., 2006. An outbreak of wound infection in cardiac surgery patients caused by *Enterobacter cloacae* arising from cardioplegia ice. J. Hosp. Infect. 64, 124–128. https://doi.org/10.1016/j.jhin.2006.06.015.

DHSC, 2016. HTM 04-01: Safe Water in Healthcare Premises.

DHSC, 2021. Health Building Note 00-09: Infection Control in the Built Environment 47.

DHSC Safe water in healthcare premises (HTM 04-01), 2017. Safe Water in Healthcare Premises (HTM 04-01). [WWW Document]. GOV.UK. URL: https://www.gov.uk/government/publications/hot-and-cold-water-supply-storage-and-distribution-systems-for-healthcare-premises. (Accessed 5 January 2021 (Accessed 4 April 2019).

EPIA, 2021. European Packaged Ice Association—Ice You Can Trust.

Graman, P.S., Quinlan, G.A., Rank, J.A., 1997. Nosocomial legionellosis traced to a contaminated ice machine. Infect. Control Hosp. Epidemiol. 18, 637–640. https://doi.org/10.1086/647689.

Iroh Tam, P.-Y., Kline, S., Wagner, J.E., Guspiel, A., Streifel, A., Ward, G., Messinger, K., Ferrieri, P., 2014. Rapidly growing *Mycobacteria* among pediatric hematopoietic cell transplant patients traced to the hospital water supply. Pediatr. Infect. Dis. J. 33, 1043–1046. https://doi.org/10.1097/INF.0000000000000391.

Kanwar, A., Domitrovic, T.N., Koganti, S., Fuldauer, P., Cadnum, J.L., Bonomo, R.A., Donskey, C.J., 2017. A cold hard menace: a contaminated ice machine as a potential source for transmission of carbapenem-resistant *Acinetobacter baumannii*. Am. J. Infect. Control 45, 1273–1275. https://doi.org/10.1016/j.ajic.2017.05.007.

Kanwar, A., Cadnum, J.L., Xu, D., Jencson, A.L., Donskey, C.J., 2018. Hiding in plain sight: contaminated ice machines are a potential source for dissemination of gram-negative bacteria and *Candida* species in healthcare facilities. Infect. Control Hosp. Epidemiol. 39, 253–258. https://doi.org/10.1017/ice.2017.321.

Laussucq, S., Baltch, A.L., Smith, R.P., Smithwick, R.W., Davis, B.J., Desjardin, E.K., Silcox, V.A., Spellacy, A.B., Zeimis, R.T., Gruft, H.M., 1988. Nosocomial *Mycobacterium fortuitum* colonization from a contaminated ice machine. Am. Rev. Respir. Dis. 138, 891–894. https://doi.org/10.1164/ajrccm/138.4.891.

MAFF, 1995. The Food Safety (General Food Hygiene) Regulations 1995.

Millar, B.C., Moore, J.E., 2020. Hospital ice, ice machines, and water as sources of nontuberculous mycobacteria: description of qualitative risk assessment models to determine host-nontuberculous mycobacteria interplay. Int. J. Mycobacteriol. 9, 347–362. https://doi.org/10.4103/ijmy.ijmy_179_20.

Mills, J.P., Zhu, Z., Mantey, J., Hatt, S., Patel, P., Kaye, K.S., Gibson, K., Cassone, M., Lansing, B., Mody, L., 2019. The devil is in the details: Factors influencing hand hygiene adherence and contamination with antibiotic-resistant organisms among healthcare providers in nursing facilities. Infect. Control Hosp. Epidemiol. 40, 1394–1399. https://doi.org/10.1017/ice.2019.292.

Newsom, S.W.B., 1968. Hospital infection from contaminated ice. Lancet 292, 620–622. https://doi.org/10.1016/S0140-6736(68)90708-3.

Nikoletti, S., Hyde, S., Shaw, T., Myers, H., Kristjanson, L.J., 2005. Comparison of plain ice and flavoured ice for preventing oral mucositis associated with the use of 5 fluorouracil. J. Clin. Nurs. 14, 750–753. https://doi.org/10.1111/j.1365-2702.2005.01156.x.

Panwalker, A.P., Fuhse, E., 1986. Nosocomial *Mycobacterium gordonae* pseudoinfection from contaminated ice machines. Infect. Control. 7, 67–70. https://doi.org/10.1017/s0195941700063918.

PHE, 2020. Examining Food, Water and Environmental Samples From Healthcare Environments—Microbiological Guidelines. https://assets.publishing.service.gov.uk/government/uploads/system/uploads/attachment_data/file/865369/Hospital_F_W_E_Microbiology_Guidelines_Issue_3_February_2020__1_.pdf.

Pinho Reis, C., Sarmento, A., Capelas, M., 2018. Nutrition and Hydration in the End-of-Life Care: Ethical Issues. pp. 36–40, https://doi.org/10.21011/apn.2018.1507.

Pisegna, J.M., Langmore, S.E., 2018. The ice chip protocol: a description of the protocol and case reports. Perspect. ASHA Special Interest Groups 3, 28–46. https://doi.org/10.1044/persp3.SIG13.28.

Prasad, R.M., Raziq, F., Kemnic, T., Nabeel, M., Ghaffar, M., 2020. *Legionella* causing lung abscess in an immunocompetent patient. Am. J. Med. Case Rep. 9, 18–21. https://doi.org/10.12691/ajmcr-9-1-6.

Ravn, P., Lundgren, J.D., Kjaeldgaard, P., Holten-Anderson, W., Højlyng, N., Nielsen, J.O., Gaub, J., 1991. Nosocomial outbreak of cryptosporidiosis in AIDS patients. BMJ 302, 277–280. https://doi.org/10.1136/bmj.302.6771.277.

Schlesinger, N., Detry, M.A., Holland, B.K., Baker, D.G., Beutler, A.M., Rull, M., Hoffman, B.I., Schumacher, H.R., 2002. Local ice therapy during bouts of acute gouty arthritis. J. Rheumatol. 29, 331–334.

Schuetz, A.N., Hughes, R.L., Howard, R.M., Williams, T.C., Nolte, F.S., Jackson, D., Ribner, B.S., 2009. Pseudo-outbreak of *Legionella pneumophila* serogroup 8 infection associated with a contaminated ice machine in a bronchoscopy suite. Infect. Control Hosp. Epidemiol. 30, 461–466. https://doi.org/10.1086/596613.

Stout, J.E., Yu, V.L., Muraca, P., 1985. Isolation of *Legionella pneumophila* from the cold water of hospital ice machines: implications for origin and transmission of the organism. Infect. Control. 6, 141–146. https://doi.org/10.1017/s0195941700062937.

Wilson, I.G., Hogg, G.M., Barr, J.G., 1997. Microbiological quality of ice in hospital and community. J. Hosp. Infect. 36, 171–180. https://doi.org/10.1016/S0195-6701(97)90192-4.

WRAS, 1999. The Water Supply (Water Fittings) Regulations 1999.

Yorioka, K., Oie, S., Hayashi, K., Kimoto, H., Furukawa, H., 2016. Microbial contamination of ice machines is mediated by activated charcoal filtration systems in a city hospital. J. Environ. Health 78, 32–35.

Yu, V.L., Stout, J.E., 2010. *Legionella* in an ice machine may be a sentinel for drinking water contamination. Infect. Control Hosp. Epidemiol. 31, 317. author reply 318 https://doi.org/10.1086/651067.

18 Dirty utility rooms

What is a dirty utility room?

Dirty utility rooms usually contain a hopper (also known as a slop hopper) for the disposal of liquid waste and either a macerator or bedpan washer disinfector and a separate sink for cleaning of equipment. Dirty utility rooms require a hand wash station, ideally located near the exit to help enable hand hygiene of staff leaving the area. The functionality of the dirty utility room and handling of liquid waste and the use of water results in the contamination of the surrounding area including the surfaces, water outlets, hand wash basins, sinks, and drains. This section deals with a number of areas including the design, use, location, size, storage of equipment drainage, and use of water in the dirty utility room.

Dirty utility rooms should be designed to enable:

- safe and hygienic disposal of clinical waste, such as the contents of bedpans, urine bottles, etc. These fluids should be assumed to be highly contaminated and may often contain antibiotic-resistant organisms.
- decontamination of specific items of equipment, e.g., bedpans, urine bottles, catheter bag carriers, commodes, slipper bedpans (used to hold disposable pulp bedpans).

Location

The location of the dirty utility room may impact on patient safety in the following ways.

- In most hospitals, there is often a considerable distance between the point of care and the dirty utility room, and local infection prevention rules do not permit the discarding of fluids into hand wash basins because of the risk of introducing contaminants and nutrients that support microbial growth. In addition, intensive care unit side rooms usually lack facilities for the safe disposal of body fluids, unlike most ward side rooms which tend to have ensuite bathrooms with a WC. As a consequence, fluids are discarded into hand wash basins, and these practices have been implicated in outbreaks (Breathnach et al., 2012; Mathers et al., 2018).
- The transportation of these body fluids and water used for patients' personal hygiene runs the risk of spillages leading to slip hazards, as well as increasing the likelihood of cross contamination.
- It is good practice to site the dirty utility room as the end point of a cold water pipe, i.e., as the last facility to promote high water usage and turnover in these rooms and to pull the water along the pipework to alleviate stagnation. However, this type of arrangement is likely to conflict with points 1 and 2 described above, especially in larger wards. Therefore, alternative methods for ensuring water movement should be considered that do not compromise staff and patient safety, such as siting staff toilets at the end of pipe runs and/or installing self-flushing outlets at the ends of pipe runs to ensure that the water does not stagnate.

Safe Water in Healthcare. https://doi.org/10.1016/B978-0-323-90492-6.00030-6

Design

This should take into account;

- Location (see above)
- Size of the ward and the number and type of patients
- Space for storage of items such as bed pans and urine bottles
- Layout to allow effective process flow, segregating dirty activities/equipment from clean
- Contents including infrastructure and equipment for the room to function
- Design of supporting services, i.e., water supply, drainage
- Minimizing cross-contamination on accessing room

Note: Good design of the dirty utility room by itself is not sufficient as it also requires staff awareness training to ensure their interaction with water services and the environment are safe.

The design of dirty utility rooms has long been an issue of concern. As far back as 1959, Professor J.M. Smellie considered the important attributes of a dirty utility room and concluded the following: "Sluice room—the design of the sluice room or utility room illustrates our attempt to avoid cross-infections while employing relatively unskilled nursing staff. It has two doors and traffic passes through the dirty side of the sluice room, where the sluice and bedpan washer are situated and where there should be an incinerator for napkins, to the clean side where sterilized bedpans are stored and where the nurse washes her hands before leaving the room" (Smellie, 1929).

Unfortunately, such design considerations, including process flow and sufficient space for segregation of dirty and clean equipment within the dirty utility room (Fig. 1), have not become embedded into standard practice. The importance of the

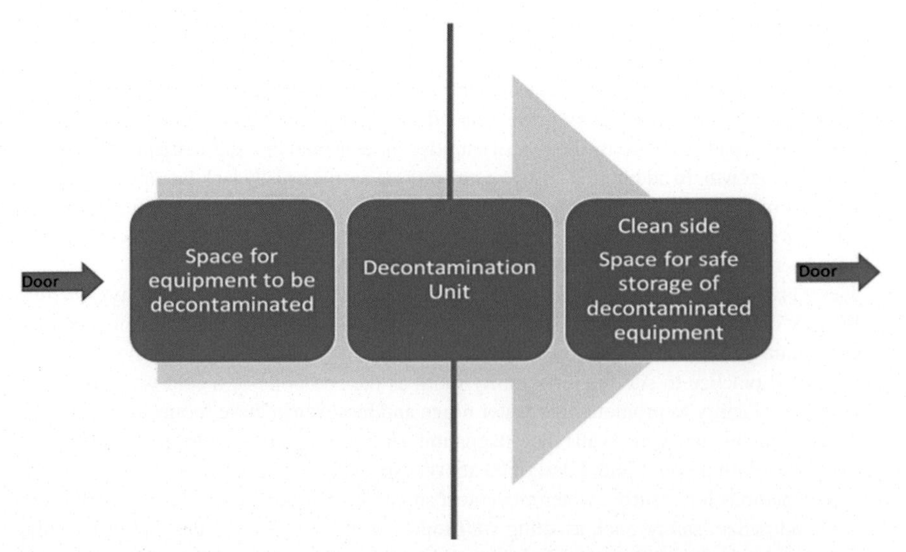

Fig. 1 Schematic of dirty to clean area workflow though rooms are often small, and demarcation may assist.

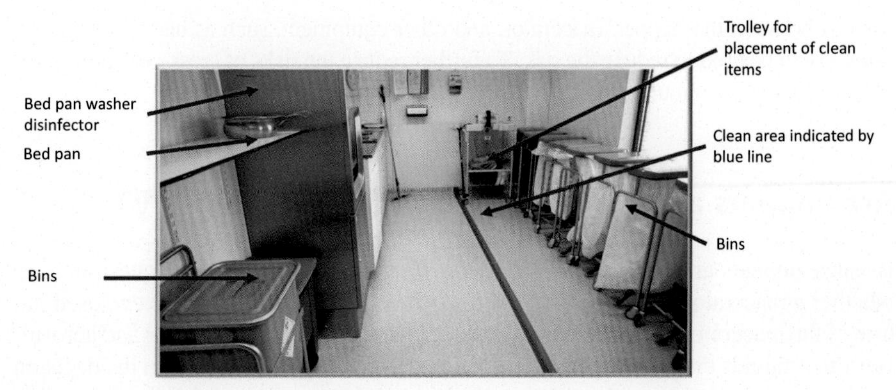

Fig. 2 Achieving zoning in an existing dirty utility room demonstrating demarcation with tape.

risks associated with poor design and lack of storage space is not being managed appropriately by impacting those with influence. Dirty utility rooms are therefore often not subject to purposeful design. Rather, they appear as an afterthought, molded by whatever space is left over when everything else is accounted for on a ward.

Poor original design of a dirty utility room can be improved by the demarcation of clean and dirty zones as demonstrated in Fig. 2 showing red (dirty) and blue (clean) zoning lines, respectively (Fig. 2) with a one-way flow through system (dirty to clean) to reduce the risk of cross-contamination. This type of arrangement is aligned with the guidance given in HBN 00-09, i.e., to designate clean and dirty areas and implement clearly defined workflow patterns (DHSC, 2021). In Fig. 2, dirty bedpans are placed on the shelf on the left (red area) prior to placement in the washer disinfector. Once removed, the clean bedpans are stored in the clean blue zone (not shown in the image).

Size

Although in terms of design considerations size, flow/layout and contents are listed as discrete entities, but in practice they can be interdependent. If a dirty utility room is too small, it is very difficult to ensure the correct layout and process flow, which are necessary to maintain effective infection control practices.

Impact of design on splash contamination

A complicating factor is the risk from splashing and aerosols. Studies have shown that droplets emanating from sinks and hoppers can contaminate the surrounding area and items of equipment within 1–2 m radius (Mathers et al., 2018). Therefore, to reduce the spread of contamination from splashing when emptying the contents of containers into the hopper or macerator, care should be taken to minimize splashing. For example, installing screens between the hopper, macerator, and other equipment, such as handwashing stations, could be used to reduce the risk. To further reduce the risks of cross-contamination, ideally, equipment should be able to be operated hands-free.

Are hoppers and separate cleaning sinks necessary?

Because hoppers are very often co-located with macerators, it begs the question as to whether a macerator can fulfill the function of the hopper; certainly, the enclosed nature of the macerator during operation reduces the risk of fluids splashing and aerosolization of liquids once the lid has been closed. Some hospitals have taken the decision to dispense with hoppers altogether and dispose of contaminated fluids using either a macerator or bedpan washer disinfector. However, where this is the case, consideration should be given to the length of the macerator operating cycle to avoid an accumulation of material awaiting disposal in the dirty utility room. As such those involved in design need to consider the volume of usage.

It should also be remembered that large dialysis bags often contain many liters of fluid and take several minutes to empty and their size, weight, and shape make disposal of the fluids they contain into a macerator very difficult to accomplish without the risk of significant splashing. For this reason, some hoppers are designed with supports to hold the weight of the bags during emptying. It may not, therefore, be feasible for all hoppers to be replaced by macerators, and local teams will need to consider this on a case-by-case basis.

Some dirty utility rooms also have a separate sink, intended for washing of equipment. Occasionally, these sinks are also used for hand washing, but this should be avoided because they are not of an appropriate design for hand washing in a clinical setting. In addition, these sinks are likely to become contaminated when they are used to clean dirty equipment, and they are often in close proximity to other sources of contamination (Fig. 3).

In some designs, the sink unit is integrated with a hopper (Fig. 4). Their proximity, relative to one another, is likely to facilitate the transfer of microorganisms between both devices, and as such most surfaces are likely to be contaminated. As can be seen in Fig. 4, the hopper and decontamination sink are within 1–2 m of the hand wash sink. The children's toys and drug preparation trays that are being cleaned in the decontamination sink often do not go through a further decontamination procedure and could become contaminated. The general clutter on and in this scenario, including the toilet brush, used to clean the hopper, presents a real challenge to infection control, and items not in use should be decontaminated before being stowed safely elsewhere.

Layout of the dirty utility room

Fig. 5 shows the layout of a dirty utility room as described in HBN 00 03 (DHSC, 2013, 2021; Wales NHS, 2006). However, if this layout is assessed for its potential to allow cross-contamination by splashing, it is clear that there is little in the way of dirty

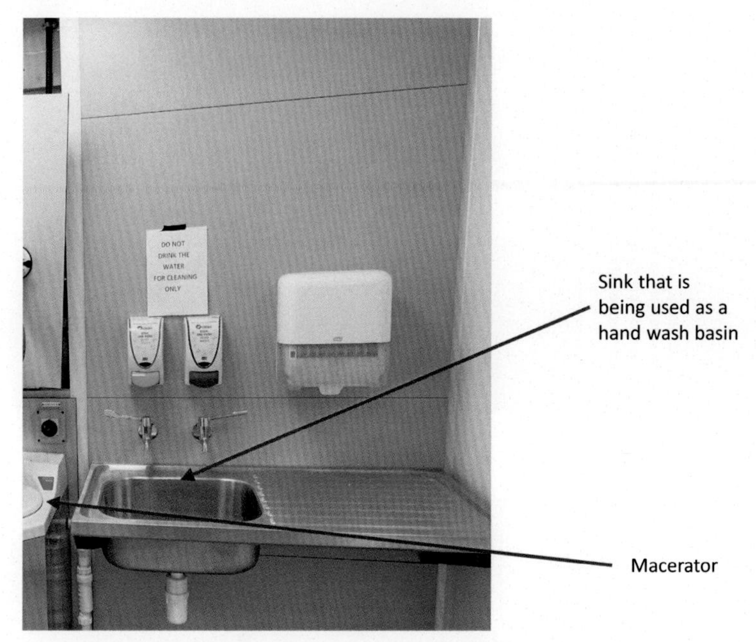

Fig. 3 Inappropriate use of a cleaning sink for hand washing purposes.

Fig. 4 Integrated sink and hopper.

to clean zone segregation, and no provision made for storage for equipment. If you were conducting a risk assessment for the use of this area, how could it be improved to enhance infection control and reduce risks from splashing and recontamination of areas when handling dirty products? Figs. 6 and 7 demonstrate alternative designs with a flow through from dirty to clean (Fig. 6) as well as completely segregated dirty and clean areas (Fig. 7) where there is the inclusion of a pass-through washer disinfector. In this example of a completely segregated clean area where staff would be washing hand before entering, then only alcohol gel would be required as no contamination should be present.

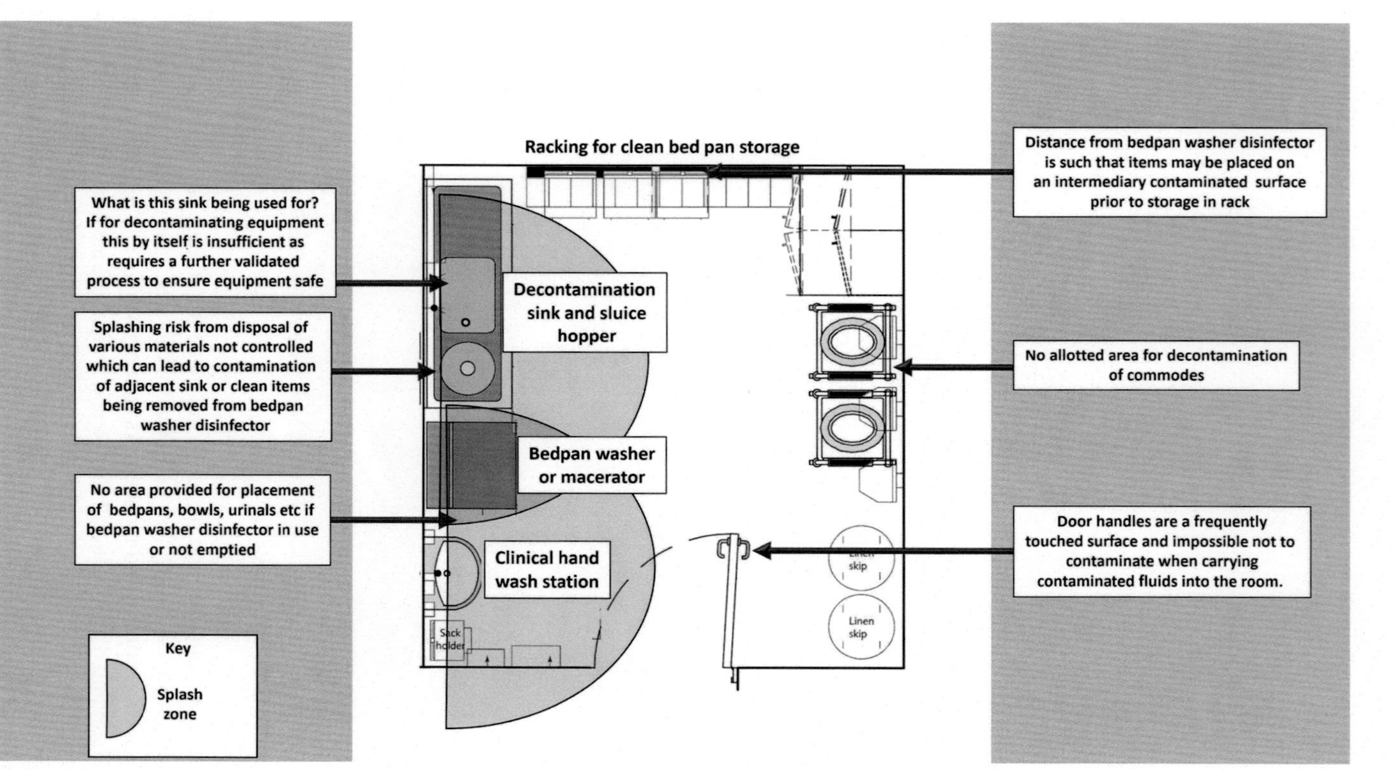

Fig. 5 Space requirements for dirty utility rooms for bed pan reprocessing (as per HBN 00-03).

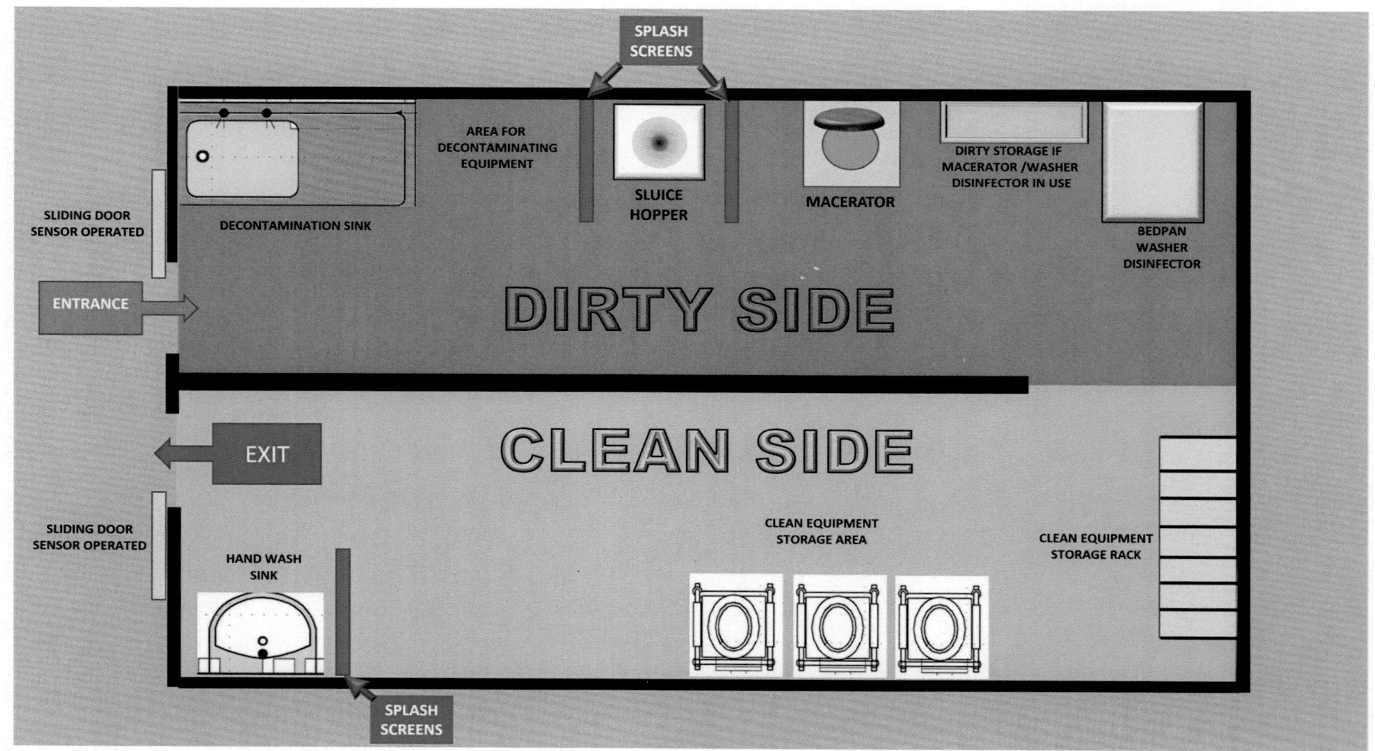

Fig. 6 Revised design taking into account deficiencies in original HBN 00-03 design—inclusion of physical walk through separation for dirty to clean.

Poor ergonomic design of the dirty utility room is often a problem, including the siting of clinical hand wash stations. If these are not positioned conveniently (and ideally) next to the entrance/exit door, staff can be discouraged from washing their hands when leaving the area, which may contribute to the spread of contamination from the dirty utility to other ward areas (Figs. 5–7) (Fornara et al., 2010).

The lack of identified zoning of dirty to clean areas increases the likelihood of cross-contamination in a number of ways.

(1) **Failing to segregate clean and dirty areas**, e.g., storing clean bedpans and urine bottles immediately above, and close to, the hopper, can result in cross-contamination of clean equipment (Fig. 8).

(2) **Failing to provide designated areas for placement of dirty equipment when a macerator or bedpan washer disinfector is in use**. For example, in the scenario presented in the following figures, contaminated bedpans were placed on top of the washer disinfector (Fig. 9) because it was already in use, and there was no designated temporary storage area.

To add to the problem, the height of most bedpan washer disinfectors means that some staff are unable to inspect the top surfaces of the washer disinfector and to clean them if necessary (Fig. 10). In this case, the top of the bedpan washer disinfector was contaminated with fecal material due to the incorrect placement of the dirty bedpans (Fig. 11). In this example, signage was displayed to discourage items from being placed on top of the washer disinfector (Fig. 12), but more thoughtful design, such as incorporating a sloped top surface, would avoid this situation in the first place.

(3) **Failing to provide a designated area for storage of clean equipment following removal from the bedpan washer** disinfector resulted in cleaned items being removed and placed on the top of the contaminated bedpan washer disinfector, recontaminating the cleaned bedpans with feces. A designated clean storage area for decontaminated equipment coming out of the bedpan washer disinfector should be close-by, otherwise it is unlikely to be used appropriately. This area might also require clean storage racks to encourage correct placement of items and as to aid drainage of any excess water.

There is often a degree of confusion regarding responsibilities for cleaning dirty utility rooms and the equipment they contain, i.e., does the responsibility lie with cleaning staff or ward staff? This uncertainty often means that cleaning of hoppers, macerators, etc., falls between two stools (pun intended), and it can be the case that these items are not cleaned frequently and regularly and to the required standard to ensure that risks to patients are minimized. Hospitals should ensure that these roles and responsibilities are defined clearly in the water safety plan and detailed in local codes or written procedures. Cleaning procedures need to be developed that consider any manufacturers' instructions and cleaning and descaling products should be such that they can be applied effectively without the need for excessive physical force, e.g., by using a toilet brush that could disperse contaminating particles onto other surfaces. Audits of cleaning procedures should be undertaken periodically to provide assurance that they are being adhered to.

Drainage

For those members of the water safety group involved in the design of the dirty utility room design, the waste pipe leading from the macerator or sluice hopper needs to be of sufficient diameter to accommodate the material going to waste and 90 degree bends

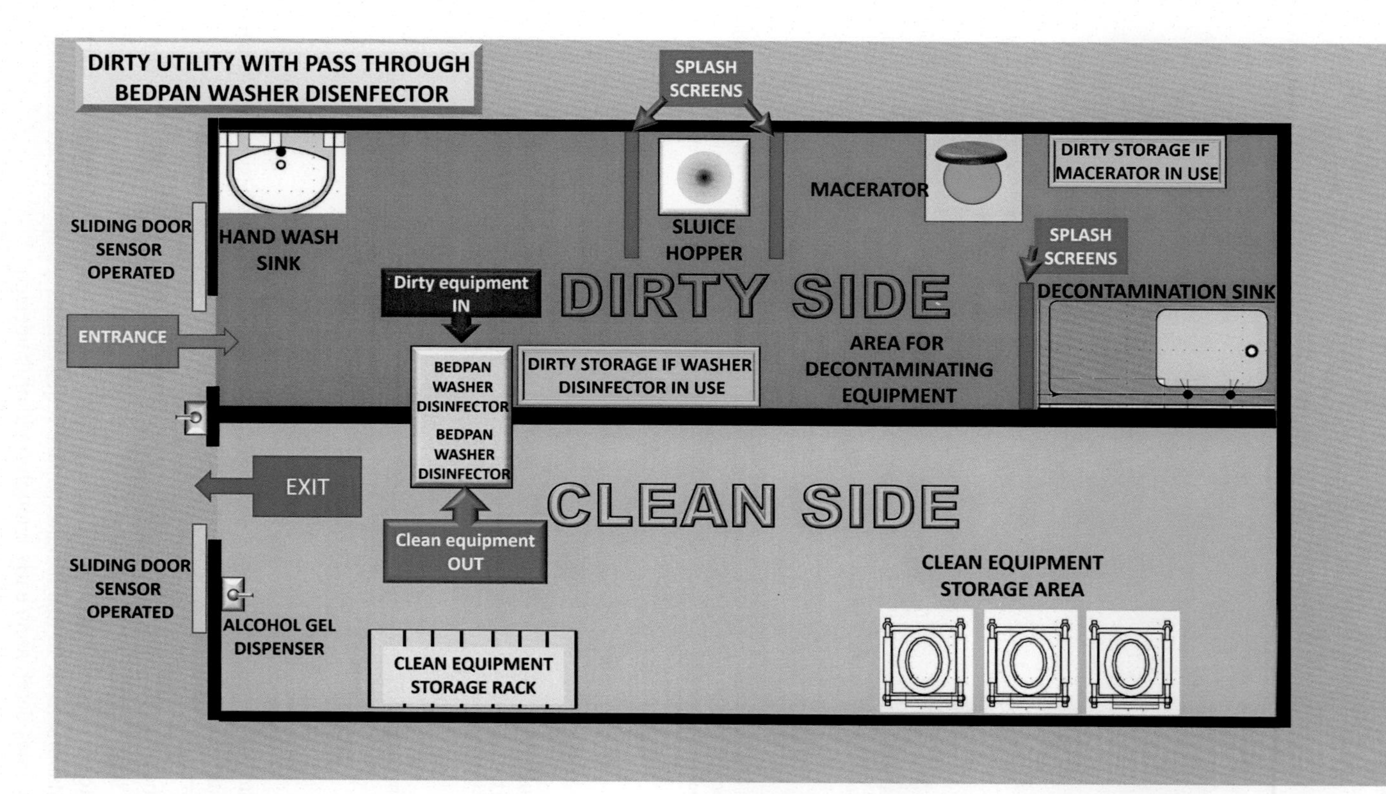

Fig. 7 Revised design taking into account deficiencies in original HBN 00-03 design.

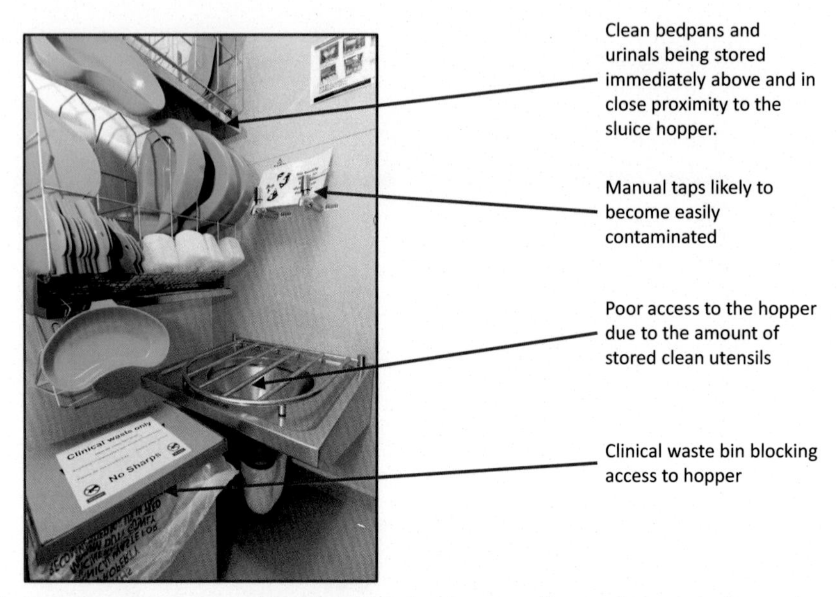

Fig. 8 Poor design of dirty utility room with the clean utensils stored above the hopper.

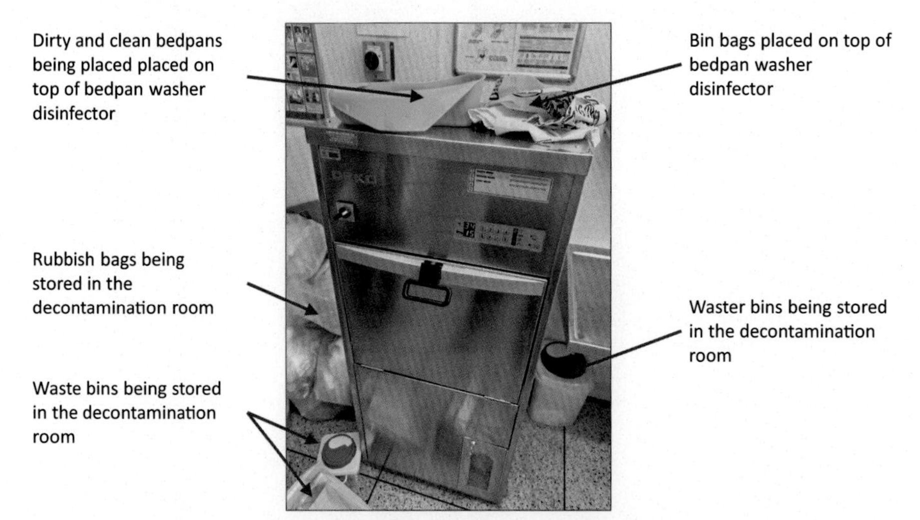

Fig. 9 Storage of bedpans on top of washer disinfector as well as rubbish bags and clinical waste bins being stored at the side of the washer disinfector.

should be avoided to minimize the chances of blockages occurring (Figs. 13 and 14). Fig. 14 shows a double 90 degrees bend that was installed despite instructions to the contractor to ensure there was a single, gradual, swept bend, as specified by the macerator manufacturer. In this instance, space requirements for appropriate drainage had not been considered at an early enough stage of the design.

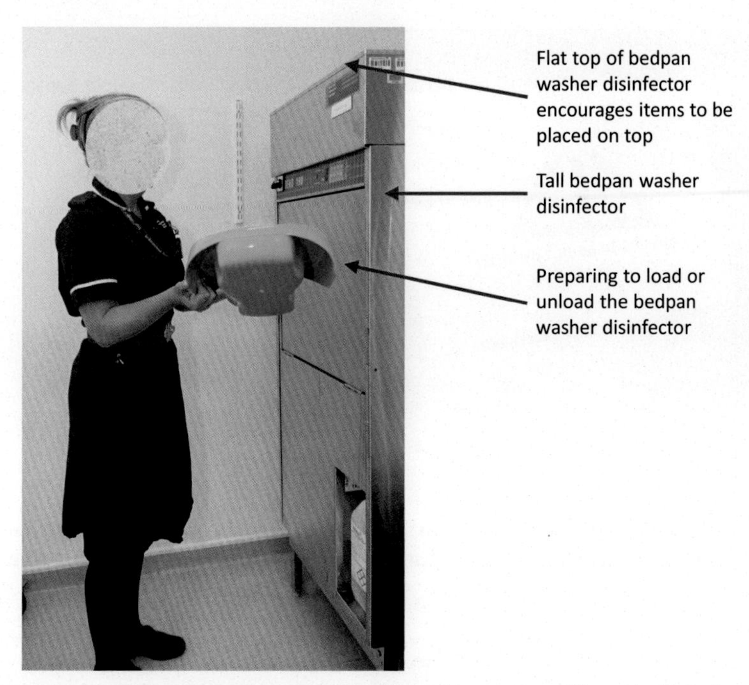

Fig. 10 Bedpan washer disinfectors are tall, and therefore many staff are unable to inspect and/or clean the surfaces.

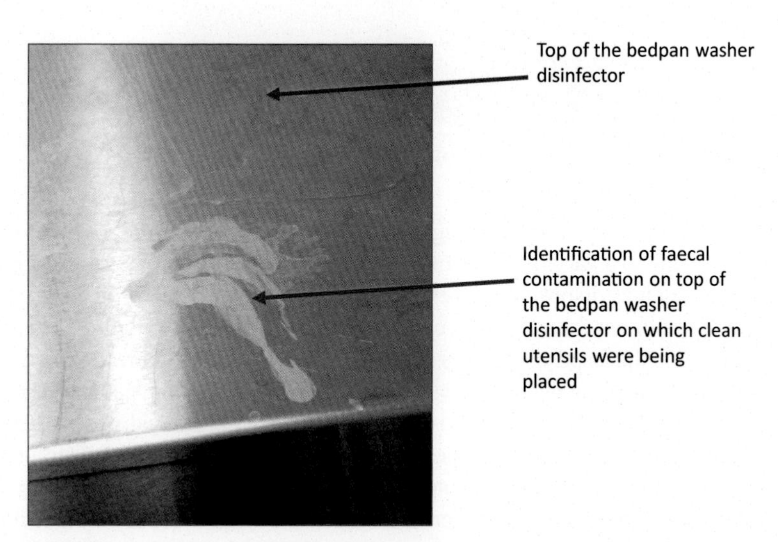

Fig. 11 Identification of fecal contamination from the top of the bedpan washer disinfector from Fig. 8.

Fig. 12 Placement of signage to indicate to staff that items should not be placed or stored on top of the washer disinfector—a physical barrier or sloping top would be more effective.

Fig. 13 Removal of the access panel to inspect drainage pipes of macerator.

Even in systems with good drainage, problems due to blockages in dirty utility areas can arise due to a variety of reasons, but commonly include inappropriate placement of items, such as disposable wipes, in macerators and hoppers (Fig. 15).

It is vitally important for staff to be educated in the appropriate disposal of items including prohibiting the disposal of wipes in hoppers and macerators (there are some wipes that are designed for placement in a macerator, and these should be validated for their intended use); wipes should be disposed of via the conventional clinical waste stream.

Blockages often lead to backflow (i.e., flow of water in the opposite direction to that intended so that dirty, contaminated water can flow backward) from the drains to other connected items such as hand wash stations, clinical sinks, and patient showering facilities.

This must be avoided to prevent transmission of harmful organisms to patients. As a precaution, no other service should drain into the drainage pipe running from the

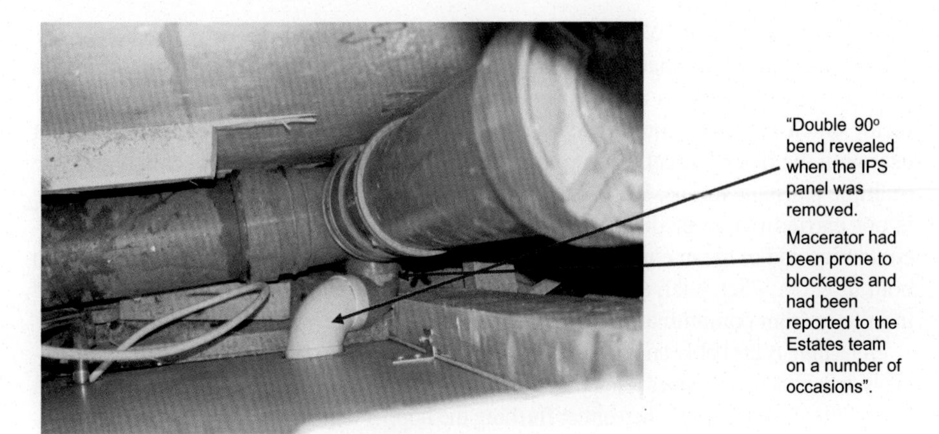

Fig. 14 Demonstration of double 90 degrees° bend in the drainage pipework from the macerator, which was shown to cause blockages.

Fig. 15 Photos showing inside of macerator blocked by inappropriate items (left hand) and on the right items removed from the blocked drain connected to the macerator.

macerator/bedpan washer disinfector to the main sewage stack. This minimizes the risks from backflow into patient areas. Additional considerations are the location of rodding points to clear any blockages that do occur.

Where blockages do occur, then the area must be a cleaned and decontaminated and all accessories decontaminated and single use items disposed of. Where macerators are installed, it is important to make sure you are not placing unsuitable materials that may cause blockages further down the drainage.

The most significant causes of contamination of dirty utility areas involve the handling of patient-derived material containing harmful microorganisms that can lead to contamination of surrounding environment and equipment during disposal (Breathnach et al., 2012).

There are several opportunities for cross-contamination during disposal. As staff empty patient waste, there will be a high likelihood that their hands or gloves will become contaminated. Where gloves are worn, it may not be recognized that the gloved hand is contaminated, resulting in the spread of environmental contamination of, e.g., on door handles or when manually opening/closing a bedpan washer disinfector. Such practices could lead to this contamination being transferred to another user with the risk of infection from potentially infectious material, which may include multiple resistant strains. Similarly, if hands do become contaminated, if hand hygiene is not carried out immediately, then any further hand contact could result in cross-contamination of clean equipment. This could cause infection by cross-contamination both directly, when such equipment is used or comes into contact with patients, and indirectly from contamination passed on via staff contact.

Splashing is probably an underestimated and largely unrecognized risk in this area especially when using a hopper. This may occur when discarded materials are dropped from a height into the hopper or when either flushing the hopper or using tap outlets located above.

Storage

It is extremely important that dirty utility rooms are used for their intended purpose and nothing else.

Note: A dirty utility room is not a safe area for storage of equipment other than for that that is re-processed in the room. Due to general constraints on space, the dirty utility room is often incorrectly used for the purpose of storing equipment either in the short or long term (Figs. 14–16).

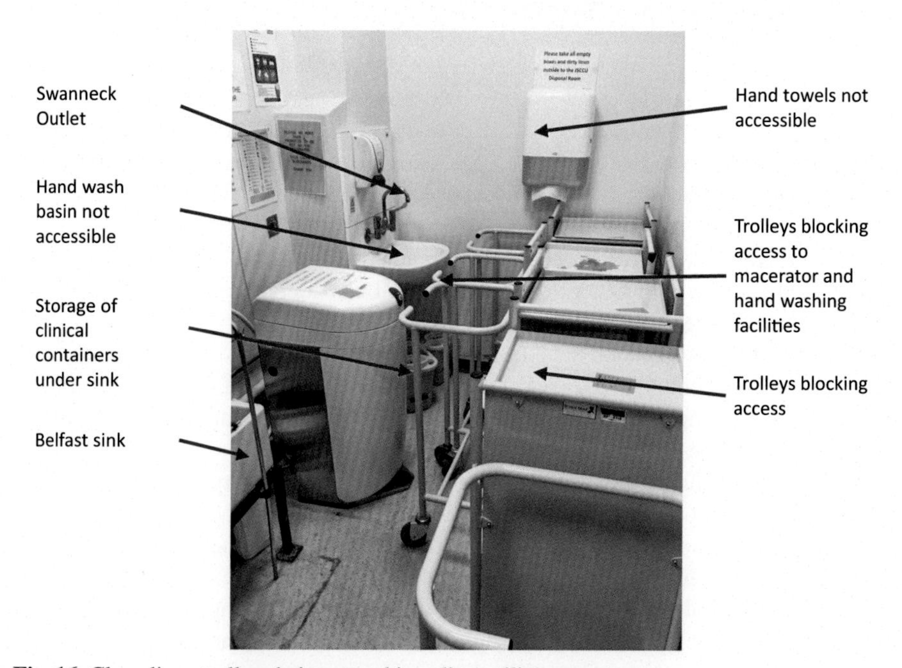

Fig. 16 Clean linen trolleys being stored in a dirty utility room.

However, space for storage of equipment and consumables is almost always under-provided in most hospitals, and this often leads to an accumulation of inappropriate items in dirty utility rooms. These items are placed unnecessarily at risk from contamination within the dirty utility environment.

Due to the presence of equipment being stored in the dirty utility room, the hand wash station may not be accessible and will be underused. This could present a microbial risk due to water stagnation in the supply pipework and proliferation of harmful bacteria (Figs. 16 and 17). As such, the potential for water supplied to the dirty utility area to contain opportunistic waterborne pathogens, including *Legionella* spp., nontuberculous mycobacteria (NTM), and *Pseudomonas aeruginosa* (Figs. 16 and 17). In addition, the use of the swan neck taps placed high above the basin will lead to an increased microbial risk due to promotion of splashing and aerosol production. Medical equipment stored in the vicinity of the dirty utility room would also become contaminated through splashing when the fluids are disposed into the hopper (Fig. 18).

Similarly, when hoppers and macerators become blocked, or if they breakdown for other reasons, they are sometimes left in a nonoperating state for lengthy periods while they wait to be fixed. This can lead to stagnation in the water supply to the fixture and the accumulation of biofilm, increasing the potential for bacteria such as *Legionella* spp. and *P. aeruginosa* to multiply within the supply pipework and outlets. This can increase risks of exposure during repair and when first used following repair, and also has the potential for biofilm to slough off and contaminate other parts of the system.

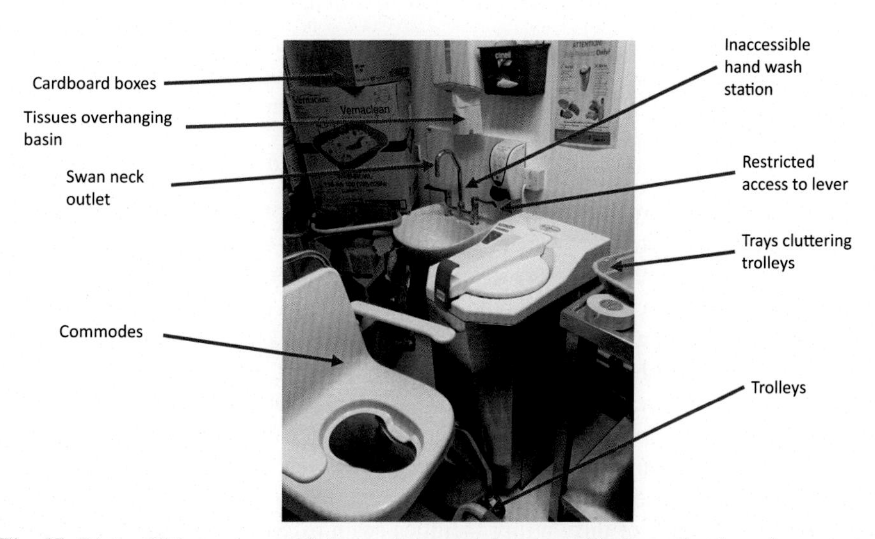

Fig. 17 Dirty utility rooms containing equipment and consumable items that impede access to the hand washing station.

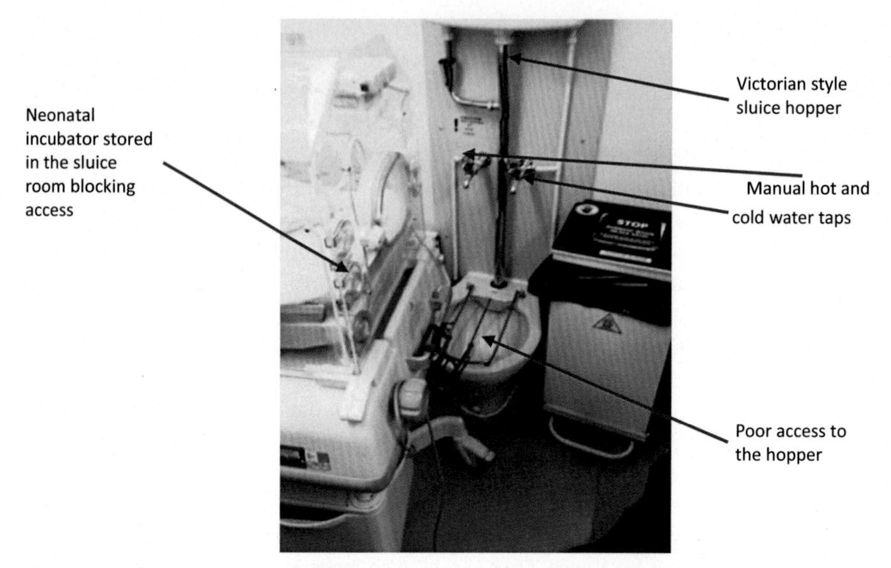

Fig. 18 Victorian-style hopper area being used to store neonatal incubators, the surfaces of which could become contaminated in this designated dirty area.

Outbreaks and incidents associated with dirty utility rooms and hoppers

Problems with dirty utility rooms were identified during the investigation into the *Clostridium difficile* outbreak at Stoke Mandeville Hospital in 2006 (Healthcare Commission, 2006). The report that was published identified that:

- Nurses had difficulties entering "dirty utility-rooms" and using the hand wash basin in these rooms, because linen, waste bags, and rubbish had not been collected.
- In one ward, the clinical waste bin lid in the dirty utility room touched paper hand towels when raised, leading to cross-contamination.
- In another ward, the dirty utility room did not have a dedicated hand wash basin

It is therefore important to keep a clean and tidy environment and to escalate any issues with the use of the space where possible. Items that do not need to be stored in this area should be stored elsewhere in the ward/area. Patient hygiene equipment such as dry wipes or mouth care items should be kept in a clean area, not the dirty utility.

Breathnach and coworkers also described an outbreak of a multidrug-resistant *P. aeruginosa* in two hospitals (Breathnach et al., 2012). Extensive environmental sampling in each outbreak yielded the same strain only from the wastewater systems. Inspection of the environment and estates records revealed many factors that may have contributed to contamination of clinical areas, including a faulty sink, shower and toilet design, and cleaned items stored near hoppers. Those inspections also identified poor design and usage of dirty utility facilities (with excessive splashing, leading to the potential for contamination of cleaned equipment). Records also revealed that an extensive number of blocked hoppers had been reported in the hospital each year, with suggestions that these were due to an accumulation of paper towels and/or patient wipes. The authors' recommendations included avoidance of inappropriate storage of clean items in dirty utility room areas.

In another outbreak, transmission networks were studied in adult intensive care units, and it was found that the majority of transmissions could not be explained by patient-to-patient events (Mathers et al., 2018). As a consequence, the investigators looked at the role of the environment in multispecies hospital acquired outbreaks of carbapenemase-producing *Enterobacteriaceae* (CPE). In a 15-month study, covers were installed on all (60) hoppers in adult intensive care units of a university hospital. In addition, in 23 surgical ICU sink traps, heating and vibration devices were also installed to pasteurize the trap and to remove accumulated biofilm. Patient acquisitions of antibiotic-resistant strains such as *Klebsiella pneumoniae* carbapenemase-producing organisms (KPCOs) for admissions were compared for 18-month preintervention and intervention periods. Fifty-six new multispecies, KPCO acquisitions occurred preintervention compared with 30 during the intervention, which was a statistically significantly difference. The effect of the sink trap devices alone could not be determined, although the proportion of sink drain cultures that tested positive for KPCO decreased from 80% to 5%. The authors concluded that interventions targeting wastewater plumbing fixtures, principally the installation of hopper covers, demonstrated a decrease in patient KPCO acquisitions, which they considered may be critical in reducing infections.

What monitoring should be considered?

Audits should be carried out to assess how the dirty utility rooms are used and how they are cleaned, and that suitable effective segregation is in place to prevent cross-contamination. Audits should also check that dirty utility rooms are not used for temporary storage of waste items or surplus equipment. Water outlets should be part of risk assessment, and water sampling for waterborne pathogens should be considered during investigations of cases or outbreaks when the epidemiology indicates a link. However, advice should be sought from microbiologists prior to sampling.

Governance

The hospital executive team/boards need to ensure there are documented processes in place within the water safety plan to ensure there is appropriate oversight of all dirty utility/decontamination areas and equipment to minimize the risk of cross-contamination to patients and staff. Common problems associated with the misuse of dirty utility areas are caused by a lack of awareness of the potential risks of causing hospital acquired infection by a lack of awareness of the associated risks and training for all those who might use the area and equipment therein. There should be clear lines of responsibility, accountability, and communication for the cleaning, operation, trouble shooting, and maintenance of the area and any equipment in. Clear governance is particularly important where the area is shared between wards or different clinical teams, for example.

- All staff should receive training during the ward/unit induction on how to use the area and equipment, the procedures in place, and how to avoid cross-contamination from equipment and within the area. Regular training updates should be provided and recorded.
- Documented escalation procedures should be in place for predictable untoward events, which identify the key personnel to be informed of any failures, the mitigation measures to be implemented, and safeguards to protect patients and staff. For example, if a macerator is not working or drain has become blocked.
  - The contact details for who should be informed and the authorized competent personnel to carry out remedial measures, including for out of hours.
  - What alternative facilities can be used and the measures that need to be taken into account to prevent the transfer of cross-contamination?
  - How to manage waste in the interim, for example, disposal of contaminated material into clinical waste bags with gel?
  - Any measures that should be put in place to prevent access, if required.
  - What should be done before the dirty utility is put back into normal operation?
  - Who is responsible for signing off that the area is safe and can be reinstated?
  - Who is responsible for reviewing these procedures and the time interval for review?
  - Who is responsible for regular audit and the reporting mechanisms for audit findings?
  - Who is responsible for service level agreements for equipment to be serviced and repaired in a timescale based on clinical need, as determined by the Water Safety Group?
- The hospital executive team/boards need to ensure there are processes in place within the Water Safety Plan to ensure there is appropriate consultation with the intended users, Water Safety Group members and subject matter experts to ensure the design, implementation, or refurbishment of dirty utility areas minimizes the risk of infection to both patients and staff by;
  - incorporating sufficient space including for equipment such as commodes and providing additional storage outside the dirty utility area for items such as cleaning equipment and consumables, e.g., disposable items such as paper towels, etc.
  - facilitating one-way workflow patterns from dirty to clean with sufficient set down and storage space.
  - ensuring the responsibilities for cleaning dirty utility rooms and their equipment are clearly defined in local policies and procedures and that auditing and monitoring are in place to demonstrate these activities have been undertaken effectively to reduce the risk to patients.

Recommendations for dirty utility rooms and facilities

Despite the best intentions of Government guidance, there is still a requirement for the infection control team and the Water Safety Group to risk assess their current facilities to determine if what is in place is fit for purpose. This should include reviewing whether the dirty utility room assists with the clean to dirty requirement to reduce contamination as well as some of the additional considerations identified below.

Design

1. Location—the location of the dirty utility rooms should minimize travel distances for staff from patient areas to reduce the risk of spillages and cross-contamination and to increase working efficiencies.
2. Access—consider hands-free options to minimize cross contamination of door handles.
3. The dirty utility room should be of sufficient size to allow safe segregation of equipment, a flow from dirty to clean, and sufficient space for storage (only items that are reusable and decontaminated in this area) and decontamination of equipment (such as commodes).
4. There should be visual use of color flooring (or alternatively tape for existing rooms, see Fig. 2) and preferably physical barriers to demarcate the dirty from clean areas.
5. The drains to the macerator or bedpan washer disinfector should have swept bends and have no other drains feeding into the pipework prior to entry into the main sewage stack. It is essential that architects/design teams review any documentation provided with equipment to ensure it is installed in accordance with manufacturer's advice.
6. Ensure there are adequate rodding points, which are accessible in case of blockages.
7. A clinical hand wash station should be sited next to the door.
8. In choosing macerator or bedpan washer disinfector, go for hand-free operation.
9. Consider whether a hopper is required.
10. Where a hopper is required avoid design, where there is (i) a sink built into the same unit and (ii) a sink drainage area on which items are placed.
11. Where required, consider splash screens to minimize risk of cross-contamination via splashing from hoppers and sinks (Fig. 19).
12. Consider whether a decontamination sink is required in the dirty utility room.
13. Consider whether a support should be added to the top of the sink to prevent items being placed in the sink (Fig. 19).

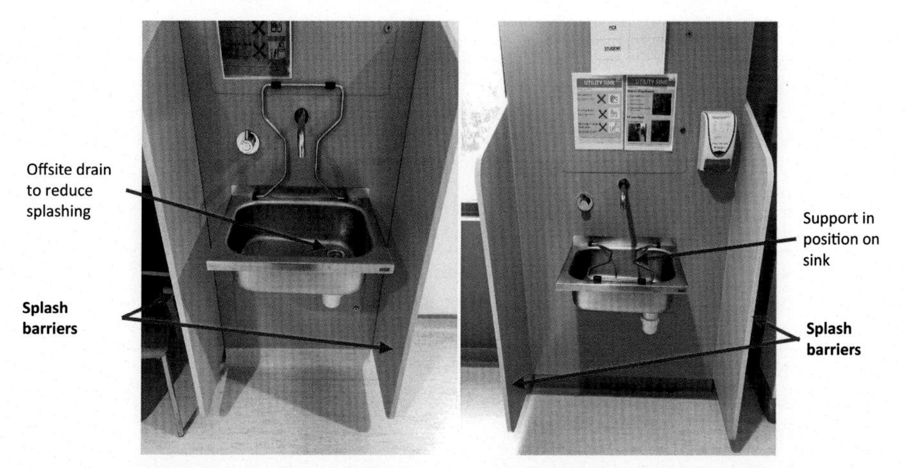

Fig. 19 Use of splash barriers to prevent dispersal of contaminant and support for heavy items on the sink. Good practice where a screen has been used to protect items from splashing in the local vicinity, and a support rail has been designed to prevent items being placed on the bottom of the sink.
Images courtesy of Infection Prevention and Control Department, Norfolk, and Norwich University Hospital NHS Foundation Trust.

14. There should be appropriately designed and demarcated areas for placement of dirty equipment prior to placement into macerator or bedpan washer disinfector (when these are already in use.
15. Closed racks or shelving or storage of clean equipment should be placed in the clean area away from any potential risks of contamination, including splashing.

Operational issues

1. Staff should be trained on correct operation and use of equipment, what issues to look for, and when to report them.
2. There should be a program of routine maintenance.
3. There should be a service level agreement with manufacturer in case of breakdowns, which stipulates the minimum response time.

Considerations for risk assessing your dirty utility rooms

Risk assessment criteria	Yes	No
Do you know where the dirty utility rooms are located in the hospital to ensure that they have all been risk assessed?		
Can dirty utility rooms be accommodated within the wards so that staff do not have to transport contaminated materials unduly long distances?		
Does the hopper/macerator drain have an air gap installed to prevent backflow?		
Do any of your dirty utility rooms have Victorian style hoppers?		
Has an infection control risk assessment been carried out for all dirty utility rooms/facilities?		
Have the appropriate microbiological risk assessments for water supply and usage in the dirty utility room been carried out, e.g., for *Legionella* and *Pseudomonas aeruginosa*?		
Has the scheme of control been agreed by the water safety group?		
Are there service-level agreements in place with the manufacturers or suppliers of equipment to ensure timely repairs and replacement based on clinical needs?		
Has there been appropriate training on the use of all equipment such as macerators/bedpan washers, etc., so that staff are aware of what can and can't be processed?		
Are inappropriate materials placed into the macerator that could lead to blockages?		
Are there documented procedures agreed by the water safety group for predictable untoward events, such as blockages?		
Are your dirty utility rooms used for storage of equipment that should not be there?		

Risk assessment criteria	Yes	No
Has a risk assessment been carried out on the dirty utility room, taking into account the potential for cross-contamination from items brought into and stored within the room (including the potential for cross-contamination within the dirty utility that could be transferred to patient areas such as on wheels of commodes, etc.)?		
Is there a storage area demarcated for dirty equipment prior to placement in the macerator or bedpan washer disinfector when it is already in use?		
Are cleaned and decontaminated items removed and stored in a safe (i.e., away from splashing from sinks, soap dispensers, etc., in the clean area) dedicated area.		
Is there a wash hand station at the entrance/exit door?		
Who is responsible for cleaning the wash hand stations in the dirty utility room?		
Is access restricted to authorized staff only		
Is there clear demarcation of clean and dirty areas and are these adhered to?		
Does the design ensure there is flow from dirty to clean?		
Is it possible to access the dirty utility without touching door handles, etc.?		
Are antisplash screens fitted between the hopper, macerator, and hand wash station to prevent splash contamination?		
Are items placed on top of equipment such as the washer disinfector that could cause cross-contamination?		
Is it clear who is responsible for cleaning and for cover for holidays? sick leave, etc.?		
Is the frequency of cleaning monitored?		
Is the efficacy of cleaning monitored?		
Are audits carried out to assess that the dirty utility room is used and cleaned appropriately?		
Have the staff been trained to use the area and do they understand the risks from cross-contamination, splashing, and not disposing of clinical waste safely?		
Is it clear who is responsible for this area and to whom problems should be reported to (especially when a dirty utility room is used by more than one clinical team)?		

Guidance and further reading for dirty utility rooms

Health Building Note 00-03: Clinical and clinical support spaces (DHSC, 2013).

HTM 04-01 Safe water in healthcare (DHSC, 2017).

HTM 01-01Part D—"Management and decontamination of surgical instruments: washer-disinfectors" (DHSC, 2016).

HBN 00-09 Infection control in the built environment (HBN 00-09). https://www.gov.uk/government/publications/guidance-for-infection-control-in-the-built-environment (Health Building Note 00-09: Infection control in the built environment, 2013).

Water Supply (Water Fittings) Regulations (WRAS, 1999).

The Health and Social Care Act 2008: "Code of Practice on the prevention and control of infections and related guidance" (the HCAI Code of Practice).

References

Breathnach, A.S., Cubbon, M.D., Karunaharan, R.N., Pope, C.F., Planche, T.D., 2012. Multidrug-resistant *Pseudomonas aeruginosa* outbreaks in two hospitals: association with contaminated hospital waste-water systems. J. Hosp. Infect. 82, 19–24. https://doi.org/10.1016/j.jhin.2012.06.007.

DHSC, 2013. Health Building Note 00-03: Clinical and Clinical Support Spaces.

DHSC, 2016. Decontamination of surgical instruments (HTM 01-01). [WWW Document]. GOV.UK. URL: https://www.gov.uk/government/publications/management-and-decontamination-of-surgical-instruments-used-in-acute-care. (Accessed 16 April 2019).

DHSC, 2017. HTM 04-01: Safe Water in Healthcare Premises. https://www.gov.uk/government/publications/hot-and-cold-water-supply-storage-and-distribution-systems-for-healthcare-premises. (Accessed 5 January 2021).

DHSC, 2021. Health Building Note 00-09: Infection Control in the Built Environment. https://www.england.nhs.uk/publication/infection-control-in-the-built-environment-hbn-00-09/.

Fornara, F., Bonaiuto, M., Bonnes, M., 2010. The relationship between the humanization of inpatient areas and the satisfaction and perceived affective qualities of hospital users. In: 3rd HaCIRIC International Conference "Better Healthcare through Better Infrastructure", pp. 19–27.

Anon., 2013. Health Building Note 00-09: Infection control in the built environment. 47.

Healthcare Commission, 2006. Investigation into outbreaks of *Clostridium difficile* at Stoke Mandeville Hospital. Buckinghamshire Hospitals NHS Trust.

Mathers, A.J., Vegesana, K., German Mesner, I., Barry, K.E., Pannone, A., Baumann, J., Crook, D.W., Stoesser, N., Kotay, S., Carroll, J., Sifri, C.D., 2018. Intensive care unit wastewater interventions to prevent transmission of multispecies *Klebsiella pneumoniae* carbapenemase–producing organisms. Clin. Infect. Dis. 67, 171–178. https://doi.org/10.1093/cid/ciy052.

Smellie, J., 1929. Section of pediatrics. Proc. Natl. Soc. Med. XLII, 631–635.

Wales NHS, 2006. Health Technical Memorandum 64; Building Component Series—Sanitary assemblies. http://www.wales.nhs.uk/sites3/documents/254/htm%2064%203rded2006.pdf.

WRAS, 1999. The Water Supply (Water Fittings) Regulations 1999.

Laundry rooms

19

Background

All reusable textile items require cleaning and disinfection via laundry services in healthcare, and laundry installations are typically supplied with mains water that is used both in the cleaning process and to rinse the laundered items after the disinfection stage.

Reusable items will include bed linen: blankets, counterpanes, cot sheets and blankets, duvets, duvet covers, pillowcases, and sheets (woven, knitted, half sheets, draw, and slide sheets). However, reusable linen may be soiled by a range of bodily fluids including human blood, sweat, wound exudate, sputum, saliva, urine, vomit, or feces, all of which may be contaminated by pathogens. Linen contaminated with bacterial, viral, or fungal pathogens including *Staphylococcus aureus*, *Bacillus cereus*, *Clostridium difficile*, *Pseudomonas aeruginosa,* and fungi including *Trichophyton mentagrophytes* (athlete's foot) and Mucorales could then pose a risk to others (Balm et al., 2012; Bloomfield et al., 2011; Michael et al., 2016a,b, 2017a,b; Overcash and Sehulster, 2021; Shunmugarajoo et al., 2020; Sundermann et al., 2021; Tarrant et al., 2018; Tsai et al., 2021).

While rates of incidence are low, healthcare staff need to be aware of the transmission of healthcare-associated infections (HAIs) from reusable healthcare textiles (Overcash and Sehulster, 2021).

The presence of pathogens can effectively be eliminated through standard hospital laundering methods using either thermal (≥65°C) or chemical disinfection (Bockmühl et al., 2021; NHS England, 2016; Owen and Laird, 2020).

- Microorganisms will be physically lifted and removed from the linen by the detergent and rinse water during the washing and rinsing cycle.
- Thermal disinfection will be considered satisfactory when the washer load temperature has reached 65°C and held for a minimum period of 10 min or at a temperature of 71°C for a period of not less than 3 min. Alternative time/temperature relationships may be used as long as the efficacy of the process chosen is equal to or exceeds that of the 65°C or 71°C processes. There is also an obligatory mixing time that needs to be added to ensure heat penetration and assure disinfection. For conventionally designed machines and those with a low degree of loading (<0.056 kg/L), 4 min should be added to these times to allow for adequate mixing. For a heavy degree of loading (i.e., >0.056 kg/L), it is necessary to add 8 min (NHS England, 2016)
- For some heat-labile materials, chemical disinfection is essential. A range of disinfectants can be used; however, hypochlorite should not be used on fabrics treated for fire retardance. It is essential that the entire laundry process is validated (including washing, dilution, and disinfection) to ensure that the disinfection process is capable of passing the specified microbiological tests. (NHS England, 2016)
- All sections of the machines that follow the disinfection process, which do not normally receive disinfection, should receive a disinfection cycle.

Safe Water in Healthcare. https://doi.org/10.1016/B978-0-323-90492-6.00027-6

Microbiological issues related to laundry

The chemical, microbial quality and volumes of the rinse water used after disinfection could have an adverse impact on the quality of the processed linen (Owen and Laird, 2020). The quality of the final rinse water during the disinfection stage is specified in HTM 01-04, and from a bioburden perspective, there should be no pathogens present, and the total viable count should be <100 CFU/mL, i.e., equivalent to that suggested by the Textile Services Association (see below):

Other issues related to laundry and water

Any vent pipes associated with laundry machines processing infectious linen should be vented to a safe point of discharge outside the building and away from any windows or ventilation plant inlets (to avoid contaminated air being reintroduced into the wards).

Due to potential for the effluent to contain microbial pathogens, the laundry machine drains must be plumbed into the waste system to prevent backflow of harmful microorganisms (WRAS, 1999). Backflow is when stagnant or contaminated water gets sucked back into and contaminates the mains water supply. Fluid category 5 is the high risk level as it represents the most serious health hazard because of the presence of pathogens and is particularly relevant to laundry departments. In these circumstances, a Class 5 backflow prevention device, incorporating a Type AB air gap or equivalent, is required (Fig. 1).

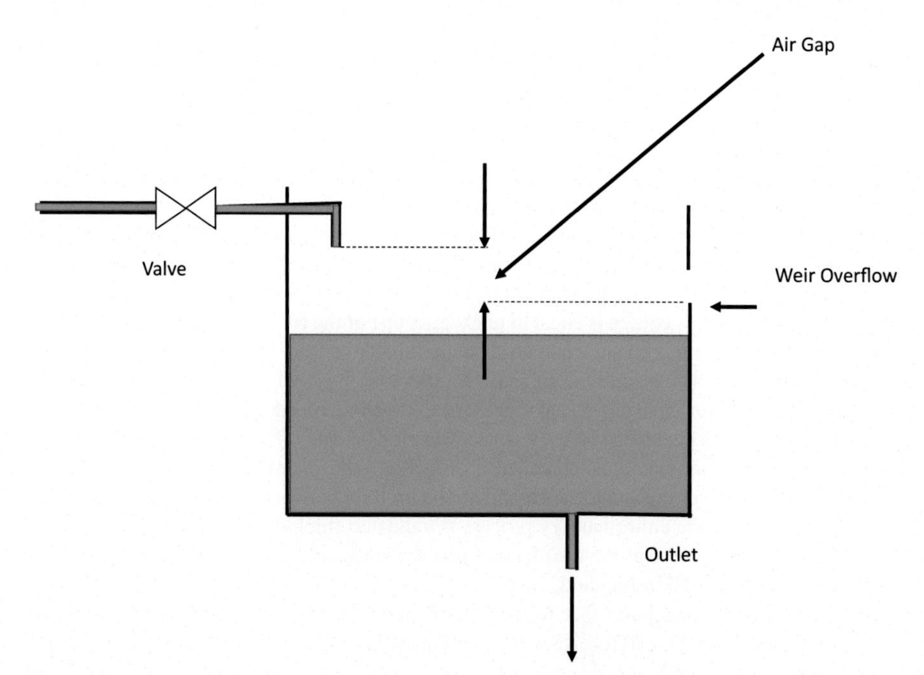

Fig. 1 Demonstration of a Class 5 fluid category risk with a Type AB air gap between the waste pipe and wastewater level to prevent backflow of microbial contamination.

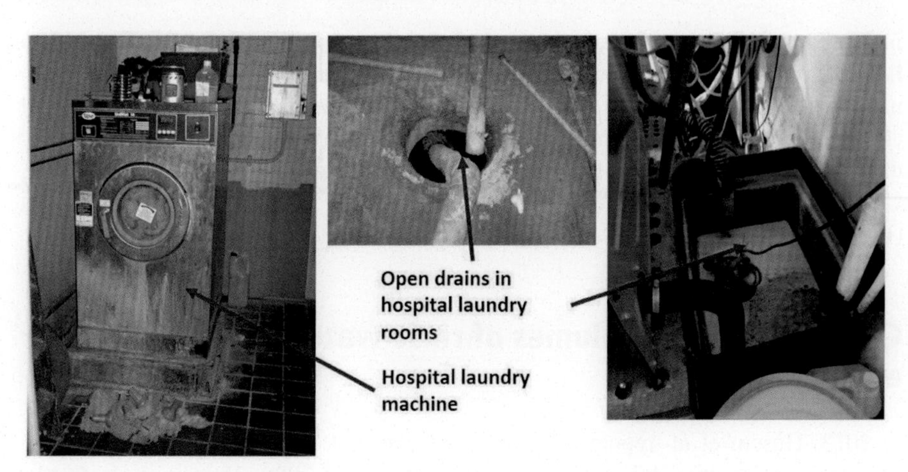

Fig. 2 Laundry machine with open drains in the floor and in a sump.
Photos courtesy of Jackie Hook (Industrial Chemist).

If the laundry machine drains to an open sump or pit (Fig. 2) immediately below the machine drain valve, the sump or pit should be covered and sealed to reduce the risk of bacteria being spread by the aerosol effect when water is pumped from the machine.

When the chemical disinfectant is intended to be discharged to drain, the drainage system should be trapped, sealed, and vented to a safe position.

The drainage system should be checked to ensure that it is not possible for toxic materials to be vented into any other part of the laundry. The maximum permitted concentration and the method of detection and analysis will depend on the chemical being used.

Case study—Multisite *B. cereus* outbreak related to laundry management

B. cereus is a Gram-positive bacillus that forms endospores and is distributed throughout various environments. It rarely causes disease in humans except for immunocompromised individuals in which it can cause sepsis. A number of cases of blood stream infections associated with *B. cereus* were recorded in a range of wards, including an acute intensive care unit and a hematology unit, across two different hospital sites. Laundry from both hospitals was being carried out in the same continuous batch or industrial tunnel washer. A serious of process failures led to a significant backlog of used soiled laundry, which was stored in sealed bags, in the open air during warm summer months. During this time, the number of blood stream infections rose 1 or 2 per months to 16 patients. Processing delays and warm summer temperatures would have led to the growth of microorganism on the laundry materials including beds sheets, baby linen, patient gowns, and theater scrubs all of which were found to be positive for *B. cereus*. The spores were resistant to the thermal controls, and it was found that low-volume laundry processes were not sufficient for these to be diluted out. The outbreak was resolved when the water volume through the tunnel washer was increased, there were improvements to the collection and storage of soiled linen, baby linen was processed in a washer extractor, disposable theater scrubs were used for neonatal theaters, and all linen rooms were emptied, cleaned, and restocked. Additional chemical treatment of the tunnel washer was undertaken using peracetic acid to reduce the microbial counts to acceptable levels. We are grateful to Karren Staniforth, United Kingdom, for allowing us to include this outbreak and the associated images (Figs. 3–5).

Case study—Low volumes of rinse water to improve water efficiency

In 2013, Hosein et al. reported that 65% of neonatal umbilical swabs in one UK hospital were positive for *B. cereus* in the summer of 2009 (Hosein et al., 2013). Environmental samples found that the linen was positive for *B. cereus*. The number of

Fig. 3 Industrial tunnel washer.
Photo courtesy of Karren Staniforth.

Fig. 4 Soiled laundry stacked outside in the warm summer temperatures.
Photo courtesy of Karren Staniforth.

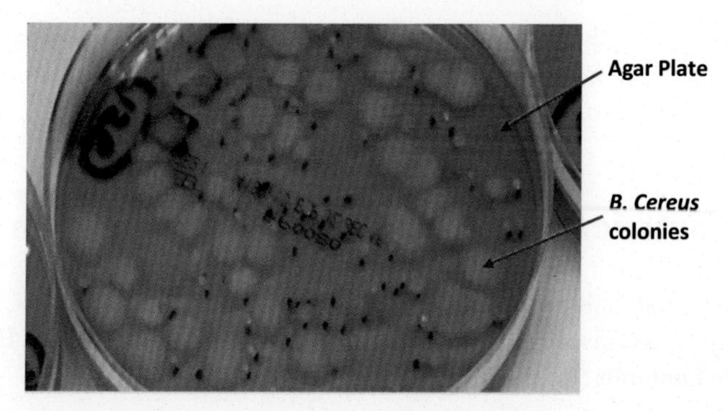

Fig. 5 Agar plate with *Bacillus cereus* colonies—after direct impression with freshly a laundered sheet.
Photo courtesy of Karren Staniforth.

positive umbilical and linen cultures approached zero during the autumn and winter and increased the following summer. It was hypothesized that *B. cereus* was replicating and sporulating on damp and soiled linen at higher summer temperatures and that the low volumes of water (at an attempt to improve water efficiency) flow through washer extractor and was unable to remove *B. cereus* spores from the linen. Washing the linen in the washer extractor at a higher dilution rates than normally used for the continuous tunnel washer coincided with lowering of detectable *B. cereus* in unused washed linen and colonization of babies.

Case study—Impact of unsatisfactory bacterial levels in incoming water supply

A UK hospital that employed a chemical disinfectant for the decontaminating of cleaning cloths and mops experienced issues relating to the Essential Quality Requirement disinfection test between 2017 and 2019. Annual testing in accordance with HTM01-04 for chemical disinfectants is a requirement for all laundries utilizing chemical disinfectants. The basic essential quality requirement test relies on maintaining the sterility of a swatch introduced into the disinfection process to demonstrate nontransference of the bacterial contamination. On all occasions, the sterile swatch retained a low level CFU (<50 CFU per swatch) following the process. The Best Practice test, which evaluates the disinfection performance of the laundry process, introduces known challenged swatches into the load. Heat-resistant bacteria *Enterococcus hirae*, *Enterococcus faecium,* or *Enterococcus faecalis* are the test organisms. At the end of the process, a ≥log 5 reduction should be achieved. While the process satisfied the requirements for ≥log 5 reduction for the best practice microbiological test, the presence of high CFU (>1000 CFU/mL) in the incoming cold supply impacted on the basic essential quality disinfection test. Tests conducted on the water supply revealed an unsatisfactory TVC value and a positive test for *Pseudomonas* spp. This was only resolved when a microbial filtration system was introduced to the incoming water supply, decreasing the bacterial load significantly and below <100 CFU/mL.

Requirements for consideration in laundry of linen

Ensure that the total viable count of the final rinse water is <100 CFU/mL.

The quality of the rinse water should comply with the specifications in HTM 01-04.

Healthcare laundries should consider (Hosein et al., 2013):

- Sampling laundered linen in tunnel washers for *B. cereus* in summertime.
- Setting a concentration of <5 CFU/10 cm^2 *B. cereus* in washed linen samples, assessed by total elution in liquid.
- Using *B. cereus* selective agar impression plates, ready-to-use linen at hospitals should be negative or just have occasional low positivity (<1% of washed linen items could have *B. cereus* and number of colonies per agar plate should be <3).

From a guidance perspective, the HTM 01-04 recommends for those laundries employing chemical disinfectants instead of thermal that an annual disinfection test is conducted (NHS England, 2016). This takes the form of a basic test (Essential Quality Requirements EQR), in which a sterile swatch is introduced into the disinfection process. The swatch should remain sterile following the process; however, a limit of <10 CFU is more practical. The second and probably more important test is the Best Quality Disinfection Test. This uses challenged swatches in the load and requires a ≥log 5 reduction to qualify the process and may also give an indication of the cleanliness of the equipment. Poor incoming water quality can impact the EQR test significantly demonstrating that filtration of the incoming water supply is required. However, while this can impact the end result for thermal disinfection process, there is no requirement under HTM 01-04 for this to take place for thermal disinfection.

The NHS is committed to tackling climate change by reducing emissions to "net zero" to be the world's first "net zero" national health service (NHS England, 2020). This of course must include opportunities for emissions reductions in the secondary and primary care estates respectively, with significant opportunities seen in energy use in buildings, waste and water, and new sources of heating and power generation. There are clearly a wide range of interventions that need to be focused on air conditioning and cooling, building fabric, space heating, ventilation, and water usage that could all be applied to laundry and could be rolled out throughout the secondary care estate over the next 5–10 years, saving some £250 million per year. Laundry and linen services are considered to account for 13% of a hospital's carbon footprint, and some trusts have invested for the future to reduce carbon emissions by at least 18% and source the majority of our water from local boreholes (NHW WSH, 2021).

Risk assessing microbial contamination in hospital laundry units

Risk assessment criteria	Yes	No
Has a legionella risk assessment been carried out for the laundry department?		
Is there a surveillance system in place that would indicate microbial infections related to contaminated laundry?		
Are the laundry machines on the asset register?		
Is an annual disinfection test carried out?		
Where necessary is any of the laundry monitored for microbial contamination?		
Are there proper and appropriate storage facilities for laundry at all times and particularly during busy periods?		
Are the storage periods of the linen monitored during the summer months when the weather is warmer?		
Are the drains sealed?		
Are there Class 5 fluid category Type AB air gaps present between the waste pipe and wastewater level to prevent backflow of microbial contamination?		
Are all vent pipes vented to a safe point of discharge to prevent exposure to toxic materials?		

Guidance relevant to laundry of linen

HTM 01-04 Decontamination of linen for health and social care (HTM 01-04) (NHS England, 2016).

HTM 04-01 Safewater in Healthcare (DHSC, 2016, pp. 04-01).

References

Balm, M.N.D., Jureen, R., Teo, C., Yeoh, A.E.J., Lin, R.T.P., Dancer, S.J., Fisher, D.A., 2012. Hot and steamy: outbreak of *Bacillus cereus* in Singapore associated with construction work and laundry practices. J. Hosp. Infect. 81, 224–230. https://doi.org/10.1016/j.jhin.2012.04.022.

Bloomfield, S., Exner, M., Carlo, S., Scott, E., 2011. The infection risks associated with clothing and household linens in home and everyday life settings, and the role of laundry. International Scientific Forum on Home Hygiene, pp. 1–43.

Bockmühl, D.P., Schages, J., Rehberg, L., 2021. Laundry and textile hygiene in healthcare and beyond. Microb. Cell 6, 299–306. https://doi.org/10.15698/mic2019.07.682.

DHSC, 2016. HTM 04-01: Safe Water in Healthcare Premises.

Hosein, I.K., Hoffman, P.N., Ellam, S., Asseez, T.-M., Fakokunde, A., Silles, J., Devereux, E., Kaur, D., Bosanquet, J., 2013. Summertime *Bacillus cereus* colonization of hospital newborns traced to contaminated, laundered linen. J. Hosp. Infect. 85, 149–154. https://doi.org/10.1016/j.jhin.2013.06.001.

Michael, K.E., No, D., Dankoff, J., Lee, K., Lara-Crawford, E., Roberts, M.C., 2016a. *Clostridium difficile* environmental contamination within a clinical laundry facility in the USA. FEMS Microbiol. Lett. 363, fnw236. https://doi.org/10.1093/femsle/fnw236.

Michael, K.E., No, D., Roberts, M.C., 2016b. Methicillin-resistant *Staphylococcus aureus* isolates from surfaces and personnel at a hospital laundry facility. J. Appl. Microbiol. 121, 846–854. https://doi.org/10.1111/jam.13202.

Michael, K.E., No, D., Daniell, W.E., Seixas, N.S., Roberts, M.C., 2017a. Assessment of environmental contamination with pathogenic bacteria at a hospital laundry facility. Ann. Work Expo. Health 61, 1087–1096. https://doi.org/10.1093/annweh/wxx082.

Michael, K.E., No, D., Roberts, M.C., 2017b. vanA-positive multi-drug-resistant *Enterococcus* spp. isolated from surfaces of a US hospital laundry facility. J. Hosp. Infect. 95, 218–223. https://doi.org/10.1016/j.jhin.2016.10.017.

NHS England, 2016. HTM 01-04 Decontamination of Linen for Health and Social Care.

NHS England, 2020. Delivering a 'Net Zero' National Health Service. https://www.england.nhs.uk/greenernhs/wp-content/uploads/sites/51/2020/10/delivering-a-net-zero-national-health-service.pdf.

NHW WSH, 2021. Multimillion Pound Refresh of St Richard's Hospital Laundry. Western Sussex Hospitals.

Overcash, M.R., Sehulster, L.M., 2021. Estimated incidence rate of healthcare-associated infections (HAIs) linked to laundered reusable healthcare textiles (HCTs) in the United States and United Kingdom over a 50-year period: do the data support the efficacy of approved laundry practices? Infect. Control Hosp. Epidemiol. 1–2. https://doi.org/10.1017/ice.2021.274.

Owen, L., Laird, K., 2020. The role of textiles as fomites in the healthcare environment: a review of the infection control risk. PeerJ 8, e9790. https://doi.org/10.7717/peerj.9790.

Shunmugarajoo, A., Mustakim, S., Sfkhan, S., Azmel, A., Rosli, R., Govindasamy, T.R., Chellapan, K., Mokhtar, A.S., Noor, N.H.M., Rmnn, N., Ang, E.L., 2020. Outbreak of *Bacillus bacteremia* from contaminated hospital linens. Int. J. Infect. Dis. 101, 315. https://doi.org/10.1016/j.ijid.2020.09.822.

Sundermann, A.J., Clancy, C.J., Pasculle, A.W., Liu, G., Cheng, S., Cumbie, R.B., Driscoll, E., Ayres, A., Donahue, L., Buck, M., Streifel, A., Muto, C.A., Nguyen, M.H., 2021. Remediation of Mucorales-contaminated Healthcare Linens at a Laundry Facility Following an Investigation of a Case Cluster of Hospital-acquired Mucormycosis. Clinical Infectious Diseases ciab638., https://doi.org/10.1093/cid/ciab638.

Tarrant, J., Jenkins, R.O., Laird, K.T., 2018. From ward to washer: The survival of *Clostridium difficile* spores on hospital bed sheets through a commercial UK NHS healthcare laundry process. Infect. Control Hosp. Epidemiol. 39, 1406–1411. https://doi.org/10.1017/ice.2018.255.

Tsai, A.-L., Hsieh, Y.-C., Chen, C.-J., Huang, K.-Y., Chiu, C.-H., Kuo, C.-Y., Lin, T.-Y., Lai, M.-Y., Chiang, M.-C., Huang, Y.-C., 2021. Investigation of a cluster of *Bacillus cereus* bacteremia in neonatal care units. J. Microbiol. Immunol. Infect. https://doi.org/10.1016/j.jmii.2021.07.008.

WRAS, 1999. The Water Supply (Water Fittings) Regulations 1999. https://www.legislation.gov.uk/uksi/1999/1148/contents/made.

Endoscope washer disinfectors

20

Introduction

Several nosocomial outbreaks have been associated with the contamination of flexible endoscopes caused by a failure in the reprocessing and decontamination procedures and the water used in these processes. The most frequently reported failures contributing to incidents are inappropriate disinfection, inadequate cleaning and rinsing, and faulty or contaminated endoscope washer disinfector (EWD) and contaminated rinse water.

Microbial contamination of endoscopes is usually associated with biofilm formation in the lumen channels associated with either endogenous flora (patients' own flora) or exogenous flora related to previous patient or contaminated reprocessing equipment, e.g., rinse water (Cottarelli et al., 2020; Kovaleva, 2017). Following disinfection of lumened instruments, it is important that the endoscope is not microbiologically recontaminated by waterborne microorganisms and that residues are flushed with rinse water of known microbiological (microbial free) and chemical quality (DHSC, 2016). Despite this, final rinse water has been shown to be contaminated by a wide range of microorganisms including *Pseudomonas* spp., *Legionella* spp., environmental mycobacteria, *Serratia* spp., *Legionella* spp., *Aspergillus* as well as endotoxins (Bennett et al., 1994; Khalsa et al., 2014; Marek et al., 2014; Vandenbroucke-Grauls et al., 1993; Walker et al., 2022; Willis, 2006)

More endoscopes have been associated with hospital-acquired infections than any other device (Rutala and Weber, 2008) including upper GI endoscopes (Kovaleva, 2017), ERCP (Cryan et al., 1984), colonoscopes (Muscarella, 2010), bronchoscopes (Schuetz et al., 2009), and laryngoscopes (Cottarelli et al., 2020).

What are endoscopes and how are they cleaned and disinfected?

Flexible endoscopes and their accessories are classified as medical devices under the Medical Devices Directive (93/42/EEC7 and 2007/47/EEC8). The Medical Device Regulations 20029 (MDR) implemented the EC Medical Devices Directives into UK law.

Flexible endoscopes are used for invasive clinical procedures and have channels and lumens that are required to be manually cleaned and decontaminated in an endoscope washer disinfector after each and every use (Fig. 1).

Effective cleaning and disinfection of endoscopes between clinical procedures are required to ensure that cross-contamination from one patient to another is prevented. Cleaning involves several stages including a bedside clean immediately after use

Safe Water in Healthcare. https://doi.org/10.1016/B978-0-323-90492-6.00012-4

Fig. 1 Endoscope being placed in a washer disinfector.

preferably by the person using the endoscope, manual cleaning stage using a brush, to remove all organic matter and other debris, followed by processing in an EWD where all lumens and channels are cleaned and disinfected and finally rinsed with water that has less than 10 CFU/100 mL.

Flexible endoscopes should undergo the following cleaning and disinfection regimens. Follow the manufacturers' instructions for reprocessing at all times.

Bedside cleaning—this preclean is carried out at the patient bedside before the endoscope is disconnected from its light source/video system to remove most of the protein immediately after use and ensure that the channels are not blocked. The working channel should be sucked through with water and low foaming neutral detergent and the insertion tube wiped down using a single-use moist lint-free wipe/sponge. Excess fluid should then be expelled from the channels using air prior to transporting the endoscope for manual cleaning.

During transportation, all endoscopes should be placed in a lined tray for protection and appropriately covered to show that it is contaminated to prevent accidental reuse. Within a suitably secure transportation system if being reprocessed off site, it should then be transported to the decontamination suite according to the unit policies and procedures.

Water bottles should be changed after every endoscopy session (i.e., 3 h) to prevent microbial contamination. They should be detached, emptied, and cleaned (as per manufacturer's instructions), then sent for steam sterilization. The bottles should only be filled with fresh sterile water immediately prior to use. The sterilization of the water bottles and the sterile water used should be tracked for purposes of traceability. Single-use water bottles and connectors should be used.

Manual cleaning—This is undertaken in a sink in the dirty section of the decontamination area and includes brushing with a specific single-use brush that is sized appropriately for each channel, exposing all external and accessible internal components to a low foaming neutral detergent and rinsing with water as recommended by the endoscope manufacturer. Detergent and rinse water must be replaced for each endoscope for the manual process. Two separate sinks (one for clean, one for rinse) are required and should not be used for any other purpose (Fig. 2).

Sinks should be height adjustable and should be cleaned thoroughly between each manual clean. The environment surrounding the sinks should be clear of clutter with no items stored nearby that could be contaminated by splashes and droplets (see shelves in Fig. 2). The sinks should have a designated volume indicator. Stickers or labels in sinks may be difficult to clean. Monitoring of the water supply should be in place to ensure that the cold water supply to that outlet is below 20°C within 2 min of running the cold tap, and that the hot water reaches 50°C within 30 s and 55°C within 1 min of that outlet being opened. If detachable parts are to be reused (e.g., air/water and suction valves/pistons, i.e., not single use), then they should be reprocessed together with the corresponding endoscope as a unique set in order to allow traceability.

To support cleaning, reprocessing needs to take place within 3 h of patient use to reduce the potential for the lumen to dry, which would make removal of protein material more difficult (DHSC, 2016).

Automated cleaning—this is carried out in the EWD and includes water sprays and pulsed liquid flowing down the lumens. The EWD must be used in accordance with the manufacturer's instructions for use and must be capable of irrigating all the channels of the endoscopes, including auxiliary channels. Only connector systems applicable for specific endoscopes should be used to ensure that all lumens and channels are cleaned accordingly (Figs. 3 and 4). Connectors should be stored vertically to precent water stagnation.

Automated disinfection—this process should always be verified by a validation protocol in line with the basic testing requirements identified in HTM 01-06 Part D—"Validation and verification" and BS EN ISO 15883 Parts 1 and 4 and is followed by rinsing and drying the endoscope (BS EN ISO, 2014; DHSC, 2016).

The process of decontamination should be concluded with further rinsing with purified water of a defined microbiological quality, e.g., RO water (Fig. 5).

Fig. 2 Manual washing sinks.

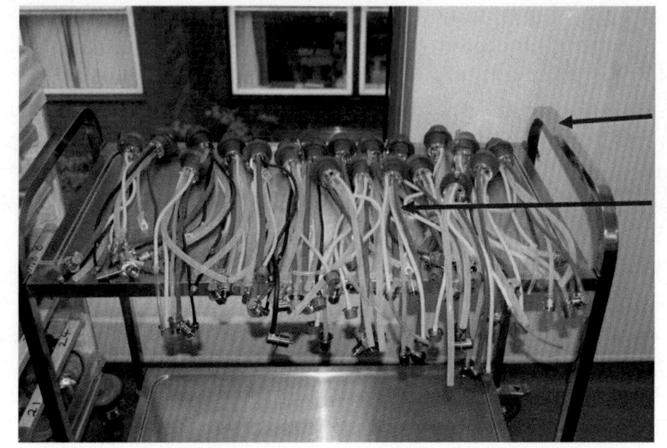

Fig. 3 Connectors for each endoscope.

Fig. 4 Endoscope connected and loaded into the endoscope washer disinfector tray to enable all lumens to be cleaned.

What are the sources of contamination of endoscopes?

It is critical that the quality of water used in reprocessing is controlled and free of microorganisms. However, there are a number of steps where the water used in the reprocessing of endoscopes can lead to contamination of the endoscope and/or surrounding area. This can occur during the bedside cleaning, manual cleaning, and during the automated cleaning and disinfection stages I.

During manual cleaning, it is important that two sinks are used, one for cleaning and one for rinsing and to ensure that splashing of water is kept to a minimum to reduce the dispersal of droplets and aerosols across the decontamination room and to reduce exposure of the staff in that area (DHSC, 2016; HSE, 2014).

The EWD must be suitable and appropriate for the types of endoscopes to ensure that all lumens and channels are cleaned appropriately. If specific lumens or channels

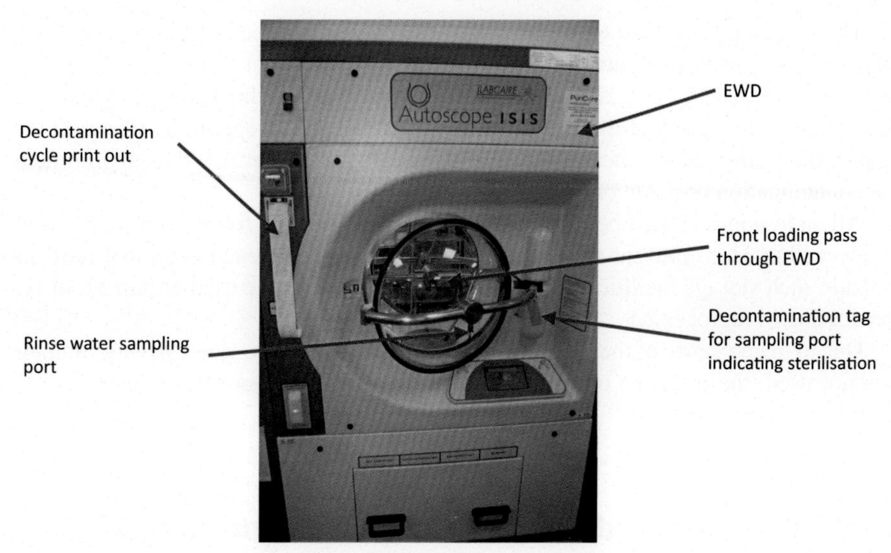

Fig. 5 Washer disinfector.

are not connected to the EWD, then they will not be cleaned. The quality of the final rinse water is critically important as regardless of how much care and quality control have been implemented and followed previously. Water used in an EWD should be free from particulate and chemical contamination and microorganisms. The final rinse water quality is often achieved either by using bacteria retention filters or by other purification systems (e.g., reverse osmosis). A water-softening and/or treatment system may be needed to prevent contamination of the EWD with limescale, biofilms, and microorganisms. It is recommended that rinse water is not reused. If the final rinse water is already contaminated by microbial pathogens, then the endoscope lumens will be contaminated. Reverse osmosis systems are used to purify the water to achieve a quality of water that is acceptable from a chemical and microbiological aspect and further details are available in Chapter 24.

What are the transmission routes with endoscopes?

As flexible endoscopes may not be fully dry on completion of the decontamination process, then there is a possibility that droplets of any remaining water may enable bacterial growth and biofilm to develop. This can lead to the presence of water borne (from contaminated rinse water) or patient-derived microorganisms (due to an incomplete or failure of the decontamination process) to the extent that cross-infections could occur when the endoscope is used on the next patient. As a consequence following cleaning and disinfection, endoscopes are then placed in controlled environmental storage cabinets, which assist in the drying and control of recontamination (from the air) (DHSC, 2016).

These cabinets are used to dry and store flexible endoscopes after processing and before use to prevent recontamination. The cabinet should be configured in line with BS EN 16442. (BSI, 2015) and should be designed to deliver high-efficiency particulate air (HEPA) to each of the individual lumens in all endoscopes to be stored, at the appropriate temperature (including ambient) and flow rate to ensure that environmental contamination does not occur.

All endoscopes should be reprocessed as soon as possible following use, but routinely within 3h to prevent recontamination. Any processed endoscope that remains outside such storage facilities or is unwrapped will need to be used within 3h of reprocessing or reprocessed

Due to the presence of the long narrow lumens and channels, where drying cabinets are not used, the presence of moisture and any viable microorganisms may lead to recontamination.

Outbreaks and incidents associated with endoscopes and endoscope washer disinfectors

Zhang et al. observed an increase in the number of positive *Pseudomonas aeruginosa* bronchoalveolar lavage fluid (BALF) samples, and *P. aeruginosa* was also found during routine microbiologic monitoring of bronchoscopes in this period (Zhang et al., 2020). Bronchoscopes were performed primarily on outpatients but also on patients on the intensive care units. Following a review of patients' samples, the number of positives were found to increase from a baseline of 23%–50% for BALF samples. Environmental samples were taken from bronchoscopes, rinse waters, swabs of connecting tube to the bronchoscope, sinks, and associated tap outlets. *P. aeruginosa* was cultured from filtered final rinsing water, from swabs of connecting tube to bronchoscope, and from swabs of drainpipes. A novel multilocus sequence type (ST) of *P. aeruginosa* was identified and cultured from bronchoscopes and the connecting tube in the manual reprocessing cleaning equipment. One strain from a patient was indistinguishable from the clones obtained from the bronchoscope and connecting tubing as revealed by pulsed-field gel electrophoresis. Two strains from two patients from the burns intensive care unit were identical and highly related to two other strains from the burns intensive care unit. The persistence of *P. aeruginosa* in bronchoscopes, connecting tubes, and final rinse water was terminated by replacement of the connecting tube. In conclusion, the authors reported a pseudo-outbreak of *P. aeruginosa* associated with bronchoscope, for which the connecting tube was the hidden reservoir for contaminating the bronchoscopes. This study highlighted that effective measures are needed to process the connecting tubing through the EWD and to control the bacterial load in final rinsing water to protect reusable equipment from contamination in reprocessing and cleaning.

Bennett et al. investigated an endoscopy-related pseudo-infection *Mycobacterium xenopi*. Typically, *M. xenopi* accounts for less than 0.3% of all clinical mycobacterial isolates; however, over a particular period, it accounted for 35% of isolates (Bennett et al., 1994). Hospital, laboratory, and bronchoscopy records were reviewed to determine case characteristics, develop a case series, and calculate procedure-specific *M. xenopi* isolation rates. A case-control study was conducted to elucidate aspects of the bronchoscopy procedure associated with *M. xenopi* isolation. Bronchoscope cleaning procedures were reviewed. Mycobacterial sampling was carried out from hot and cold water tanks in the three buildings making up the hospital complex and from the hot and cold water taps in the room where the bronchoscopes were cleaned. In addition, the bronchoscope cleaning brush, bronchoscope disinfectant solution were sampled and sterile water flushed through bronchoscopes after disinfection and before use. Four isolates were from three patients with disease attributable to *M. xenopi*. Of the other isolates, specimens obtained by bronchoscopy were more likely to yield *M. xenopi* than were specimens obtained by other routes (relative risk, 9.7; 950/0 confidence intervals, 3.2, 29.6). Bronchoscopes were disinfected in a 0.13% glutaraldehyde-phenate and tap water bath and then were rinsed in tap water. Water from the hot water tank supplying this area yielded *M. xenopi* and *Mycobacteria* were cultured from bronchoscopes after disinfection. The investigation indicated that inadequate disinfection of bronchoscopes, rinsing bronchoscopes with contaminated water, and inadequate drying time between bronchoscope cleaning and reuse were possible factors contributing to bronchoscope contamination. The authors concluded that *M. xenopi* in the tap water appears to have contaminated the bronchoscopes during cleaning and advised that adequate disinfection of contaminated bronchoscopes and careful collection of specimens to avoid contamination with contaminated water are essential, both for limiting diagnostic confusion caused by mycobacterial pseudoinfections and for reducing risks of disease transmission.

In 2004, a hospital in Northern Ireland announced that the arrangements for cleaning and disinfecting one of their endoscopes were not fully in accordance with necessary standards (Doherty et al., 2004). Therefore, the hospital involved undertook a patient notification exercise in relation to over 400 patients, which involved (i) identifying patients who had undergone an endoscopy using this scope and (ii) contacting these patients to invite them for testing for blood-borne virus infection. The results of the notification exercise have showed no cause for concern. In response to this incident, the Department of Health, Social Services and Public Safety (DHSSPS, Northern Ireland) initiated a formal audit involving a detailed observational assessment of all endoscopes in use at hospitals. The audit identified a number of issues in relation to the decontamination of endoscopes (16 in total, including gastroscopes, duodenoscopes, and colonoscopes) at four hospitals. The specific issues in relation to the cleaning and disinfection of these endoscopes fall were that (1) in a small number of endoscopes, one narrow channel on the endoscope was not fully cleaned or disinfected despite going through the normal cleaning and disinfection process; and (2) in a second group, all the channels in the endoscope had been fully cleaned, but one channel may not have been disinfected despite going through the normal cleaning and disinfection process. Following this response to this incident, the DHSSPS convened a regional team to advise on the risk assessment for the transmission of blood-borne viruses associated with the endoscopy cleaning and the disinfection issues identified. To reassure the public, the local regional team concluded that all patients who were examined with endoscopes from the second group should be contacted, made aware of the situation, and offered reassurance and advice. In addition, a regional 24-h helpline was established and was supported by local helplines setup in each of the four hospitals. Over 1700 patients were included in a patient notification exercise; this includes patients involved in the notification exercise at the first hospital. As a consequence the NHS (England), the national assembly of Wales and the Scottish Executive Health Department all asked for an urgent review of practice. This publication concluded that while the risk of acquiring any form of infection from an endoscope is very low, evidence suggests that the main reasons for transmission appear to be improper cleaning and disinfection procedures; the contamination of endoscopes by automatic washers/reprocessors; and an inability to decontaminate endoscopes, despite the use of standard disinfection techniques, because of their complex channel and valve systems.

Monitoring water associated with the decontamination suite

Water quality management can be difficult to understand and control for decontamination processes as there are many areas that can cause problems. Many factors contribute to poor water quality results, and users need to be aware of the water supply and distribution system and how the quality is managed at the point of use. Problems can manifest themselves within the EWDs or within the water treatment equipment in the vicinity of the decontamination suite as opposed to the actual incoming water supply. Regular monitoring is of prime importance so that users can see if standards are being maintained at a level that is safe for patients (Walker et al., 2022).

The EWD must be subjected to weekly cleaning verification tests using a validated process challenge device and weekly testing for the total viable count of the final rinse water (BS EN ISO, 2014; ISO, 2016; DHSC, 2016). Sampling water from the EWD must be done aseptically such that there is no contamination of the sample and staff undertaking sample should be trained. This will mean training and following protocols to ensure the sampling port or bottles for water collection are not contaminated by personnel undertaking this task (Willis, 2006).

RO systems need to be tested weekly, serviced and maintained to ensure that the final rinse water is microbially safe to use (Figs. 6 and 7).

Microbial retention filters should be checked and changed in accordance with the manufacturer's instructions or more often if the water quality is poor (as suggested by frequent clogging of filters). Hard water can cause a deposit of limescale on internal pipe work. Advice may need to be taken from 0 specialists in water treatment and from the local Authorizing Engineer (Decontamination).

The final rinse water should contain <10 CFU/100 mL (Richards et al., 2002) (Table 1). Identification, e.g., using MALDI-TOF will be useful for the risk assessment, particularly if unusual strains, which may be clinically relevant, are recovered. However, users should be aware that even MALDI-TOF may not identify environmental waterborne organisms if they are not already in the database or if the colonies

Fig. 6 Sampling port for microbiological assessment of the endoscope washer disinfector rinse water.

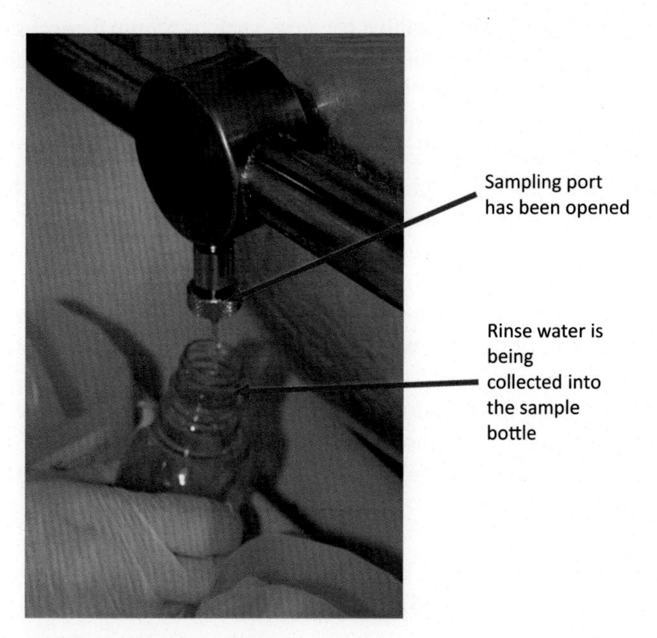

Fig. 7 Aseptic sampling of the rinse water.

are not pure. There should be quarterly testing for atypical mycobacteria and *P. aeruginosa*, with culture plates incubated at 30°C and 37°C for atypical mycobacteria. However, in units carrying out clinical procedures in augmented care, more frequent (e.g., weekly) testing for atypical mycobacteria and *P. aeruginosa* may be prudent. Annual testing for endotoxin has been suggested, but there is no real evidence to support this additional step in nonsterile endoscopy practice, and it is no longer included in HTM 01-06.

Table 1 Interpretation of the results based on satisfactory and unsatisfactory results relevant to risk (Walker et al., 2022).

Aerobic colony count in 100 mL	Interpretation	Action
<1 CFU/100 mL	Satisfactory (green)	No action required
1–9 CFU/100 mL repeatedly	Acceptable (yellow)	Indicates bacterial number are under reasonable level of control, no action required
10–100 CFU/100 mL	Unsatisfactory (orange)	Risk assessment required to investigate potential problems. Superchlorinate or repeat EWD self-disinfect
>100 CFU/100 mL or >0 CFU/100 microorganisms of significance	Unacceptable (red)	Risk assessment required, consider taking EWD out of service until water quality improved

The actions taken for Gram-positive bacteria, such as *Staphylococcus epidermidis* (skin flora possibly indicating microbial contamination during sampling handling), will be different to those taken for Gram-negative bacteria such as *P. aeruginosa*.

So what should the reader be doing if the water results are above this? Taking out of action, seeking advice from IPC teams, etc.

Some patients, for example, those who are immunosuppressed, may be more susceptible to postendoscopic infections, and some endoscopes may present a higher risk than others, e.g., those endoscopes used for invasive procedures such as ERCP or endoscopies that breach the mucous membranes.

Endoscopes used in invasive procedures require different management where microbial quality of the final rinse water has been inadequate (NHS Scotland, 2019). Upon receiving either unsatisfactory or unacceptable results (10–100 CFU/100 mL) for high-risk endoscopes, the endoscope washer disinfector may not be appropriate for reprocessing of high-risk endoscopes but may still be appropriate for low-risk endoscopes during the time when the corrective actions are undertaken (NHS Scotland, 2019; Walker et al., 2022).

Recommendations for rinse waters

Ensure that the water quality of the water supply to the endoscopy decontamination unit is appropriate.

Determine whether the existing hot and cold water system in the hospital is capable of supplying the new unit, without starving any other department in the hospital of their supplies (this could have legionellae implications).

Assess the type of water treatment plant required to provide the final rinse water.

Design the decontamination unit such that the water treatment plant, e.g., RO system and associated equipment is located within or adjacent to the unit.

Sinks used for manual cleaning should be supplied with water directly from the main supply.

Ensure the treated water is routinely monitored including physical, chemical, and microbiological parameters.

Water used in an EWD should be free from particulate and chemical contamination and microorganisms.

The endoscope washer disinfector has the capability of flushing the elevator wire channel with detergent, disinfectant, and rinse water.

The decontamination of endoscopy equipment is a specialized procedure and should only be carried out by personnel who have been trained, are competent and experienced, and who understand the principles involved.

Trend analysis of the results assists in the management of the process to identify situation where control of the water quality if failing.

Consider the requirement for more frequent testing for *P. aeruginosa,* e.g., weekly rather than quarterly as this will provide clinically relevant information.

The use of flow charts or algorithm is recommended.

When adverse microbiological results have been obtained, e.g., where microbiological counts are significantly greater than normal and *P. aeruginosa* has been identified, then a microbiologist should decide on the corrective actions, e.g., taking the EWD out of use and starting a look back exercise.

Considerations for risk assessing endoscope washer disinfectors and rinse waters

Risk assessment criteria	Yes	No
Are robust governance criteria in place for the endoscopy decontamination unit?		
Are all the endoscopy washer disinfectors included in the asset register?		
Is there a decontamination risk group, an endoscopy users group, and or do you have representation on the water safety group?		
Has the water treatment plant for the endoscopy unit been designed appropriately and is it fit for purpose?		
Has each location where endoscopes are used and decontaminated in the hospital been identified?		
Do all employees handling and using endoscopes undergo rigorous training and assessment with validation and annual revalidation to ensure they are competent?		
Are the protocols and facilities for manual cleaning appropriate and risks assessments in place to safeguard staff from waterborne infections		
Has the temperature of the hot and cold water from the sink outlets been assessed?		
Is each EWD capable of connecting to each and every channel of each endoscope?		

Risk assessment criteria	Yes	No
Is reprocessing being carried out within 3 h, particularly for endoscopes that are used clinically off site or out of hours		
If reprocessing is delayed beyond 3 h, what practices are employed to keep endoscopes moist during that time?		
Are all endoscopes cleaned and flushed immediately after use, using clean safe water by the person using the scope and before it can be placed in the transport container?		
Is the rinse water from the EWD achieving both the microbiological and chemical requirements as set out in standards and guidelines?		
Is the rinse water free from microbial pathogens including *Pseudomonas aeruginosa*?		
Are the controlled environment storage cabinets within specification providing the appropriate safe and clean environment for the endoscopes?		

Regulations and guidance relevant to rinse waters and EWDs

HTM 0106 Management and decontamination of flexible endoscopes (HTM 01-06) (DHSC, 2016).

BS EN ISO 15883 Parts 1 and 4 (BS EN ISO, 2014).

ISO 13485 Medical devices—Quality management systems—Requirements for regulatory purposes (ISO, 2016).

ISO 14971 Medical devices. Application of risk management to medical devices (ISO, 2016).

Health and Social Care Act (2015).

HTM 01-01 (2016) Decontamination of surgical instruments (DHSC, 2016).

References

Bennett, S.N., Peterson, D.E., Johnson, D.R., Hall, W.N., Robinson-Dunn, B., Dietrich, S., 1994. Bronchoscopy-associated *Mycobacterium xenopi* pseudoinfections. Am. J. Respir. Crit. Care Med. 150, 245–250. https://doi.org/10.1164/ajrccm.150.1.8025757.

BS EN ISO, 2014. 15883—Washer-disinfectors. BSI British Standards., https://doi.org/10.3403/BSENISO15883.

BSI, 2015. Controlled Environment Storage Cabinet for Processed Thermolabile Endoscopes., https://doi.org/10.3403/30248649.

Cottarelli, A., De Giusti, M., Solimini, A.G., Venuto, G., Palazzo, C., Del Cimmuto, A., Osborn, J., Marinelli, L., 2020. Microbiological surveillance of endoscopes and implications for current reprocessing procedures adopted by an Italian teaching hospital. Ann. Ig. 32, 166–177. https://doi.org/10.7416/ai.2020.2340.

Cryan, E., Falkiner, F.R., Mulvihill, T.E., Keane, C.T., Keeling, P.W., 1984. *Pseudomonas aeruginosa* cross-infection following endoscopic retrograde cholangiopancreatography. J. Hosp. Infect. 5. https://doi.org/10.1016/0195-6701(84)90004-5.

DHSC, 2016. Management and Decontamination of Flexible Endoscopes (HTM 01-06). GOV.UK.

Doherty, L., Mitchell, E., Smyth, B., 2004. Investigation of the decontamination arrangements for endoscopes in Northern Ireland. Weekly Releases (1997–2007) 8, 2499. https://doi.org/10.2807/esw.08.27.02499-en.

HSE, 2014. HSG 274 Legionnaires' Disease—Technical Guidance Part 2: The Control of Legionella Bacteria in Hot and Cold Water Systems Technical Guidance. http://www.hse.gov.uk/pubns/books/hsg274.htm. (Accessed 5 January 2021).

ISO, 2016. ISO 13485:2016. Medical devices—Quality management systems—Requirements for regulatory purposes. https://www.iso.org/cms/render/live/en/sites/isoorg/contents/data/standard/05/97/59752.html.

Khalsa, K., Smith, A., Morrison, P., Shaw, D., Peat, M., Howard, P., Hamilton, K., Stewart, A., 2014. Contamination of a purified water system by *Aspergillus fumigatus* in a new endoscopy reprocessing unit. Am. J. Infect. Control 42, 1337–1339. https://doi.org/10.1016/j.ajic.2014.08.008.

Kovaleva, J., 2017. Endoscope drying and its pitfalls. J. Hosp. Infect. 97, 319–328. https://doi.org/10.1016/j.jhin.2017.07.012.

Marek, A., Smith, A., Peat, M., Connell, A., Gillespie, I., Morrison, P., Hamilton, A., Shaw, D., Stewart, A., Hamilton, K., Smith, I., Mead, A., Howard, P., Ingle, D., 2014. Endoscopy supply water and final rinse testing: five years of experience. J. Hosp. Infect. 88, 207–212. https://doi.org/10.1016/j.jhin.2014.09.004.

Muscarella, L.F., 2010. The study of a contaminated colonoscope. Clin. Gastroenterol. Hepatol. 8, 577–580.e1. https://doi.org/10.1016/j.cgh.2010.04.025.

NHS Scotland, 2019. Guidance for the interpretation and clinical management of endoscopy final rinse water. V1.0. National Services Scotland.

Richards, J., Spencer, R., Fraise, A., Lee, J., Parnell, P., Cookson, B., Hoffman, P., Phillips, G., Brown, N., 2002. Rinse water for heat labile endoscopy equipment. J. Hosp. Infect. 51, 7–16. https://doi.org/10.1053/jhin.2002.1172.

Rutala, W.A., Weber, D.J., 2008. Guideline for Disinfection and Sterilization in Healthcare Facilities, 2008. CDC. p. 163.

Schuetz, A.N., Hughes, R.L., Howard, R.M., Williams, T.C., Nolte, F.S., Jackson, D., Ribner, B.S., 2009. Pseudo-outbreak of *Legionella pneumophila* serogroup 8 infection associated with a contaminated ice machine in a bronchoscopy suite. Infect. Control Hosp. Epidemiol. 30, 461–466. https://doi.org/10.1086/596613.

Vandenbroucke-Grauls, C.M., Baars, A.C., Visser, M.R., Hulstaert, P.F., Verhoef, J., 1993. An outbreak of *Serratia marcescens* traced to a contaminated bronchoscope. J. Hosp. Infect. 23, 263–270. https://doi.org/10.1016/0195-6701(93)90143-n.

Walker, J.T., Bak, A., Marsden, G., Spencer, W., Griffiths, H., Stanton, G.A., Williams, C., White, L.J., Ross, E., Sjogren, G., Bradley, C.R., Garvey, M., 2022. Final rinse water quality for flexible endoscopy to minimize the risk of post-endoscopic infection. Report from Healthcare Infection Society Working Party. J. Hosp. Infect. 0. https://doi.org/10.1016/j.jhin.2022.02.022.

Willis, C., 2006. Bacteria-free endoscopy rinse water—a realistic aim? Epidemiol. Infect. 134, 279–284. https://doi.org/10.1017/S0950268805005066.

Zhang, Y., Zhou, H., Jiang, Q., Wang, Q., Li, S., Huang, Y., 2020. Bronchoscope-related *Pseudomonas aeruginosa* pseudo-outbreak attributed to contaminated rinse water. Am. J. Infect. Control 48, 26–32. https://doi.org/10.1016/j.ajic.2019.06.013.

Cardiac heater coolers

21

Introduction to cardiac heater coolers

What is a cardiac heater cooler?

Heater cooler devices are used during heart and lung bypasses (cardiothoracic surgeries) to warm or cool the temperature of the patient's blood during cardiac, vascular, or transplant surgery. They are an essential part of the equipment needed for these types of procedures, and the machine uses warm or cool water from tanks to regulate the temperature of patients' blood as it is circulated outside of the patient's body. The water in the device does not come into direct contact with the patient or the patient's blood (Fig. 1).

Background

Not all medical equipment that requires water is permanently plumbed into the water system. Sometimes such equipment is not even considered a microbial risk. In the case of heater coolers, these units contain several water tanks for the basic operation of the equipment. Prior to 2012, no one had a made link between the heater coolers and microbial infections occurring in patients having undergone major heart surgery. Invasive *Mycobacterium chimaera* infections were first diagnosed in 2012 in two heart surgery patients on extracorporeal circulation, and the groundbreaking publication from Sax et al., provided the microbiological and epidemiological evidence for the airborne transmission of *M. chimaera* from contaminated heater cooler unit water tanks to patients during open-heart surgery (Sax et al., 2015).

The Swiss investigation implicated the Sorin (now LivaNova) 3T heater cooler unit (HCU) of the cardiopulmonary bypass equipment, with the transmission of bacteria to the surgical site by aerosolization of contaminated water from within the unit. The LivaNova device was widely used both in the United Kingdom and internationally.

This was not an isolated infection occurring in one or two hospitals but was in fact a worldwide outbreak of 120 cases where airborne bioaerosols generated by the 3T heater-cooler unit (HCU) used during cardiac bypass surgical procedures infected the patients while they were going through lengthy operations (Fig. 2) (Ghodousi et al., 2020; Schreiber et al., 2021). These infections were difficult to diagnose because of the long latency period (up to several years), with no specific symptoms and a highly specialized microbiological diagnosis with a high mortality rate of around 50%. Heater coolers while being complex machines could at the same time be considered as a small water system with tanks, tubing, moving water, and a mechanism of releasing aerosols, albeit in this case a problem of sealing the tanks and tubing (Walker et al., 2017).

Safe Water in Healthcare. https://doi.org/10.1016/B978-0-323-90492-6.00025-2

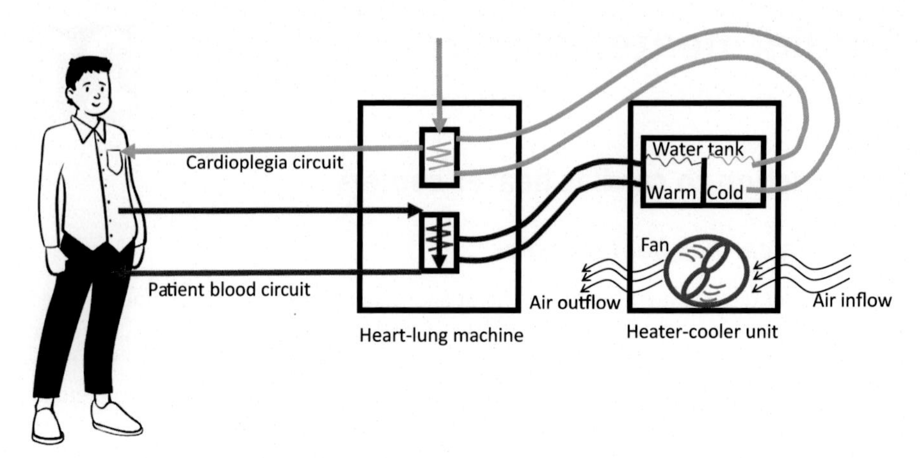

Fig. 1 Schematic of a heater cooler in operation.

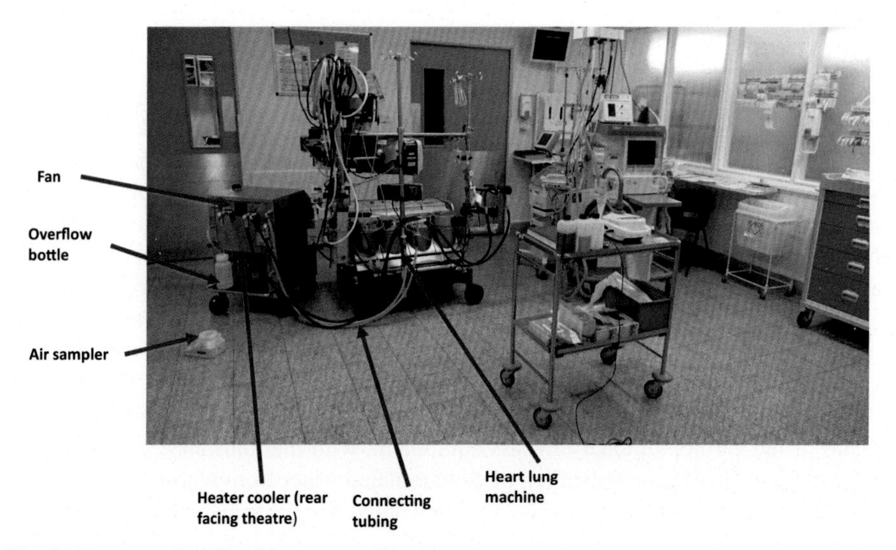

Fig. 2 Heater cooler placed in an operating theatre. (Image supplied by UKHSA.)

Investigations at PHE involved environmental tests of heater coolers both in hospitals and laboratory investigations where a number of heater coolers were dismantled and tested for dispersal of microorganisms (Chand, 2017; Walker et al., 2017).

Investigators found that heater coolers in use in hospitals and those that were newly produced at the manufacturing site were both contaminated with the same *M. chimaera*. Upon further examination using whole genome sequencing, the data suggested that the strains originated from the manufacturers site and hence indicated a point source outbreak where the units had been contaminated during production (Sommerstein et al., 2017).

Manufacturers have very strict decontamination policies, and the onus is on hospitals to ensure that the heater cooler disinfection polices are adhered to and that on site sampling is undertaken to ensure that vulnerable patients do not succumb to infections. However, decontamination was very complex, and even when trusts were carrying out the manufacturer's instructions to the best of their ability, they were unable to control the presence of microorganisms in the unit without replacing parts of the internal tubing, which was covered in biofilm (Garvey et al., 2016).

The very nature of heater coolers having to indirectly warm and cool the patient's blood requires the presence of water in tanks inside the unit. When the outside metal coverings are removed from the HCU, one can see the top of the two or three interconnected cold and warm water tanks, which function in combination to deliver temperature controlled water to external heat exchangers associated with cardioplegia and oxygenation equipment as well as the patient heating blanket (pump/stirrer identified by arrow).

The heater cooler units are placed in the operating theater and are adjacent to the heart lung machine as they are connected by extensive lengths of tubing (Fig. 2). The unit is built on wheels for mobility. There is an overflow bottle at the back of the unit.

When the outside covers are removed and the unit is viewed from above, the top of the water tanks can been with the stirring motors (Fig. 3).

When the top of one of the tanks was removed, there was extensive visible biofilm on some of the insulated piping material inside the tanks (Fig. 4). This insulation material obviously provided a good substratum for biofilm growth as well as making disinfection challenging (image supplied by Simon Parks).

The water tanks had an overflow pipe at the side of the tank that was connected to the overflow bottle (Fig. 2). There was an additional overflow tubing, which was available for use when greater volumes of water were stored in the units, but which was capped. Both these overflow lines, in use or not, were found to have extensive visible amounts of debris and biofilm (Fig. 5).

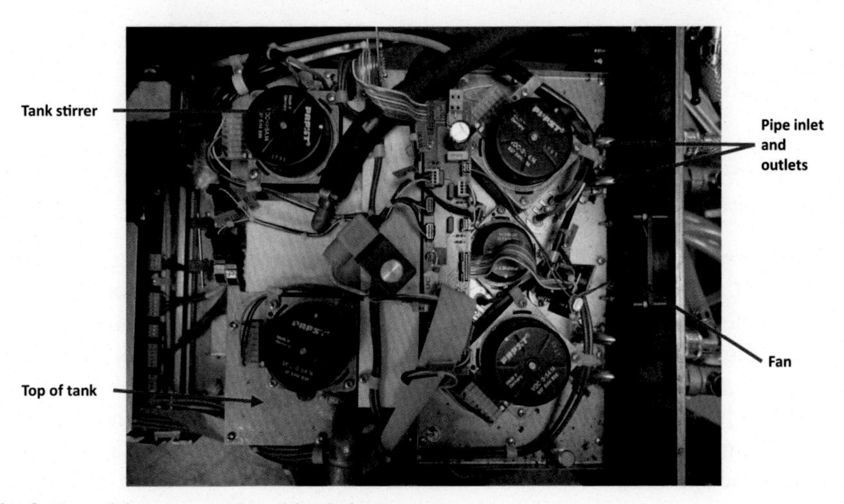

Fig. 3 Top of the water tanks with stirring motors.

Fig. 4 Extensive biofilm present inside the tanks.
(Image supplied by Simon Parks.)

Fig. 5 Identification of overflow tubing with biofilm debris.

When the top of the tanks was examined, several gaps were found to be present near the inlet and outlet pipes that would have allowed air to escape from the tanks while the water was circulating. When manufactured, the top of the tanks would have been sealed in place. These gaps suggested that the top plate of the tank had shifted during use.

When the heater cooler was switched on and the water circulating, the fan at the back of the unit (circled) would extract the air expelled from the tanks and disperse that air into the operating theater (arrow indicating direction for flow from left to right) (Fig. 6).

Fig. 6 Demonstration of smoke from inside the heater cooler being blown outside by the rear-facing fan.
(Image supplied by Simon Parks.)

When microbiological sampling was carried out on air from the the back of the heater cooler using an Anderson sampler, it was evident that the water from the tanks was extensively colonized, and the sampling indicated that the particle sizes ranged from small to larger aerosols and droplets that would have extended across an operating theater (Fig. 7).

Following the identification of the gaps in water tank top plates, moldable putty (Fig. 8) was used to close the gaps in the top of the tanks, and this successfully prevented the release of microbials-contaminated particles.

Microbial contamination of cardiac heater coolers

However, heater coolers have been found to be heavily contaminated with the water circulating containing extremely high heterotrophic plate counts (>10^8 CFU/L) comprising a range of environmental Gram-negative organisms including *P. aeruginosa*, *Sphingomonas paucimobilis*, *Stenotrophomonas maltophilia,* and *Brevundimonas vesicularis* have been reported (Chand et al., 2017; Garvey et al., 2017). Other opportunistic pathogens including *Mycobacterium chelonae*, *Aspergillus flavus,* and non-aspergillus molds (*Paecilomyces* spp.) have been identified. The presence of *Legionella pneumophila* has also been detected (Chand et al., 2017; Diaz-Guerra et al., 2000; Dupont et al., 2016; Garvey et al., 2016). The aerosolization of *L. pneumophila* could represent a risk to users and surrounding staff (Weitkemper et al., 2002). However, in the United Kingdom, no cases of Legionnaires' disease associated with HCUs have been identified (Chand et al., 2017). The presence of high numbers and

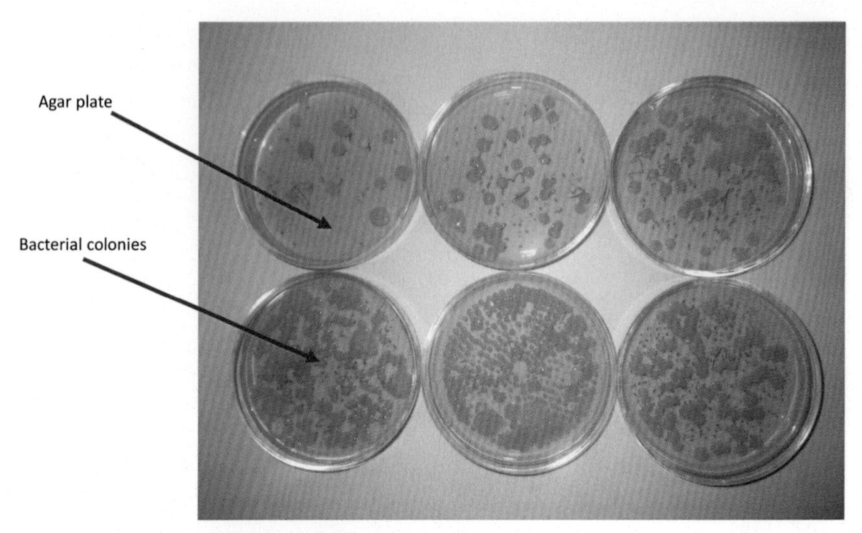

Fig. 7 Sequential agar plates from an Anderson Sampler demonstrating extensive microbial growth from each of the impactor stages.

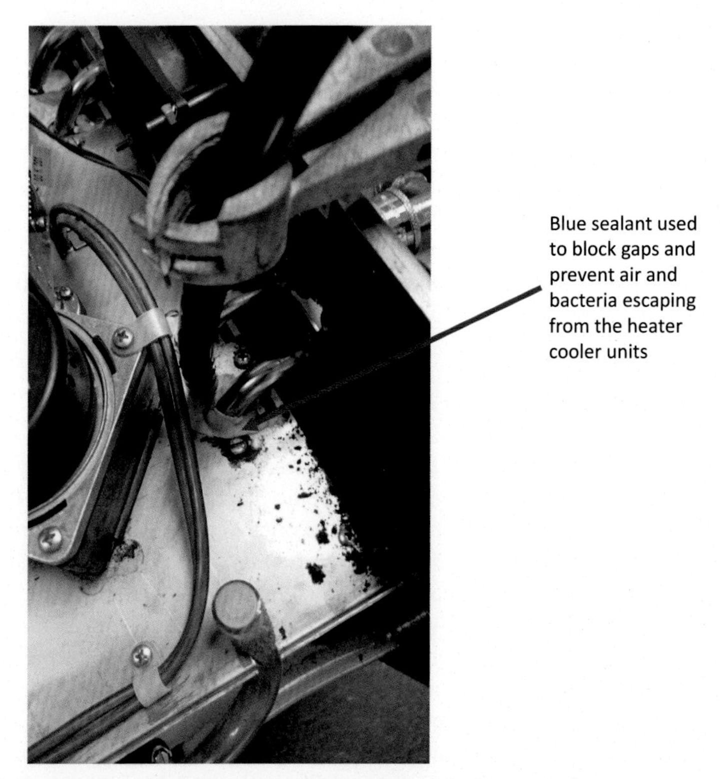

Fig. 8 Sealant used to close the gaps in the top plate of the heater cooler. (Image supplied by Simon Parks.)

a diverse range of microorganisms within an HCU is likely to facilitate the formation of extensive biofilms on tubing associated with the water tanks (Fig. 3) (Chand et al., 2017; Garvey et al., 2016).

Due to the risks involved, several countries have produced guidance over the years including regular microbiological monitoring. However, disinfection of the heater coolers is complex, and reliable disinfection is difficult to achieve to the extent that eradication of *M. chimaera* from a contaminated heater cooler has been difficult to achieve (Garvey et al., 2016). As a result, some hospitals separated the heater cooler from the operating room air to ensure patient safety. This required some ingenious engineering solutions by the engineers and clinical teams involved having the heater cooler outside the operating theater but having to connect the equipment with long lines of tubing, which created other risks (Walker et al., 2017). While our understanding of the causes and the extent of the *M. chimaera* outbreak is growing, several aspects of patient management, device handling, and risk mitigation still require clarification.

What monitoring should be carried out for cardiac heater coolers?

Devices should be monitored according to the instructions from the manufacturer (LivaNova, 2020).

At least once a month, monitor the tanks' water for bacteria and NTM (LivaNova, 2020) (Tables 1 and 2).

Table 1 Interpreting total viable counts for the heater cooler 3T (LivaNova, 2020).

Criteria	Count	Action
Acceptable	<100 CFU/mL	None
Unacceptable	>100 CFU/mL	Remove the unit from use and investigate probable cause

Table 2 Interpreting NTM results for the heater cooler 3T (LivaNova, 2020).

Criteria	Count	Action
Acceptable	<1 CFU/100 mL	None
Unacceptable	>1 CFU/100 mL	Remove the unit from use and investigate probable cause

Risk assessing heater coolers

Risk assessment criteria for heater cooler	Yes	No
Have the current recommendations from manufacturer's and regulators been reviewed?		
Have the cleaning and disinfection instructions provided in the manufacturer's heater cooler device labelling been strictly followed?		
Has only sterile water or water that has been passed through a filter of less than or equal to 0.22 microns been used for the heater cooler?		
If ice is needed for use in the heater cooler, has the water been passed through a filter of less than or equal to 0.22 microns and has the ice been tested for pathogens?		
Are the heater cooler devices' exhaust vents pointed away from the surgical/sterile fields and toward an operating room exhaust vent during device setup and surgical procedures as well as after use to mitigate the risk of aerosolized heater cooler tank water reaching the sterile field?		
Have regular cleaning, disinfection, and maintenance schedules been established for heater cooler devices according to the manufacturer's instructions to minimize the risk of device contamination and patient infection?		
Have the manufacturer's instructions been followed for storage of heater cooler devices and accessories when not in use, which may include removing all water from the device and tubing?		
Are risk assessments available for performing environmental, air, and water sampling and monitoring if heater cooler contamination is suspected?		
Has consideration been taken to positioning the unit outside of the HCU outside the theater?		

Recommendations for heater coolers

Do undertake a full risk assessment of the units used and the environment in which they are used.

Do follow instructions from the relevant manufacturer, including Field Safety Notices, and should seek assistance from the manufacturer as required to effect changes. However, disinfection of contaminated units may be difficult to achieve.

Do microbiological monitoring of the HCU should be carried out to assess the decontamination regimens.

Do position the HCU outside the theater. This has been achieved in a number of European hospitals.

If HCUs have to be in the theater, attention should be given to positioning of the unit such that the rear is not facing the operating area; the HCUs should be retrospectively fitted with a vacuum pump, which reduces exhaust emissions from the rear of the machine by attaching the vacuum pump to the hospital suction system.

Do use only sterile water or water that has been passed through a filter of less than or equal to 0.22 microns.

Ultraclean ventilation (UCV) systems are validated as resisting the ingress of particles as part of the theater's commissioning procedure. Therefore, while some may consider that UCV would provide a safer environment, there is evidence that the airflow from the HCU exhaust can compromise the effect of the UCV airflow pattern.

DO NOT use tap water to rinse, fill, refill, or top-off heater cooler water tanks since this may introduce NTM organisms.

Guidelines for heater coolers

NHS—Heater cooler devices used in cardiac surgery—risk of infection with Mycobacterium species update (NHS, 2016).

NHS Scotland Guidance for Decontamination and testing of Cardiac Heater Cooler Units (HCUs) (NHS Scotland, 2019).

PHE *Mycobacterium chimaera*: infections linked to heater cooler units (PHE, 2021).

FDA—Recommendations for the Use of Any Heater Cooler Device (CDC, 2020).

Acknowledgment

Acknowledgment and thanks must go to many colleagues at PHE including Allan Bennett, Simon Parks (for images 3, 4, 6, 7, 8 and 9), and Ginny Moore. A dedicated and inquisitive set of scientists who have an ability to investigate outbreaks and through due diligence identify problems and routes of transmission of waterborne pathogens. I am indebted to them for their work and for the above images.

References

CDC C for D and R, 2020. Recommendations for the Use of Any Heater Cooler Device. FDA. Available at: https://www.fda.gov/medical-devices/what-heater-cooler-device/recommendations-use-any-heater-cooler-device. (Accessed 29 December 2020).

Chand, M., 2017. Infections Associated With Heater Cooler Units Used in Cardiopulmonary Bypass and ECMO: 20.

Chand, M., Lamagni, T., Kranzer, K., et al., 2017. Insidious risk of severe *Mycobacterium chimaera* infection in cardiac surgery patients. Clin. Infect. Dis. 64 (3), 335–342. https://doi.org/10.1093/cid/ciw754.

Diaz-Guerra, T.M., Mellado, E., Cuenca-Estrella, M., et al., 2000. Genetic similarity among one *Aspergillus flavus* strain isolated from a patient who underwent heart surgery and two environmental strains obtained from the operating room. J. Clin. Microbiol. 38 (6), 2419–2422.

Dupont, C., Terru, D., Aguilhon, S., et al., 2016. Source-case investigation of *Mycobacterium wolinskyi* cardiac surgical site infection. J. Hosp. Infect. 93 (3), 235–239. https://doi.org/10.1016/j.jhin.2016.03.024.

Garvey, M.I., Ashford, R., Bradley, C.W., et al., 2016. Decontamination of heater-cooler units associated with contamination by atypical mycobacteria. J. Hosp. Infect. 93 (3), 229–234. https://doi.org/10.1016/j.jhin.2016.02.007.

Garvey, M.I., Phillips, N., Bradley, C.W., et al., 2017. Decontamination of an extracorporeal membrane oxygenator contaminated with *Mycobacterium chimaera*. Infect. Control Hosp. Epidemiol. 38 (10), 1244–1246. https://doi.org/10.1017/ice.2017.163.

Ghodousi, A., Borroni, E., Peracchi, M., et al., 2020. Genomic analysis of cardiac surgery-associated *Mycobacterium chimaera* infections in Italy. PLoS One 15 (9). https://doi.org/10.1371/journal.pone.0239273. Public Library of Science: e0239273.

LivaNova, 2020. Heater-Cooler System 3T Operating Instructions. Available at: https://livanovamediaprod.azureedge.net/livanova-media/livanova-public/media/resources01/cp_ifu_16-xx-xx_usa_021.pdf?ext=.pdf.

NHS, 2016. Heater-Cooler Devices Used in Cardiac Surgery—Risk of Infection With Mycobacterium Species Update. *GOV.UK*: NHSScotland Guidance for Decontamination and testing of Cardiac Heater Cooler Units (HCUs). Available at: https://www.gov.uk/drug-device-alerts/heater-cooler-devices-used-in-cardiac-surgery-risk-of-infection-with-mycobacterium-species-update. (Accessed 29 May 2019).

NHS Scotland, 2019. Guidance for Decontamination and Testing of Cardiac Heater Cooler Units (HCUs). v1.0. *National Services Scotland*. Available at: https://www.nss.nhs.scot/publications/nhsscotland-guidance-for-decontamination-and-testing-of-cardiac-heater-cooler-units-hcus-v10/. (Accessed 14 April 2022).

PHE, 2021. *Mycobacterium chimaera*: Infections Linked to Heater Cooler Units. *GOV.UK*. Available at: https://www.gov.uk/government/collections/mycobacterial-infections-associated-with-heater-cooler-units. (Accessed 14 April 2022).

Sax, H., Bloemberg, G., Hasse, B., et al., 2015. Prolonged outbreak of *Mycobacterium chimaera* infection after open-chest heart surgery. Clin. Infect. Dis. 61 (1), 67–75. https://doi.org/10.1093/cid/civ198.

Schreiber, P.W., Kohl, T.A., Kuster, S.P., et al., 2021. The global outbreak of *Mycobacterium chimaera* infections in cardiac surgery—a systematic review of whole-genome sequencing studies and joint analysis. Clin. Microbiol. Infect. 27 (11), 1613–1620. https://doi.org/10.1016/j.cmi.2021.07.017.

Sommerstein, R., Schreiber, P.W., Diekema, D.J., et al., 2017. *Mycobacterium chimaera* outbreak associated with heater-cooler devices: piecing the Puzzle together. Infect. Control Hosp. Epidemiol. 38 (1), 103–108. https://doi.org/10.1017/ice.2016.283.

Walker, J., Moore, G., Collins, S., et al., 2017. Microbiological problems and biofilms associated with *Mycobacterium chimaera* in heater–cooler units used for cardiopulmonary bypass. J. Hosp. Infect. 96 (3), 209–220. https://doi.org/10.1016/j.jhin.2017.04.014.

Weitkemper, H.H., Spilker, A., Knobl, H.J., et al., 2002. The heater-cooler unit—a conceivable source of infection. J Extra Corpor Technol 34 (4), 276–280.

Dental chairs and dental unit water lines

22

Dental chairs and dental unit water lines

Dental chairs are designed not just to ensure that the patient is seated comfortably but also to ensure the dentist has the required equipment at hand for rinsing, drilling, descaling, and polishing the patient's teeth. A series of narrow bore tubing supplies water to all of the dental chair supplied instruments, cup-filler and bowl-rinse outlets. Water from the dental chair is also used for oral rinsing by patients (water supplied via the cup filler outlet) and to wash out the dental chair spittoon, or cuspidor, after oral rinsing (water supplied via the bowl rinse outlet). Dental chairs are categorized as medical devices under the European Union Medical Devices Directive (Anon, 1993).

While we can all see that the dental chair is visually clean, we have very little knowledge on the microbial contamination of the water and tubing used in the dental chair. Dental practices can either be a sole practitioner with one chair, a number of practitioners and with a number of chairs (~5) and healthcare premises and dental hospitals can contain a multitude of dental chairs (more than 10).

Potable water is a fundamental component of the dental chair, which contains a wide range of narrow bore dental unit water lines (~6 mm) through which the water flows to devices such as the three-in-one hand piece syringe, handpiece.

Dental unit water is both swallowed and inhaled by the patient and staff. When the dental equipment such as the handpiece and the ultrasonic scalers are operated, they can result in significant dispersal of water droplets and aerosols in the vicinity of the patient and the dental staff (Bennett et al., 2000). It is the dental staff who are exposed to these microorganisms on a regular basis.

It must also be considered that dentists are undertaking surgery when treating patients, and the water from dental chair devices should be of a quality that wound infections are avoided not by luck but by appropriate management. As such for dental surgical procedures surgical flaps or other access into body cavities involving irrigation, the use of sterile water or sterile isotonic saline provided from a separate single-use source is recommended (DHSC, 2013).

Dental chairs are equipped with a suction system that has a variety of purposes. Primarily, the suction system is used to remove oral fluids and debris from the oral cavity during dental procedures and also to minimize aerosol release into the dental clinic environment during the use of high-energy dental tools.

A number of dental chairs are equipped with heaters that warm water lines so that the water is more comfortable for the patient. However, heating DUWL water to >20°C may result in the proliferation of Legionella bacteria present in DCU supply water within DUWLs or within the DCU water heater itself. Naturally occurring

Safe Water in Healthcare. https://doi.org/10.1016/B978-0-323-90492-6.00020-3

Legionella pneumophila readily multiply at 25–45°C (DHSC, 2016). It is also feasible that heating DCU water may promote the proliferation of human bacteria that have entered the water lines.

What are the microbial contamination sources of dental chairs and associated equipment?

Dental chairs can be supplied by water from a number of different sources within a practice, and the responsible person needs to ensure that the different sources of water used in that practice are described within the written water line management scheme and also any written scheme plan from their legionella risk assessment.

Dental care professionals (dental nurses, hygienists, dentists) who use the chair need to be aware of the type of water that is used for the different purposes within their practices:

The different sources of water in a practice can include:

- domestic systems (will either be fed from a tank in the loft or direct mains)
- dental water lines to handpieces, 3-in-1, scalers (bottled water system)
- mains water servicing dental chairs (cup filler/spittoon)
- other risk systems (e.g., water purification including reverse osmosis)
- sterile water used for surgery.

While the water should be free of pathogens, there will still be microorganisms entering the tubing in the dental unit water lines.

Dental chairs will either be fed directly from the potable water supply, cisterns or tanks, bottled water, and/or sterile water from a sterile source. In a recent survey in France, majority of units were fed by mains water (91.0%), and one-third (33.6%) of the units had an independent water bottle reservoir (Baudet et al., 2019).

While the concept of using sterile water in a dental chair may appeal the water being used and will only be sterile at the point of use. As soon as that sterile water enters the dental unit water lines, then the water will quickly become contaminated.

In addition, dental chair water systems (regardless of the type of water used) will quickly and easily become contaminated with a range of environmental and skin flora (Lizzadro et al., 2019; O'Donnell et al., 2011; Volgenant and Persoon, 2019; Walker et al., 2000).

The dental unit water lines can also be contaminated from aspiration of patient secretions through handpieces into the water lines, and therefore, antiretraction device should be fitted to equipment to prevent this. The detection of oral bacterial species and other human-derived microorganisms in DUWL output water has provided convincing evidence for likely failure of antiretraction devices (Berlutti et al., 2003; Montebugnoli et al., 2004; Petti and Tarsitani, 2006; Tuttlebee et al., 2002).

Where potable water is being supplied to the dental chair, it would typically already have traveled some way from the original mains supply and will contain microorganisms. Like most waterborne microorganisms, the majority will be harmless Gram-negative environmental bacteria. However, the dental chair water may be

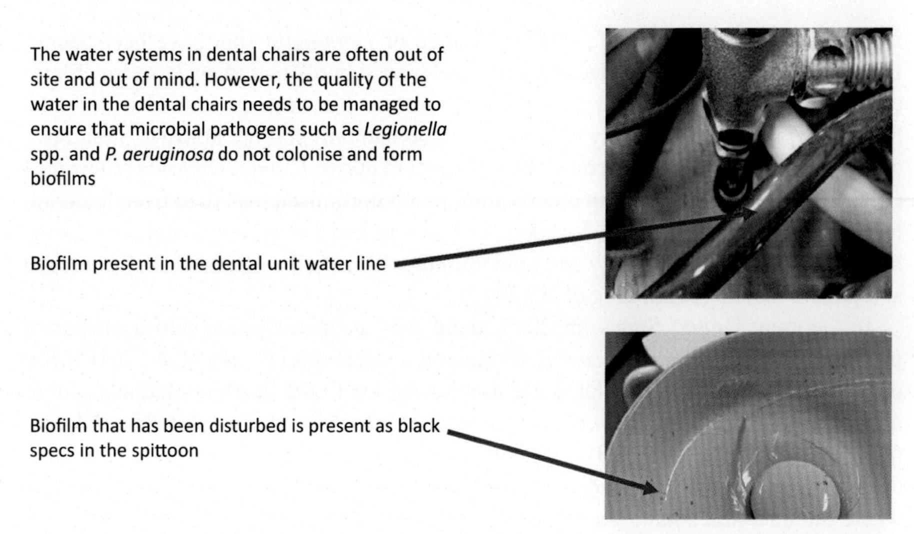

Fig. 1 Demonstration of biofilm in the dental chair feed line and presence of biofilm debris in the spittoon.
Images courtesy of Mary Henderson.

contaminated by a low level range of microbial pathogens including *Legionella* spp. and *Pseudomonas aeruginosa,* among others, including bacteria, fungi, and viruses.

These microorganisms once present in the dental unit water line will form biofilms that will perpetually contaminate the tubing. The small-bore narrow lumens present a high surface-area-to-volume ratio, there is a relatively slow flow rate, and the entire column of water in the tubing can remain stagnant for long periods of time. As the water in the lines is used only intermittently, then the temporary stagnation will result in ideal conditions for biofilm formation.

It is not uncommon for there to be a relatively high heterotrophic plate count in excess of 10^5 CFU/mL, which can equate to a high endotoxin concentration that can result in mild or severe inflammatory responses leading to diseases such as periodontitis and systemic, septic shock (Pankhurst et al., 2017). An association has been demonstrated between the onset of occupational asthma in dentists and the concentration of bacteria in their DUWLs in a cross-sectional multicenter DUWL survey (Pankhurst et al., 2005) (Fig. 1).

What are the transmission sources?

It is the dispersal of aerosols that is such a risk for microorganisms such as *Legionella* spp. The presence of *Legionella* spp. should be considered a risk due to the high energy devices (hand pieces and ultrasonic devices) producing aerosols as a number of dental patients may be in the at-risk categories including males, middle-aged, smokers, alcohol consumers, and even those who are immunocompromised. In one recent study, 36% of samples from a dental hospital facility were positive for *Legionella* spp. (Sedlata Juraskova et al., 2017). A range of *Legionella* spp. have been identified in water sampled from dental chairs in dental schools including *L. anisa* (Fleres et al., 2018). Transmission of Legionnaires' disease has been associated with dental treatments. In 2011, an elderly (82) woman who had only left her house to attend two dental appointments died as a result of an infection due to *L. pneumophila* serogroup 1 (Ricci et al., 2012). In 2012, an elderly immunocompromised man died from legionellosis at a hospital in Uppsala, Sweden. The patient had visited the dental ward at the hospital, and the microbiology and epidemiology identified a common environmental and patient isolate (Schönning et al., 2017).

In Georgia, United States, in 2015, there was an investigation into a cluster of pediatric *Mycobacterium abscessus* odontogenic infections (Peralta et al., 2016). The dental practice's lack of regular water monitoring led to the unintentional use of municipal water that had a high concentration of *M. abscessus* during pediatric pulpotomies (Hatzenbuehler et al., 2017). Other studies have also reported facial infections of dental origin, due to an infections with *M. fortuitum*, *M. abscessus,* and *M. peregrinum* where the infection source was the dental unit waterlines (DUWLs), which were colonized with NTM (Pérez-Alfonzo et al., 2020).

There is also evidence that dental unit waterlines colonized with *P. aeruginosa* may be associated with infections (Martin, 1987; O'Donnell et al., 2011).

Monitoring dental chair water systems

Apart from situations where there are taste or odor problems, microbiological monitoring for total viable counts (TVCs) is not considered to be necessary.

CDC recommends that dental unit water used in nonsurgical procedures measure less than or equal to 500 CFU/mL (≤500 CFU/mL) of water, the standard set for drinking water by the Environmental Protection Agency (EPA).

In the EU, the limits for total viable counts at 37°C vary per country. The upper limit set is <500 CFU/mL (Denmark), <200 CFU/mL (Holland), and <100 CFU/mL (Sweden and Germany).

In the United Kingdom, the HTM 01-05 recommends that where monitoring is undertaken that the TVC (preferably at 22°C) should be expected to lie in the range 100–200 CFU/mL (DHSC, 2013).

In HTM 04-01, dental facilities should have *Legionella* spp. expressed as colony counts per liter and *P. aeruginosa* colony counts per 100 mL, and the presence of these pathogens in the dental chair water system would be unacceptable.

Requirements for dental practices and dental chairs

All water systems require a risk assessment; however, not all systems will require elaborate control measures.

All premises are required to have a written waterline management scheme and legionella risk assessment. These schemes should be written by experienced and competent people. A competent person is someone with the necessary skills, knowledge, and experience to carry out this function.

The registered manager must ensure that all the recommendations of the written scheme and risk assessment are implemented.

Water and air lines must be fitted with antiretraction valves in accordance with EU regulations.

It is mandatory to control *Legionella* spp. within the dental waterline system, but there is no one single system of treatment, which is 100% effective.

Recommended practice for dental practices and dental chairs

To reduce microbial accumulation, run (flush) water through the water lines for 2–3 min at the start of each session and 20–30 s between every patient. Checklists are useful to ensure compliance and provide auditable evidence.

At the end of the day, the bottle(s) should be disconnected, emptied, rinsed, and stored inverted clean and dry overnight.

Isolate the water supply from the mains water by using an independent bottled water system on the unit.

Fill the bottle with freshly distilled/reverse osmosis water at the start of each day (if bottled water is used, this must be from a previously unopened bottle).

Do not fill bottles with tap water as this will introduce opportunistic respiratory bacteria into the waterline and rapidly lead to biofilm formation.

Do not attach dental handpieces or dental instruments to dental unit waterlines that have not been cleaned or disinfected per the manufacturer's instructions.

Guidelines for dental chairs, dental unit water lines, and spittoons

ALCOP L8 (The control of legionella bacteria in water systems).

HTM 04-01 (The control of Legionella, hygiene, "safe" hot water, cold water and drinking systems).

HTM 01-05 (Decontamination in primary care dental practices).

CQC (https://www.cqc.org.uk/guidance-providers/dentists/dental-mythbuster-5-legionella-dental-waterline-management).

FDA—dental unit water lines (https://www.fda.gov/medical-devices/dental-devices/dental-unit-waterlines).

Council Directive 93/42/EEC of 14 June 1993 concerning medical devices. OJEU L169, 1–43.

Acknowledgment

Many thanks to Mary Henderson for her comments and use of images.

References

Anon, 1993. Council Directive 93/42/EEC of 14 June 1993 concerning medical devices. OJEU L169. pp. 1–43.

Baudet, A., Lizon, J., Martrette, J.-M., Camelot, F., Florentin, A., Clément, C., 2019. Dental unit waterlines: a survey of practices in Eastern France. Int. J. Environ. Res. Public Health 16. https://doi.org/10.3390/ijerph16214242.

Bennett, A.M., Fulford, M.R., Walker, J.T., Bradshaw, D.J., Martin, M.V., Marsh, P.D., 2000. Microbial aerosols in general dental practice. Br. Dent. J. 189, 664–667. https://doi.org/10.1038/sj.bdj.4800859.

Berlutti, F., Testarelli, L., Vaia, F., Luca, M.D., Dolci, G., 2003. Efficacy of anti-retraction devices in preventing bacterial contamination of dental unit water lines. J. Dent. 31, 105–110. https://doi.org/10.1016/S0300-5712(03)00004-6.

DHSC, 2013. Decontamination in Primary Care Dental Practices (HTM 01-05). GOV.UK.

DHSC, 2016. HTM 04-01: Safe Water in Healthcare Premises.

Fleres, G., Couto, N., Lokate, M., van der Sluis, L.W.M., Ginevra, C., Jarraud, S., Deurenberg, R.H., Rossen, J.W., García-Cobos, S., Friedrich, A.W., 2018. Detection of *Legionella anisa* in water from hospital dental chair units and molecular characterization by whole-genome sequencing. Microorganisms 6. https://doi.org/10.3390/microorganisms6030071.

Hatzenbuehler, L.A., Tobin-D'Angelo, M., Drenzek, C., Peralta, G., Cranmer, L.C., Anderson, E.J., Milla, S.S., Abramowicz, S., Yi, J., Hilinski, J., Rajan, R., Whitley, M.K., Gower, V., Berkowitz, F., Shapiro, C.A., Williams, J.K., Harmon, P., Shane, A.L., 2017. Pediatric dental clinic-associated outbreak of *Mycobacterium abscessus* infection. J. Pediatr. Infect. Dis. Soc. 6, e116–e122. https://doi.org/10.1093/jpids/pix065.

Lizzadro, J., Mazzotta, M., Girolamini, L., Dormi, A., Pellati, T., Cristino, S., 2019. Comparison between two types of dental unit waterlines: how evaluation of microbiological contamination can support risk containment. Int. J. Environ. Res. Public Health 16. https://doi.org/10.3390/ijerph16030328.

Martin, M.V., 1987. The significance of the bacterial contamination of dental unit water systems. Br. Dent. J. 163, 152–154. https://doi.org/10.1038/sj.bdj.4806220.

Montebugnoli, L., Chersoni, S., Prati, C., Dolci, G., 2004. A between-patient disinfection method to control water line contamination and biofilm inside dental units. J. Hosp. Infect. 56, 297–304. https://doi.org/10.1016/j.jhin.2004.01.015.

O'Donnell, M.J., Boyle, M.A., Russell, R.J., Coleman, D.C., 2011. Management of dental unit waterline biofilms in the 21st century. Future Microbiol. 6, 1209–1226. https://doi.org/10.2217/fmb.11.104.

Pankhurst, C.L., Coulter, W., Philpott-Howard, J.N., Surman-Lee, S., Warburton, F., Challacombe, S., 2005. Evaluation of the potential risk of occupational asthma in dentists exposed to contaminated dental unit waterlines. Prim. Dent. Care 12, 53–59. https://doi.org/10.1308/1355761053695176.

Pankhurst, C.L., Scully, C., Samaranayake, L., 2017. Dental unit water lines and their disinfection and management: a review. Dent. Update 44, 284–285. 289–292 10.12968/denu.2017.44.4.284.

Peralta, G., Tobin-D'Angelo, M., Parham, A., Edison, L., Lorentzson, L., Smith, C., Drenzek, C., 2016. Notes from the field: *Mycobacterium abscessus* infections among patients of a pediatric dentistry practice—Georgia, 2015. MMWR Morb. Mortal. Wkly Rep. 65, 355–356. https://doi.org/10.15585/mmwr.mm6513a5.

Pérez-Alfonzo, R., Poleo Brito, L.E., Vergara, M.S., Ruiz Damasco, A., Meneses Rodríguez, P.L., Kannee Quintero, C.E., Carrera Martinez, C., Rivera-Oliver, I.A., Da Mata Jardin, O.J., Rodríguez-Castillo, B.A., de Waard, J.H., 2020. Odontogenic cutaneous sinus tracts due to infection with nontuberculous mycobacteria: a report of three cases. BMC Infect. Dis. 20, 295. https://doi.org/10.1186/s12879-020-05015-5.

Petti, S., Tarsitani, G., 2006. Detection and quantification of dental unit water line contamination by oral *Streptococci*. Infect. Control Hosp. Epidemiol. 27, 504–509. https://doi.org/10.1086/504500.

Ricci, M.L., Fontana, S., Pinci, F., Fiumana, E., Pedna, M.F., Farolfi, P., Sabattini, M.A.B., Scaturro, M., 2012. Pneumonia associated with a dental unit waterline. Lancet (London England) 379, 684. https://doi.org/10.1016/S0140-6736(12)60074-9.

Schönning, C., Jernberg, C., Klingenberg, D., Andersson, S., Pääjärvi, A., Alm, E., Tano, E., Lytsy, B., 2017. Legionellosis acquired through a dental unit: a case study. J. Hosp. Infect. 96, 89–92. https://doi.org/10.1016/j.jhin.2017.01.009.

Sedlata Juraskova, E., Sedlackova, H., Janska, J., Holy, O., Lalova, I., Matouskova, I., 2017. *Legionella* spp. in dental unit waterlines. Bratisl. Lek. Listy 118, 310–314. https://doi.org/10.4149/BLL_2017_060.

Tuttlebee, C.M., O'Donnell, M.J., Keane, C.T., Russell, R.J., Sullivan, D.J., Falkiner, F., Coleman, D.C., 2002. Effective control of dental chair unit waterline biofilm and marked reduction of bacterial contamination of output water using two peroxide-based disinfectants. J. Hosp. Infect. 52, 192–205. https://doi.org/10.1053/jhin.2002.1282.

Volgenant, C.M.C., Persoon, I.F., 2019. Microbial water quality management of dental unit water lines at a dental school. J. Hosp. Infect. 103, e115–e117. https://doi.org/10.1016/j.jhin.2018.11.002.

Walker, J.T., Bradshaw, D.J., Bennett, A.M., Fulford, M.R., Martin, M.V., Marsh, P.D., 2000. Microbial biofilm formation and contamination of dental-unit water systems in general dental practice. Appl. Environ. Microbiol. 66, 3363–3367. https://doi.org/10.1128/AEM.66.8.3363-3367.2000.

Drains and wastewater

Introduction

Sanitation was developed by ancient civilizations including the Romans (De Feo et al., 2014) who recognized the importance of segregating water supplies from human waste (Yannopoulos et al., 2017); however, these innovations were not always encompassed by the every society.

The Industrial Revolution brought change to many aspects of society. The large influx of the rural population into cities lacking safe water and sanitation resulted in conditions where infectious disease rampaged. Cesspits at the bottom of houses became common, the association between the ensuing stench and disease in Victorian minds promoting the miasma theory, that is, the spread of infection through the air (van Oosten, 2016). Sir Edwin Chadwick, a great social reformer, acting in good faith ordered that all cesspits should drain into the river Thames (Collins, 1924). Up until then, this had been a relatively safe source of drinking water. Although the consequences to water safety were disastrous, the other unforeseen outcome "the great stink" drove sanitation. The Great Stink was an event in Central London in July and August 1858, during which the hot weather exacerbated the smell of untreated human waste and industrial effluent that was present on the banks of the River Thames (Porter, 2001). Parliament, which was on the banks of the river Thames, was keen to act to reverse the situation and commissioned Joseph Bazalgette to design the service system for London, which is still in use today. Coincidentally, 1858 was the year that Florence Nightingale published her work on hospitals in which she stated so eloquently "it may seem a strange principle to enunciate as the very first requirement in a hospital that it should do the sick no harm" (Loveday, 2020).

Moving on more than a century later, Joachim Kohn published in the Lancet the first thermally disinfecting waste trap (Kohn, 1967). Kohn believed (correctly as time has shown) that drain was frequently the source from where organisms such as *Pseudomonas aeruginosa* originated when producing contamination of water outlets. However, his views were not generally accepted or welcomed.

Moving on 25 years later, an outbreak in Manchester changed the way the areas would be designed (Farwell, 1995). Contaminated total parenteral nutrition resulted in pediatric sterile pharmaceutical preparation deaths. The source of the contamination was identified as a sink in the sterile preparation area. Splashing from the sink had resulted in contamination of the feeds during preparation. The resultant Farwell enquiry recommended that all water services be removed from such facilities. Their view was water was so dangerous that staff should wash their hands before entering the facility and if need be, use alcohol hand rub. Thus, this was the first area within hospitals to become water-free.

Safe Water in Healthcare. https://doi.org/10.1016/B978-0-323-90492-6.00015-X

Drain-associated transmission of infections

Publications by Hota (Hota et al., 2009) and Breathnach (Breathnach et al., 2012) were regarded as landmark. Hota et al. described a multidrug-resistant *P. aeruginosa* outbreak in an ITU in Toronto, Canada. The source of the outbreak could be traced to a sink located in close proximity to a work surface where drugs would be prepared. Water from the sink outlet tested negative. However, the sink drain contained the outbreak strain. By placing fluorescein dye in the drain, and running the outlet, the authors could demonstrate splashing of drain contents up to a meter from the sink. Breathnach et al. described a hospital-wide outbreak again with a multidrug-resistant strain of *P. aeruginosa*. It took 4 years to uncover the source of the outbreak, but given it took 45 years for the United Kingdom to finally accept Joachim Kohn's assertion that water from outlets could contain *P. aeruginosa*, this was pretty good going. The source of the outbreak was a wastewater system—blocked sinks, drains, showers, and toilets throughout the hospital. What this publication demonstrated was that the wastewater system could act as a superhighway for the spread of organisms around a healthcare facility.

But Breathnach et al. also made another interesting statement (Breathnach et al., 2012). A multidrug-resistant organism naturally stands out and attracts everyone's attention. Could it be that sensitive organisms were being transmitted by the same route, but because they blend into the background of other infections, the link to sensitive organisms is not made?

The advent of Carbapenemase-producing Enterobacteriaceae (CPE) has driven renewed interest in drains/wastewater systems to the extent that it is hard to read an infection control journal without a report describing their transmission from drains (Jamal et al., 2019; Tang et al., 2020). Although ESBL-producing organisms could also be linked to transmission from drains, the number of reports would appear to be small compared with CPEs (Roux et al., 2013; Wolf et al., 2014). However ESBLs were not regarded with the same fervor as CPEs, presumably because the world felt safe as the carbapenems seemed to be the answer to infection with these organisms. ESBLs drove the world to the brink, as the response was to substantially increase the use of carbapenems, and should we be surprised that resistance to these agents resulted (Bonomo et al., 2018; D'Angelo et al., 2016)? However, from the perspective of drains, is it that carbapenemase-producing organisms have a special propensity to spread from the source, or is it that this is not the case, they merely attracting our attention (Jung et al., 2020; Kearney et al., 2021)?

In 2017, Hopman et al. published on the "reduced rate of intensive care unit acquired Gram-negative bacilli after removal of sinks and introduction of 'water free' patient care" (Hopman et al., 2017). An intractable outbreak with a multidrug-resistant organism drove this group to introduce the concept of "water-free" patient care. Apart from two sinks on the ITU (one in the central area of the unit principally for surgical scrub) and the other in the dirty sluice, all other sinks were removed. Not only did this eradicate the outbreak but produced an overall reduction in acquisition of all Gram-negative organisms. This would suggest that drains are a source of a wide variety of organisms, but our surveillance systems are primarily geared to detecting transmission events with multidrug-resistant organisms as these are easier to recognize. Highly resistant organisms are merely highlighting the well-trodden pathways used by other organisms, which until now have mostly gone unrecognized.

How are drains contaminated?

It is a regular occurrence to dispose of liquids in the sink or to flush the toilet and to assume that is the end of the matter (Feng et al., 2020). However, this is not the case. Wastewater systems provide a superhighway for microorganisms to travel within the building and eventually be transmitted to patients (Smismans et al., 2019). As the human gut is the reservoir for many of the Gram-negative organisms, which threaten the end of the antibiotic era (i.e., Carbapenemase-Producing Enterobacteriaceae (CPE)), wastewater systems are increasingly recognized as a major risk factor in the propagation of antibiotic resistance (Breathnach et al., 2012; Carling, 2018).

The combination of a number of mechanisms facilitates this situation including:

1. Ability of bacteria to defy gravity—when a toilet is flushed and water and fecal material enter the main sewage stack, gravity will cause the water and fecal material to drop down the pipe. However, trapped air is forced to escape upward, which creates a suspension of water and fecal organisms, which can travel up several floors in the building to contaminate other branches of the drainage system (Wong et al., 2021).
2. Blockage of drainage systems in hospitals is common, the build-up behind the blockage aiding spread of microorganisms across the drainage network (Vardoulakis et al., 2022).
3. Experimental models have shown that if the waste trap of one sink (connected via the drainage system to a gallery of sinks) is inoculated with a tracer organism, within a week the same organism can be found in most other waste traps (Aranega-Bou et al., 2018; Mathers et al., 2018).
4. Disposing of a carbon source down a sink drain will stimulate the growth of biofilm up the vertical section of the drain at a rate of 1 mm/h (Kotay et al., 2020). The biofilm will reach the sieve at the top of the drain. Water from an outlet directly hitting the sieve can disperse organisms in the biofilm up to 2 m away.

The above will allow dispersal of organisms across the drainage network. But how do they escape from the drainage system and find their way to a patient? Most of the work has been done in relation to hand wash stations. The purpose of a waste trap (present in any device connected to the drainage system) is to provide a water seal to prevent escape of sewer gases. The waste trap will inevitably contain bacteria (Aranega-Bou et al., 2018; Moloney et al., 2020). In a drain located directly below an outlet, bacteria can grow up the vertical section reaching the top of the drain where there is usually a sieve. When water directly hits a drain, it can be shown by placement of agar plates outside of the sink that drain organisms will be dispersed widely (up to 2 m) in the environment. A recent study has shown that even when the drain is placed at the bottom of the sink but offset, so that the main body of water does not hit it, dispersal of organisms still takes place (Aranega-Bou et al., 2018). In the United Kingdom, the recommended design is for the drain to be located at the rear of the basin. Work from Porton Down has shown that a rear drain is effective at preventing dispersal of drain organisms providing drainage is not impaired (Aranega-Bou et al., 2018, 2021). But impaired drainage with rear drains can be common because they frequently lack a sieve to prevent objects going down into the drain where they may obstruct flow (Fig. 1). Items removed from the waste trap of a blocked sink included end caps from giving sets, razor blade covers, capillary sampling devices, and intravenous connecting devices. Sieves are available to prevent items from falling into the u-bends but are thought to be a risk for extensive biofilm formation on the plastic grid (Walker et al., 2014) (Figs. 2 and 3). Clinical staff have not been taught to report poorly draining sinks, so in practice blockages only tend to be reported when they have become so bad that no drainage occurs and residual water results in microbial contamination of the hand wash basin surface.

While disposal of any fluids, including patient fluids/secretions, should not occur in hand wash stations, this does happen in practice. The design of intensive care units makes appropriate disposal of fluids more difficult as side rooms tend not to have ensuite facilities (as would be found inside rooms on a general ward), which facilitate

Fig. 1 Sieve for a rear drain, which is infrequently used due to concerns these will promote attachment of bacteria and biofilm formation.

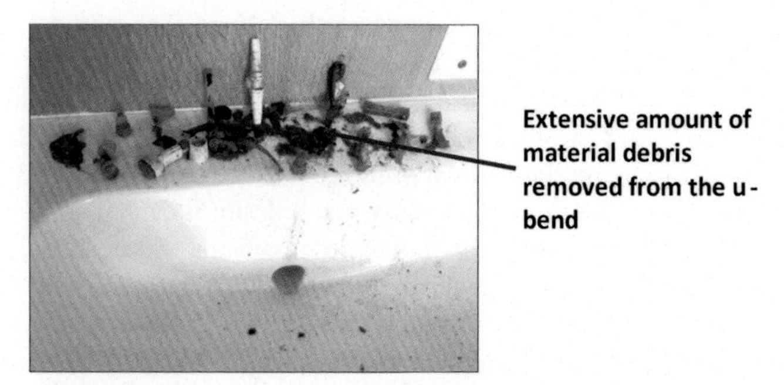

Fig. 2 Items removed from the waste trap (u-bend or p-trap) from a blocked hand wash station.

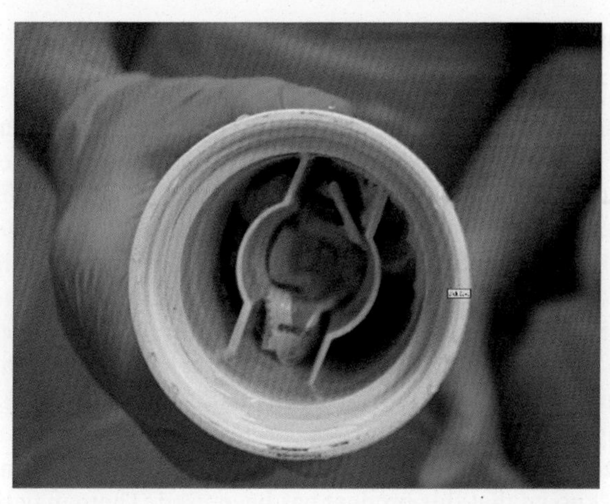

Fig. 3 Looking inside the waste trap and see the plastic wrappers and paper towel. Note how the device is compartmentalized, increasing the risk that objects will obstruct flow.

disposal. Instead, there may be one or two dirty sluices requiring staff to take the most heavily contaminated fluids out of an isolation room, in doing so often having to make contact with touch surfaces such as door handles, traverse a considerable distance before entering the dirty sluice. Once within the dirty sluice, the risks are not over—staff often drop the secretions into a sluice hopper, generating splashing, which can contaminate items incorrectly stored within the vicinity. Many sluices lack appropriate infection control governance including a flow from dirty to clean (Breathnach et al., 2012; Choudhuri et al., 2014; Mathers et al., 2018).

Our understanding of the risk from drains continues to evolve rapidly in this environment of increasing antimicrobial resistance. In general, anything that impairs wastewater drainage will increase risk of dispersal. Impaired drainage is commonly the result of inappropriate disposal of items (Fig. 3). Nonbiodegradable wipes are frequent causes of blockage of waste pipes both within and outside healthcare facilities. Incorrect disposal of items is further hindered by poor design/engineering practices such as incorporating 90 degree bends in the waste pipes from a macerator (if the manufacturer's installation instructions have been read, this would not have occurred in most instances) (Weinbren, 2020).

Reducing clinical hand wash station risks from drains

In terms of developing a strategy for mitigating risk, two other factors need to be taken into account:

1. sensitivity of surveillance in detecting transmission events.
2. effectiveness of mitigating measures.

Sensitivity of surveillance systems

If it were possible to detect and recognize the source of every transmission event linked to drainage systems, then while it is preferable to prevent such occurrences, at least once they have been identified, mitigations can be put in place. Unfortunately, the sensitivity of infection control surveillance varies globally and between organisms. A recent contamination of wipes with *P. aeruginosa* produced in one country but distributed across Europe has highlighted some interesting differences between countries. The contamination was initially detected in Norway, and following this a product recall was issued in several countries but for a variety of reasons was not totally successful. This issue came to light requiring issuing of a further product recall. At this stage, Scotland identified cases linked to the wipes still being used in clinical practice. To date, and this is early days, no cases have been linked to the use of the wipes in the remainder of the United Kingdom despite it being known that this product has been used. There may be a number of explanations for this, but at first sight it would suggest that Norway has the best surveillance system, followed by Scotland. Further evidence demonstrating that surveillance systems can miss transmission of sensitive strains comes from the whole genome sequencing study in four UK hospitals, which showed ongoing transmission of PA from water systems in three of the hospitals, which was not detected by the surveillance systems in place. As already mentioned, it is likely that transmission of sensitive organisms from drains has occurred on the whole undetected. The literature relating to water and wastewater transmission events is overrepresented by multidrug-resistant organisms—these transmission events are easier to detect as they currently stand out from the more sensitive organisms. When antimicrobial resistance becomes more common, it is likely that even multidrug-resistant organism outbreaks will blend in with the background and no longer be so readily detected.

What is the effectiveness of mitigating measures?

Kearney et al. ranked the effectiveness of interventions (Kearney et al., 2021) with the most effective intervention being eliminating the hazard (i.e., removal of sink (hand hygiene station) and associated pipework) and the least administrative controls (policy, guidelines, and education). The second most effective was to isolate a separate hazard, such as installation of physical barriers between patients and sink or ensuring no sinks were in proximity to the patient or their environment. The second least effective were engineering solutions to prevent dispersal of contamination from drains, such as rear drains, and the use of disinfectants.

The situation where the surveillance methodology lacks sensitivity combined with the knowledge that our most used mitigating measure policies, guidelines, and education are the least effective strategy is uncomfortable, carrying a high risk to patients. The solution is to either improve the surveillance methodology or move to more effective interventions (either eliminating the hazard or isolation or separation of the hazard) or a combination of both.

Mitigating risk in hand wash stations

1. **Sink removal**—is the hand wash station necessary? For example, in a drug preparation area, staff hands should already be clean (alcohol hand rub should be available), making it possible to remove the hand wash station. In a small but an increasing number of ICUs, the majority of hand wash stations are being removed. New hospitals are opening, many of which have single room accommodation, and in Holland (personal communication Joost Hopman), there are no clinical hand wash stations inside the patient room (they are not placed outside the room either). The move to single room occupancy has a major effect on the number of water services, which in turn increases the risk of water system contamination should there be insufficient water turnover.
2. Sink placement—as splashing can occur up to 2 m away from the sink, it is imperative that the patient or their environment and work preparation areas outside of the zone (Fig. 4). In the image on the left, while the risk from the quality of water has been recognized and a point-of-use filter attached, the risk from splashing has not. This is a high-risk situation with water directly hitting the sieve of the drain below. The red arrow represents 1 m. This sink should be removed. Also note that the elbow-operated lever is at the wrong angle, and its operation is also impaired by the placement of the hand detergent dispenser.

   Splash screens can be installed to mitigate against the risk of splashing and contaminating nearby equipment (Fig. 5). The amount of space above the sink has been compromised, which means that the hand towel dispenser is now directly over the sink. This runs a potential risk of paper towel going down the rear drain and impairing flow.
3. Correct installation of rear drain (Fig. 6)—the red arrow is pointing in the direction from the inside of the basin looking down the rear drain. The downpipe has been the attached to the rear drain of the sink, but the overzealous use of sealant has resulted in a dam, preventing wastewater drainage. This has allowed build-up of a biofilm. This incorrect design was implicated in transmission events on a pediatric ward.

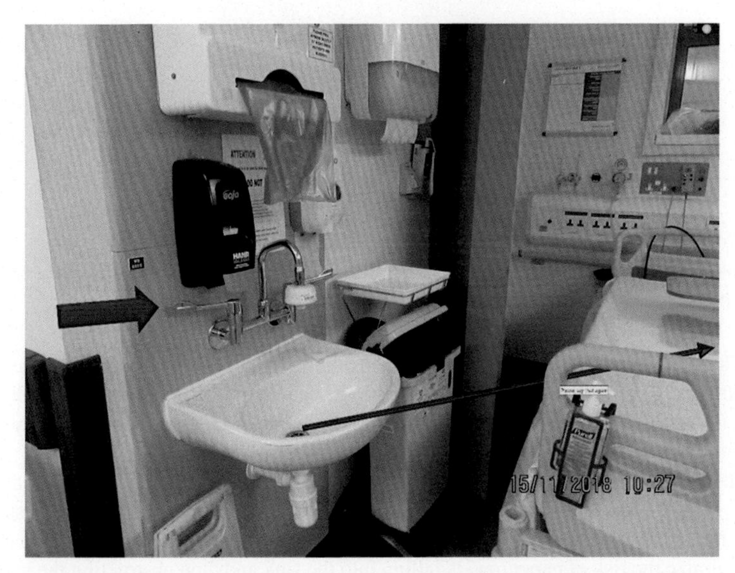

Fig. 4 Location of patient environment within 2 m of the clinical hand wash station.

Fig. 5 Positioning of screens to prevent splashing of equipment and patient environment. Image courtesy of Sarah Morter and colleagues, Infection Prevention and Control Department, Norfolk and Norwich University Hospital NHS Foundation Trust.

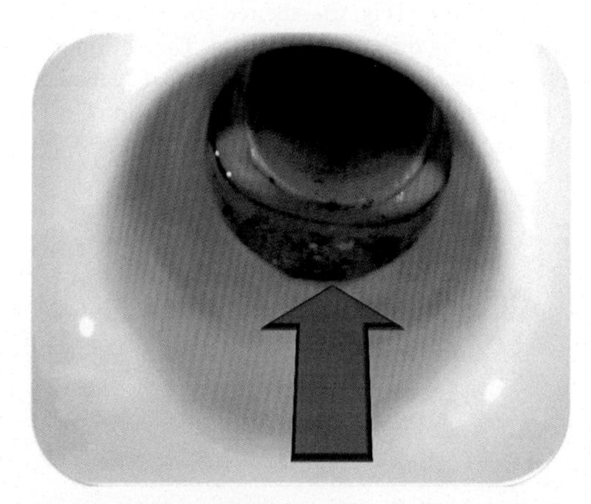

Fig. 6 Overzealous use of sealant has resulted in a dam in this rear-facing drain in a clinical hand wash basin, preventing wastewater drainage.

There are other variations in incorrect installation linked to wastewater transmission events where a trough was present between the rear drain of the sink and the downpipe (Fig. 7). As would be expected, there is cooling of wastewater and unsurprisingly biofilm formation.

4. Setup of clinical hand wash station—one might have expected that the layout of a clinical hand wash station (placement of hand towel dispenser, detergent, etc.) might be standardized, but this is not the case. The hand towel dispenser should ideally be placed to one side to minimize the risk of paper towels falling into the sink and then obstructing the rear drain. One of the reasons this is not done is because people quote a slip risk—because hands have

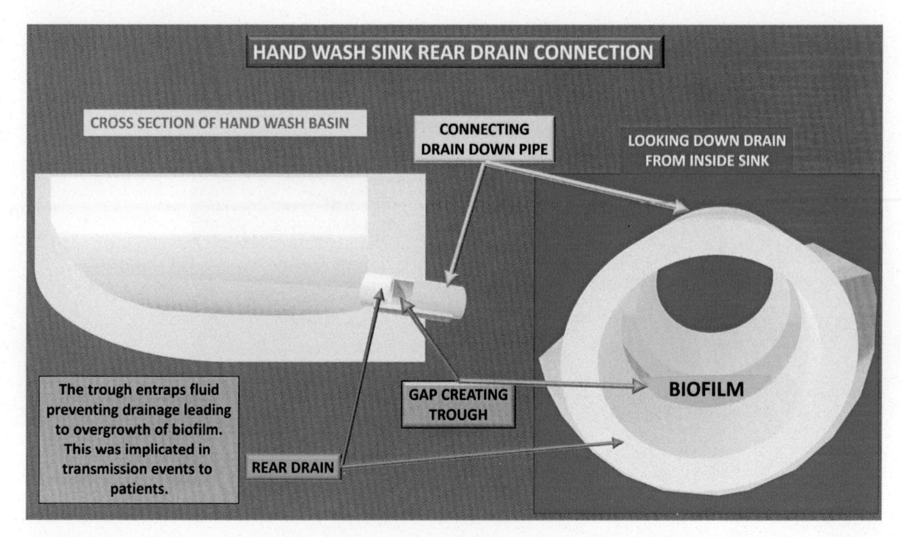

Fig. 7 Evidence of a trough that would trap biofilm between the rear drain of the sink and the downpipe.

to go to one side to get the paper towel, water droplets may fall off their hands onto the floor. My personal feeling is that this is theoretical, and that the greater risk comes from drain obstruction. If one looks at the surrounding floor after a hand wash station has been used, there is often splashes. The distance between the end of the outlet and the top of the sink basin, the activity space is important in ensuring hands either do not come into contact with the end of the outlet (or a point-of-use filter if in place) and equally to prevent hands being forced down into the basin where they may pick up wastewater organisms. The use of thermally disinfecting waste traps has to date had limited appeal. However, in new build projects, especially in augmented care areas, consideration for their installation should be made.

5. Staff discipline—Gabrowski et al. placed a video camera over a clinical hand wash station in an ITU. The camera angle was such that it would not identify individuals but could see what practices were occurring at the hand wash station. Only 4% of activities were for the correct practice, i.e., hand decontamination. To put it another way round, 96 out of every 100 activities at the hand wash station were for the wrong purpose. The inappropriate activities included using the sink as a shelf, disposal of fluids, washing items of equipment, etc. An outbreak of multidrug-resistant Pseudomonas in one hospital was linked to washing items of respiratory equipment at the clinical hand wash station. Testing the water showed this was free of Pseudomonas, but samples from the drain cultured the outbreak strain (personal communication, Dr Thekli Gee). Most staff have not received training around the risks arising from a clinical hand wash station. Their perception is that this is a place of safety.
6. Other sinks—the risk from drains/wastewater is of course not restricted to clinical hand wash stations. The design criteria to minimize risk in the clinical hand wash station are not applied to sinks used by patients, even the most vulnerable. An outbreak of CPE in Korea was linked to patients brushing their teeth—naturally the toothbrush will come into close proximity with the drain (Jung et al., 2020). Outbreaks are not limited just to sinks in the immediate vicinity of the patient. Transmission events have originated in ward kitchens and

Fig. 8 Patient water jugs being filled in a kitchen sink. The *blue* arrow is pointing at the sieve of the drain, which is obscured by the jug. In one CPE outbreak, water directly hitting the kitchen sink drain could be shown to be dispersing the outbreak strain.

the main hospital kitchens. There are risks arising from cleaning staff contaminating a cloth while cleaning the sink drain and then using the same cloth to clean the outlet is recognized, other practices that carry a similar risk are not. Patient water jugs being filled in a kitchen sink can be a risk (Fig. 8) (Decraene et al., 2018). In one CPE outbreak, water directly hitting the kitchen sink drain could be shown to be dispersing the outbreak strain. The route of transmission, although unproven, was thought to be the filling of the water jugs. Contact between the base of the jug and the sieve is likely to have contaminated the base. The jug would then be given to the patient. Other practices run a similar risk—one often sees on the wards nice disposable wash bowls, but these are filled by placing them in a sink where the base comes into contact with the drain.

7. Showers—outbreaks have been linked to the wastewater systems in showers. The inherent design, where the patient is frequently asked to stand in contact with the drain located immediately beneath the shower, is not commensurate with good infection control! It would seem to be a simple change in design to offset the drain so the patient does not have to come into contact with it. As a general principle when drainage is impaired, the risk of transmission of organisms to patients increases substantially. Perversely some of the most high-risk patients increase the risk from the environment. The administration of chemotherapy is often associated with hair loss, which leads to drain blockages. Even worse in some reports, the patients have tried to block the drains themselves, placing themselves at high risk of acquiring the organism.
8. Reporting impaired drainage/routine preventative maintenance—experience shows that staff tend to report issues with impaired drainage only at a very advanced stage, when drainage is almost completely blocked. This runs a risk as an increased likelihood of dispersal of organisms will occur long before the peaceful equipment is completely blocked. With so many demands on ward staff, whether request to report impaired drainage at an early stage will yield results remains to be seen. Domestic staff, as they often have to run appliances, might be a better source of information and compliance. In high-risk areas, for example,

patients undergoing chemotherapy or other augmented care areas, the requirement for increased routine preventative maintenance of drains should be considered. It is essential that the maintenance staff understand the risks when working around drains, the requirement for good infection control practices, and the need to ensure their equipment poses no risk (it is not carrying contamination or will be carrying contamination to the next job).

9. The water safety group—it is suggested that the following items are reviewed by the water safety group, some of which should be standing agenda items;
- review procedures for unblocking drains
- review procedures for decontaminating equipment used in unblocking drains
- review requirements for planned preventative maintenance of drains in high-risk areas
- collate data on blocked drains throughout the healthcare facility and review to see if there are underlying patterns, which could reflect poor design or poor practices (such as placement wipes into drainage system) with a view to putting in place interventions.

10. Recognizing drain transmission events—Carling published a comprehensive review of drain/wastewater outbreaks (Carling, 2018). Although there was heterogeneity in terms of the outbreaks, they tended to be protracted, transmission was often intermittent, and once recognized, standard infection control precautions failed to control the outbreak. This requires infection control teams to review data on units over a protracted period of time when conducting surveillance otherwise previous cases may not be linked. Additionally understanding that outbreaks can emanate from main kitchens, meaning that the traditional way of analyzing surveillance result by ward environment may miss hospital-wide outbreaks.

11. A number of control strategies are discussed elsewhere [cross-reference to the "Control of water borne microorganisms in healthcare"] but may include a range of physical and chemical control strategies as well as removal of the clinical hand wash basin and/or sink unit where a risk to patients has been identified (Coleman et al., 2020; Jones et al., 2020; Livingston et al., 2018a,b; Tang et al., 2020).

Summary

Although the risk from drains was highlighted in 1967 by Kohn, this was mostly ignored until the new millennium. The advent of CPEs has rascally increased the recognition of drain transmission events. However, there is nothing to suggest that CPEs have any special adaptation to spread from wastewater systems. It is thought that they merely attract our attention. The vigor with which CPE outbreaks have been investigated was not applied to ESBL-producing organisms, presumably because with the latter it was felt we always had effective antibiotics, the carbapenems. However, when people did investigate outbreaks, ESBL's transmission events could be linked to wastewater systems.

The publications in recent years linked to wastewater systems has on the whole not translated into new guidance. This is of concern as the prevention of transmission events from wastewater systems has an important bearing on the prevention of spread of antimicrobial resistance. In England, a large new hospital building programme (48 new hospitals planned) is underway. Failure to improve wastewater systems at the design stage becomes extremely expensive to rectify once a building is complete. There is an urgent need to produce new guidance and wastewater systems.

References

Aranega-Bou, P., George, R.P., Verlander, N.Q., Paton, S., Bennett, A., Moore, G., Aiken, Z., Akinremi, O., Ali, A., Cawthorne, J., Cleary, P., Crook, D.W., Decraene, V., Dodgson, A., Doumith, M., Ellington, M., Eyre, D.W., George, R.P., Grimshaw, J., Guiver, M., Hill, R., Hopkins, K., Jones, R., Lenney, C., Mathers, A.J., McEwan, A., Moore, G., Neilson, M., Neilson, S., Peto, T.E.A., Phan, H.T.T., Regan, M., Seale, A.C., Stoesser, N., Turner-Gardner, J., Watts, V., Walker, J., Sarah Walker, A., Wyllie, D., Welfare, W., Woodford, N., 2018. Carbapenem-resistant *Enterobacteriaceae* dispersal from sinks is linked to drain position and drainage rates in a laboratory model system. J. Hosp. Infect. https://doi.org/10.1016/j.jhin.2018.12.007.

Aranega-Bou, P., Cornbill, C., Verlander, N.Q., Moore, G., 2021. A splash-reducing clinical handwash basin reduces droplet-mediated dispersal from a sink contaminated with gram-negative bacteria in a laboratory model system. J. Hosp. Infect. 114, 171–174. https://doi.org/10.1016/j.jhin.2021.04.017.

Bonomo, R.A., Burd, E.M., Conly, J., Limbago, B.M., Poirel, L., Segre, J.A., Westblade, L.F., 2018. Carbapenemase-producing organisms: a global scourge. Clin. Infect. Dis. 66, 1290–1297. https://doi.org/10.1093/cid/cix893.

Breathnach, A.S., Cubbon, M.D., Karunaharan, R.N., Pope, C.F., Planche, T.D., 2012. Multidrug-resistant *Pseudomonas aeruginosa* outbreaks in two hospitals: association with contaminated hospital waste-water systems. J. Hosp. Infect. 82, 19–24. https://doi.org/10.1016/j.jhin.2012.06.007.

Carling, P.C., 2018. Wastewater drains: epidemiology and interventions in 23 carbapenem-resistant organism outbreaks. Infect. Control Hosp. Epidemiol. 39, 972–979.

Choudhuri, J.A., Chan, J.D., Schreuder, A.B., Hafermann, M.J., Fulton, C., Melius, E., McNamara, E., Pergamit, R.F., Lynch, J.B., Dellit, T.H., 2014. Shared hoppers: a novel risk factor for the transmission of *Clostridium difficile*. Infect. Control Hosp. Epidemiol. 35, 1314–1316. https://doi.org/10.1086/678077.

Coleman, D.C., Deasy, E.C., Moloney, E.M., Swan, J.S., O'Donnell, M.J., 2020. 7—Decontamination of hand washbasins and traps in hospitals. In: Walker, J. (Ed.), Decontamination in Hospitals and Healthcare, second ed. Woodhead Publishing, pp. 135–161, https://doi.org/10.1016/B978-0-08-102565-9.00007-8. Woodhead Publishing Series in Biomaterials.

Collins, W.J., 1924. The life and doctrine of Sir Edwin Chadwick. Hosp. Health Rev. 3, 50–52.

D'Angelo, R.G., Johnson, J.K., Bork, J.T., Heil, E.L., 2016. Treatment options for extended-spectrum beta-lactamase (ESBL) and AmpC-producing bacteria. Expert. Opin. Pharmacother. 17, 953–967. https://doi.org/10.1517/14656566.2016.1154538.

De Feo, G., Antoniou, G., Fardin, H.F., El-Gohary, F., Zheng, X.Y., Reklaityte, I., Butler, D., Yannopoulos, S., Angelakis, A.N., 2014. The historical development of sewers worldwide. Sustainability 6, 3936–3974. https://doi.org/10.3390/su6063936.

Decraene, V., Phan, H.T.T., George, R., Wyllie, D.H., Akinremi, O., Aiken, Z., Cleary, P., Dodgson, A., Pankhurst, L., Crook, D.W., Lenney, C., Walker, A.S., Woodford, N., Sebra, R., Fath-Ordoubadi, F., Mathers, A.J., Seale, A.C., Guiver, M., McEwan, A., Watts, V., Welfare, W., Stoesser, N., Cawthorne, J., Group, the T.I, 2018. A large, refractory nosocomial outbreak of *Klebsiella pneumoniae* carbapenemase-producing *Escherichia coli* demonstrates carbapenemase gene outbreaks involving sink sites require novel approaches to infection control. Antimicrob. Agents Chemother. 62. https://doi.org/10.1128/AAC.01689-18.

Farwell, J., 1995. Aseptic dispensing for NHS patients [Farwell report]. Department of Health, London.

Feng, Y., Wei, L., Zhu, S., Qiao, F., Zhang, X., Kang, Y., Cai, L., Kang, M., McNally, A., Zong, Z., 2020. Handwashing sinks as the source of transmission of ST16 carbapenem-resistant *Klebsiella pneumoniae*, an international high-risk clone, in an intensive care unit. J. Hosp. Infect. 104, 492–496.

Hopman, J., Tostmann, A., Wertheim, H., Bos, M., Kolwijck, E., Akkermans, R., Sturm, P., Voss, A., Pickkers, P., vd Hoeven, H., 2017. Reduced rate of intensive care unit acquired Gram-negative bacilli after removal of sinks and introduction of 'water-free' patient care. Antimicrob. Resist. Infect. Control 6, 59. https://doi.org/10.1186/s13756-017-0213-0.

Hota, S., Hirji, Z., Stockton, K., Lemieux, C., Dedier, H., Wolfaardt, G., Gardam, M.A., 2009. Outbreak of multidrug-resistant *Pseudomonas aeruginosa* colonization and infection secondary to imperfect intensive care unit room design. Infect. Control Hosp. Epidemiol. 30, 25–33. https://doi.org/10.1086/592700.

Jamal, A., Brown, K.A., Katz, K., Johnstone, J., Muller, M.P., Allen, V., Borgia, S., Boyd, D.A., Ciccotelli, W., Delibasic, K., Fisman, D., Leis, J., Li, A., Mataseje, L., Mehta, M., Mulvey, M., Ng, W., Pantelidis, R., Paterson, A., McGeer, A., 2019. Risk factors for contamination with carbapenemase-producing Enterobacteriales (CPE) in exposed hospital drains in Ontario, Canada. Open Forum Infect. Dis. 6, S441. https://doi.org/10.1093/ofid/ofz360.1091.

Jones, L.D., Mana, T.S.C., Cadnum, J.L., Jencson, A.L., Silva, S.Y., Wilson, B.M., Donskey, C.J., 2020. Effectiveness of foam disinfectants in reducing sink-drain Gram-negative bacterial colonization. Infect. Control Hosp. Epidemiol. 41, 280–285. https://doi.org/10.1017/ice.2019.325.

Jung, J., Choi, H.-S., Lee, J.-Y., Ryu, S.H., Kim, S.-K., Hong, M.J., Kwak, S.H., Kim, H.J., Lee, M.-S., Sung, H., Kim, M.-N., Kim, S.-H., 2020. Outbreak of carbapenemase-producing *Enterobacteriaceae* associated with a contaminated water dispenser and sink drains in the cardiology units of a Korean hospital. J. Hosp. Infect. 104, 476–483. https://doi.org/10.1016/j.jhin.2019.11.015.

Kearney, A., Boyle, M.A., Curley, G.F., Humphreys, H., 2021. Preventing infections caused by carbapenemase-producing bacteria in the intensive care unit—think about the sink. J. Crit. Care 66, 52–59.

Kohn, J., 1967. *Pseudomonas* infection in hospital. Br. Med. J. 4, 548. https://doi.org/10.1136/bmj.4.5578.548.

Kotay, S.M., Parikh, H.I., Barry, K., Gweon, H.S., Guilford, W., Carroll, J., Mathers, A.J., 2020. Nutrients influence the dynamics of *Klebsiella pneumoniae* carbapenemase producing enterobacterales in transplanted hospital sinks. Water Res. 176, 115707. https://doi.org/10.1016/j.watres.2020.115707.

Livingston, S., Cadnum, J.L., Gestrich, S., Jencson, A.L., Donskey, C.J., 2018a. Efficacy of automated disinfection with ozonated water in reducing sink drainage system colonization with *Pseudomonas* species and *Candida auris*. Infect. Control Hosp. Epidemiol. 39, 1497–1498. https://doi.org/10.1017/ice.2018.176.

Livingston, S.H., Cadnum, J.L., Gestrich, S., Jencson, A.L., Donskey, C.J., 2018b. A novel sink drain cover prevents dispersal of microorganisms from contaminated sink drains. Infect. Control Hosp. Epidemiol. 39, 1254–1256. https://doi.org/10.1017/ice.2018.192.

Loveday, H.P., 2020. Revisiting Florence Nightingale: International Year of the Nurse and Midwife 2020. J. Infect. Prev. 21, 4–6. https://doi.org/10.1177/1757177419896246.

Mathers, A.J., Vegesana, K., German Mesner, I., Barry, K.E., Pannone, A., Baumann, J., Crook, D.W., Stoesser, N., Kotay, S., Carroll, J., Sifri, C.D., 2018. Intensive care unit wastewater interventions to prevent transmission of multispecies *Klebsiella pneumoniae* carbapenemase–producing organisms. Clin. Infect. Dis. 67, 171–178. https://doi.org/10.1093/cid/ciy052.

Moloney, E.M., Deasy, E.C., Swan, J.S., Brennan, G.I., O'Donnell, M.J., Coleman, D.C., 2020. Whole-genome sequencing identifies highly related *Pseudomonas aeruginosa* strains in multiple washbasin U-bends at several locations in one hospital: evidence for trafficking of potential pathogens via wastewater pipes. J. Hosp. Infect. 104, 484–491.

Porter, D.H., 2001. The Great Stink of London: Sir Joseph Bazalgette and the Cleansing of the Victorian Metropolis (review). Vic. Stud. 43, 530–531. https://doi.org/10.1353/vic.2001.0074.

Roux, D., Aubier, B., Cochard, H., Quentin, R., van der Mee-Marquet, N., HAI Prevention Group of the Réseau des Hygiénistes du Centre, 2013. Contaminated sinks in intensive care units: an underestimated source of extended-spectrum beta-lactamase-producing *Enterobacteriaceae* in the patient environment. J. Hosp. Infect. 85, 106–111.

Smismans, A., Ho, E., Daniels, D., Ombelet, S., Mellaerts, B., Obbels, D., Valgaeren, H., Goovaerts, A., Huybrechts, E., Montag, I., Frans, J., 2019. New environmental reservoir of CPE in hospitals. Lancet Infect. Dis. 19, 580–581. https://doi.org/10.1016/S1473-3099(19)30230-0.

Tang, L., Tadros, M., Matukas, L., Taggart, L., Muller, M., 2020. Sink and drain monitoring and decontamination protocol for carbapenemase-producing *Enterobacteriaceae* (CPE). Am. J. Infect. Control 48, S17. https://doi.org/10.1016/j.ajic.2020.06.132.

van Oosten, R., 2016. The Dutch Great Stink: The End of the Cesspit Era in the Pre-Industrial Towns of Leiden and Haarlem. Eur. J. Archaeol. 19, 704–727.

Vardoulakis, S., Espinoza Oyarce, D.A., Donner, E., 2022. Transmission of COVID-19 and other infectious diseases in public washrooms: a systematic review. Sci. Total Environ. 803, 149932. https://doi.org/10.1016/j.scitotenv.2021.149932.

Walker, J.T., Jhutty, A., Parks, S., Willis, C., Copley, V., Turton, J.F., Hoffman, P.N., Bennett, A.M., 2014. Investigation of healthcare-acquired infections associated with *Pseudomonas aeruginosa* biofilms in taps in neonatal units in Northern Ireland. J. Hosp. Infect. 86, 16–23. https://doi.org/10.1016/j.jhin.2013.10.003.

Weinbren, M.J., 2020. Dissemination of antibiotic resistance and other healthcare waterborne pathogens. The price of poor design, construction, usage and maintenance of modern water/sanitation services. J. Hosp. Infect. 105, 406–411. https://doi.org/10.1016/j.jhin.2020.03.034.

Wolf, I., Bergervoet, P.W.M., Sebens, F.W., van den Oever, H.L.A., Savelkoul, P.H.M., van der Zwet, W.C., 2014. The sink as a correctable source of extended-spectrum β-lactamase contamination for patients in the intensive care unit. J. Hosp. Infect. 87, 126–130. https://doi.org/10.1016/j.jhin.2014.02.013.

Wong, S.-C., Yuen, L.L.-H., Chan, V.W.-M., Chen, J.H.-K., To, K.K.-W., Yuen, K.-Y., Cheng, V.C.-C., 2021. Airborne transmission of severe acute respiratory syndrome coronavirus 2 (SARS-CoV-2): what is the implication of hospital infection control? Infect. Control Hosp. Epidemiol. 1–2. https://doi.org/10.1017/ice.2021.318.

Yannopoulos, S., Yapijakis, C., Kaiafa-Saropoulou, A., Antoniou, G., Angelakis, A.N., 2017. History of sanitation and hygiene technologies in the Hellenic world. J. Water Sanit. Hyg. Dev. 7, 163–180. https://doi.org/10.2166/washdev.2017.178.

Reverse osmosis systems

24

What are reverse osmosis units and what are they used for?

Reverse osmosis (RO) units are used to purify supply water and remove microorganisms and impurities in the water. Drinking water standards are inadequate for high-dependency patients in hemodialysis units or for the use in washer disinfectors. RO units are used in a number of hospitals to supply purified water to haemodialysis units or equipment such as endoscopy repressors, washer disinfector, and dental chairs. However, due to their continued use, they can result in microorganisms breaching the membranes and contamination of the water source they are supposed to protect.

RO is a physical water purification process (FDA, 2019). Osmosis is the process by which semipermeable membranes are used to remove unwanted pollutants, e.g., ions, molecules, large particles, and bacteria from drinking water. This normal osmotic flow can be reversed (hence the name reverse osmosis) by applying pressure to the more concentrated (contaminated) solution to produce purified water.

In essence, RO units remove dissolved inorganic and organic contaminants, bacterial endotoxins, and microorganisms by passing the feed water, under pressure, through the semipermeable membrane against an osmotic gradient to produce the product water. RO units can be fitted with a final 0.2 μm filter to control bacterial numbers (Fig. 1).

The purified water has a low Total Organic Carbon (TOC) level and microbial population and thus provides water quality suitable for diluting washer disinfectors (surgical instruments or for endoscopy), chemicals, and final rinse.

Design and operation of RO units

RO is part of the water treatment plant that may consist of a cold water break tank (to water supply regulation standards), water-softening plant, prefilters; granular-activated carbon (GAC) filters for chlorine and chloramine removal, and a final RO treatment unit (Figs. 2 and 3).

Measures are required to maintain the microbial quality of water during storage and distribution as there will be no residual chemicals to control the microbial growth. The retention of this water quality requires a high level of maintenance.

The quality of the source of the water supply (e.g., local tank supply or directly from town mains) needs to be assessed as the level of residual chlorine can otherwise destroy RO membranes.

Safe Water in Healthcare. https://doi.org/10.1016/B978-0-323-90492-6.00029-X

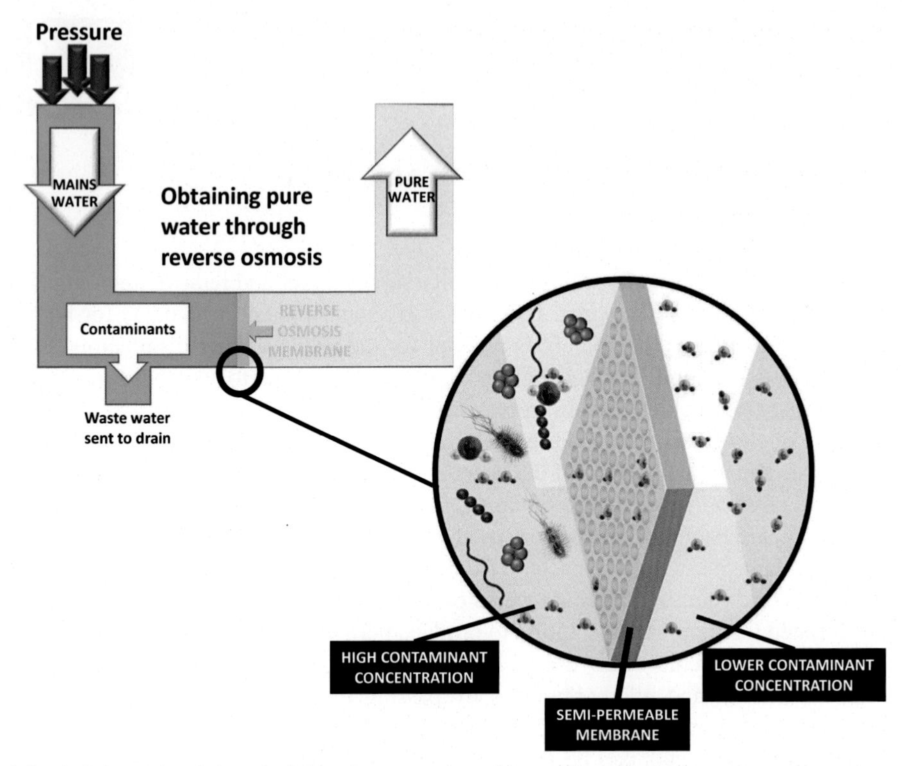

Fig. 1 Schematic of the principles of reverse osmosis.

Fig. 2 Prefiltration and carbon filters.
Image courtesy of Peter Brown.

Fig. 3 Reverse osmosis unit suppling decontamination unit for endoscope washer disinfectors. Image courtesy of Peter Brown.

What is renal dialysis water?

For hemodialysis, water purity must meet the minimum standards for regular water quoted by the European Pharmacopoeia (European Pharmacopoeia 3rd Edition 1997 (Supplement), 2000). For hemodiafiltration, the water quality must achieve ultra-pure standards (European Renal Association-European Dialysis and Transplant Association's (ERA-EDTA) "European best practice guidelines for haemodialysis") by, for example, a "double pass reverse osmosis (RO)" may be required, and this will have an effect on the space allocated to the water treatment room.

What are the sources of microbial contamination?

Where manufacturers have wet tested components in their manufacturing plants, then they should ensure that components have been disinfected to control any residual microbial biofilm growth.

The water supply can be a source of contamination for RO units. The RO may be positioned far downstream of the incoming mains supply and so the quality of the water supplied to the RO unit needs to be assessed locally to the unit.

What are the routes of dispersal of microorganisms from RO units?

Following production of the RO water, the quality of the water will need to be managed and maintained to ensure the water does not deteriorate. Where microorganisms have penetrated the RO membranes and/or the downstream pipework becomes contaminated, then biofilms will start to form. These biofilms will lead to a deterioration of the produced water and contamination of downstream equipment such as hemodialysis units, dental chairs, surgical instrument washer disinfectors, and endoscopy washer disinfectors.

What are the microbiological risks with RO units?

Naturally occurring Gram-negative bacteria can colonize, multiply, and be disseminated from RO water units resulting in contaminated pipework downstream including *Delftia acidovorans* (https://pubmed.ncbi.nlm.nih.gov/31952870/), *Ralstonia mannitolilytica* (Said et al., 2020), *Stenotrophomonas maltophilia* (Thet et al., 2019), *Aspergillus fumigatus* (Khalsa et al., 2014).

Stagnant water in pipes downstream of the membrane is one of the major sources of bacteria and endotoxin in the product water (FDA, 2019).

Incidents and outbreaks associated with haemodialysis units

Said et al. investigated an outbreak that resulted in *R. mannitolilytica* bacteraemia in 16 affected patients in a hemodialysis unit in which there was one fatality. All infected patients underwent hemodialysis within the hospital hemodialysis unit. Environmental sampling identified that the RO unit was positive for *R. mannitolilytica,* which were identical to the patient strain. The pump unit of the RO system was found to be leaking. Water within the RO system was found to be positive for *Cupriavadus pauculus* and *R. pickettii,* and *Sphingomonas paucimobilis* was also detected after the UV light. The authors speculated that *Ralstonia* species may have survived the UV light and occurred in small numbers, likely below the limit of detection of culture, hence the reason that it was cultured from patient specimens but not from the water samples collected in the wards. Following an overhaul of the RO units, there were no further infections

Microbiological monitoring for RO units

Drinking water standards for dialysis patients are inadequate as patients are exposed to many thousands of liters of dialysis fluid annually. Water to be used for dialysis needs to be treated appropriately to remove impurities. For normal hemodialysis, water purity must meet the minimum standards for regular water quoted by the European Pharmacopoeia (European Pharmacopoeia 3rd Edition 1997 (Supplement), 2000).

A continuous supply of water of the specified chemical and microbial quality is essential for a number of hospital departments including hemodialysis, washer disinfector, and endoscopy washer disinfectors. Water that is too hard or has too high a concentration of total organic carbon may impair the activity of detergents (or require the use of increased quantities of chemical additives) and cause deposits, scaling, or corrosion of the items being processed. Ensure conductivity readings, endotoxin levels, and TVCs are satisfactory (HTM 01-06, part E) (Table 1).

Table 1 Periodic final rinse-water tests: satisfactory results.

Water test	Satisfactory results	Frequency
Total viable count	<10 CFU/100 mL acceptable	Weekly
Environmental *Mycobacteria*	Nondetected in 100 mL	Quarterly
Electrical conductivity	<40 μS/cm at 25°C	Weekly
Pseudomonas aeruginosa	Nondetected in 100 mL	Quarterly
Total organic carbon	<1 mg/L	Annually
Appearance	Clear, bright and colorless	Annually
pH	5.5–8.0	Annually
Hardness	<50 mg/L $CaCO_3$	Weekly (if appropriate)

Modified from HTM 01-06 (DHSC, 2016. Management and Decontamination of Flexible Endoscopes (HTM 01-06). GOV.UK., pp. 01–06).

Guidance for reverse osmosis units

HTM 04-01 Safe water in healthcare premises.
HTM 01-06 Management and decontamination of flexible endoscopes.
Health Building Note 07-02: Main renal unit: planning and design.
ISO 13959: "Water for haemodialysis and related therapies"; or • AAMI RD-52 2004 (Association for the Advancement of Medical Instrumentation) standards.
European Pharmacopoeia 3rd Edition 1997 (Supplement) (2000). Water for diluting concentrated haemodialysis solutions. European Directorate for the Quality of Medicines, Council of Europe, Strasbourg, France.
ISO 13959:2002. Water for hemodialysis and related therapies. International Organization for Standardization, Geneva, Switzerland.

Requirements and recommendations for RO units

Ensure that there has been an audit of mains water supply from point of entry to hospital to point of use where the RO unit is situated.

Determine the quality of feed water supply (e.g., local tank supply or directly from town mains) and have records on the hardness and chlorine concentrations as the level of residual chlorine is important as it can destroy RO membranes.

Determine where the RO will be placed to ensure that there is sufficient space for maintenance and servicing the unit, whether there is sufficient space for the water storage tank? Will the floor stand the weight?

If RO water is to be distributed to several units, specify the type and quality of pipe material as the RO distribution system and associated pipework may need to be sanitized regularly.

Determine how will treated RO water be routinely monitored to determine if the RO system is fouling, suffering from scale formation, or if the RO membranes are deteriorating.

Feed water and product water should be monitored for microbiological quality. The system should be disinfected when microbiological quality levels are exceeded.

The chemical and microbial quality of water should be tested at predetermined intervals during a production cycle. In-line conductivity probes should be installed at key points for continuous monitoring of water quality.

Design water sample draw-off points at convenient locations within the system for monitoring the supply and produced RO water with sampling ports that are easy to access.

Where microbial filters are used post RO, determine how often the filters will be serviced and maintained.

Acknowledgment

We are grateful to Peter Brown for supplying the images and for his comments on the reverse osmosis section.

References

European Pharmacopoeia 3rd Edition 1997 (Supplement), 2000. Water for Diluting Concentrated Haemodialysis Solutions. European Directorate for the Quality of Medicines, Council of Europe, Strasbourg, France, 2000.

FDA, O. of R, 2019. Reverse Osmosis. FDA.

Khalsa, K., Smith, A., Morrison, P., Shaw, D., Peat, M., Howard, P., Hamilton, K., Stewart, A., 2014. Contamination of a purified water system by *Aspergillus fumigatus* in a new endoscopy reprocessing unit. Am. J. Infect. Control 42, 1337–1339. https://doi.org/10.1016/j.ajic.2014.08.008.

Said, M., van Hougenhouck-Tulleken, W., Naidoo, R., Mbelle, N., Ismail, F., 2020. Outbreak of *Ralstonia mannitolilytica* bacteraemia in patients undergoing haemodialysis at a tertiary hospital in Pretoria, South Africa. Antimicrob. Resist. Infect. Control 9, 117. https://doi.org/10.1186/s13756-020-00778-7.

Thet, K., Pelobello, M.L.F., Das, M., Alhaji, M.M., Chong, V.H., Khalil, M.A.M., Chinniah, T., Tan, J., 2019. Outbreak of nonfermentative Gram-negative bacteria (*Ralstonia pickettii* and *Stenotrophomonas maltophilia*) in a hemodialysis center. Hemodial. Int. Int. Symp. Home Hemodial. 23, E83–E89. https://doi.org/10.1111/hdi.12722.

Remote monitoring of water temperature and biofilms

Why you should monitor your water system

The importance of maintaining water temperatures outside of the recognized growth temperature ranges of most opportunistic waterborne pathogens is well recognized (DHSC, 2016; HSE, 2014). In healthcare premises, hot water should be heated to 60°C and distributed at a minimum of 55°C with the cold water stored and distributed below 20°C (DHSC, 2016; HSE, 2014). These are important control measures and, as such, their application must be assessed and monitored to provide assurances that they remain effective and water systems are safe to use.

Most healthcare water systems are large and often complex, either because of their original design or due to modifications of the systems over time. This means that maintaining water temperatures within safe limits in both the hot and cold storage and distribution systems can be difficult to achieve. As such extensive monitoring is required to identify problem areas so that effective remedial actions can be taken to resolve temperature issues. Historically, and still the case in most healthcare buildings, monitoring of water storage and distribution temperatures has been performed manually. This requires a person to measure water temperatures using a thermometer at calorifiers, in cold water storage tanks, and at selected outlets (or the supply pipework to TMVs) on distribution networks. For large and complex water systems, this represents a significant task and takes many person hours to achieve. Resources that are available to healthcare service providers are finite, with many competing priorities. This means that the extent and frequencies of these checks, when performed manually, are likely to be similarly limited.

Current national guidance on water temperature monitoring is based on manual monitoring, which means that the limitations inherent in this type of approach are reflected in the level of assurances that can be afforded (DHSC, 2016; HSE, 2014). Monthly checks of calorifier flow and return temperatures and hot water temperatures at sentinel and outlets and return legs on subordinate loops are recommended, as are annual cold water storage and monthly temperature checks at sentinel cold water outlets. These checks reflect the performance of the water system at a particular point in time and may be accurate at that moment in time. However, because water systems in a hospital building are in an almost constant state of flux, a temperature measurement taken at the same location at a different time may be significantly different. Indeed, it is well known that hot water temperatures measured at outlets can be low if, for example, a lot of hot water has been drawn recently from the system and the calorifier is in the process of recovering from meeting the recent demand.

Safe Water in Healthcare. https://doi.org/10.1016/B978-0-323-90492-6.00003-3

Although some healthcare buildings operate Building Management Systems (BMSs), which report on a range of important operational parameters, including some that cover water (e.g., stored hot and cold water temperatures and calorifier flow and return temperatures), these are insufficient to cover the water system in its entirety. So, even if a BMS is used, manual temperature measurements over the majority of the system are required also.

Novel approaches to hot and cold water temperatures

In recent years, a novel approach has been developed that allows temperature measurements to be taken at strategic point across water systems with minimal human intervention. By deploying temperature sensors attached to the hot and cold water supplies to water outlets (see Fig. 1), temperature measurements can be automatically recorded at frequencies far exceeding those possible by manual measuring. Similarly, much more frequent measurements of stored cold water temperatures and at calorifier flow and return pipework are also possible.

If sufficient sensors are deployed, a comprehensive temperature profile of the entire hot and cold water system can be obtained and monitored in real time.

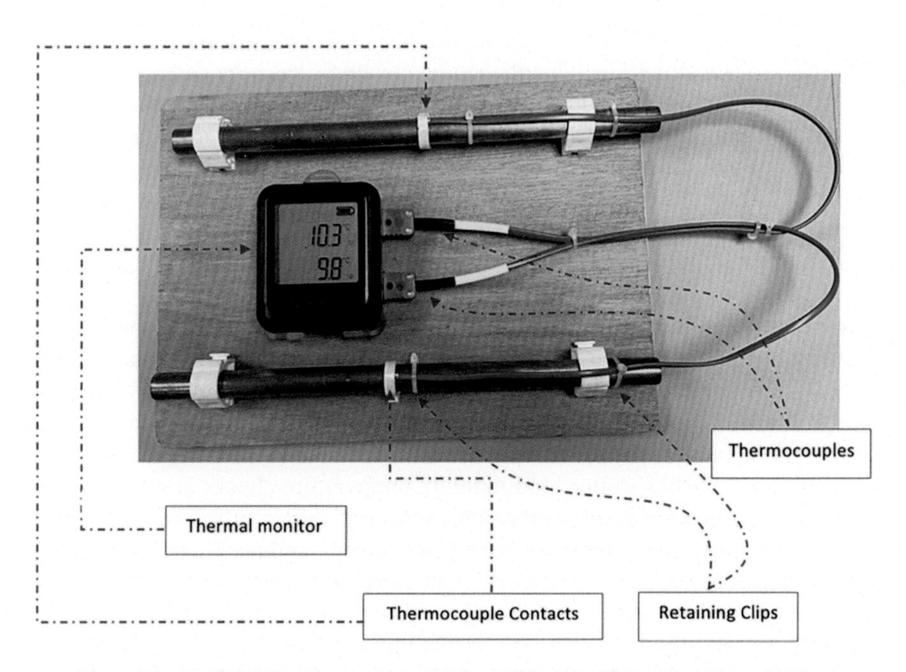

The monitor itself clips directly on to the cold pipe and 'stands off' the pipe with a suitable air gap so that there is no physical contact between the monitor and the pipe. The monitors are battery operated and the battery life under typical conditions is 15 to 18 months and is relatively easy to achieve by an engineer using a portable battery pack.

Fig. 1 A typical sensor system attached to the hot and cold supply pipework to an outlet.

Advantages of remote monitoring

The water safety group can provide their own targets for the flow and return temperatures across the estate, and the installation of monitors at strategic points allows a daily picture of performance. The monitoring system provides alerts, via text messages, to key personnel, when any of the set points fall outside of the prescribed ranges. In this way, confidence is provided in the performance of the boilers, calorifiers, or plate heat exchangers in the knowledge that if a fault arises, they will receive a "report" within a short period of time, normally of the unspecified event.

Identifying infrequently used outlets

The technology was originally developed to solve a perennial problem; to identify outlets that were either not used or used at frequencies that might allow risks to develop in water stagnating in pipework leading to outlets.

The concept is simple: when a water outlet is opened, the temperature of the supply pipework will change from the ambient temperature of the room to reflect the temperature of the water flowing through the pipework. Accordingly, if a hot tap is activated, the pipework temperature will increase from ambient. If a cold water outlet is activated, the pipework temperature will fall from ambient. If an outlet is served by a TMV, the hot and cold supplies to the TMV will change similarly (see Fig. 2). The technology

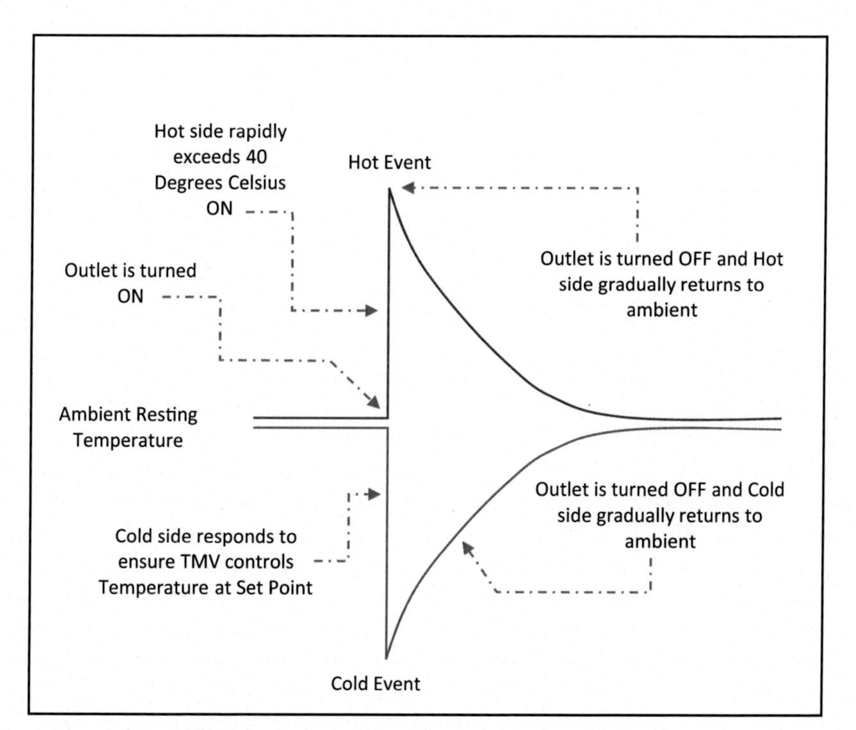

Fig. 2 Showing the changing temperatures in the supply pipework serving a TMV.

has been used successfully in some hospital wards to provide unequivocal evidence of outlets that are used less frequently than had been genuinely believed by ward staff.

Reports that are generated automatically (see Fig. 3), and which are sent to the ward, identify only those outlets that require proactive flushing, providing valuable information to ward staff regarding actual usage of outlets and avoiding unnecessary precautionary flushing of all outlets, which is often the case in some high-risk wards (e.g., augmented care areas) (Weinbren et al., 2018).

Benefits for infection control, facilities, management, and estates teams

There are additional benefits for (i) infection prevention and control practitioners who gain important information on hand washing habits of clinical workers, and for (ii) soft facilities management and cleaning contractors where the technology accurately identifies outlets that have not been cleaned to prescribed frequencies because the outlet has not been activated during the cleaning process.

After the reports have been sent, and remedial flushing has been undertaken, subsequent reports can then provide assurances that the flushing has actually taken place because the sensors will detect the changes in outlet supply temperatures. These reports then provide the records that appropriate remedial action has been taken, as required by law.

Monitoring water temperatures at outlets

In addition to determining whether or not an outlet has been activated within a prescribed timeframe, the thermometers attached to the probes can measure the temperature of the hot and cold water within the pipework. This means that supply temperatures to that outlet can be measured in much the same way as they would be if a person, using a contact probe thermometer, might measure the temperatures at that outlet manually. The main difference is that, using remote sensors, temperature measurements can be taken much more frequently, e.g., at 10 s intervals.

When measured manually, the person doing the measuring would normally open the outlet for a prescribed period. This is to check whether the hot water at the supply reached 55°C within 1 min, and that cold supply reached a temperature below 20°C within 2 min, as recommended in national guidance. These types of measurements provide information on the capability of the water system supplying that outlet to meet the performance criteria deemed necessary to control growth of pathogens in the distributed water. However, during normal operation, not all outlets are opened for 2 min when they are used, which means that the temperatures measured using remote sensors are often measured over a shorter timeframe of operation. Remote monitoring has advantages over the traditional approach because it provides information on how the water system supplying the outlet actually performs day to day, rather than how it is capable of performing under artificial conditions (see Fig. 4).

Omnia-Klenz
Intelligent Flushing & Outlet Monitoring

INTELLIGENT FLUSHING REPORT

ALL OUTLETS ARE USED / FLUSHED ON A DAILY BASIS AS AN AID TO INFECTION CONTROL PROCEDURES.

FLUSHING REPORT		LOCATION	EXAMINATION PERIOD			
DATE	15/01/2022	Ward ABC	START	14/01/2022	END	14/01/2022
USAGE MINIMUM TARGET HOT	280	USAGE MINIMUM TARGET COLD	280	TRUNCATED MAX USAGE TIME	260	

ZONE 1	TRUNCATED USAGE		MAX TEMP	MIN TEMP	ACTION REQUIRED	
LOCATION	HOT	COLD	HOT	COLD	HOT	COLD
WABC 5-144 SNK 25	520 (2)	260 (1)	62.5	11.5	--	--
WABC 5-144 SNK 26	1570 (16)	400 (6)	52.6	11.1	--	--
WABC 5-144 SNK 27	510 (2)	510 (2)	61.0	10.9	--	--
WABC 5-144 SNK 28	520 (2)	520 (2)	67.8	10.4	--	--
ZONE 2	**TRUNCATED USAGE**		**MAX TEMP**	**MIN TEMP**	**ACTION REQUIRED**	
LOCATION	HOT	COLD	HOT	COLD	HOT	COLD
WABC 5-144 SNK 29	510 (6)	140 (4)	68.5	10.3	--	FLUSH
WABC 5-144 SNK 30	330 (2)	310 (2)	70.0	10.5	--	--
WABC 5-144 SNK 31	1360 (18)	1360 (11)	69.8	10.5	--	--
WABC 5-144 SNK 32	280 (2)	270 (2)	59.3	10.1	--	--
WABC 5-144 SNK 33	260 (1)	260 (1)	58.1	10.8	--	--
WABC 5-144 SNK 34	440 (8)	290 (2)	58.5	10.7	--	--
WABC 5-144 SNK 35	0 (0)	0 (0)	22.3	20.1	FLUSH	FLUSH
WABC 5-144 SNK 36	260 (1)	260 (1)	51.9	11.2	--	--
ZONE 3	**TRUNCATED USAGE**		**MAX TEMP**	**MIN TEMP**	**ACTION REQUIRED**	
LOCATION	HOT	COLD	HOT	COLD	HOT	COLD
WABC 5-144 SNK 37	520 (2)	520 (2)	70.3	10.9	--	--
WABC 5-144 SNK 38	400 (3)	390 (4)	64.3	10.2	--	--
WABC 5-144 SNK 39	680 (3)	660 (3)	63.7	10.4	--	--
WABC 5-144 SNK 40	330 (3)	1180 (14)	64.2	10.9	--	--
ZONE 4	**TRUNCATED USAGE**		**MAX TEMP**	**MIN TEMP**	**ACTION REQUIRED**	
LOCATION	HOT	COLD	HOT	COLD	HOT	COLD
WABC 5-144 SNK 41	380 (2)	370 (2)	68.5	9.9	--	--
WABC 5-144 SNK 42	60 (3)	330 (5)	55.4	14.0	FLUSH	--
WABC 5-144 SNK 43	500 (3)	500 (4)	66.5	9.9	--	--
WABC 5-144 SNK 44	260 (1)	290 (3)	65.6	9.8	--	--
WABC 5-144 SNK 45	840 (10)	870 (10)	53.5	12.9	--	--
WABC 5-144 SNK 46	260 (1)	260 (1)	51.8	11.0	--	--
WABC 5-144 SNK 47	850 (5)	900 (5)	61.5	10.8	--	--
WABC 5-144 SNK 48	260 (1)	260 (1)	53.8	10.7	--	--

Fig. 3 Shows a typical report generated by the remote sensors located in a ward area and which may be sent to the ward manager.

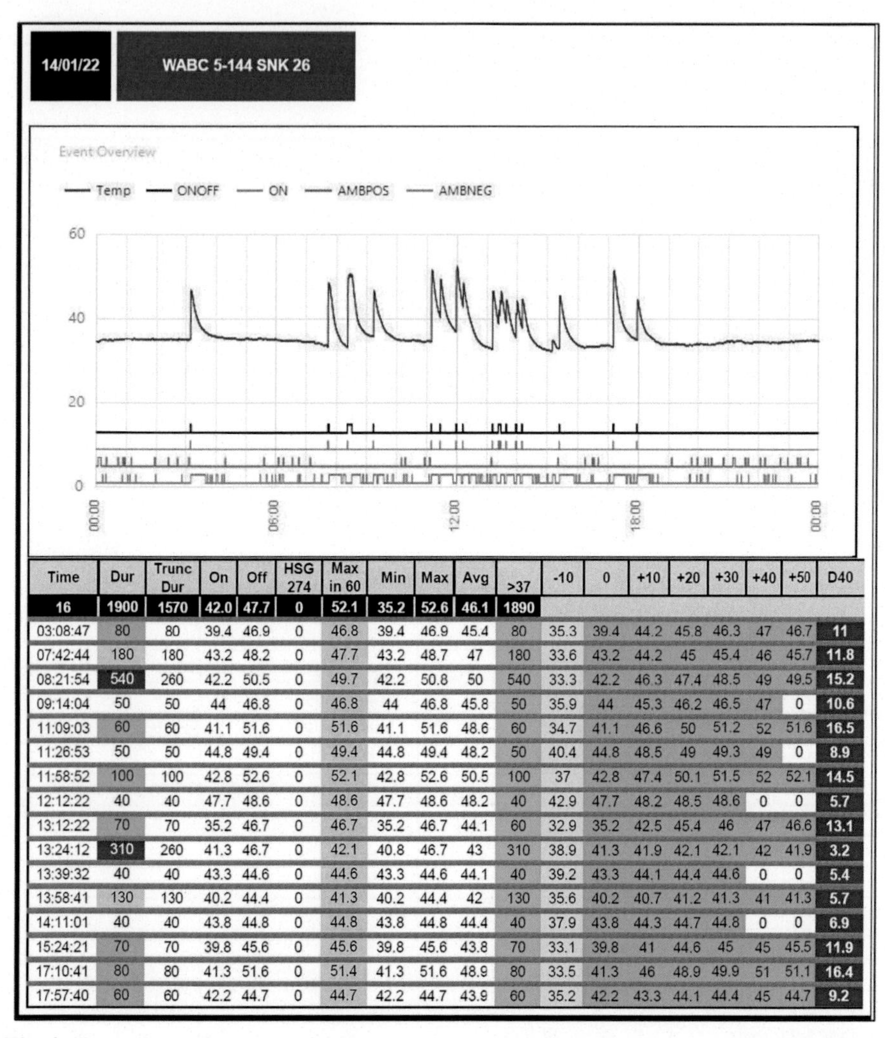

Time	Dur	Trunc Dur	On	Off	HSG 274	Max in 60	Min	Max	Avg	>37	-10	0	+10	+20	+30	+40	+50	D40
16	1900	1570	42.0	47.7	0	52.1	35.2	52.6	46.1	1890								
03:08:47	80	80	39.4	46.9	0	46.8	39.4	46.9	45.4	80	35.3	39.4	44.2	45.8	46.3	47	46.7	11
07:42:44	180	180	43.2	48.2	0	47.7	43.2	48.7	47	180	33.6	43.2	44.2	45	45.4	46	45.7	11.8
08:21:54	540	260	42.2	50.5	0	49.7	42.2	50.8	50	540	33.3	42.2	46.3	47.4	48.5	49	49.5	15.2
09:14:04	50	50	44	46.8	0	46.8	44	46.8	45.8	50	35.9	44	45.3	46.2	46.5	47	0	10.6
11:09:03	60	60	41.1	51.6	0	51.6	41.1	51.6	48.6	60	34.7	41.1	46.6	50	51.2	52	51.6	16.5
11:26:53	50	50	44.8	49.4	0	49.4	44.8	49.4	48.2	50	40.4	44.8	48.5	49	49.3	49	0	8.9
11:58:52	100	100	42.8	52.6	0	52.1	42.8	52.6	50.5	100	37	42.8	47.4	50.1	51.5	52	52.1	14.5
12:12:22	40	40	47.7	48.6	0	48.6	47.7	48.6	48.2	40	42.9	47.7	48.2	48.5	48.6	0	0	5.7
13:12:22	70	70	35.2	46.7	0	46.7	35.2	46.7	44.1	60	32.9	35.2	42.5	45.4	46	47	46.6	13.1
13:24:12	310	260	41.3	46.7	0	42.1	40.8	46.7	43	310	38.9	41.3	41.9	42.1	42.1	42	41.9	3.2
13:39:32	40	40	43.3	44.6	0	44.6	43.3	44.6	44.1	40	39.2	43.3	44.1	44.4	44.6	0	0	5.4
13:58:41	130	130	40.2	44.4	0	41.3	40.2	44.4	42	130	35.6	40.2	40.7	41.2	41.3	41	41.3	5.7
14:11:01	40	40	43.8	44.8	0	44.8	43.8	44.8	44.4	40	37.9	43.8	44.3	44.7	44.8	0	0	6.9
15:24:21	70	70	39.8	45.6	0	45.6	39.8	45.6	43.8	70	33.1	39.8	41	44.6	45	45	45.5	11.9
17:10:41	80	80	41.3	51.6	0	51.4	41.3	51.6	48.9	80	33.5	41.3	46	48.9	49.9	51	51.1	16.4
17:57:40	60	60	42.2	44.7	0	44.7	42.2	44.7	43.9	60	35.2	42.2	43.3	44.1	44.4	45	44.7	9.2

Fig. 4 Shows the performance of the water system supplying an outlet over a 24 h period.

Because the data gathered using remote monitoring are different from those obtained using the conventional approach, thought must be given to the way that the data are processed. It is possible for these types of systems to be set up so as to allow alerts to be sent to an appropriate individual (e.g., the Authorized Person in an Estates Department). However, careful consideration must be given to deciding on the tolerances within the system and set points that trigger an alert to avoid inundating that person with messages that might reflect brief and/or transient changes in water temperatures that are not critical to the safe operation of the water system. Sophisticated algorithms based on risk analysis have been created to identify when issues need to be reported.

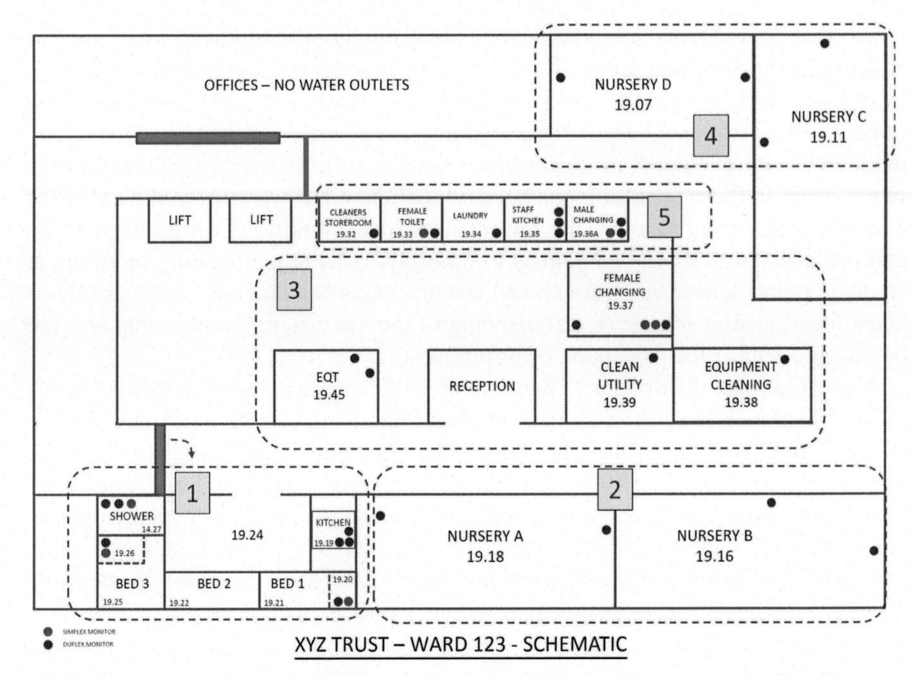

Fig. 5 Site-specific ward map indicating where outlet is located for personnel to identify specific locations.

Reports can also be accompanied by a specific ward diagram to assist those tasked with the flushing of those outlets identified as being under-utilized to easily identify the specific locations (Fig. 5).

Monitoring the performance of other parts of the hot and cold water system

Remote temperature sensors can be located almost anywhere on a water system, provided that access to that location is possible. Accordingly, remote sensors can be installed at the incoming supply, cold water supply tanks, and at hot water calorifiers to provide assurances that hot and cold water is at the correct temperature before it begins its journey around the building; if these temperatures are not correct, then temperatures elsewhere in the building may present an opportunity for microbial growth (HSE, 2014).

As discussed in Chapter 4 [cross-reference to the hot water section], most hot water systems in hospital buildings circulate the hot water supplied from the calorifier in loops (principal loops) around individual floors of the building, traveling through risers between the floors. It is often that case that smaller loops (subordinate loops) are fed from this main loop, as water is distributed within certain areas on that floor, e.g., a ward area. There may also be further smaller loops (tertiary loops) that are fed from subordinate loops that supply other discrete areas, such as en-suite facilities in side rooms. It is often that, as water passes from principal loop to subordinate loop and then into tertiary loops, changes in water pressure result in poor water flow as the size of the loop and the diameter of the pipework within the loop diminish. Poor flow performance usually results in a drop in temperature of the water in these smaller loops because sufficient hot water from the calorifier does not pass through them. This can result in local conditions within these loops that can allow waterborne pathogens to multiply. Placing temperature monitors at strategic points within subordinate and tertiary loops can diagnose these problems and allow remedial actions (e.g., balancing of the hot water system in that area) to be taken to reduce the likelihood of proliferation.

Most cold water distribution systems in hospitals are "single pipe" [cross-reference to the cold section in Chapter 4], meaning that water is supplied in a network of pipework running from the cold water tank but not recirculated back to that cold water tank. This type of design means that water often travels significant distances through ductwork before it reaches an outlet. While it is traveling to the outlet, there is potential for the cold water to gain temperature and exceed 20°C, a situation that is often exacerbated by the proximity often of hot water distribution pipework in the same ducting but may also occur near heating panels (HSE, 2014).

Placing temperature sensors on these long pipe runs can provide vital information on the performance of the cold water distribution system, allowing measures to be taken to reduce heat gain, if necessary, e.g., investigating frequency of outlet use, increasing flushing activities to draw more cold water through the pipework, or improving standards of thermal insulation (DHSC, 2016; HSE, 2014).

Alternatives may be to install a recirculating cold water system (Baillie, 2020; Kemper, 2022).

Additional benefits of remote monitoring

Clearly, the use of remote temperature monitoring affords more frequent and extensive monitoring of the performance of water systems and allows, for the first time, an effective means of monitoring the frequency and duration of outlet usage. Targeted flushing of only those outlets that have not been used within a frequency prescribed by the water safety group means that the task of proactive flushing is much less onerous, and there are significant cost savings that can be made in the number of person hours required alone. It also means that busy ward staff who may be given this task have less outlets to run, leaving their time free to provide care to patients.

Carbon footprint and the economics of water resources

Perhaps more importantly, from a cost saving perspective, is the reduction in the amount of water that might otherwise be run needlessly to drain. It should be remembered that in addition to the intrinsic cost of the water itself, a significant amount of hospital water is also heated, softened, and/or dosed with biocides, which adds to the overall value of the water that is lost during flushing activities.

It is relatively straightforward to calculate the costs of flushing in terms of water used, water costs, manpower costs, energy costs, and the combined Carbon Footprint associated with the process. From the experience of the those who have devised and implemented the remote monitoring systems in large number of applications, they estimate that a system operating with an "Intelligent flushing" program in place can reduce the number of outlets to be flushed compared with a blanket flushing approach by around 80%, and the subsequent savings in costs and Carbon Footprint are substantial (Figs. 6 and 7).

CURRENT BLANKET FLUSHING PROCEDURES:		CARBON FOOTPRINT / 1000 OUTLETS / YEAR			
WATER SUPPLY	EFFLUENT TREATMENT	ENERGY			TOTAL kg CO_2e
kg CO_2e	kg CO_2e	kg CO_2e	kg CH_4	kg N_2O	
619.84	1131.52	18520.39	25.28	10.11	20271.75

Fig. 6 Carbon footprint of large healthcare building with the installation of remote monitoring.

INTELLIGENT FLUSHING CASE:		CARBON FOOTPRINT / 1000 OUTLETS / YEAR			
WATER SUPPLY	EFFLUENT TREATMENT	ENERGY			TOTAL kg CO_2e
kg CO_2e	kg CO_2e	kg CO_2e	kg CH_4	kg N_2O	
137.48	250.97	3789.27	5.17	2.07	4177.72

Fig. 7 Estimates regarding potential reductions in carbon footprint and cost savings provided through the use of remote sensors and a targeted approach to flushing infrequently used outlets.

Impact of remote monitoring on reducing the risk to patients

Finally, another benefit of using remote monitoring centers on infection prevention and control. Traditional monitoring approaches necessitate individuals entering ward areas to open outlets to take water temperature measurements. This traffic, particularly through ward areas that provide care to highly vulnerable patients, introduces risks of contamination and cross-contamination from operatives, their clothing, and the equipment that they use. Remote monitoring effectively eliminates these risks and should help reduce infection risks in all clinical areas.

Linking remote monitoring to automated flushing events

While the conventional method of flushing has been manual using human resources within hospitals, a number of trusts are now connecting their remote monitoring systems to "intelligent automatic flushing" regimes that ensure that the flushing is carried out effectively and efficiently to reduce the risk of microbial growth and as recorded by the remote monitoring.

Use of remote monitoring to assess biofilm development

Other types of monitoring have also now been bolted on to existing monitoring systems including biofilm sensors placed in wash hand basins and drains (Baillie, 2020.

On-site biofilm detection methods

Unlike flow and temperature monitoring, biofilm monitoring devices provide definitive evidence that biofilm formation is taking place. Such real-time online monitoring system contamination enables profiles to be established and action to remediate the presence of the biofilm.

A new method of biofilm detection developed by Angel Guard Ltd. utilizes the current response gap across an in-line electrode in a system (Baillie, 2020). In a report by the University of York (Biofilm detection for Angel Guard Ltd., 2019), tests were taken using cyclic voltammetry type methods over an electrode in a variety of biofilms demonstrating that biofilms can be easily detected using an electrode. These systems are now being installed and biofilm monitoring undertaken.

The biofilm monitoring box known as "Angel Guard Clarence" containing microcontrollers and wireless connections requires a stable power supply. When the electrode that is placed in the water begins to develop biofilm, its conductive properties change. By monitoring the voltage and current changes across the electrodes, a pattern of biofilm development can be established over time. When this occurs, software within the "Clarence monitoring box" provides alerts to a nominated member of the water safety team. Through using other measurements such as hot and cold water temperatures, risk profiles are established and algorithms are used to identify "high-risk" events. As such, actions can be taken by the water safety team to determine a safe course of action, e.g., to clean or remove particular components (Oliver, 2021).

Conclusions

The days of personnel walking round the hospital to manually record temperatures from the water tanks to the outlets are clearly numbered. The advance of remote monitoring is now enabling up to the minute updates and accurate outputs to enable the hot and cold water systems including outlets to be managed to reduce the risk to patients.

In addition, the information gathered from remote monitoring allows all of the costs to be determined, which will assist many hospitals in meeting their carbon footprint strategies.

With this additional information, estates department and infection control departments are able to utilize live accurate information to maintain and support the water distribution system to ensure that the patients are safe from waterborne infections and that systems are run cost-effectively.

Acknowledgment

We would like to acknowledge Steve and Ross Finch from Omnia-Klenz for their support and enthusiasm in sharing with us their experiences with remote monitoring and for giving permission for using their input, results, and graphics. In addition, we would also like to thank Jonathan Waggot from Angel Guard Ltd. for discussions and for literature related to their biofilm detection technologies.

References

Baillie, J., 2020. Radical washbasin and new 'pipe within a pipe' system. Health Estates J., 42–46 (Feb.).

DHSC, 2016. HTM 04-01: Safe Water in Healthcare Premises.

HSE, 2014. HSG 274 Legionnaires' Disease—Technical Guidance Part 2: The Control of Legionella Bacteria in Hot and Cold Water Systems Technical Guidance. http://www.hse.gov.uk/pubns/books/hsg274.htm. (Accessed 5 January 2021).

Kemper, 2022. Cold Water Circulation System. https://www.kemper-uk.com/building-technology/product-information/khs-drinking-water-hygiene/cold-water-circulation-khs-coolflow/?L=0.

Oliver, M., 2021. Use of Screen Printed Electrode as a Real IMTE Biosensor. Angel-Guard Ltd, East Kilbride, pp. 3–5.

Weinbren, M.J., Scott, D., Bower, W., Milanova, D., 2018. Observation study of water outlet design from a cross-infection/user perspective: time for a radical re-think? J. Hosp. Infect. https://doi.org/10.1016/j.jhin.2018.11.007.

Waterborne pathogens in healthcare water systems

26

Introduction

Infections caused by waterborne pathogens are of a major public health concern worldwide, not only by the morbidity and mortality that they cause, but by the high cost of their prevention and treatment and their ability to spread antibiotic resistance. The aim of this section is to discuss the ecology and epidemiology of infections that arise from water systems, which are colonized with pathogens within biofilms in healthcare settings. Waterborne opportunistic pathogens of major concern include: Gram-negative pathogens such as *Legionella* spp., *Pseudomonas aeruginosa*, *Cupriavidus pauculus*, *Stenotrophomonas maltophilia*, but also protozoa, fungi, and nontuberculous mycobacteria, which are normally considered Gram-positive.

Opportunistic pathogens

It is important to understand that water supplies are not sterile and contain a range of microorganisms, which do not cause harm to the general population, even within the healthcare environment. In incoming supplies, which comply with drinking water regulations, waterborne opportunistic pathogens are likely to be detected in low numbers, many in a viable but nonculturable form (VBNC) and of low virulence due to the low temperature, especially in waters below 20°C. However, once within a building when the temperature and availability of nutrients increase, they can very quickly colonize and grow to levels that can cause high morbidity and mortality in the most susceptible patients, especially those who are immunocompromised as a result of illness or treatment. Opportunistic pathogens are, therefore, natural inhabitants of water, not contaminants; this is their natural environment. They possess a number of key characteristics when growing within biofilms or protozoa including changes in their physiology, which increases their tolerance to the levels of disinfectants and temperatures that are normally advised within guidelines for controlling microbial growth in water systems.

Microorganisms termed "opportunistic" pathogens may be considered to be of relatively low virulence for most people in the general population, but able to cause infection when provided with "opportunities" to cause disease in certain patient groups. Patients who provide such opportunities include those with impaired immune systems such as premature neonates, cystic fibrosis patients, and those with breaches in their skins' integrity including burns patients (patients with significant burns are also immunocompromised) and those with indwelling medical devices, including arterial and venous lines and catheters. Waterborne infections are becoming increasingly

Safe Water in Healthcare. https://doi.org/10.1016/B978-0-323-90492-6.00031-8

important as the numbers of immunocompromised patients are increasing, and they are often discharged into the community with no risk assessment and little or no education as to how to avoid infection in the home environment. As such, any area where these patients are exposed to water or aerosols derived from water should be considered by the WSG and the risks of infection assessed.

Advances in the identification of microorganisms

From the laboratory perspective, most of these organisms tend to be metabolically inert. Up until recently, most microbiological culture and identification methods were intended for the identification of clinical isolates. When environmental isolates are cultured and isolates tested for identification purposes, usually against a panel of biochemical reagents, many may be able to metabolize only one or two of the standard metabolic test reagents. Unlike their respective clinical isolates, they are taken from a relatively nutrient starved environment to a nutrient-rich environment in culture media so are subjected to nutrient shock, which may both affect both their ability to recover and identify them. This can result in either no identification or a misidentification of the isolates from water sources, sometimes with alarming results. e.g., a misidentification of *Yersinia pestis* from an experimental laboratory test rig, the wrong identification was due to the database being intended for clinical isolates. The atmospheric conditions during incubation can also determine how well they can grow some, e.g., some *Legionella* species grow, better when incubated with low levels of CO_2 (2%–5%).

Recent technological advances including matrix-assisted laser desorption ionization-time of flight (MALDI-ToF) have only relatively recently been introduced into medical and environmental microbiology laboratories. This technology has a number of advantages including rapid analysis of colonies in hours, though bacteria still have to be cultured on agar, as well as improved sensitivity and specificity in the identification of microorganisms, rapid turnaround of large numbers of isolates, and significantly reduced costs per isolate. However, the extent of the database used may still influence the ability to correctly identify environmental isolates.

One of the reasons that novel pathogens such as *Cupriavidus* spp. reports seem to be increasing is the availability of MALDI-ToF (Cheng et al., 2016; Popović et al., 2017). Prior to the use of MALDI-ToF, such organisms may have been reported as a "non-fermenting environmental organism of dubious clinical significance." The important aspect is that until this technology was integrated into laboratories, clinical infections involving the transmission of these rare opportunistic pathogens would not have been identified and the source of such infections not sought or understood to prevent further cases.

MALDI-ToF was developed primarily for clinical isolates. Its lack of databases for environmental isolates and the large capital cost and ongoing servicing costs mean it has been slower to be integrated into water testing laboratories. Hence, there has been an untimely lag in the ability to identify waterborne microorganisms (Ashfaq et al., 2022; Pinar-Méndez et al., 2021; Popović et al., 2017). From the water safety group perspective, it is important to know whether the laboratory being used for water testing has this or other validated technology including molecular methods such as 16S rRNA or whole-genome sequencing, when investigating clinical infections, which may be associated with exposure to water, especially in the highly immunocompromised. Most competent water testing laboratories should be able to culture and identify *P. aeruginosa* and *Legionella* without MALDI-ToF technology, though care needs to be taken to ensure that the laboratory chosen preforms well in external quality assurance programs and isolates may still need to be sent away to reference laboratories for confirmation and typing with clinical strains during outbreak investigations, However, in a competent and accredited laboratory, the confidence and speed of reporting are significantly increased where MALDI-ToF is used allowing interventions to prevent further cases sooner than with traditional methods.

Conventional methods of identification involve: morphological examination (observing the color, size, smooth, or roughness of the isolate), microscopy, culture using a range of nonselective and selective enrichment media, and culture conditions, followed by biochemical confirmation, which can be unreliable, and for many of these organisms, there is no validated and accredited test for their detection in water. This has required laboratories to innovate or revert to out-of-date biochemical identification texts and the use un-validated novel tests to support outbreak investigations. For example, in the *Cupriavidus* spp., waterborne outbreak in Glasgow, the organism was detected in water samples by processing the sample initially using a nonselective method used for enumerating total viable counts (TVCs) in water (Inkster et al., 2022). This allowed the recovery of organisms, which were subsequently identified using MALDI-ToF (Inkster et al., 2022). Though as in all things microbiological, microbiology is a very inexact science, identification may not be absolute, Ferone et al. (2020) report the misidentification of closely related bacteria such as *Shigella* and *Escherichia coli* using MALDI-ToF (Ferone et al., 2020).

Microorganisms growing in water do not generally grow in pure culture, if you have a mixed population on a culture plate, how many picks from each phenotype are needed to ensure you have not missed your target? One answer may be based on the calculation of probability, and therefore, up to 30 colonies may have to be selected and analyzed. In the one outbreak, this strategy allowed the laboratory to be able to demonstrate *Cupriavidus* spp. in the water sample, which was later matched to the patient isolate (Inkster et al., 2021, 2022).

However, it must be stated that, where new technologies such as MALDI-ToF are unable to provide a statistically reliable identification, there is still a use for simple tests such as Gram-stain, biochemical test panels, various molecular methods such as qPCR, the use of antibody testing (Parraga-Nino et al., 2018), and antibiotic resistance profiling (Berg et al., 1999).

Legionella

Legionella is one of the most high-profile waterborne pathogens, with respect to causing nosocomial waterborne infection. *Legionella* are bacteria that occur naturally in freshwater environments, and once introduced into building water systems and equipment that use water, they can multiply and colonize the surfaces within them unless meticulous water management practices are in place throughout the whole lifecycle of the system. Transmission primarily occurs via inhalation of aerosols developed from water droplets containing *Legionella*, which have been dispersed into the air. Infectious aerosols are too small to be seen and have a negative falling velocity so can hang around for hours if the environmental conditions allow, these aerosol particles of $<5\,\mu m$ are then inhaled deep into the lungs and in susceptible patients can result in the development of Legionnaires' disease. For certain patient groups, particularly those with swallowing difficulties, aspiration of *Legionella* directly into the lungs when drinking water or sucking ice can also be a significant mode of infection. There are currently more than 60 known species and 80 serogroups, the majority of which have only been isolated from environmental, rather than from clinical, sources. *L. pneumophila* is the predominant species responsible for the majority of both community-acquired and nosocomial infections (ECDC, 2019; Scaturro et al., 2020). Though there is likely to be an underestimation of legionellosis caused by non-*L. pneumophila* serogroup 1 and also nonpneumophila species as the clinical diagnostic technique most often used;, urinary antigen, only reliably detects *L. pneumophila serogroup 1* (Sivagnanam et al., 2017; Miyata et al., 2017); in addition, few laboratories *use* culture techniques to detect *Legionella* (ECDC, 2019).

In England and Wales, Legionnaires' disease is notifiable under the Health Protection (Notification) Regulations 2010 and in Scotland under the Public Health (Notification of Infectious Diseases) (Scotland) Regulations 1988 (HSE, 2014). Legionnaires' disease is considered preventable if the water system is managed correctly, and duty holders including employers or those in control of premises must ensure the health and safety of their employees or others who may be affected by their undertaking. Therefore, they must take suitable precautions to prevent or control the risk of exposure to legionella (HSE, 2014).

Microbiological characteristics

Legionella are small, aerobic, are weakly staining Gram-negative but only, nonmotile bacteria, with no capsule, are weakly catalase-positive, weakly oxidase-positive.

Legionellae are fastidious in their growth requirements and use amino acids rather than carbohydrates as their preferred energy source and require l-cysteine. Iron is required for both growth and pathogenicity (Cianciotto, 2015). As a result, it took many years to develop a medium to isolate colonies (Edelstein, 1981) Legionellae can be difficult to identify when other colonies present (Fitzgeorge and Dennis, 1983) especially as they are slow growing (taking around 3 days to see on a culture plate) as they can be masked by overgrowth of faster-growing background bacteria. Some *P. aeruginosa* can inhibit *Legionella* growth on a culture plate potentially leading to a false-negative result. When grown on buffered charcoal yeast extract (BCYE) agar on culture plates, the colonies may vary in color depending on species and age, with young colonies of *Legionella* pneumophila varying from pink/purple or blue/green turning to a typically gray-white in color as they age with a textured almost mottled "ground glass"-like appearance on (Figs. 1 and 2). Some species, such as the typically slower-growing

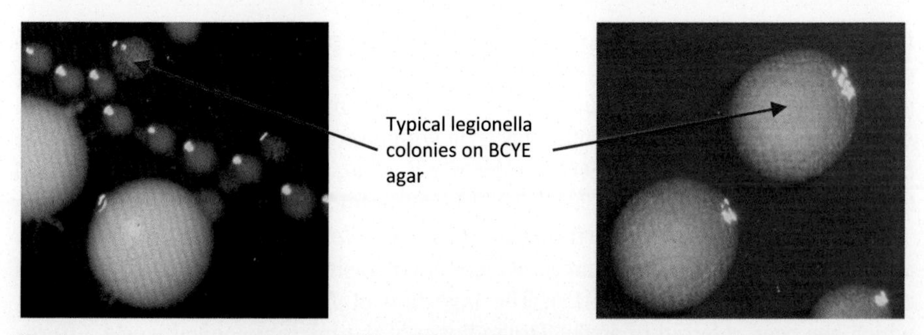

Fig. 1 *Legionella pneumophila* on buffered charcoal yeast extract demonstrating the presence of classical mottled/glassy-like colonies. Other species exhibit different colors including red, blue, green, brown, and some, including clinically relevant species, may fluoresce under UV light.

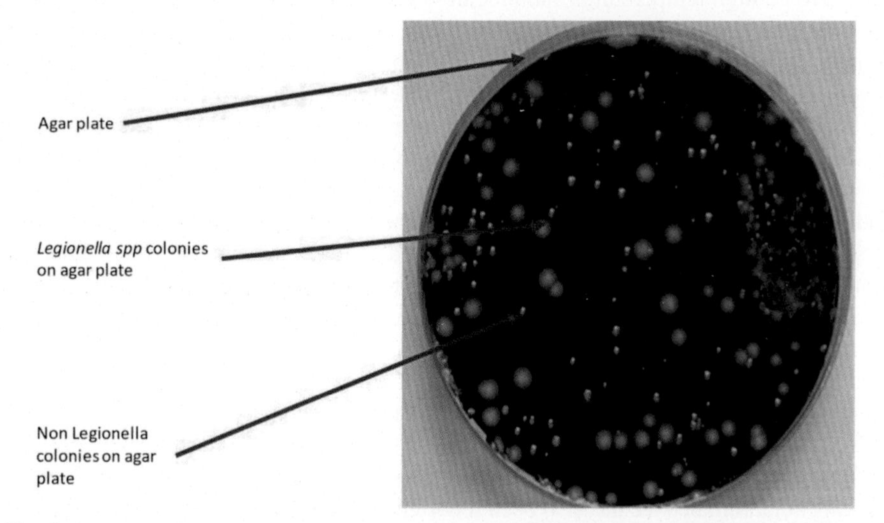

Fig. 2 Presence of *Legionella pneumophila* in a water sample grown on GVPC showing the presence of background bacteria resistant to the supplements added to BYCE limit background growth from environmental samples.

nonpneumophila species, including some of which can rarely cause human infections (primarily in immunocompromised patients), fluoresce under UV light e.g., *L.* bozemanii, *L. anisa, L. dumoffii, L. gormanii, and L. parisiensis.* Experienced microbiologists can usually predict there will be positive isolates in the incubator as they have a characteristic odor when grown on specific agar (Burillo et al., 2017).

Ecology

Legionella are widespread worldwide in natural water sources such as rivers, lakes, reservoirs albeit in low numbers. In these natural environments, they usually pose insignificant risks to health in a healthy population in cold water because of their low numbers probably because of natural grazing by protozoa and competition with other natural aquatic microbes. Risks to health increase when the bacteria enter water systems in the built environment because the conditions there tend to favor microbial growth. *Legionella* depend on other bacteria within biofilms for growth (Surman-Lee et al., 2007; Rogers et al., 1994; van der Kooij et al., 2020), relying on amino acid from coinhabitants. They are also able to grow rapidly within protozoa including *Acanthamoeba Vermamoeba* (formerly *Hartmannella vermiformis*) and *Naegleria* also common in freshwater sources, which they parasitize and rapidly grow in warm conditions (Boamah et al., 2017). The ingestion of *Legionella* by these protozoa, which has been referred to as the Trojan horse of the microbiological world, affords the bacterium protection from control strategies including heating and disinfection (Barker and Brown, 1994) and enhances virulence (Cirillo et al., 1999).

In some circumstances, the bacteria can be protected from environmental challenges (e.g., high temperatures and biocides) when the host protozoan reacts to these external influences and forms a protective cyst with *Legionella* inside. The cyst can remain dormant until such time as the unfavorable conditions have passed, and then reactivate, allowing the *Legionella* to continue to multiply until lysis.

While they will survive at temperatures lower than 20°C, most pathogenic species multiply between 25°C and 42°C, thrive at 35°C, a frequently occurring temperature in man-made systems. They will be found where water is stagnant, where scale or sediment is found, and in dead legs and blind ends of plumbing systems.

Legionella are typically found in constructed water systems such as hot and cold water systems, spa pools and hot tubs, cooling towers, evaporative condensers where optimal temperatures and other factors, such as the availability of suitable nutrient sources enable them to proliferate (Ezzeddine et al., 1989). Outbreaks associated with healthcare premises sources include: hot and cold water systems (Hugo Johansson et al., 2006; Fisher-Hoch et al., 1981), cooling towers (Brundrett, 1991; Fisher-Hoch et al., 1981), respiratory therapy equipment (Bou and Ramos, 2009; Mastro et al., 1991; Palmore et al., 2009), decorative fountains (Haupt et al., 2012), aspiration while sucking contaminated ice has also been reported (Graman et al., 1997) as well as from contaminated water used to flush nasogastric tubes.

Historical perspectives

Legionella was first identified following a large outbreak of severe pneumonia among members of the American Legion in the Stratford hotel in Philadelphia in the United States in 1976 (Stout and Yu, 1997). It has since been associated with outbreaks linked to water systems including cooling towers or evaporative condensers, hot and cold water systems in buildings and whirlpool spas before and after the outbreak in Philadelphia.

Transmission routes

Legionella multiplies in man-made water systems where it is transported through the water and is disseminated at the point of use in aerosols from water systems, which are then inhaled. Most water sources that produce aerosols have been associated with transmission of *Legionella* and include air-conditioning cooling towers, hot and cold water systems, humidifiers, and whirlpool spas (Allegra et al., 2016). Infection can also occur by aspiration of contaminated water or ice, particularly in susceptible hospital patients, and by exposure of babies during water births. Direct human-to-human transmission is rare with only one reported case of a probable but not proven case (Borges et al., 2016).

Disease symptoms

Legionellosis is the description given to the pneumonic and nonpneumonic forms of infection due to *Legionella*. The nonpneumonic form, known as Pontiac fever, has an incubation period up to 48 h is an acute, self-limiting influenza-like illness lasting 2–5 days and is nonfatal with symptoms including fever, chills, headache, malaise, and muscle pain (myalgia) (Hamilton et al., 2018).

The pneumonic form, Legionnaires' disease, has an incubation period of 2–10 days but may last up to 16 days (Beauté, 2017). Symptoms include mild cough, fever, loss of appetite, headache, malaise, lethargy, muscle pain, diarrhea, and confusion. Blood-streaked phlegm or haemoptysis occurs in about one-third of the patients. The severity of disease ranges from a mild cough to a rapidly fatal pneumonia. Death occurs through progressive pneumonia with respiratory failure and/or shock and multiorgan failure and sometime loss of limbs and extremities due to circulatory shutdown (Taylor, 2016).

Early diagnosis and prompt antimicrobial therapy improve outcomes. Untreated Legionnaires' disease usually worsens during the first week, and the most frequent complications of legionellosis are respiratory failure, shock, and acute kidney and multiorgan failure. Recovery requires antibiotic treatment but will take several weeks or months. For those that do recover, there are usually life-changing consequences (Lettinga et al., 2002).

Extent of the disease

Legionnaires' disease fatality rate depends on the severity of the disease, the appropriateness of initial antimicrobial treatment, the setting where *Legionella* was acquired, the virulence and number of bacteria, and host factors. For example, the disease is usually more serious in patients with immuno-suppression. The death rate may be as high as 40%–80% in untreated immuno-suppressed patients and may be reduced to 5%–30% through appropriate and rapid case management and depending on the severity of the clinical signs and symptoms. Overall the death rate is usually within the range of 5%–10%.

In Europe and the United States, there are about 20–25 cases detected per million population per year with 75%–80% reported cases being over 50 years and 60%–70% are male. Risk factors for community-acquired and travel-associated legionellosis include smoking, heavy drinking, pulmonary-related illness, immuno-suppression, and chronic respiratory, or renal illnesses.

According to ECDC in the 5 years up to 2021, notification rates have nearly doubled in the EU/EEA, from 1.4 in 2015 to 2.2 per 100,000 population. Four countries, France, Germany, Italy, and Spain, accounted for 71% of all notified cases (ECDC, 2019).

In the United States, CDC has reported that from 2000 through 2017, the national incidence rate increased 5.5-fold from 0.42 per 100,000 persons in 2000 to 2.29 per 100,000 persons in 2017 (CDC, 2021), and *Legionella* has been identified as the primary cause of all potable water-related outbreaks (Benedict, 2017).

Hospital-acquired pneumonia risk factors include recent surgery, intubation, mechanical ventilation, aspiration, nasogastric tubes, and the use of respiratory therapy equipment. However, the most susceptible patients are those who are immunocompromised, including organ transplant recipients and cancer patients and those receiving corticosteroid treatment.

Death from Legionnaires' disease is influenced by delay in diagnosis and administration of appropriate antibiotic treatment, increasing age, and presence of coexisting diseases.

Methods of diagnosis for *Legionella* infection include urinary antigen testing, serology, culture, and nucleic acid testing (Herwaldt and Marra, 2018). Particular concern should be raised when non-*L. pneumophila* sero-group 1 infections are identified as the environmental sources may be more difficult to identify.

Clinical treatments

Currently, there is no vaccine available for Legionnaires' disease. The nonpneumonic infection (Pontiac fever) is self-limiting and does not typically require medical interventions, including antibiotic treatment. However, patients with Legionnaires' disease do require specific antibiotic treatments.

International treatment recommendations for patients with *Legionella* infection often provide differing guidelines and regimens (Lim et al., 2009). While antibiotic resistance in *Legionella* has not been of concern, there have been recent reports of the genes encoding a macrolide efflux pump and single-point mutations in *L. pneumophila* 23S rRNA (Massip et al., 2017; Shadoud et al., 2015), which have resulted in intermediate to high levels of macrolide resistance. A novel tetracycline resistance gene in *L. longbeachae* (this organism is mainly associated with soil and nonpeat-based potting composts) is the most common source of legionellosis in Australia and New Zealand, though cases have occurred elsewhere including in the United States, Europe, and Japan (Matsushita et al., 2017; Currie and Beattie, 2015; Han et al., 2015), which highlights the need for caution and research at an international level (Forsberg et al., 2015; Portal et al., 2021).

Control mechanisms in water systems

Despite awareness of Legionnaires' disease since the 1970s and numerous national and international guidance documents, *Legionella* outbreaks still frequently occur. and it is if often due to the failure to design and manage the water system appropriately with lack of risk assessment and lack of effective schemes of control with poor management, communication, and training often found during investigations (DHSC, 2016; HSE, 2014). Various guidance documentations address control mechanisms through the implementation of water safety plans by authorities responsible for building safety or water system safety (Bartram, 2007; BSI, 2022; DHSC, 2016; HSE, 2014). Such plans can only be specific to a particular site, the users within, its water systems, other water reservoirs, the types of water use, and the susceptibility of those who might be exposed, and no two situations are identical. Strategies must include ensuring the competence of all involved in water management and treatment, identification of all inherent hazards and risks in the water system, validation of control measures, regular and accurate monitoring (preferably multiple remote monitoring points) of the water system flow and throughput, temperature, and where appropriate biocide concentrations, to reduce the risks to vulnerable users.

Prevention of Legionnaires' disease not only depends on the efficacy of control measures to reduce the proliferation of *Legionella* within the water system per se but also in controlling the equipment that results in the dissemination of aerosols. These measures, which are described in other sections, include other aspects as follows:

- Educating and training staff including within the water safety group, to ensure they are competent and the regular retraining takes place.
- Regular maintenance, cleaning, and disinfection of evaporative cooling towers together with frequent or continuous addition of biocides and the installation of close-fitting efficient drift eliminators to reduce dissemination of aerosols from cooling towers.
- The hot and cold water systems should be clean, maintained, serviced, and monitored (remotely). For example, the hot water needs to be maintained well above 50°C on the return to the calorifier and above 55°C at the last point of use before the calorifier in healthcare buildings. This will require water leaving the heating unit at or above 60°C and the cold water being 20°C at the point of use, ideally below 18°C influent to the building. Where these cold temperatures cannot be maintained and a risk has been identified, then there may an option of installing a chiller unit prior to the insulated cold water storage tank or treating the cold water with a suitable biocide to limit growth particularly in healthcare buildings. Global warming is creating challenges to the delivery of cold water at temperatures less than 20°C.
- Evidence-based choices on the type of materials used in pipework and components, flexible hoses are kept to minimum and short as possible and inappropriate jointing compounds are not used.
- Reducing stagnation by adequate flushing of unused or low used outlets in buildings on a daily/weekly basis (remote monitoring and automated flushing) should be considered. Such practices may be helpful at the end of a long run, but the complexity and material choices in touchless auto-flushing taps need to be risk-assessed based on manufacturer's long-term testing. Focus should be on removing these high-risk outlets as a longer-term solution.
- Identification and removal of dead legs and blind ends—continuous and relentless maintenance program to keep on top of this, particularly in large buildings.
- Maintaining appropriate concentrations of biocides in water systems, for example, by using remote monitoring at predetermined points.
- The development of an asset register is a requirement in the United Kingdom to determine those items that require a documented risk assessment for Legionella, scheme of control and/or maintenance (HSG 274). This should be extended in healthcare premises to include all assets, which also pose a risk of harm from other waterborne pathogens of concern to ensure they are risk-assessed, managed, and where appropriate there are maintenance contracts for plumbed in equipment, for example, water/drink dispensers, assisted baths, footbaths, whirlpool baths, ice machines, washing machines, dishwashers.

- Maintenance and servicing of water-related components, including regular cleaning, disinfection, and applying other physical (temperature) or chemical measures (biocide) to minimize growth.

Controls described above will help reduce the risk of *Legionella* colonization and growth and may reduce the occurrence of sporadic cases and outbreaks. As identified in other sections, precautions may be required for water and water-related products such as ice provided to highly susceptible patients in hospitals including those at risk of aspiration.

Effective clinical surveillance may also assist in the identification of cases in healthcare (National Services Scotland, 2019).

Summary of Legionella

- The causative bacterium of Legionnaires' disease was named after an outbreak of pneumonia at an American Legion convention in 1976 (*Legionella* after the legionnaires infected, and *pneumophila,* which is Greek for lung loving).
- Legionnaires' disease is a notifiable disease in the United Kingdom, meaning local health authorities must be informed by doctors if a case is diagnosed to help with tracing the source of a potential outbreak.
- *Legionella* spp. is a waterborne pathogen that multiplies in all water systems including healthcare water systems between 20°C and 50°C and will survive at less than 20°C and will survive at temperatures above 50°C for relatively short periods of time (Cazals et al., 2022; Dennis et al., 1984; Henry, 2017).
- Water outlets including taps and showers, evaporative cooling towers, fountains, spa pools and hot tubs, sinks, and ice have all been sources of Legionnaires' disease outbreaks.
- Legionnaires' disease is contracted by inhaling small droplets or water, or aerosols containing *Legionella* spp., or by aspiration of water or ice containing legionellae.
- Vulnerable and immuno-compromised patients including the elderly, newborns (water-births), heavy smokers, those suffering from heart, lung, or kidney disease are most at risk from contracting Legionnaires' disease. Males are more at risk than females.
- *Legionella* spp. are most likely to be present in water systems as biofilm on a wide range of plumbing materials.

Pseudomonas aeruginosa

P. aeruginosa is ubiquitous in the environment and an opportunistic pathogen responsible for both acute and chronic infections in wards/units treating immunocompromised patients such as intensive care, transplant, and cystic fibrosis units. Its name is derived from *Pseudomonas* meaning "false unit," and *aeruginosa,* which refers to the blue-green color (pyocyanin, see Fig. 2) of laboratory cultures of the species. infections can be of rapid onset and progression, difficult to diagnose, and have limited treatment options due to increasing antibiotic resistance; in 2008, it was reported that antibiotic resistance was present in up to one-third of all isolates in European surveys (Souli et al., 2008).

P. aeruginosa is able to take advantage of a number of scenarios, for example, water outlets can cause infection in vulnerable patients such as preterm babies (Coppry et al., 2020; Fleiszig et al., 2020; Spencer et al., 2021). It has a natural resistance to many antibiotics, and its ability to form biofilm often renders both environmental treatments and host defense mechanisms ineffective.

Microbiological characteristics

P. aeruginosa is a Gram-negative, rod-shaped, with a single flagellum about 1–5 μm long and 0.5–1.0 μm wide. *P. aeruginosa* uses aerobic respiration (with oxygen) as its optimal metabolism although can also respire anaerobically on nitrate or other alternative electron acceptors. *P. aeruginosa* produces a wide range of organic molecules, including organic compounds such as benzoate and is recognizable on agar plates due to the production of pyocyanin, which results in the typical characteristic green color (Fig. 2).

Ecology

P. aeruginosa is found naturally in a wide range of environments including soil and water. It can also be found colonizing domestic water systems, including those in healthcare. Unlike *Legionella*, *P. aeruginosa* is often carried by humans and animals and is also a colonizer and pathogen of plants, which can, in turn, contaminate healthcare water systems. These organisms typically grow at temperatures of between 25°C and 42°C and grow optimally at 37°C. Due to their simple nutritional requirements, they can utilize an amazing range of organic food sources, including plasticizers in plastic pipework, and can even grow in distilled water, scavenging nutrients from the air, but can also grow in high-nutrient environments such as the human gut. They are commonly found in healthcare water systems and the associated environment, including items left damp such as facecloths, cleaning equipment such as mops, cloths, and sanitizers, shaving and toothbrushes, bath toys especially where there are nutrients present (Matz et al., 2008). They will proliferate in the water phase and as biofilms where water is stagnant, scale is present, sediment is found, and in dead legs of plumbing systems (BSI, 2022; Moloney et al., 2020; Reynolds and Kollef, 2021). They are also found associated with tap outlets, therapy pools, ice machines, drinking water dispensers, and also with daily bathing in bed.

P. aeruginosa isolates demonstrate three types of colonies. Natural isolates from soil or water typically are a small, rough colony, while clinical isolates are likely smooth colony mucoid types, occasionally with a fried-egg appearance that is large, smooth, with flat edges, and an elevated appearance. Isolates from respiratory and urinary tract secretions may show a mucoid type (alginate slime).

Historical perspectives

P. aeruginosa was first described as a distinct bacterial species at the end of the 19th century, after the development of sterile culture media by Pasteur. In 1882, the first scientific study on *P. aeruginosa*, entitled "On the blue and green coloration of

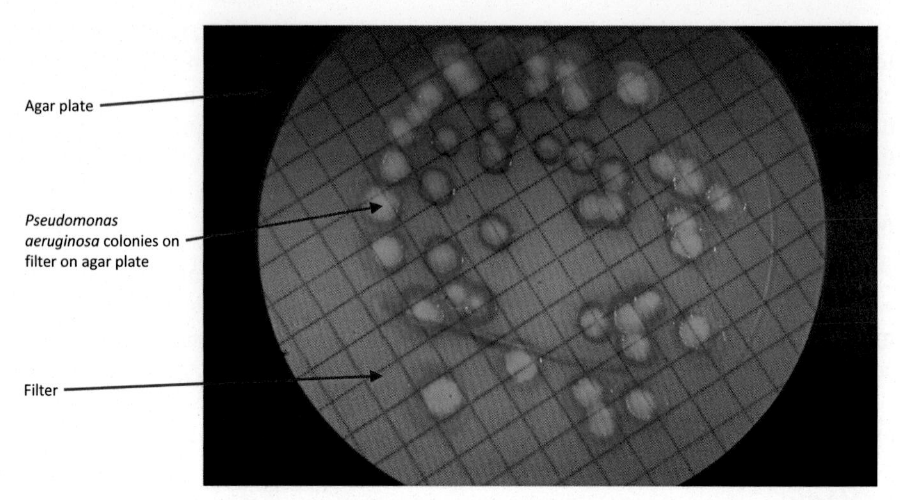

Fig. 3 *Pseudomonas aeruginosa* colonies recovered on a filter and cultured on a selective agar.
Photo courtesy of Zak Prior, UKHSA.

bandages," was published by a pharmacist named Carle Gessard, who demonstrated *P. aeruginosa's* characteristic pigmentation on exposure to ultraviolet light. This fluorescent blue-green pigmentation was later attributed to pyocyanine, a derivative of phenazine, which is also reflected in historical names: *Bacillus pyocyaneus*, *Bakterium aeruginosa*, *P. polycolor*, and *P. pyocyaneus* (Fig. 3). There are many *P. aeruginosa* strains, including PA01, PA7, UCBPP-PA14, a2192, which were isolated based on their distinctive grape-like odor of aminoacetophenone, pyocyanin production, and the colonies' structure on agar media.

Transmission routes

Transmission of *P. aeruginosa* may result in colonization and infection of patients. Transmission routes may include direct and indirect contact, direct and indirect ingestion, inhalation and aspiration, indirect contact (from hands, surfaces, and/or equipment), aerosolization (from contaminated water or body fluids), and aspiration (contaminated drinking water in the airways) (National Services Scotland, 2019). Ventilation increases the risk of *P. aeruginosa* infections (Gerardin et al., 2006; Odeh and Quinn, 2000).

Some patients may be colonized with the organism before admission. It is this that causes the greatest amount of problems in distinguishing between carriage and what may reflect acquisition on a unit (Halstead et al., 2021; Martak et al., 2022). No baseline has been set for endogenous carriage.

Endogenous routes include antibiotic selective pressures due to overprescription of the wrong type of antibiotics.

Exogenous routes include cross-contamination from hands plus environmental sources including water, damp materials, surfaces, and drains.

Disease symptoms

P. aeruginosa rarely causes infections in healthy people, except where it can gain access to areas not usually accessible, e.g., pseudomonas folliculitis, swimmers' ear, and eye infections as a result of swimming/bathing in contaminated water. Some patient groups are more at risk of acquiring infection than others including immunocompromised patients, especially those with chemotherapy-induced neutropenia who are at particular risk of pseudomonal blood-stream infections as well as cystic fibrosis patients where colonization leading to infection is the most common cause of death. Colonization of *P. aeruginosa* can lead to a wide range of infections including bacteraemia, pneumonia, urosepsis, wound, and surgical intervention site infection as well as secondary infection of burns.

Extent of disease

In the United States, *P. aeruginosa* is the sixth most common cause of healthcare-associated infections, accounting for 7.1% of all hospital infections and 32,600 infections among hospitalized patients and 2700 estimated deaths in the United States (CDC, 2019a; Magill et al., 2014). The most common infection is pneumonia, and prevalence is increasing (Williams et al., 2010). The increasing incidence of antibiotic-resistant strains is a cause of global concern, in 2017, multidrug-resistant *P. aeruginosa* caused an estimated 32,600 infections among hospitalized patients and 2700 estimated deaths in the United States (Asempa et al., 2019).

However, the prevalence rate among immunocompromised hosts is evident; studies that have demonstrated that in ICU patients, *P. aeruginosa* represented 16.2% of patient infections and was the cause of 23% of all intensive care unit-acquired infections, with a respiratory source being the most common site of *P. aeruginosa* infection (Vincent et al., 2020).

According to UKHSA, rates of *P. aeruginosa* infections in ICUs have also gradually increased year on year between April 2017 and March 2021 (UKHSA, 2021a). An increase was observed from April 2019/March 2020 to April 2020/March 2021, with the rate rising from 0.20 to 0.29 cases per 1000 ICU bed days greater than 2 days.

Between April 2020 and March 2021, reported *P. aeruginosa* cases peaked at 391 cases, and most cases were hospital-onset (54%), which is the first time since the start of mandatory surveillance where there were more hospital-onset cases than community-onset cases. These increases coincided with the COVID-19 pandemic and may have been as a result of the change in hospital population, long-term stays in ICUs, and changes in practice during this period. In addition, during the pandemic, the dominant primary focus of *P. aeruginosa* infection was respiratory infection compared with urinary tract infection in previous years (UKHSA, 2021b). Most cases occur in adults aged 45 and over, and in terms of gender, the proportion of *P. aeruginosa* cases was greater in males of all age groups compared with their female counterparts.

Pseudomonas spp. is one of the leading causes of hospital-acquired bacteremia, accounting for 4% of all cases and the third leading cause of Gram-negative blood stream infections (Magill et al., 2014). While the source of infection in patients remains unknown in up to 40% of the cases, the most common afflictions are respiratory

(25%) and urinary tract (19%) followed by central venous catheter and skin and soft tissue infections.

Why is *P. aeruginosa* able to cause such severe and wide-ranging infections? *P. aeruginosa* can attach to cells using its flagella and uses its export systems to secrete and release virulence factors such as proteases, elastases, and phenazine pigments (including pyocyanin and rhamnolipids). These type III secretion systems inject the proteins directly into the cytoplasm of the host cells, which has a profound and devastating effect on epithelial barrier function and wound healing.

Some subtypes of *P. aeruginosa* are resistant to nearly all antibiotics, including carbapenems. A small percentage (2%–3%) of carbapenem-resistant *P. aeruginosa* carry a mobile genetic element, which makes a carbapenemase enzyme that destroys these important drugs. In addition, mobile genetic elements are easily shared between bacteria, thereby rapidly spreading resistance.

The ability of *P. aeruginosa* to form biofilm is important for its persistence in humans and in drinking water systems where the production of alginate and lipopolysaccharide provides a protective niche for the survival of this microorganism (Flemming and Wingender, 2010; Vander Elzen et al., 2019). This biofilm mode of growth and its tolerance to biocides and resistance to antibiotics are associated with the pathogenesis of burn wounds, chronic lung infections in patients with cystic fibrosis and bronchiectasis as well as infections associated with indwelling devices and other prosthetic material.

Clinical treatment

P. aeruginosa infections are generally treated with antibiotics. Unfortunately, in healthcare settings such as hospitals or nursing homes, *P. aeruginosa* infections are becoming more difficult to treat because of increasing antibiotic resistance within these settings, possibly due to poor antibiotic strategies.

To identify the best antibiotic to treat a specific infection, healthcare providers need to correctly collect and send a specimen to the laboratory and test any bacteria that grow against a set of antibiotics to determine which are active against the isolate. Healthcare professionals will then select an antibiotic based on the activity of the antibiotic and other factors, such as potential side effects or interactions with other drugs. For some multidrug-resistant types of *P. aeruginosa*, treatment options might be limited (Bassetti et al., 2018).

Control mechanisms in the hospital environment

Effective clinical surveillance, i.e., monitoring the prevalence of *P. aeruginosa* is essential to monitor trends and identify early infections in areas previously clear, and where there are exceedances over the level normally seen in a particular area, this is especially important to identify multidrug-resistant (MDR) strains among high-risk patients and acting accordingly to identify and remove or mitigate reservoirs of cross-transmission. The effective management of rinse water quality during decontamination of instruments is also needed, for example, bronchoscopes (Bou et al., 2006)

Where clusters of infections due to *P. aeruginosa* are detected, reservoirs including water (e.g., the water system, any other uses of water including, water in bottles, containers, taps, showers, baths, and water-based equipment, rinse water during instrument decontamination) all medical solutions such as IV fluids, TPN, thickened drinks, contact lens solution, sterile water, and disinfectants (products and utensils) should be screened as well as damp items and those left in the splash zone from an outlet or drain.

Adherence to standard infection control guidelines will assist in limiting the spread of *P. aeruginosa*, and such measures to prevent the transmission of MDR *P. aeruginosa* in healthcare facilities should include hand hygiene (with the appropriate use of alcohol-based solutions following hand washing), contact precautions, patient isolation (single room or cohort), environmental design, cleanliness, awareness training, including for staff and visitors in high-risk areas and surveillance.

However, hands may become contaminated during hand washing in a sink with a contaminated aerator, faucet, or drain or slow moving water within a wash hand basin or sink must be rectified without delay to the reflux of drain water contaminating the wash hand basin or sink. Therefore, hand disinfection using antiseptic agents (e.g., chlorhexidine or alcohol-based disinfectants) should be undertaken prior to patient contact.

Clinical hand wash basins, sinks, and drains should be maintained, serviced, cleaned, and disinfected regularly as part of routine hospital environmental cleaning.

Remediation of outbreaks due to hand wash station or sink contamination may require disassembly, physical cleaning followed by disinfection and/or replacement of plumbing/drain components and redirection of the water flow. Preferably, wash stations and sinks should be removed or replacement of the sink and/or rearrangement of the surrounding space such that all items, objects, or patients are not within 2 m to reduce splash contamination.

Use sterile fluids and sterile water for medical equipment including nebulizers, humidifiers, and for rinsing tracheal suction catheters.

Do not reuse containers for liquid soap or disinfectants or use previously opened vial of water or sodium chloride solution for injection. Don't have soap and towel dispensers located above the hand wash basin/sink to avoid contamination.

Summary

- *Pseudomonas* means "false unit," and *aeruginosa* refers to the blue-green color of laboratory cultures of the species.
- *P. aeruginosa* is not a common part of the normal transient flora.
- *P. aeruginosa* is one of the most frequent and severe causes of hospital-acquired infections, particularly affecting immunocompromised (especially neutropenic) and intensive care unit (ICU) patients including patients with burns and foot ulcers and associated with significant morbidity and mortality.
- Strains are resistant to most antibiotics including carbapenem and multidrug resistance is increasing.
- Outbreaks have been traced to contaminated solutions (tracheal irrigate, mouthwash, IV fluids), water, disinfectants, and inadequately disinfected or sterilized endoscopes, ventilators or contaminated mesh grafts in burn patients, but have also been linked to direct and indirect transmission, e.g., via the hands of hospital personnel washed under a contaminated aerator (Inglis et al., 2010).

Stenotrophomonas maltophilia

Stenotrophomonas maltophilia is a globally emerging environmental Gram-negative multidrug-resistant organism that is commonly associated with respiratory infections in humans. This opportunistic pathogen is commonly found in and around water and causes serious infections in humans. Risk factors include prior exposure to antimicrobials (especially broad-spectrum antibiotics), mechanical ventilation, and prolonged hospitalization. It may also affect the lungs of patients with cystic fibrosis. Infections in previously healthy patients are unusual. It is recognized as one of the underestimated important multidrug-resistant organisms in hospitals by the World Health Organization (Brooke, 2012).

Microbiological characteristics

S. maltophilia is an aerobic, nonfermentative, catalase positive, oxidase negative, Gram-negative bacillus possessing four polar flagella in a multitrichous formation and naturally lives in the rhizosphere,

While *S. maltophilia* is an aerobe, it can still grow using nitrate as a terminal electron acceptor in the absence of oxygen.

They form small colonies after 24 h at 37°C on solid culture media, smooth, convex colonies about 3 mm in diameter after 48 h usually producing a yellow pigmentation. While they have optimal growth at 35°C, they do not grow at 4°C or 41°C but do demonstrate thermotolerance at higher temperatures.

Ecology

S. maltophilia is an environmental organism that has been isolated from aqueous sources both inside and outside the hospital/clinical setting. *S. maltophilia* has been recovered from soils and plant roots, animals, invertebrates, water treatment and distribution systems, wastewater plants, sinkholes, lakes, rivers, biofilms on fracture surfaces in aquifers, washed salads, haemodialysis water and dialysate samples, taps, tap water, bottled water, contaminated chlorhexidine-cetrimide topical antiseptic, hand-washing soap, contact lens solutions, ice machines, and sink drains.

A significant feature of *S. maltophilia* is its ability to adhere to surfaces and form biofilms in intravenous cannulae, prosthetic devices, dental unit waterlines, and nebulizers.

Historical perspectives

S. maltophilia was first isolated in 1943 as *Bacterium bookeri* and then named *P. maltophilia*; later, rRNA analysis determined that it was more appropriately named *Xanthomonas maltophilia*.

Transmission routes

S. maltophilia does not readily spread between patients. Apparent outbreaks attributed to *S. maltophilia* are frequently caused by multiple strains, implying acquisition from environmental sources as opposed to interpatient spread.

Indirect transmission may occur through hospital tap water faucets, sinks, shower outlets, air-cooling systems, ice-making and soda fountain machines, disinfectant solutions, intravenous fluids, catheters, ethylenediaminetetraacetic acid (EDTA) containing blood collection tubes, blood gas analyzers, dialysis machines, intra-aortic balloon pumps, nebulizers, oxygen humidifiers, breathing circuits, scopes, dental equipment, lens care systems, and the hands of healthcare workers (Said et al., 2022).

Disease symptoms

Although human infection is relatively rare, it most commonly causes respiratory infections particularly in cystic fibrosis patients and chronic obstructive pulmonary disease. *S. maltophilia* is also able to cause bacteraemia, meningitis, cellulitis, long line infections, and eye infections. *S. maltophilia* may be associated with polymicrobial infections or grow slowly in the host, resulting in some difficulty in isolating this bacterium. It is occasionally misidentified as *Burkholderia cepacia* (previously known as *P. cepacia*), which is also a common cystic-fibrosis-associated pathogen.

Extent of disease

S. maltophilia is the most common carbapenem-resistant cause of bloodstream infections in US hospitals as it is inherently resistant to carbapenems and causes about 1% of nosocomial bacteremia cases (Ryan et al., 2009). Infection rates are estimated to be from 5.7 to 37.7 cases per 10,000 hospital discharges and has increased progressively since the 1970s (Patterson et al., 2020). Crude mortality rates range from 14% to 69% in patients with bacteremia.

Risk factors include chronic respiratory diseases, especially cystic fibrosis, hematologic malignancy, chemotherapy-induced neutropenia, organ transplant patients, human immunodeficiency virus (HIV) infection, hemodialysis patients, and neonates (Abbott et al., 2011).

Increased infection rates are believed to be primarily due to the increase in the number of immunocompromised patients and the wide use of broad-spectrum antibiotics.

S. maltophilia has been shown to possess powerful virulence factors that are believed to be present in all environmental isolates without any specific evolutionary branching, similar to *P. aeruginosa*. Virulence factors include:

- ability to form biofilms to both animate surfaces (respiratory epithelial tissue) and inanimate surfaces (ventilation tubes and circuits).
- intrinsically resistant to multiple and broad-spectrum antibiotic agents including most beta-lactams.
- using virulence exoenzymes for tissue invasion and escaping the host immunity.
- ability to form small-colony variants, which are slowly growing and sometimes challenging to detect.

Clinical treatments

S. maltophilia is resistant to many antibiotic classes such as cephalosporins, carbapenems, and aminoglycosides. This means that treatment options are relatively limited. However, most strains remain susceptible to co-trimoxazole, which is regarded as the drug of choice for treating infections.

All interprofessional healthcare team members need to have good communication regarding the treatment plan for *S. maltophilia.* This includes the clinicians (including mid-level practitioners), residents, infectious diseases consultants, nurses, pharmacists, and the infection prevention team. By coordinating treatment efforts and sharing information about the case, patient outcomes will be improved with fewer adverse events.

Antibiotic stewardship programs are also critical in preventing and treating *S. maltophilia* infections by limiting the unnecessary use of broad-spectrum empiric antimicrobials to the minimum.

Control mechanisms in the healthcare environment

Strict infection prevention measures must be implemented and stressed to all team members, including hand hygiene, central venous line insertion precautions, appropriate disposal of potentially contaminated solutions, and proper handling and effective disinfection of medical equipment.

Environmental sampling, especially in outbreak situations, is important to identify potential sources (National Services Scotland, 2019). Maintenance of the water supplies, water filtration, and disinfection of plumbing systems disinfection is recommended to limit the transmission. Disinfection is a problem unless plumbing components are physically cleaned first to disrupt the biofilm as like many waterborne pathogens *S. maltophilia* will be tolerant to disinfectants.

Summary

- *S. maltophilia* is the only species of the Stenotrophomonas genus that infects man.
- *S. maltophilia* has previously been referred to as *P. maltophilia* and *X. maltophilia.*
- *S. maltophilia* is an opportunistic pathogen commonly found in and around water, both in the wider environment and in hospitals. It survives in minimal nutrient concentration and forms biofilms on the surfaces of medical devices including nebulizers and cannulas. It has optimal growth at 35°C, no growth at 4°C or 41°C but can survive refrigeration and at higher temperatures.
- Human infection has been relatively rare but has become more prevalent especially in hospital settings where there are more immunocompromised patients and increased use of antibiotics. It most commonly causes respiratory infections in humans including cystic fibrosis patients as well as obstructive lung cancer patients. *S. maltophilia* is also able to cause other infections including bacteremia, meningitis, cellulitis, and eye infections.
- *S. maltophilia* may be associated with polymicrobial infections or grow slowly in the host and is occasionally misidentified as *Burkholderia cepacia* (previously known as *P. cepacia*), which is also a common cystic-fibrosis-associated pathogen.

Cupriavidus pauculus

C. pauculus is an emerging environmental bacterium that causes waterborne associated infections in vulnerable patients in healthcare including immunocompromised patients, neonates, those with extracorporeal membrane oxygenation or in intensive care areas (Inkster et al., 2022; Valdés-Corona et al., 2021; Yahya and Mushannen, 2019).

Microbiological characteristics

The genus *Cupriavidus* is a member of the family *Burkholderiaceae. C. pauculus* is an aerobic oxidase and catalase-positive, Gram-negative bacillus (rod shaped), non-fermentative, motile bacterium with peritric flagella (multiple flagella distributed randomly around the cell) isolated from water, ultrafiltration systems, and bottled mineral water (Fig. 4). Colonies are round, smooth, convex, and nonpigmented.

Ecology

Cupriavidus species are primarily environmental organisms found in soil but have been isolated from wastewater treatment plants, groundwater remediation systems, water systems, bottled mineral water, and the International Space Station (Inkster et al., 2022; Manaia et al., 1990; Mora et al., 2016; Schmidt and Schlegel, 1989; Suenaga et al., 2015).

C. pauculus, strains have also been isolated from various clinical sources including hydrotherapy pools (Aspinall and Graham, 1989), nebulizers (Oie et al., 2006), ultra-filtrated water, thermos-regulator reservoir water, extra-corporeal membrane

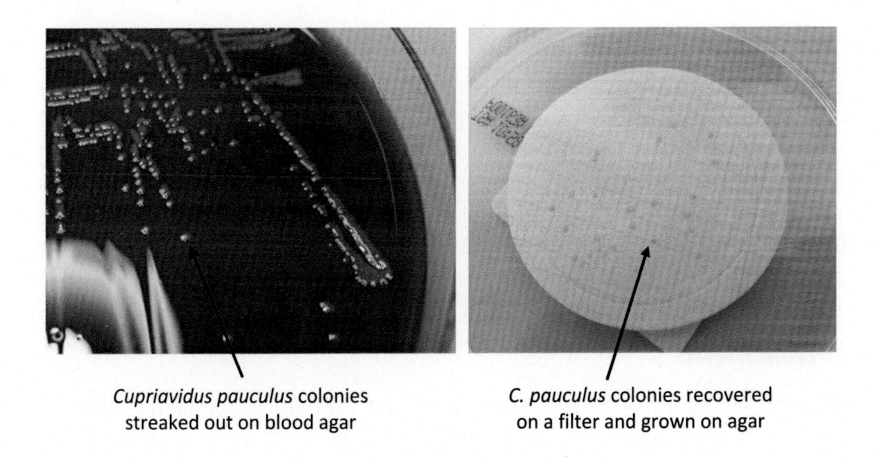

Fig. 4 *Cupriavidus pauculus* colonies streaked out on blood agar and isolated on filters grown on agar.
Images courtesy of BMS Gareth Wilson.

oxygenation system, as well as from environmental samples of pool water, groundwater, tap water, soil, and bottled mineral water (Almasy et al., 2016). One study found *C. pauculus* to be present in 40% of UK hospitals' water systems, and a key area for investigation is the drain/sink trap (Butler et al., 2022; Inkster et al., 2022).

Historical perspectives

In 1995, the genus *Ralstonia,* a genus of proteobacteria, was established to include species formerly known as *Alcaligenes eutrophus*, *Burkholderia solanacearum*, and *Burkholderia pickettii* (Yabuuchi et al., 1995). In 1999, Vandamme et al. described it as *Ralstonia paucula* (Vandamme and Coenye, 2004); the description is identical to the one given by Vaneechoutte et al. in 2004 as *Wautersia paucula* (*Vaneechoutte* et al., 2004), and they were finally renamed *C. pauculus* (Vandamme and Coenye, 2004).

Transmission routes

Transmission has involved splashing or contact via hospital water systems, drainage systems, patients undergoing extracorporeal membrane oxygenation, sterile swabs prewetted in tap water (Balada-Llasat et al., 2010; Inkster et al., 2021; Stovall et al., 2010).

Disease symptoms

There have been reported sepsis cases, such as ventilator-associated airway infections, peritonitis, catheter infections, corneal ulcer, tenosynovitis, and meningitis, caused by this pathogen (Bianco et al., 2018; Fiel et al., 2020).

Extent of disease

Although this pathogen occasionally causes serious human infections, especially in immunocompromised patients, thus far *C. pauculus* has rarely been identified as a pathogen in patients. A few reports have described *C. pauculus* infection in children and neonates, and several cases have reported positive cultures from intubated intensive care unit patients (Almasy et al., 2016; Balada-Llasat et al., 2010; Fiel et al., 2020; Huda et al., 2020). Additional use of update MALDI-ToF database may reduce the risk of misidentification.

Clinical treatments

There is a limited of information concerning the optimal regimen for therapy of *C. pauculus* and clinical advice should be obtained. Wide susceptibility to fluoroquinolones, ceftazidime, and imipenem had been observed, as well as reports of meropenem-and imipenem-susceptible bacteremia (Vay et al., 2007) and colistin-sensitivity (Aydın et al., 2012). However, resistance to carbapenems has also been described in a child on extracorporeal membrane oxygenation

(Salar et al., 1998; Uzodi et al., 2014). There has been comments in the literatures that isolation of *C. pauculus* from patients should be viewed as potential contamination; however, surveillance should be adopted to determine if there is an ongoing environmental presence (Inkster et al., 2021, 2022).

Control mechanisms in healthcare

For effective control mechanisms to be implemented, then the source has to be identified. This may involve investigating a range of water systems and outlets as well as medical equipment in which water is used such as extracorporeal membrane oxygenation (Uzodi et al., 2014). In one recent outbreak, there was extensive colonization of the water system in which microbial retention filters were used to control the outbreak (Inkster et al., 2021).

Summary

- *C. pauculus* is an emerging environmental bacteria;
- has resulted in a number of waterborne associated infections in vulnerable patients in healthcare with reports of infections in immunocompromised patients, neonates, and in those with extracorporeal membrane oxygenation or in intensive care areas;
- is an aerobic oxidase and catalase-positive, Gram-negative rod that is nonfermentative and is motile via multiple flagella distributed randomly around the cell;
- species are primarily environmental organisms found in soil but have also been isolated from a wastewater treatment plant, groundwater remediation systems, bottled mineral water, and even the International Space Station;
- have been isolated from various clinical sources including hydrotherapy pools, nebulizers, ultra-filtrated water, extra-corporeal membrane oxygenation system, as well as hospital water systems, drains/traps;
- causes a range of infections including peritonitis, corneal ulcer, tenosynovitis, bacteremia, and meningitis particularly in immunocompromised patients;
- due to the wide susceptibility to fluoroquinolones, ceftazidime, and imipenem resistance profiles should be determined.

Nontuberculous mycobacteria

Nontuberculous mycobacteria (NTM) are mycobacteria other than *M. tuberculosis* (the cause of tuberculosis) and *M. leprae* (the cause of leprosy). NTM are also referred to as atypical mycobacteria, mycobacteria other than tuberculosis (MOTT), or environmental mycobacteria. NTM are responsible for causing chronic lung diseases and are associated with substantial morbidity and mortality. NTM are environmental organisms that can be found in soil, dust, and water including natural water sources (such as lakes, rivers, and streams) and healthcare water systems. They are tolerant to disinfectants and thermal disinfection used in water systems due to their thick lipid and hydrophobic cell wall, which ensures that they are protected in or outside of biofilms and also highly resistant to antibiotics (Falkinham, 2018).

Microbiological characteristics

NTM are members of the *Mycobacterium* bacterial genus that now includes over 200 distinct species (Waman et al., 2019). NTM are acid-fast bacteria that are ubiquitous in the environment and can colonize soil, dust particles, water sources, and food supplies. NTM are oligotrophic, able to grow at low organic matter concentrations and over a wide range of temperatures, and even at low oxygen concentrations

They are divided into rapidly growing mycobacteria such as *M. fortuitum, M. chelonae*, and *M. abscessus* as well as slowly growing species such as *M. avium, M. kansasii*, and *M. marinum.*

All *Mycobacterium* spp. strains can be distinguished from other bacteria by virtue of the fact that they are "acid-fast," i.e., mycobacterial cells retain red dye carbol fuchsin even after decolorizing with acidic alcohol, due to the high concentration of lipid in the outer membrane (Fig. 5).

Evidence suggests that some NTM species can alternate between smooth and rough colony morphotypes, but it remains unclear whether these phenotypic differences alter adherence on medical devices, aerosol survival, and/or susceptibility to disinfectants and antibiotics. Recent publications confirm that NTM can survive extensive periods of desiccation and still be viable (Malcolm et al., 2017)

Ecology

NTM are environmental organisms found in soil, dust, and water (Falkinham, 2015). *M. avium* complex (MAC) is the most common pathogen in most areas followed by *M. abscessus* complex (MABC) and *M. kansasii* (Hoefsloot et al., 2013).

NTM form biofilms, which are difficult to eliminate with standard disinfectants and antibiotics and as such are able to survive in water systems and cause a number of long-term illnesses in patients, typically lasting more than 5 years.

Historical perspectives

NTM date back to 1868 when "tuberculosis" was first described in chickens, and by 1890, NTM was recognized to be distinct from *M. tuberculosis*, the cause of

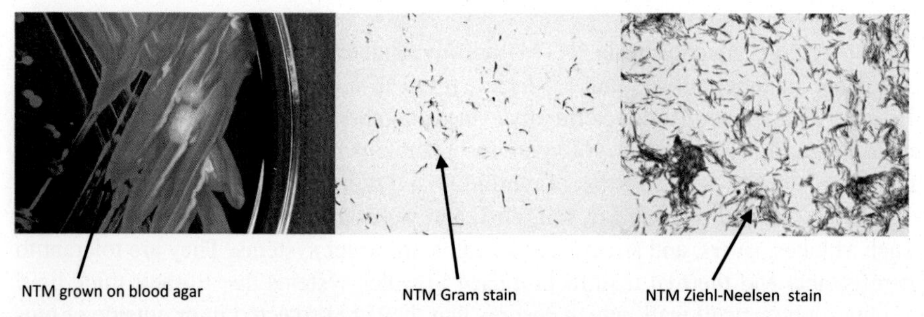

Fig. 5 Nontuberculous mycobacteria growing on blood agar from a blood culture in patient with line infection who was contaminated via the water from a shower. Initial Gram-stain and Ziehl-Neelsen stain to demonstrate the presence of the acid-fast bacteria.

tuberculosis. The organism that caused disease in chickens was later identified to be *M. avium*. Because these organisms did not cause characteristic disease when injected into guinea pigs, they were recognized as distinct from *M. tuberculosis* and were not believed to cause disease in humans.

In the 1930s, NTM were recognized to cause lung disease in humans due to the *M. avium* complex (MAC), which was described in 1943 in a man with underlying silicosis (a form of lung disease).

Pulmonary disease due to NTM became more commonly recognized in the 1950s and accounted for approximately 1%–2% of admissions to tuberculosis hospitals.

Interest in NTM increased when human immunodeficiency virus (HIV)-infected patients began to develop infections distributed throughout the body (disseminated) due to various NTM species, particularly *M. avium*. As antiretroviral drugs (medications for the treatment of HIV) became available, the incidence of disseminated MAC decreased in these patients.

However, there is now an increasing number of cases of pulmonary disease due to NTM.

Transmission routes

Unfortunately, waterborne NTM infection affects susceptible patients through common, preventable exposure routes such as water outlets and showers. Patient infections can also occur by exposure to NTM contaminated medical devices through a range of exposure routes such as uncovered central venous catheters (CVCs), wound exposure, and contamination during surgical procedures has been reported. Exposure through open heart surgery has been demonstrated as heater coolers were contaminated and were distributing NTM aerosols in the operating theatres in a global outbreak (Sax et al., 2015).

Disease symptoms

Clinical symptoms vary but commonly include chronic cough, often with purulent sputum and coughing up blood. Systemic symptoms include malaise, fatigue, enlarged lymph nodes, and weight loss in advanced disease (Winthrop et al., 2020).

Soft-tissue disease due to NTM infection includes posttraumatic abscesses (caused by rapid growers after injection), swimming pool granuloma (caused by *M. marinum*), and *Buruli ulcer* (caused by *M. ulcerans* or *M. shinshuense*).

Diagnosis of opportunistic NTM is made through repeated isolation and identification of this opportunistic pathogen with compatible clinical and radiological features. Similar to *M. tuberculosis*, most NTM can be detected microscopically and grow on Löwenstein-Jensen medium. NTM are traditionally hard to culture, and the results take weeks to be reported. Hence, nucleic-acid-based molecular methods are now being more frequently used including detecting sequence differences in the gene coding for 16S ribosomal RNA to identify the species.

Diagnosis of NTM lung disease requires review of the clinical, radiographic, and microbiological data, but, ultimately, diagnosis can be confirmed by (1) at least two positive cultures from sputum, (2) one positive culture in the case of bronchoscopic wash or lavage, or (3) a transbronchial or other lung biopsy with a positive culture for

NTM or compatible histopathological features such as granulomatous inflammation or stainable acid-fast bacilli (AFB), and one positive sputum or bronchial wash culture for NTM regardless of the mycobacterial strain (Ryu et al., 2016).

Extent of disease

Pulmonary disease caused by NTM is most often seen in low BMI, postmenopausal tall women, and patients with underlying lung disease such as cystic fibrosis, bronchiectasis, and prior tuberculosis. Pulmonary NTM can also be found in individuals with AIDS and malignant disease. It can be caused by many NTM species, which depends on region, but most frequently MAC and *M. kansasii*.

Disseminated mycobacterial disease was common in United States and European AIDS patients in the 1980s, though the incidence has declined in developed nations since the introduction of highly active antiretroviral therapy. It can also occur in individuals after having renal transplantation.

In the United States (1997–2007), data estimated that NTM lung disease has increased at an annual rate of 8.2% and also significantly increased by 7.5% between 2008 and 2015. Women and people aged 65 years or older had consistently higher incidence and prevalence rates than men and people aged less than 65 years, respectively (Winthrop et al., 2020).

Clinical treatments

Treatment is lengthy and varies by species and therefore a challenge. Treatment may be complicated by potential toxicity with discouraging outcomes, which are not always positive. The decision to start treatment for NTM lung disease is not easy and requires careful individualized analysis of risks and benefits.

For NTM lung disease, macrolides (clarithromycin and azithromycin) are used for treatment. The standard optional regimen includes a rifamycin (rifampin or rifabutin), ethambutol, and a macrolide administered for 18–24 months, including 12 months of sputum culture negativity. Potential toxicity and intolerance also must be considered.

Control mechanisms in water

Effective prevention strategies will require both medical and environmental health expertise, and interprofessional cooperation will optimize these efforts.

Many NTM species are significantly more resistant to chlorine than *E. coli*; for example, *M. avium* is more than 500-fold more resistant than *E. coli* to chlorine and is even more resistant when grown in water (Le Dantec et al., 2002). The worldwide heater cooler outbreak was a classic example where a lack of appropriate disinfection and treatment regimen led to fouling and macro-biofilm formation inside the heater cooler tanks. As the equipment was used, aerosols were generated through the flow of the water and small fans in the equipment resulted in the transmission of aerosols across the operating theater and contamination of the open wounds during surgery. These outbreaks were only brought to a halt by the complete overhaul of the heater coolers including replacement of the many parts that were biofouled, extensive biocide treatment as well as new designs and disinfection protocols from the manufacturer.

For water systems, point-of-use filters may be one of the limited options that can be put in place to prevent patient exposure to NTM.

Pseudo outbreaks with NTM are typically due to contaminated final rinse water resulting in a positive NTM following the clinical use of bronchoscopes (Abdolrasouli et al., 2021).

Monitoring can also be used as a control strategy, and there have been recent moves to use molecular techniques when commissioning endoscopy units rather than wait for many week for the culture results.

Summary of NTM

- NTM are mycobacteria other than *M. tuberculosis* (the cause of tuberculosis) and *M. leprae* (the cause of leprosy). NTM are referred to as atypical mycobacteria, mycobacteria other than tuberculosis, or environmental mycobacteria.
- Although anyone can get an NTM infection, NTM are opportunistic pathogens placing some patient groups at increased risk, including those with underlying lung disease or depressed immune systems. These pathogens are typically not transmitted person to person.
- NTM are environmental organisms that can be found in soil, dust, and water including natural water sources (such as lakes, rivers, and streams) and municipal water sources (such as water that people drink or shower in). NTM can form difficult-to-eliminate biofilms, which are collections of microorganisms that stick to each other and adhere to surfaces in moist environments, such as the insides of plumbing in buildings.
- NTM are typically tolerant of chemical disinfectants and resistant to a range of antibiotics.

Protozoa and amoeba

Free-living amoebae belonging to the genera *Acanthamoeba*, *Balamuthia*, *Sappinia pedata*, *Hartmonella,* and *Naegleria* are ubiquitous in the environment, and although they rarely cause disease in humans, they are responsible for a number of infections including eye and skin infections, inflammation of the lungs or sinuses, and have the potential to spread to the brain. However, as well as causing infections, the vegetative (trophozoites) form of these free-living amoeba feed mainly by phagocytosis and are thus able to ingest waterborne bacteria including pathogens. In doing so, the protozoa afford the bacterium protection from control strategies including heating and disinfection.

Microbiological characteristics

Amoebae are microscopic, single-celled organisms that are highly motile and are able to move around through what is known as pseudopodia. Free-living amoebae are aerobic, mitochondriate, and eukaryotic microorganisms and examples include:

- *Naegleria fowleri,* which has three stages in its life cycle: cysts, trophozoites, and flagellated forms. The trophozoites replicate by promitosis (nuclear membrane remains intact). *N. fowleri* is the only species of *Naegleria* known to infect people.

- *Acanthamoeba* has only two stages, cysts and trophozoites, in its life cycle. No flagellated stage exists as part of the life cycle. The trophozoites replicate by mitosis (nuclear membrane does not remain intact). The trophozoites are the infective forms, although both cysts and trophozoites gain entry into the body through various means.
- *Balamuthia mandrillaris* is a free-living amoeba that is known to cause the rare but deadly neurological condition known as granulomatous amoebic encephalitis

Ecology

Free-living amoebae are naturally present in soil and water environments. There are more than 11,300 species of amoebae currently identified, with only a few known to be hazardous to humans (Loret and Greub, 2010). Water commonly contains free-living amoebae including genera of *Acanthamoeba*, *Harmannella*, *Naegleria,* and *Vahlkampfia* (Delafont et al., 2013).

N. fowleri is a heat-loving (thermophilic), free-living amoeba (single-celled microbe), commonly found around the world in warm freshwater (such as lakes, rivers, and hot springs) and soil.

Acanthamoeba spp. are ubiquitous in the environment and have been found in a variety of sites, including soil; fresh, brackish, and sea water; field-grown vegetables; sewage; swimming pools; contact lens supplies; medicinal pools; dental treatment units; dialysis machines; heating, ventilating, and air-conditioning systems; and tap water; mammalian cell cultures; and vegetables. *Acanthamoeba* growth is inhibited by temperatures above 35°C–39°C, although recent evidence suggests that *Acanthamoeba* may be able to grow at least inefficiently at higher temperatures.

Free-living amoeba (FLA) as bacterial hosts

Other than health risks associated from direct exposure and infection, waterborne amoebae can serve as protective hosts for pathogenic bacteria, such as *Legionella*, that hide inside amoeba hosts to effectively avoid inactivation from disinfectants or other water treatment works. A significant association has been established between FLA and the presence of pathogens including *Legionella*, *P. aeruginosa,* and *Mycobacterium* species (Loret and Greub, 2010; Scheid, 2018; Shaheen and Ashbolt, 2018). *Legionella* rapidly recolonize domestic water systems, sometimes immediately after increased disinfectant concentrations, which may be due to the constant supply of the bacteria from free-living amoeba hosts already established in the system.

Historical perspectives

The name Amoeba is derived from a-mioba, i.e., without form, and was first described in 1875 by Fedor Lösch in the first proven case of amoebic dysentery in St. Petersburg, Russia. Free-living amoebae began to be recognized in the early 1960s as opportunistic human pathogens, capable of causing infections of the central nervous system in both immunocompetent and immunocompromised hosts.

Transmission routes

Diseases caused by protozoa and amoeba are transmitted by contaminated drinking water. The portal of entry of the amoeba is not always known but is likely to be through the nostrils, contact with mucous membranes (eyes), or through breaks in the skin with hematogenous spread.

Disease symptoms

Infections through free-living amoeba affecting the brain and spinal cord can activate the immune system, which leads to inflammation. These diseases, and the resulting inflammation, can produce a wide range of symptoms, including fever, headache, seizures, and changes in behavior or confusion. In extreme cases, these can cause brain damage, stroke, or even death.

Acanthamoeba spp. infection typically shows up as inflammation of the lungs or sinuses, and/or skin infections but has the potential to spread to the brain. Granulomatous amoebic encephalitis has a chronic onset that progressively worsens over weeks and months. Signs and symptoms are typical of meningoencephalitis and encephalitis, which involves varying degrees of neurological impairment. There are very few known survivors of granulomatous amoebic encephalitis.

Acanthamoeba keratitis was first described in 1974. Corneal ulceration and scarring can occur if not treated. Although highly associated with contact lens use, cases in the absence of contact lenses have occurred.

Cutaneous acanthamebiasis presents as a single or disseminated chronic skin lesions, which are most commonly crusted or ulcerated, they may be indurated or have an eschar. Skin infections caused by *Acanthamoeba* can appear as reddish nodules, skin ulcers, or abscesses in the skin. The lesions may be mistaken for fungal or mycobacterial skin infection, cutaneous amebiasis caused by *Entamoeba histolytica* or cutaneous leishmaniasis. These may occur with or without concurrent central nervous system disease.

N. fowleri infections are *characterized* by CNS dysfunction with degeneration caused by haemorrhagic-necrotizing meningoencephalitis. Onset of symptoms occurs quickly following infection (1–9 days; median 5 days after swimming or other nasal exposure to *Naegleria*-containing water). Symptoms are like bacterial meningitis, for which it is often mistaken, with deteriorating neurological function and complications. The case fatality rate is extremely high.

Extent of the disease

Data are limited on the number of infections caused by free-living amoeba globally. In 2003, Seal estimated an annual incidence rate of one case of *Acanthamoeba keratitis* per 30,000 soft contact lens wearers for Europe and Hong Kong, based on several cohort studies and surveys (Seal, 2003) and has increased to reach 1 in 21,000 in 2015 in the Netherlands (Randag et al., 2019). Worldwide, the annual incidence ranges from 0.15 per million to 1.4 per million.

Clinical treatments

Eye and skin infections caused by *Acanthamoeba* spp. are generally treatable. Topical use of 0.1% propamidine isethionate (Brolene) plus neomycin-polymyxin B-gramicidin ophthalmic solution has been a successful approach; keratoplasty is often necessary in severe infections. Although most cases of brain (CNS) infection with *Acanthamoeba* have resulted in death, patients have recovered from the infection with proper treatment.

Control mechanisms in water

Acanthamoeba has been found to survive a 2-h exposure to 100 mg/L of chlorine and is resistant to UV light and thermal treatment strategies. A combination of treatments, which include the use of filtration for physical removal of the amoeba, is expected to be the most broadly effective. Controlling the proliferation of free-living amoeba is an important strategy to minimize health effects, which is best accomplished by reducing organic matter and biofilms in the treated water distribution and storage system. However, in order to determine if this strategy is working, then sampling and testing will have to be carried out. Very few laboratories carry out testing for FLA with many of the protocols based on previously published methods (Shaheen and Ashbolt, 2018).

Summary of free living amoeba

- Free-living amoebae are single-celled organisms that are highly mobile and are frequently found in water and soils.
- *Acanthamoeba* spp. and *Balamuthia mandrillaris* cause granulomatous amoebic encephalitis (GAE), which usually presents as a mass, while *N. fowleri* causes primary amoebic meningoencephalitis (PAM).
- *Acanthamoeba* spp. can also cause keratitis, and both *Acanthamoeba* spp. and *Balamuthia mandrillaris* can cause lesions in skin and respiratory mucosa.
- These amoebae can be difficult to diagnose clinically as these infections are rare and, if not suspected, can be misdiagnosed with other more common diseases.
- Clinical treatment requires a combination of antibiotics and antifungals and, even with prompt diagnosis and treatment, the mortality for neurological disease is extremely high.
- Control in the water may require a combination of treatments, including the use of chemicals and point-of-use filters to physically remove the amoeba from the water source.

Fungi

Water systems worldwide have been shown to be colonized with pathogenic molds such as *Aspergillus* spp., *Mucorales,* and *Fusarium* spp. Fungi are increasingly recognized in hospital water systems where they become part of the biofilm. Fungal spores and hyphal fragment are released into the water and dispersed through showerheads, taps, toilet cisterns, or via aspiration or drinking of contaminated water may lead to patient exposure. Even areas where there has been water leakage from the water system will indirectly lead to fungal growth and potential patient exposure from contaminated absorbent materials.

The incidence of nosocomial mold infections continues to increase despite the widespread use of air filtration systems, suggesting that other hospital sources for molds exist, such as the water system.

Microbiological characteristics

Aspergillus fumigatus is an opportunistic fungal pathogen. The genus *Aspergillus* exceeds 100 species. *Aspergillus* are filamentous, made up mainly of chain cells that in turn form a structure known as hypha, which have an approximate diameter of between 2.6 and 8.0 microns. One of its distinctive characteristics is its morphology, as it is made up of conidiophores that end in apical vesicles that can produce up to 500,000 conidia when in contact with air. Viewed under the microscope, the hyphae are uniform and have a tree-like branching pattern. Importantly, the branches are dichotomous.

The colonies that are obtained by culturing in the laboratory are of various colors. At first, they are white, but later that color can vary to yellow, brown, green, or even black. This will depend on the species of *Aspergillus* that is being cultivated. When it comes to the texture of the colonies, they look like cotton or velvet. The fungi belonging to the genus *Aspergillus* contemplate both sexual and asexual reproduction in their life cycle. This is one of the fungi of the genus *Aspergillus* that has been most studied since it constitutes an important pathogen for humans. It is the cause of numerous respiratory tract infections, mainly due to its inhalation.

Fusarium typically produce both macro- and microconidia from slender phialides. Macroconidia are hyaline, two to several-celled, fusiform to sickle-shaped, mostly with an elongated apical cell and pedicellate basal cell.

Ecology

Aspergillus spp. are widespread in the environment, growing on plants, decaying organic matter, and in soils, air/bioaerosols, in/on animal systems, and in freshwater and marine habitats. *Aspergillus* are also found in indoor environments (surfaces of buildings, air, household appliances, etc.) and in drinking water and dust. *Aspergillus fumigatus*, a thermotolerant/thermophilic fungus, is capable of growing over a temperature range from 12°C to 53°C with optimal growth occurring at approximately 40°C, a temperature may be inhibitory to the growth of most other saprophytic fungi (they will survive at higher temperatures).

Mucorales colonize all kinds of wet, organic materials and represent a permanent part of the human environment. They are economically important as fermenting agents of soybean products and producers of enzymes, but also as plant parasites and spoilage organisms.

Fusarium are commonly distributed widely in the soil they are saprophytic fungi known to associate with plants, causing a wide range of plant diseases. This is because of their ability to produce mycotoxins especially in cereal crops, which can cause

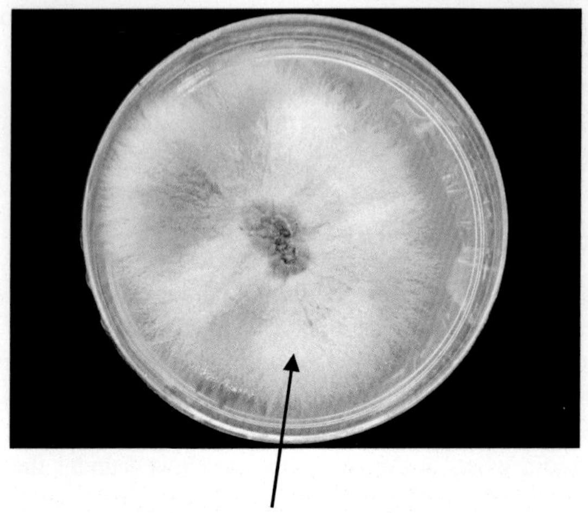

Fusarium grown on an agar plate

Fig. 6 Fusarium isolated and grown on agar.

disease in human and animal hosts if ingested. *Fusarium* spp. majorly produces fumonisins and trichothecenes mycotoxins (Fig. 6).

Aspergillus was first named in 1729 by the Italian priest and biologist Pier Antonio Micheli, who described the molds as resembling an aspergillum, a holy water sprinkler (from Latin Spargere to sprinkle). In 1863, the species *Fumigatus* was first described by physician Georg W. Fresenius. Subsequent cases of *Aspergillus* infections over the next several years showed the most frequent species of *Aspergillus* responsible for human infection was *Aspergillus fumigatus* (Plaignaud, 1781). From 1920 to 1965, cases of disseminated aspergillus infections were implicated in the heart and CNS in addition to the sinuses and lungs.

Disease associated with *Mucorales* spp. was first described when by Fürbinger in 1876 in Germany after a patient died of cancer and the right lung showed a hemorrhagic infarct with fungal hyphae and sporangia (Fürbringer, 1876). In 1885, Arnold Paltauf published the first case of disseminated mucormycosis, which he named "Mycosis mucorina" (Paltauf, 1885).

The genus *Fusarium* was first described in Link (1809) and later by Wollenweber (1935).

Transmission routes

Traditionally site renovation and construction have been found to be responsible for disturbing *Aspergillus*-contaminated dust resulting in the production of bursts of airborne fungal spores that have been associated with clusters of healthcare-acquired infections in immunocompromised patients (CDC, 2019b).

The principal portal of entry for fungi is the airways, followed by the skin at site of tissue breakdown and possibly the mucosal membranes (Nucci et al., 2004). In one study in the United States, results indicated that a geographically widespread clonal lineage was responsible for >70% of all clinical isolates, and these strains were genetically similar to those isolated from the water systems of three US hospitals (O'Donnell et al., 2004), further supporting the risk of nosocomial waterborne fusariosis.

Absorbent building materials (e.g., wallboard) also serve as an ideal substrate for the proliferation of this organism when they become and remain wet, thereby increasing the numbers of fungal spores in the area (Inkster and Weinbren, 2021a).

Disease and infections

The presence of aspergilli in the healthcare facility environment is a substantial extrinsic risk factor for opportunistic invasive aspergillosis (invasive aspergillosis being the most serious form of the disease). Respiratory tract infections are caused by various species of Aspergillus, especially *Aspergillus fumigatus.* Since its entry into the body occurs mainly through inhalation, the tissues that are affected are those of the respiratory tract.

However, aspergillosis can present in several clinical forms: allergic bronchopulmonary aspergillosis, chronic pulmonary aspergillosis, and invasive aspergillosis (Warris et al., 2001).

Fusarium cause disease in humans if ingested. *Fusarium* spp. produces fumonisins and trichothecenes mycotoxins. Invasive infections, such as sinusitis, pneumonia, deep cutaneous infections, and disseminated infections, present in immunocompromised patients and most commonly manifest as fever not responding to antimicrobial medications (Nucci et al., 2004). In addition, immunocompetent patients present more frequently with superficial infections, such as keratitis and onychomycosis (Chang et al., 2006). They cause opportunistic infections in immunocompromised patients.

Extent of the disease

In the United States, allergic bronchopulmonary aspergillosis (ABPA) likely affects between 1% and 15% of cystic fibrosis patients (CDC, 2019b, c). Studies have calculated that 2.5% of adults who have asthma also have ABPA, which is approximately 4.8 million people worldwide. Of these 4.8 million people who have ABPA, an estimated 400,000 also have chronic pulmonary aspergillosis (CPA). Another 1.2 million people are estimated to have CPA after having tuberculosis, and over 70,000 people are estimated to have CPA as a complication of sarcoidosis (CDC, 2019b).

There are limited data on the incidence of Mucorales, since there are few population-based studies, but multiple studies in France, Belgium, and India have shown that it is increasing, and the prevalence of mucormycosis in India is about 80 times the prevalence in developed countries (Skiada et al., 2020).

The prognosis of Fusarium disease is poor and is determined largely by degree of immunosuppression and extent of infection, with virtually a 100% death rate among persistently neutropenic patients with disseminated disease. The importance of T-cell defenses against *Fusarium* is illustrated by the occurrence of disseminated fusariosis in nonneutropenic hematopoietic stem cell transplant (HSCT) recipients where the overall incidence of fusariosis is ~6 per 1000 cases HSCTs (Nucci et al., 2004).

Diagnosis

The diagnosis of Aspergillosis includes high-resolution computer tomography (CT) scans and serum *Aspergillus* galactomannan antigen tests.

The hallmark of mucormycosis is tissue necrosis resulting from angioinvasion and thrombosis, and histopathology of affected tissue is key to obtaining a definitive diagnosis.

The diagnosis of fusariosis depends on the clinical form of the disease. For diagnosing keratitis, cultures of corneal scrapings or tissue biopsy is usually required for a definitive diagnosis.

Treatment of infections

Drugs used to treat fungal infections include Amphotericin B, itraconazole, Posaconazole, Echinocandins, and vorconazoles.

Keratitis is usually treated with topical antifungal agents, and natamycin is the drug of choice (Dóczi et al., 2004). More recently, successful treatment has been achieved with topical and oral voriconazole has been reported (Bunya et al., 2007).

Control in the environment

Infection-control strategies and engineering controls (filtration), when consistently implemented, have been shown to be effective in preventing opportunistic, environmentally related infections in immunocompromised populations (CDC, 2019b).

However, there are limited guidelines for preventing exposure to *Aspergillus* spp. found in hospital water. General recommendations include minimizing the exposure of high-risk patients to potential sources of *Aspergillus* spp., for example, by using microbial retention filters, as well as activities that may aerosolize *Aspergillus* spp. Furthermore, the source of *Aspergillus* spp. should be eliminated These recommendations imply that water precautions need to be introduced in those hospitals where water is contaminated with *Aspergillus* spp. such as replacing components and using point-of-use water filters. Systemic chemical and thermal activities have very limited effectiveness.

Since the airways are the principal portal of entry for *Fusarium* species, the placement of patients at high risk (prolonged and profound neutropenia and allogeneic HSCT recipients) in rooms with HEPA filter and positive pressure may decrease the risk of nosocomial acquisition of fungal infections. In addition, since the water may be a source of *Fusarium* species in the hospital, every effort should be made to prevent patient exposure (e.g., by avoiding contact with reservoirs of *Fusarium* spp., such as tap water, and/or cleaning showers prior to use by high-risk patients during periods at risk.

Summary of fungal infections related to water

- Water systems worldwide have been shown to be colonized with pathogenic molds such as *Aspergillus* spp., *Mucorales,* and *Fusarium* spp.

- Fungi are increasingly recognized in hospital water systems where they become part of the biofilm and can be released into the water and dispersed through showerheads, taps, drains, toilet cisterns, or via aspiration or drinking of contaminated water.
- The incidence of nosocomial mold infections continues to increase despite the widespread use of air filtration systems, suggesting that other hospital sources for molds exist, such as the water system.
- Fungi primarily cause respiratory disease in immunocompromised patients, e.g., particularly transplant patients.
- The extent of fungal disease indicates that rates are increasing, partly through improved recognition and diagnosis.
- Environmental and engineering controls have been shown to reduce and minimize fungal infections in patients at risk.
- Where exposure is thought to occur through water, then reducing patient exposure to splashes and aerosols may help limit the extent of infections. That removes the source or use microbial retention filters to protect patients.

Biofilm

Biofilms are one of the oldest forms of a living entity. Where moisture collects or water flows, then bacteria will attach to surfaces. For millennia, streams have flowed and biofilm grown on rocks and surfaces. Biofilm is a consortium of microorganisms that preferentially attach to surfaces where they are able to harvest nutrients from the environment. As the biofilm develops, then cells will act in collaboration for the benefit of the biofilm or slough in the flow and be carried downstream.

There are numerous examples of biofilms in our bodies, and plaque on our teeth is one of the simplest examples. Over time, bacteria are able to invade our bodies either through poor dental hygiene, urinary catheters, or through breakages in our skin membranes, and as such those bacteria are able to form biofilms that are detrimental to our heath.

In healthcare, biofilms have developed niches on plumbing components that have seeded the water and contaminated the product that we all take for granted. However, those often unseen bacterial biofilms result in hospital-acquired infections in immunocompromised patients be they preterm babies, transplant, or burn patients and exact a cost in patient mortality.

Microbiological characteristics

A biofilm is a collection of microorganisms associated with a surface and enclosed in a matrix of polysaccharide materials and is mainly composed of water (Costerton, 1995; Flemming and Wingender, 2010). Depending on the environment a range of materials including blood, proteins, minerals, crystals, corrosion products, clay or silt particles may also be found in the biofilm. Biofilm microorganisms differ from those in the planktonic phase (suspended) with respect to the genes that are transcribed (Oggioni et al., 2006).

Ecology

Biofilms form on a wide variety of surfaces from healthcare to industrial situations, including living tissues, indwelling medical devices, heater coolers used in cardiac surgery, industrial or potable water system (pipe plumbing components), or natural aquatic systems including streams and rivers (Fig. 7).

Bacteria and microorganisms including waterborne pathogens are delivered by the incoming water supply to our healthcare buildings. A wide range of waterborne pathogens including *Legionella* spp. and other Gram-negative microorganisms including *P. aeruginosa*, *S. maltophilia,* and *C. pauculus* are able to establish on pipework and plumbing components in our healthcare water systems including EPDM washers and outlet fittings (Arciola et al., 2018; Høiby et al., 2010; Moritz et al., 2010).

The variable nature of biofilms can be visualized using light microscopy or scanning electron micrographs of plumbing components from water systems or medical devices, respectively (Wilks et al., 2021). Water system biofilms are highly complex, containing corrosion products, clay material, freshwater diatoms, and filamentous bacteria (Fig. 8).

Viable but nonculturable cells

Viable-but-nonculturable (VBNC) cells have been identified in a wide range of bacteria. Entry into the VNBC state has implicated due to a number of environmental stressors, including starvation, nutrient shock, low temperature, antibiotic pressure, and oxidative stress. While the plate count technique is one of the most standard methods for the enumeration of bacteria (Li et al., 2014), the process of removing bacterial

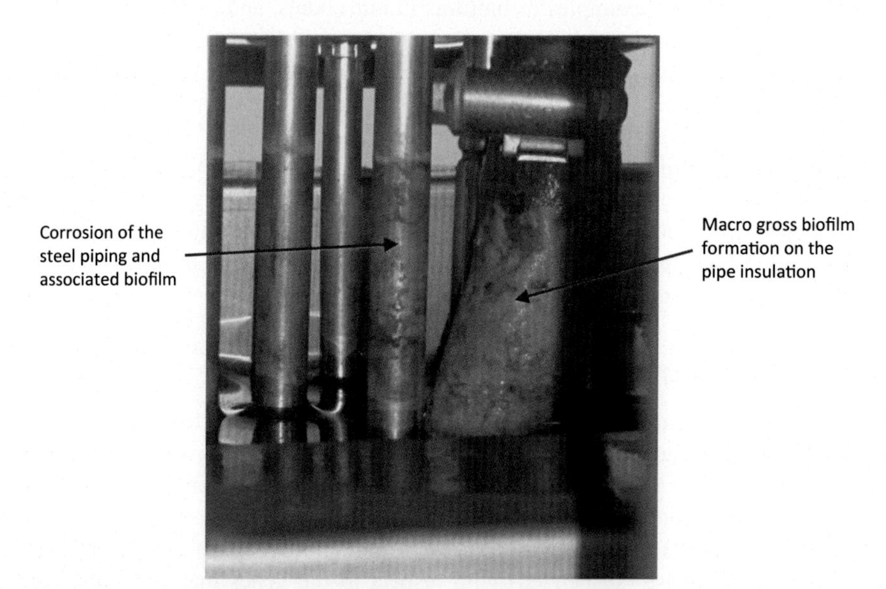

Fig. 7 Macro biofilm formation from the inside of a heater cooler. (Image courtesy of Simon Parkes UKHSA.)

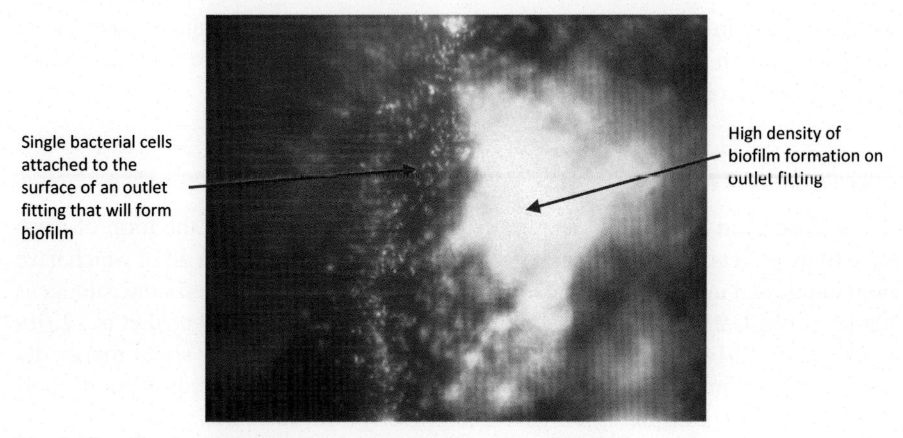

Fig. 8 Visualization of biofilm using light microscopy techniques.
Images courtesy of PHE.

biofilm cells from a surface can lead to an underestimation of the biofilm population. The presence of VBNC cells will impact on the number of biofilm cells recovered as they will not grow once placed on the agar plate. However, VBNC bacteria can still retain infectivity (Highmore et al., 2018) and may be implicated in chronic recurring infections, as VBNC is induced by the presence of antibiotics (Ayrapetyan et al., 2018). As a consequence, microscopy may be used to assess the viability of cells on a surface and can complement plate culture to understand the biofilms present on a surface (Wilks et al., 2021).

Historical aspects

The historical aspects of biofilms have been described for generations of scientists covering hundreds of years including both Leeuwenhoek (1632–1723) and Pasteur (1822–95), who described matrix adhering to surfaces or located in tissues or secretion. In modern-day disease, the detrimental human biofilm was first recognized as dental plaque. In 1914, Inglis described the following "As ordinarily understood a microbic plaque is by first a precipitation of mucin the sticky substance or solid matter in solution in saliva (Inglis, 1914). The teeth, plates, etc., being under conditions or rest as at night the mucin is deposited upon them and becomes infected with bacteria which ferment it and make it more adhesive." If ever there was a finer description of biofilm! There were very many early pioneers of biofilm research and water and its supply was a conduit through microorganisms could be transmitted to patients. One such early report was of *P. aeruginosa* infections in 24 newborn babies with nine deaths in 1947 when the milk supplied to a maternity unit was found to be contaminated from a rag that had been tied over a leak in water pipe and from this rag water dripped into the milk—there were also other issues (Hunter and Ensign, 1947). *P. aeruginosa* was isolated from the water dripping off the rag into the milk. Clearly, *P. aeruginosa* had been shown to contaminate water source in the 1940s, and it was not until the 1970s that *Legionella* was recognized as contamination water systems including those in healthcare and the role of the biofilm was paramount in the presence of this water born pathogen and its transmission route. Biofilms have now been shown to play a major role in medicine and disease.

Transmission routes

Water systems in healthcare institutes have hundreds of outlets in the form of wash hand basin outlets, showers, drinking faucets, and wastewater drains all of which have been implicated in the dispersal of biofilms and transmission of waterborne pathogens (de-las-Casas-Cámara et al., 2019; Inkster and Weinbren, 2021b; Jamal et al., 2019; Walker et al., 2014). Then there are the multitude of areas including sluice rooms, decontamination units, medical equipment including ice machines, hydrotherapy pools and heater coolers all of which have been involved in transmission of waterborne infections (Aspinall and Graham, 1989; Baudet et al., 2019; Breathnach et al., 2006; Marek et al., 2014).

Disease

Infections caused by these waterborne biofilms in healthcare are extensive and account for the deaths of many patients. It is often susceptible vulnerable immunocompromised patients who are at greatest risk. The routes of transmission whether it is direct from the water or a secondary route through splashing or aerosols ensure that the microorganisms are presented to the patients directly into wounds, the respiratory tract, through mucosal membranes or breaches in the skins integrity such as insertion points of catheters (Beatson and Bartley, 2017; Decraene et al., 2018; Jamal et al., 2018; Randag et al., 2019; RQIA, 2012). Being immunocompromised, biofilm bacteria are able to proliferate and take advantage of the inability of the host to defend against these microbial invaders.

Extent of the biofilm problem

The impact of clinical biofilm infections in healthcare is extensive with the US National Institutes of Health reported that over 80% of microbial infections are due to biofilms, which are resistant to many antimicrobial treatments and surgery is often required to physically remove chronic infection (Metcalf and Bowler, 2015). In addition, 78% of chronic wounds have been reported to be biofilm-associated (Malone et al., 2017). Financially biofilms account for a global expenditure on wounds in healthcare of $281 billion (WHO, 2017).

Water systems play a major role in encouraging the presence, growth, and transmission of biofilms to vulnerable patients in healthcare buildings. Water and wastewater biofilms result in enormous costs of $117 billion. Many products have been developed to control biofilms in water systems and combined with the time for disinfection and presentation of biofilms, the cost is estimated to be in the region of $41.5 billion (Cámara et al., 2022).

Diagnosis

Water samples are the mainstay of routine samples in water systems, and there is very little routine biofilm analysis carried out. However, during an outbreak, the incident team may decide that biofilm should be sampled either from tap outlets, pipework, equipment, or drains. A number of publications have described how best to take biofilm samples (Halstead et al., 2021; Proctor et al., 2018; Walker et al., 2014). Where biofilm needs to be sampled to identify the source of an infection, then swabs can be used to remove the biofilm from surface and the swab then analyzed for the presence of microorganisms. Alternatively sections of tubing, pipework, plumbing components, or tanks can swabbed and/or examined directly using microscopy to visualize the bacteria directly. In this way, either specific dyes or labels can be used to clarify the physiology or indeed the type of bacteria present.

Treatment of biofilm infections

Clinical biofilm infections create challenges for hosts and clinicians (Olivares et al., 2020). The structure and extracellular matrix of the biofilm can create an innate protective mechanical barrier blocking access for antimicrobials to the bacteria and fungi contained (Ciofu et al., 2022). So while many of the conventional approaches and use of antibiotics will be the first line of treatment, unfortunately, biofilms have a tolerance and resistance to antimicrobials. As such a range of antibiotics may have to be tested against the isolated microorganisms and other biofilm releasing chemical or physical treatments may be required to disrupt and disperse the biofilm to aid antimicrobial treatments.

Control of biofilms in water systems

Where water is used, then microorganisms will contaminate that environment and biofilms will form leading to contamination and infections. Training and education of those involved in the design, build, commission, and use of water in healthcare needs to be paramount to ensure that we reduce the number of hospital-acquired infections due to biofilms. The control of biofilms in water systems must start from first principals and the design of any system in which water is used. It is far easier to have a well-designed water system that will minimize the growth of biofilm with control strategies engineered into those designs to minimize the amount of water, biofilm formation, and transmission routes to patients. There are many control strategies including flow rates (high enough to reduce biofilm formation), temperature (cold water at 20°C and hot water leaving the calorifier at 60°C and returning at 55°C), filtration (point-of-use filters). and chemical treatments (their application across a water system is often not effective in the long term). However, more often than not the potential for biofilm growth and transmission is not understood by those responsible for the water system or medical equipment in which water is used. In addition, manufacturers often fail to understand the propensity of biofilm formation in their equipment or on their components. Plumbing components, which have been wet tested in production sites, should be decontaminated using a validated process and a certificate provided. Pipes and components, which have been wet tested during commission, should still supply wholesome water and monitored to ensure this is the case. The risk from biofilms is still poorly understood, and users require education in the risks of handing an everyday product such as water.

Summary of biofilms related to water

- A biofilm is a community of microorganisms including bacteria, fungi, protozoa, and viruses attached to a surface and enclosed in a matrix of polysaccharide materials.
- Biofilms have been established in natural water sources including: rivers and streams for millennia and the clinical implications of biofilm in water systems and associated equipment, fittings, and components, have only been fully recognized in recent times.
- Water system biofilms are ubiquitous and will include particulate matter, corrosion deposits, scale, debris from the environment, and the microbes within them scavenge nutrients from the water or from surfaces, including the materials of the water system pipework and associated fittings and components.
- Sources of exposure from water systems include hot and cold water systems, hand wash basin outlets, wastewater systems, showers, and medical equipment that uses water as part of its functionality, e.g., ice machines and heater coolers used in heart bypass surgery.
- Transmission routes include direct and indirect ingestion (e.g., from food prepared, irrigated, or sprayed with contaminated water), direct and indirect contact (e.g., splashing from contaminated sources, cross-contamination from hands, equipment, surfaces), inhalation, and aspiration.
- In clinical infections, the infectious hazards may colonize the skin, mucosal membranes, respiratory/digestive tracts or gain access through wounds or arterial, venous/urinary catheters, biofilms may then form and can include skin cells, hairs, red blood cells, sugars, and proteins. Foreign bodies such as medical prosthesis, e.g., catheters are susceptible to biofilm colonization. Waterborne pathogens cause an extensive range of infections from colonization to organs and tissues to sepsis and organ failure.
- Bacteria within biofilms are physiologically different to cells in the aqueous phase and due to the high impenetrability of the biofilm architecture can exhibit high tolerance to antimicrobials and antibiotics. These characteristics result in major challenges for biofilm control in both the healthcare water systems in clinical infections.
- All of those involved in the design, commissioning, build, and use of water in healthcare need to understand the risks that the development of biofilms poses, especially to vulnerable patients in critical care, transplant, and hematology/oncology units.
- There are a wide range of strategies for controlling waterborne pathogens in water systems. However, too often the use of these control strategies or the presence of the biofilm is not understood and control measures fail.
- The water sampling plan should encompass pre- and post-flush samples as appropriate. Preflush samples represent the highest risk to the person first using the outlet and so are the samples of choice for routine monitoring, postflush samples followed by disinfection of the outlet can be used to identify if the contamination is systemic or localized. Where the preflush is positive and postflush is negative, it may be indicative that there are biofilms present in the outlet and feed pipework and/or components such as TMVs.
- Education has a major role to play in preventing waterborne infections, ensuring systems as designed to deliver safe water and all those involved in the design, construction, installation, commissioning, operation, maintenance, and cleaning of water systems and associated equipment fittings and components understand their role in ensuring water at each point of use is safe for all uses and users.

References

Abbott, I.J., Slavin, M.A., Turnidge, J.D., Thursky, K.A., Worth, L.J., 2011. *Stenotrophomonas maltophilia*: emerging disease patterns and challenges for treatment. Expert Rev. Anti-Infect. Ther. 9, 471–488. https://doi.org/10.1586/eri.11.24.

Abdolrasouli, A., Gibani, M.M., de Groot, T., Borman, A.M., Hoffman, P., Azadian, B.S., Mughal, N., Moore, L.S.P., Johnson, E.M., Meis, J.F., 2021. A pseudo-outbreak of *Rhinocladiella similis* in a bronchoscopy unit of a tertiary care teaching hospital in London, United Kingdom. Mycoses 64, 394–404. https://doi.org/10.1111/myc.13227.

Allegra, S., Leclerc, L., Massard, P.A., Girardot, F., Riffard, S., Pourchez, J., 2016. Characterization of aerosols containing *Legionella* generated upon nebulization. Sci. Rep. 6, 33998. https://doi.org/10.1038/srep33998.

Almasy, E., Szederjesi, J., Rad, P., Georgescu, A., 2016. A fatal case of community acquired *Cupriavidus pauculus* pneumonia. J. Critic. Care Med. 2, 201–204. https://doi.org/10.1515/jccm-2016-0027.

Arciola, C.R., Campoccia, D., Montanaro, L., 2018. Implant infections: adhesion, biofilm formation and immune evasion. Nat. Rev. Microbiol. 16, 397. https://doi.org/10.1038/s41579-018-0019-y.

Asempa, T.E., Nicolau, D.P., Kuti, J.L., 2019. Carbapenem-nonsusceptible Pseudomonas aeruginosa isolates from intensive care units in the United States: a potential role for new β-lactam combination agents. J. Clin. Microbiol. 57. https://doi.org/10.1128/JCM.00535-19. e00535-19.

Ashfaq, M.Y., Da'na, D.A., Al-Ghouti, M.A., 2022. Application of MALDI-TOF MS for identification of environmental bacteria: a review. J. Environ. Manag. 305, 114359. https://doi.org/10.1016/j.jenvman.2021.114359.

Aspinall, S.T., Graham, R., 1989. Two sources of contamination of a hydrotherapy pool by environmental organisms. J. Hosp. Infect. 14, 285–292. https://doi.org/10.1016/0195-6701(89)90068-6.

Aydın, B., Dilli, D., Zenciroğlu, A., Okumuş, N., Ozkan, S., Tanır, G., 2012. A case of newborn with community acquired pneumonia caused by *Cupriavidus pauculus*. Tuberk. Toraks 60, 160–162.

Ayrapetyan, M., Williams, T., Oliver, J.D., 2018. Relationship between the viable but nonculturable state and antibiotic persister cells. J. Bacteriol. 200. https://doi.org/10.1128/JB.00249-18. e00249-18.

Balada-Llasat, J.-M., Elkins, C., Swyers, L., Bannerman, T., Pancholi, P., 2010. Pseudo-outbreak of *Cupriavidus pauculus* infection at an outpatient clinic related to rinsing culturette swabs in tap water. J. Clin. Microbiol. 48, 2645–2647. https://doi.org/10.1128/JCM.01874-09.

Barker, J., Brown, M.R., 1994. Trojan horses of the microbial world: protozoa and the survival of bacterial pathogens in the environment. Microbiology (Reading) 140, 1253–1259. https://doi.org/10.1099/00221287-140-6-1253.

Bartram, J., 2007. WHO Legionella and the Prevention of Legionellosis. https://www.who.int/water_sanitation_health/publications/legionella/en/.

Bassetti, M., Vena, A., Croxatto, A., Righi, E., Guery, B., 2018. How to manage *Pseudomonas aeruginosa* infections. Drugs Context 7, 212527. https://doi.org/10.7573/dic.212527.

Baudet, A., Lizon, J., Martrette, J.-M., Camelot, F., Florentin, A., Clément, C., 2019. Dental unit waterlines: a survey of practices in Eastern France. Int. J. Environ. Res. Public Health 16. https://doi.org/10.3390/ijerph16214242.

Beatson, S.A., Bartley, P.B., 2017. Diving deep into hospital-acquired *Legionella pneumophila* with whole-genome sequencing. Clin. Infect. Dis. 64, 1260–1262. https://doi.org/10.1093/cid/cix156.

Beauté, J., 2017. Legionnaires' disease in Europe, 2011 to 2015. Euro Surveill. 22. https://doi.org/10.2807/1560-7917.ES.2017.22.27.30566.

Benedict, K.M., 2017. Surveillance for waterborne disease outbreaks associated with drinking water—United States, 2013–2014. MMWR Morb. Mortal. Wkly Rep. 66. https://doi.org/10.15585/mmwr.mm6644a3.

Berg, G., Roskot, N., Smalla, K., 1999. Genotypic and phenotypic relationships between clinical and environmental isolates of Stenotrophomonas maltophilia. J. Clin. Microbiol. 37, 3594–3600.

Bianco, G., Boattini, M., Audisio, E., Cavallo, R., Costa, C., 2018. Septic shock due to meropenem- and colistin-resistant *Cupriavidus pauculus*. J. Hosp. Infect. 99, 364–365. https://doi.org/10.1016/j.jhin.2018.03.025.

Boamah, D.K., Zhou, G., Ensminger, A.W., O'Connor, T.J., 2017. From many hosts, one accidental pathogen: the diverse protozoan hosts of *Legionella*. Front. Cell. Infect. Microbiol. 7.

Borges, V., Nunes, A., Sampaio, D.A., Vieira, L., Machado, J., Simões, M.J., Gonçalves, P., Gomes, J.P., 2016. *Legionella pneumophila* strain associated with the first evidence of person-to-person transmission of Legionnaires' disease: a unique mosaic genetic backbone. Sci. Rep. 6, 26261. https://doi.org/10.1038/srep26261.

Bou, R., Ramos, P., 2009. Outbreak of nosocomial Legionnaires' disease caused by a contaminated oxygen humidifier. J. Hosp. Infect. 71, 381–383.

Bou, R., Aguilar, A., Perpinan, J., Ramos, P., Peris, M., Lorente, L., Zuniga, A., 2006. Nosocomial outbreak of Pseudomonas aeruginosa infections related to a flexible bronchoscope. J. Hosp. Infect. 64, 129–135.

Breathnach, A.S., Riley, P.A., Shad, S., Jownally, S.M., Law, R., Chin, P.C., Kaufmann, M.E., Smith, E.J., 2006. An outbreak of wound infection in cardiac surgery patients caused by *Enterobacter cloacae* arising from cardioplegia ice. J. Hosp. Infect. 64, 124–128. https://doi.org/10.1016/j.jhin.2006.06.015.

Brooke, J.S., 2012. *Stenotrophomonas maltophilia*: an emerging global opportunistic pathogen. Clin. Microbiol. Rev. 25, 2–41. https://doi.org/10.1128/CMR.00019-11.

Brundrett, G.W., 1991. Outbreaks of Legionnaires' Disease at Stafford District General Hospital. The Chartered Institution of Building Services Engineers, pp. 53–65.

BSI, 2022. BS 8580-2:2022—Risk Assessments for *Pseudomonas aeruginosa* and Other Waterborne Pathogens. Code of practice https://standardsdevelopment.bsigroup.com.

Bunya, V.Y., Hammersmith, K.M., Rapuano, C.J., Ayres, B.D., Cohen, E.J., 2007. Topical and oral voriconazole in the treatment of fungal keratitis. Am J. Ophthalmol. 143, 151–153. https://doi.org/10.1016/j.ajo.2006.07.033.

Burillo, A., Pedro-Botet, M.L., Bouza, E., 2017. Microbiology and epidemiology of Legionnaire's disease. Infect. Dis. Clin. N. Am. 31, 7–27. https://doi.org/10.1016/j.idc.2016.10.002.

Butler, J., Kelly, S.D., Muddiman, K.J., Besinis, A., Upton, M., 2022. Hospital sink traps as a potential source of the emerging multidrug-resistant pathogen *Cupriavidus pauculus*: characterization and draft genome sequence of strain MF1. J. Med. Microbiol. 71, 001501. https://doi.org/10.1099/jmm.0.001501.

Cámara, M., Green, W., MacPhee, C.E., Rakowska, P.D., Raval, R., Richardson, M.C., Slater-Jefferies, J., Steventon, K., Webb, J.S., 2022. Economic significance of biofilms: a multidisciplinary and cross-sectoral challenge. NPJ Biofilms Microbiomes 8, 1–8. https://doi.org/10.1038/s41522-022-00306-y.

Cazals, M., Bédard, E., Doberva, M., Faucher, S., Prévost, M., 2022. Compromised effectiveness of thermal inactivation of *Legionella pneumophila* in water heater sediments and water, and influence of the presence of *Vermamoeba vermiformis*. Microorganisms 10, 443. https://doi.org/10.3390/microorganisms10020443.

CDC, 2019a. *Pseudomonas aeruginosa* Infection. https://www.cdc.gov/hai/organisms/pseudomonas.html.

CDC, 2019b. Guidelines for Environmental Infection Control in Health-care Facilities: Recommendations of CDC and the Healthcare Infection Control Practices Advisory Committee (HICPAC). https://www.cdc.gov/infectioncontrol/pdf/guidelines/environmental-guidelines-P.pdf.

CDC, 2019c. Aspergillosis Statistics, Aspergillosis, Types of Fungal Diseases, Fungal Diseases.

CDC, 2021. National Outbreak Reporting System (NORS) Dashboard.

Chang, D.C., Grant, G.B., O'Donnell, K., Wannemuehler, K.A., Noble-Wang, J., Rao, C.Y., Jacobson, L.M., Crowell, C.S., Sneed, R.S., Lewis, F.M.T., Schaffzin, J.K., Kainer, M.A., Genese, C.A., Alfonso, E.C., Jones, D.B., Srinivasan, A., Fridkin, S.K., Park, B.J., Fusarium Keratitis Investigation Team, 2006. Multistate outbreak of *Fusarium keratitis* associated with use of a contact lens solution. JAMA 296, 953–963. https://doi.org/10.1001/jama.296.8.953.

Cheng, K., Chui, H., Domish, L., Hernandez, D., Wang, G., 2016. Recent development of mass spectrometry and proteomics applications in identification and typing of bacteria. Proteomics Clin. Appl. 10, 346–357. https://doi.org/10.1002/prca.201500086.

Cianciotto, N.P., 2015. An update on iron acquisition by *Legionella pneumophila*: new pathways for siderophore uptake and ferric iron reduction. Future Microbiol 10, 841–851. https://doi.org/10.2217/fmb.15.21.

Ciofu, O., Moser, C., Jensen, P.Ø., Høiby, N., 2022. Tolerance and resistance of microbial biofilms. Nat. Rev. Microbiol. 1–15. https://doi.org/10.1038/s41579-022-00682-4.

Cirillo, J.D., Cirillo, S.L.G., Yan, L., Bermudez, L.E., Falkow, S., Tompkins, L.S., 1999. Intracellular growth in Acanthamoeba castellanii affects monocyte entry mechanisms and enhances virulence of Legionella pneumophila. Infect. Immun. 67, 4427–4434. https://doi.org/10.1128/IAI.67.9.4427-4434.1999.

Coppry, M., Leroyer, C., Saly, M., Venier, A.-G., Slekovec, C., Bertrand, X., Parer, S., Alfandari, S., Cambau, E., Megarbane, B., Lawrence, C., Clair, B., Lepape, A., Cassier, P., Trivier, D., Boyer, A., Boulestreau, H., Asselineau, J., Dubois, V., Thiébaut, R., Rogues, A.-M., 2020. Exogenous acquisition of *Pseudomonas aeruginosa* in intensive care units: a prospective multi-centre study. J. Hosp. Infect. 104, 40–45. https://doi.org/10.1016/j.jhin.2019.08.008.

Costerton, J.W., 1995. Overview of microbial biofilms. J. Ind. Microbiol. 15, 137–140.

Currie, S.L., Beattie, T.K., 2015. Compost and Legionella longbeachae: an emerging infection? Perspect. Public Health 135, 309–315.

Decraene, V., Phan, H.T.T., George, R., Wyllie, D.H., Akinremi, O., Aiken, Z., Cleary, P., Dodgson, A., Pankhurst, L., Crook, D.W., Lenney, C., Walker, A.S., Woodford, N., Sebra, R., Fath-Ordoubadi, F., Mathers, A.J., Seale, A.C., Guiver, M., McEwan, A., Watts, V., Welfare, W., Stoesser, N., Cawthorne, J., Group, the T.I, 2018. A large, refractory nosocomial outbreak of *Klebsiella pneumoniae* carbapenemase-producing *Escherichia coli* demonstrates carbapenemase gene outbreaks involving sink sites require novel approaches to infection control. Antimicrob. Agents Chemother. 62. https://doi.org/10.1128/AAC.01689-18.

Delafont, V., Brouke, A., Bouchon, D., Moulin, L., Héchard, Y., 2013. Microbiome of free-living amoebae isolated from drinking water. Water Res. 47, 6958–6965. https://doi.org/10.1016/j.watres.2013.07.047.

de-las-Casas-Cámara, G., Giráldez-García, C., Adillo-Montero, M.I., Muñoz-Egea, M.C., Martín-Ríos, M.D., 2019. Impact of removing sinks from an intensive care unit on isolations by gram-negative non-fermenting bacilli in patients with invasive mechanical ventilation. Med. Clín. (English Ed.) 152, 261–263. https://doi.org/10.1016/j.medcle.2018.06.014.

Dennis, P.J., Green, D., Jones, B.P.C., 1984. A note on the temperature tolerance of *Legionella*. J. Appl. Bacteriol. 56, 349–350. https://doi.org/10.1111/j.1365-2672.1984.tb01359.x.

DHSC, 2016. HTM 04-01: Safe Water in Healthcare Premises.

Dóczi, I., Gyetvai, T., Kredics, L., Nagy, E., 2004. Involvement of *Fusarium* spp. in fungal keratitis. Clin. Microbiol. Infect. 10, 773–776. https://doi.org/10.1111/j.1469-0691.2004.00909.x.

ECDC, 2019. Legionnaires' Disease—Annual Epidemiological Report for 2019. European Centre for Disease Prevention and Control.

Edelstein, P.H., 1981. Improved semi-selective medium for isolation of *Legionella pneumophila* from contaminated clinical and environmental specimens. J. Clin. Microbiol. 14, 298–303.

Ezzeddine, H., Van Ossel, C., Delmée, M., Wauters, G., 1989. *Legionella* spp. in a hospital hot water system: effect of control measures. J. Hosp. Infect. 13, 121–131. https://doi.org/10.1016/0195-6701(89)90018-2.

Falkinham, J.O., 2015. The *Mycobacterium avium* complex and slowly growing *Mycobacteria*. In: Tang, Y.-W., Sussman, M., Liu, D., Poxton, I., Schwartzman, J. (Eds.), Molecular Medical Microbiology, second ed. Academic Press, Boston, MA, pp. 1669–1678, https://doi.org/10.1016/B978-0-12-397169-2.00094-9.

Falkinham, J.O., 2018. Challenges of NTM drug development. Front. Microbiol. 9, 1613. https://doi.org/10.3389/fmicb.2018.01613.

Ferone, M., Gowen, A., Fanning, S., Scannell, A.G.M., 2020. Microbial detection and identification methods: bench top assays to omics approaches. Compr. Rev. Food Sci. Food Saf. 19, 3106–3129. https://doi.org/10.1111/1541-4337.12618.

Fiel, D., Calca, R., Cacheira, E., Rombo, N., Querido, S., Nascimento, C., Jorge, C., Weigert, A., Birne, R., Clemente, B., Martinho, A., Toscano, C., Bruges, M., Malvar, B., Machado, D., 2020. *Cupriavius pauculus* causing a respiratory tract infection in a post-kidney-transplant patient: a firstly described rare clinical case. Pharm. Pharmacol. Int. J. 8, 34–36. https://doi.org/10.15406/ppij.2020.08.00277.

Fisher-Hoch, S.P., Bartlett, C.L., Tobin, J.O., Gillett, M.B., Nelson, A.M., Pritchard, J.E., Smith, M.G., Swann, R.A., Talbot, J.M., Thomas, J.A., 1981. Investigation and control of an outbreaks of legionnaires' disease in a district general hospital. Lancet 1, 932–936.

Fitzgeorge, R.B., Dennis, P.J., 1983. Isolation of *Legionella pneumophila* from water supplies: comparison of methods based on the guinea-pig and culture media. J. Hyg. (Lond.) 91, 179–187.

Fleiszig, S.M.J., Kroken, A.R., Nieto, V., Grosser, M.R., Wan, S.J., Metruccio, M.M.E., Evans, D.J., 2020. Contact lens-related corneal infection: intrinsic resistance and its compromise. Prog. Retin. Eye Res. 76, 100804. https://doi.org/10.1016/j.preteyeres.2019.100804.

Flemming, H.-C., Wingender, J., 2010. The biofilm matrix. Nat. Rev. Microbiol. 8, 623–633. https://doi.org/10.1038/nrmicro2415.

Forsberg, K.J., Patel, S., Wencewicz, T.A., Dantas, G., 2015. The tetracycline destructases: a novel family of tetracycline-inactivating enzymes. Chem. Biol. 22, 888–897. https://doi.org/10.1016/j.chembiol.2015.05.017.

Fürbringer, P., 1876. Beobachtungen über Lungenmycose beim Menschen. Archiv f. Pathol. Anat. 66, 330–365. https://doi.org/10.1007/BF01878266.

Gerardin, P., Farny, K., Simac, C., Laurent, A.F., Grandbastien, B., Robillard, P.Y., 2006. Pseudomonas aeruginosa infections in a neonatal care unit at Reunion Island. Arch. Pediatr. 13, 1500–1506.

Graman, P.S., Quinlan, G.A., Rank, J.A., 1997. Nosocomial legionellosis traced to a contaminated ice machine. Infect. Control Hosp. Epidemiol. 18, 637–640.

Halstead, F.D., Quick, J., Niebel, M.O., Garvey, M., Cumley, N., Smith, R., Neal, T., Roberts, P., Hardy, K., Shabir, S., Walker, J.T., Hawkey, P., Loman, N.J., 2021. *Pseudomonas aeruginosa* infection in augmented care: the molecular ecology and transmission dynamics in four large UK hospitals. J. Hosp. Infect. https://doi.org/10.1016/j.jhin.2021.01.020.

Hamilton, K.A., Prussin, A.J., Ahmed, W., Haas, C.N., 2018. Outbreaks of Legionnaires' disease and pontiac fever 2006-2017. Curr. Environ. Health Rep. 5, 263–271. https://doi.org/10.1007/s40572-018-0201-4.

Han, X.Y., Ihegword, A., Evans, S.E., Zhang, J., Li, L., Cao, H., Tarrand, J.J., El-Kweifi, O., 2015. Microbiological and clinical studies of Legionellosis in 33 patients with cancer. J. Clin. Microbiol. 53, 2180–2187.

Haupt, T.E., Heffernan, R.T., Kazmierczak, J.J., Nehls-Lowe, H., Rheineck, B., Powell, C., Leonhardt, K.K., Chitnis, A.S., Davis, J.P., 2012. An outbreak of legionnaires disease associated with a decorative water wall fountain in a hospital. Infect. Control Hosp. Epidemiol. 33, 185–191.

Henry, R., 2017. Etymologia: Legionella pneumophila. Emerg. Infect. Dis. 23, 1851. https://doi.org/10.3201/eid2311.ET2311.

Herwaldt, L.A., Marra, A.R., 2018. Legionella: a reemerging pathogen. Curr. Opin. Infect. Dis. 31, 325–333. https://doi.org/10.1097/QCO.0000000000000468.

Highmore, C.J., Warner, J.C., Rothwell, S.D., Wilks, S.A., Keevil, C.W., 2018. Viable but nonculturable *Listeria monocytogenes* and *Salmonella enterica* serovar Thompson induced by chlorine stress remain infectious. mBio 9. https://doi.org/10.1128/mBio.00540-18. e00540-18.

Hoefsloot, W., van Ingen, J., Andrejak, C., Angeby, K., Bauriaud, R., Bemer, P., Beylis, N., Boeree, M.J., Cacho, J., Chihota, V., Chimara, E., Churchyard, G., Cias, R., Daza, R., Daley, C.L., Dekhuijzen, P.N.R., Domingo, D., Drobniewski, F., Esteban, J., Fauville-Dufaux, M., Folkvardsen, D.B., Gibbons, N., Gómez-Mampaso, E., Gonzalez, R., Hoffmann, H., Hsueh, P.-R., Indra, A., Jagielski, T., Jamieson, F., Jankovic, M., Jong, E., Keane, J., Koh, W.-J., Lange, B., Leao, S., Macedo, R., Mannsåker, T., Marras, T.K., Maugein, J., Milburn, H.J., Mlinkó, T., Morcillo, N., Morimoto, K., Papaventsis, D., Palenque, E., Paez-Peña, M., Piersimoni, C., Polanová, M., Rastogi, N., Richter, E., Ruiz-Serrano, M.J., Silva, A., da Silva, M.P., Simsek, H., van Soolingen, D., Szabó, N., Thomson, R., Tórtola Fernandez, T., Tortoli, E., Totten, S.E., Tyrrell, G., Vasankari, T., Villar, M., Walkiewicz, R., Winthrop, K.L., Wagner, D., Nontuberculous Mycobacteria Network European Trials Group, 2013. The geographic diversity of nontuberculous mycobacteria isolated from pulmonary samples: an NTM-NET collaborative study. Eur. Respir. J. 42, 1604–1613. https://doi.org/10.1183/09031936.00149212.

Høiby, N., Bjarnsholt, T., Givskov, M., Molin, S., Ciofu, O., 2010. Antibiotic resistance of bacterial biofilms. Int. J. Antimicrob. Agents 35, 322–332. https://doi.org/10.1016/j.ijantimicag.2009.12.011.

HSE, 2014. HSG 274 Legionnaires' Disease—Technical Guidance Part 2: The Control of Legionella Bacteria in Hot and Cold Water Systems Technical Guidance. http://www.hse.gov.uk/pubns/books/hsg274.htm. (Accessed 5 January 2021).

Huda, S.A., Yadava, S., Kahlown, S., Jilani, M.H., Sharma, B., 2020. A rare case of ventilator-associated pneumonia caused by *Cupriavidus pauculus*. Cureus 12. https://doi.org/10.7759/cureus.8573.

Hugo Johansson, P.J., Andersson, K., Wiebe, T., Schalén, C., Bernander, S., 2006. Nosocomial transmission of Legionella pneumophila to a child from a hospital's cold-water supply. Scand. J. Infect. Dis. 38, 1023–1027.

Hunter, C.A., Ensign, P.R., 1947. An Epidemic of Diarrhoea in a New-born Nursery Caused by *Pseudomonas aeruginosa*., p. 4.

Inglis, O., 1914. New things in dental pathology. Dental Reg. 68, 512–516.

Inglis, T.J., Benson, K.A., O'Reilly, L., Bradbury, R., Hodge, M., Speers, D., Heath, C.H., 2010. Emergence of multi-resistant Pseudomonas aeruginosa in a Western Australian hospital. J. Hosp. Infect. 76, 60–65.

Inkster, T., Weinbren, M., 2021a. Water springing to life the fungal desert. J. Hosp. Infect. 111, 65–68. https://doi.org/10.1016/j.jhin.2021.02.015.

Inkster, T., Weinbren, M., 2021b. Is it time for water and drainage standards to be part of the accreditation process for haemato-oncology units? Clin. Microbiol. Infect. 27, 1721–1723. https://doi.org/10.1016/j.cmi.2021.08.011.

Inkster, T., Peters, C., Wafer, T., Holloway, D., Makin, T., 2021. Investigation and control of an outbreak due to a contaminated hospital water system, identified following a rare case of *Cupriavidus pauculus* bacteraemia. J. Hosp. Infect. https://doi.org/10.1016/j.jhin.2021.02.001.

Inkster, T., Wilson, G., Black, J., Mallon, J., Connor, M., Weinbren, M., 2022. *Cupriavidus* spp. and other waterborne organisms in healthcare water systems across the UK. J. Hosp. Infect. 123, 80–86. https://doi.org/10.1016/j.jhin.2022.02.003.

Jamal, M., Ahmad, W., Andleeb, S., Jalil, F., Imran, M., Nawaz, M.A., Hussain, T., Ali, M., Rafiq, M., Kamil, M.A., 2018. Bacterial biofilm and associated infections. J. Chin. Med. Assoc. 81, 7–11. https://doi.org/10.1016/j.jcma.2017.07.012.

Jamal, A., Brown, K.A., Katz, K., Johnstone, J., Muller, M.P., Allen, V., Borgia, S., Boyd, D.A., Ciccotelli, W., Delibasic, K., Fisman, D., Leis, J., Li, A., Mataseje, L., Mehta, M., Mulvey, M., Ng, W., Pantelidis, R., Paterson, A., McGeer, A., 2019. Risk factors for contamination with carbapenemase-producing Enterobacteriales (CPE) in exposed hospital drains in Ontario, Canada. Open Forum Infect. Dis. 6, S441. https://doi.org/10.1093/ofid/ofz360.1091.

Le Dantec, C., Duguet, J.-P., Montiel, A., Dumoutier, N., Dubrou, S., Vincent, V., 2002. Chlorine disinfection of atypical *Mycobacteria* isolated from a water distribution system. Appl. Environ. Microbiol. 68, 1025–1032. https://doi.org/10.1128/AEM.68.3.1025-1032.2002.

Lettinga, K.D., Verbon, A., Weverling, G.-J., Schellekens, J.F.P., Den Boer, J.W., Yzerman, E.P.F., Prins, J., Boersma, W.G., van Ketel, R.J., Prins, J.M., Speelman, P., 2002. Legionnaires' disease at a dutch flower show: prognostic factors and impact of therapy. Emerg. Infect. Dis. 8, 1448–1454. https://doi.org/10.3201/eid0812.020035.

Li, L., Mendis, N., Trigui, H., Oliver, J.D., Faucher, S.P., 2014. The importance of the viable but non-culturable state in human bacterial pathogens. Front. Microbiol. 5.

Lim, W.S., Baudouin, S.V., George, R.C., Hill, A.T., Jamieson, C., Jeune, I.L., Macfarlane, J.T., Read, R.C., Roberts, H.J., Levy, M.L., Wani, M., Woodhead, M.A., 2009. BTS guidelines for the management of community acquired pneumonia in adults: update 2009. Thorax 64. https://doi.org/10.1136/thx.2009.121434. iii1–iii55.

Link, J., 1809. Observationes in ordines plantarum naturales. Dissertatio Ima. Gesellschaft Naturforschender Freunde zu Berlin, Magazin 3 (1), 3–42. Biota of NZ.

Loret, J.-F., Greub, G., 2010. Free-living amoebae: biological by-passes in water treatment. In: International Journal of Hygiene and Environmental Health, First Ph.D. student seminar on "Water & Health" organized within the "Cannes Water Symposium", Cannes, July 2009 213, 167–175., https://doi.org/10.1016/j.ijheh.2010.03.004.

Magill, S.S., Edwards, J.R., Bamberg, W., Beldavs, Z.G., Dumyati, G., Kainer, M.A., Lynfield, R., Maloney, M., McAllister-Hollod, L., Nadle, J., Ray, S.M., Thompson, D.L., Wilson, L.E., Fridkin, S.K., 2014. Multistate point-prevalence survey of health care–associated infections. N. Engl. J. Med. 370, 1198–1208. https://doi.org/10.1056/NEJMoa1306801.

Malcolm, K.C., Caceres, S.M., Honda, J.R., Davidson, R.M., Epperson, L.E., Strong, M., Chan, E.D., Nick, J.A., 2017. *Mycobacterium abscessus* displays fitness for fomite transmission. Appl. Environ. Microbiol. 83. https://doi.org/10.1128/AEM.00562-17. e00562-17.

Malone, M., Bjarnsholt, T., McBain, A.J., James, G.A., Stoodley, P., Leaper, D., Tachi, M., Schultz, G., Swanson, T., Wolcott, R.D., 2017. The prevalence of biofilms in chronic wounds: a systematic review and meta-analysis of published data. J. Wound Care 26, 20–25. https://doi.org/10.12968/jowc.2017.26.1.20.

Manaia, C.M., Nunes, O.C., Morais, P.V., Costa, M.S.D., 1990. Heterotrophic plate counts and the isolation of bacteria from mineral waters on selective and enrichment media. J. Appl. Bacteriol. 69, 871–876. https://doi.org/10.1111/j.1365-2672.1990.tb01586.x.

Marek, A., Smith, A., Peat, M., Connell, A., Gillespie, I., Morrison, P., Hamilton, A., Shaw, D., Stewart, A., Hamilton, K., Smith, I., Mead, A., Howard, P., Ingle, D., 2014. Endoscopy supply water and final rinse testing: five years of experience. J. Hosp. Infect. 88, 207–212. https://doi.org/10.1016/j.jhin.2014.09.004.

Martak, D., Gbaguidi-Haore, H., Meunier, A., Valot, B., Conzelmann, N., Eib, M., Autenrieth, I.B., Slekovec, C., Tacconelli, E., Bertrand, X., Peter, S., Hocquet, D., Guther, J., 2022. High prevalence of *Pseudomonas aeruginosa* carriage in residents of French and German long-term care facilities. Clin. Microbiol. Infect. https://doi.org/10.1016/j.cmi.2022.05.004.

Massip, C., Descours, G., Ginevra, C., Doublet, P., Jarraud, S., Gilbert, C., 2017. Macrolide resistance in *Legionella pneumophila*: the role of LpeAB efflux pump. J. Antimicrob. Chemother. 72, 1327–1333. https://doi.org/10.1093/jac/dkw594.

Mastro, T.D., Fields, B.S., Breiman, R.F., Campbell, J., Plikaytis, B.D., Spika, J.S., 1991. Nosocomial Legionnaires' disease and use of medication nebulizers. J. Infect. Dis. 163, 667–671.

Matsushita, K., Hijikuro, K., Arita, S., Kaneko, Y., Isozaki, M., 2017. A case of severe Legionella longbeachae Pneumonia and usefulness of LAMP assay. Rinsho Biseibutshu Jinsoku Shindan Kenkyukai Shi 27, 57–63.

Matz, C., Moreno, A.M., Alhede, M., Manefield, M., Hauser, A.R., Givskov, M., Kjelleberg, S., 2008. Pseudomonas aeruginosa uses type III secretion system to kill biofilm-associated amoebae. ISME J. 2, 843–852. https://doi.org/10.1038/ismej.2008.47.

Metcalf, D.G., Bowler, P.G., 2015. Biofilm delays wound healing: A review of the evidence. Burns Trauma 1, 5–12. https://doi.org/10.4103/2321-3868.113329.

Miyata, J., Huh, J.Y., Ito, Y., Kobuchi, T., Kusukawa, K., Hayashi, H., 2017. Can we truly rely on the urinary antigen test for the diagnosis? Legionella case report. J. Gen. Fam. Med. 18, 139–143.

Moloney, E.M., Deasy, E.C., Swan, J.S., Brennan, G.I., O'Donnell, M.J., Coleman, D.C., 2020. Whole-genome sequencing identifies highly related *Pseudomonas aeruginosa* strains in multiple washbasin U-bends at several locations in one hospital: evidence for trafficking of potential pathogens via wastewater pipes. J. Hosp. Infect. 104, 484–491.

Mora, M., Perras, A., Alekhova, T.A., Wink, L., Krause, R., Aleksandrova, A., Novozhilova, T., Moissl-Eichinger, C., 2016. Resilient microorganisms in dust samples of the International Space Station—survival of the adaptation specialists. Microbiome 4, 65. https://doi.org/10.1186/s40168-016-0217-7.

Moritz, M.M., Flemming, H.-C., Wingender, J., 2010. Integration of *Pseudomonas aeruginosa* and *Legionella pneumophila* in drinking water biofilms grown on domestic plumbing materials. In: International Journal of Hygiene and Environmental Health, First Ph.D. student seminar on "Water & Health" organized within the "Cannes Water Symposium", Cannes, July 2009 213, 190–197., https://doi.org/10.1016/j.ijheh.2010.05.003.

National Services Scotland, 2019. Prevention and Management of Healthcare Water-associated Infection Incidents/Outbreaks. https://www.nipcm.hps.scot.nhs.uk/media/1680/2019-08-water-incidents-info-sheet-v1.pdf.

Nucci, M., Marr, K.A., Queiroz-Telles, F., Martins, C.A., Trabasso, P., Costa, S., Voltarelli, J.C., Colombo, A.L., Imhof, A., Pasquini, R., Maiolino, A., Souza, C.A., Anaissie, E., 2004. *Fusarium* infection in hematopoietic stem cell transplant recipients. Clin. Infect. Dis. 38, 1237–1242. https://doi.org/10.1086/383319.

O'Donnell, K., Sutton, D.A., Rinaldi, M.G., Magnon, K.C., Cox, P.A., Revankar, S.G., Sanche, S., Geiser, D.M., Juba, J.H., van Burik, J.-A.H., Padhye, A., Anaissie, E.J., Francesconi, A., Walsh, T.J., Robinson, J.S., 2004. Genetic diversity of human pathogenic members of the *Fusarium oxysporum* Complex Inferred from multilocus DNA sequence data and amplified fragment length polymorphism analyses: evidence for the recent dispersion of a geographically widespread clonal lineage and nosocomial origin. J. Clin. Microbiol. 42, 5109–5120. https://doi.org/10.1128/JCM.42.11.5109-5120.2004.

Odeh, R., Quinn, J.P., 2000. Problem pulmonary pathogens: Pseudomonas aeruinosa. Semin. Respir. Crit. Care Med. 21, 331–339.

Oggioni, M.R., Trappetti, C., Kadioglu, A., Cassone, M., Iannelli, F., Ricci, S., Andrew, P.W., Pozzi, G., 2006. Switch from planktonic to sessile life: a major event in pneumococcal pathogenesis. Mol. Microbiol. 61, 1196–1210. https://doi.org/10.1111/j.1365-2958.2006.05310.x.

Oie, S., Makieda, D., Ishida, S., Okano, Y., Kamiya, A., 2006. Microbial contamination of nebulization solution and its measures. Biol. Pharm. Bull. 29, 503–507. https://doi.org/10.1248/bpb.29.503.

Olivares, E., Badel-Berchoux, S., Provot, C., Prévost, G., Bernardi, T., Jehl, F., 2020. Clinical impact of antibiotics for the treatment of *Pseudomonas aeruginosa* biofilm infections. Front. Microbiol. 10.

Palmore, T.N., Stock, F., White, M., Bordner, M., Michelin, A., Bennett, J.E., Murray, P.R., Henderson, D.K., 2009. A cluster of cases of nosocomial legionnaires disease linked to a contaminated hospital decorative water fountain. Infect. Control Hosp. Epidemiol. 30, 764–768.

Paltauf, A., 1885. Mycosis mucorina. Archiv f. pathol. Anat. 102, 543–564. https://doi.org/10.1007/BF01932420.

Parraga-Nino, N., Quero, S., Uria, N., Castillo-Fernandez, O., Jimenez-Ezenarro, J., Munoz, F.X., Sabria, M., Garcia-Nunez, M., 2018. Antibody test for Legionella pneumophila detection. Diagn. Microbiol. Infect. Dis. 90, 85–89.

Patterson, S.B., Mende, K., Li, P., Lu, D., Carson, M.L., Murray, C.K., Tribble, D.R., Blyth, D.M., 2020. *Stenotrophomonas maltophilia* infections: clinical characteristics in a military trauma population. Diagn. Microbiol. Infect. Dis. 96, 114953. https://doi.org/10.1016/j.diagmicrobio.2019.114953.

Pinar-Méndez, A., Fernández, S., Baquero, D., Vilaró, C., Galofré, B., González, S., Rodrigo-Torres, L., Arahal, D.R., Macián, M.C., Ruvira, M.A., Aznar, R., Caudet-Segarra, L., Sala-Comorera, L., Lucena, F., Blanch, A.R., Garcia-Aljaro, C., 2021. Rapid and improved identification of drinking water bacteria using the drinking water library, a dedicated MALDI-TOF MS database. Water Res. 203, 117543. https://doi.org/10.1016/j.watres.2021.117543.

Plaignaud, M., 1781. Observation concerning a fungus in the maxillary sinus. J. Chir. 87, 244–251.

Popović, N.T., Kazazić, S.P., Strunjak-Perović, I., Čož-Rakovac, R., 2017. Differentiation of environmental aquatic bacterial isolates by MALDI-TOF MS. Environ. Res. 152, 7–16. https://doi.org/10.1016/j.envres.2016.09.020.

Portal, E., Descours, G., Ginevra, C., Mentasti, M., Afshar, B., Chand, M., Day, J., Echahidi, F., Franzin, L., Gaia, V., Lück, C., Meghraoui, A., Moran-Gilad, J., Ricci, M.L., Lina, G., Uldum, S., Winchell, J., Howe, R., Bernard, K., Spiller, O.B., Chalker, V.J., Jarraud, S., the ESCMID Study Group for Legionella Infections (ESGLI), 2021. Legionella antibiotic susceptibility testing: is it time for international standardization and evidence-based guidance? J. Antimicrob. Chemother. 76, 1113–1116. https://doi.org/10.1093/jac/dkab027.

Proctor, C.R., Reimann, M., Vriens, B., Hammes, F., 2018. Biofilms in shower hoses. Water Res. 131, 274–286. https://doi.org/10.1016/j.watres.2017.12.027.

Randag, A.C., van Rooij, J., van Goor, A.T., Verkerk, S., Wisse, R.P.L., Saelens, I.E.Y., Stoutenbeek, R., van Dooren, B.T.H., Cheng, Y.Y.Y., Eggink, C.A., 2019. The rising incidence of *Acanthamoeba keratitis*: a 7-year nationwide survey and clinical assessment of risk factors and functional outcomes. PLoS One 14, e0222092. https://doi.org/10.1371/journal.pone.0222092.

Reynolds, D., Kollef, M., 2021. The epidemiology and pathogenesis and treatment of *Pseudomonas aeruginosa* Infections: an update. Drugs 81, 2117–2131. https://doi.org/10.1007/s40265-021-01635-6.

Rogers, J., Dowsett, A.B., Dennis, P.J., Lee, J.V., Keevil, C.W., 1994. Influence of plumbing materials on biofilm formation and growth of *Legionella pneumophila* in potable water systems. Appl. Environ. Microbiol. 60, 1842–1851. https://doi.org/10.1128/aem.60.6.1842-1851.1994.

RQIA, 2012. Regulation and Quality Improvement Authority—RQIA Inspection Reports | Regulation and Quality Improvement Authority Standards Reports. [WWW Document]. URL: https://rqia.org.uk/reviews/review-reports/2012-2015/rqia-pseudomonas-review/. (Accessed 11 April 2019).

Ryan, R.P., Monchy, S., Cardinale, M., Taghavi, S., Crossman, L., Avison, M.B., Berg, G., van der Lelie, D., Dow, J.M., 2009. The versatility and adaptation of bacteria from the genus *Stenotrophomonas*. Nat. Rev. Microbiol. 7, 514–525. https://doi.org/10.1038/nrmicro2163.

Ryu, Y.J., Koh, W.-J., Daley, C.L., 2016. Diagnosis and treatment of nontuberculous Mycobacterial lung disease: clinicians' perspectives. Tuberc. Respir. Dis. (Seoul) 79, 74–84. https://doi.org/10.4046/trd.2016.79.2.74.

Said, M.S., Tirthani, E., Lesho, E., 2022. Stenotrophomonas maltophilia. StatPearls.

Salar, A., Carratalà, J., Zurita, A., González-Barca, E., Grañena, A., 1998. Bacteremia caused by CDC group IV c-2 in a patient with acute leukemia. Haematologica 83, 670–672.

Sax, H., Bloemberg, G., Hasse, B., Sommerstein, R., Kohler, P., Achermann, Y., Rössle, M., Falk, V., Kuster, S.P., Böttger, E.C., Weber, R., 2015. Prolonged outbreak of *Mycobacterium chimaera* infection after open-chest heart surgery. Clin. Infect. Dis. 61, 67–75. https://doi.org/10.1093/cid/civ198.

Scaturro, M., Buffoni, M., Girolamo, A., Cristino, S., Girolamini, L., Mazzotta, M., Bucci Sabattini, M.A., Zaccaro, C.M., Chetti, L., Laboratory, M.A.N., Bella, A., Rota, M.C., Ricci, M.L., 2020. Performance of Legiolert test vs. ISO 11731 to confirm *Legionella pneumophila* contamination in potable water samples. Pathogens 9, 690. https://doi.org/10.3390/pathogens9090690.

Scheid, P., 2018. Free-living amoebae as human parasites and hosts for pathogenic microorganisms. Proceedings 2, 692. https://doi.org/10.3390/proceedings2110692.

Schmidt, T., Schlegel, H.G., 1989. Nickel and cobalt resistance of various bacteria isolated from soil and highly polluted domestic and industrial wastes. FEMS Microbiol. Lett. 62, 315–328. https://doi.org/10.1016/0378-1097(89)90014-1.

Seal, D.V., 2003. *Acanthamoeba keratitis* update—incidence, molecular epidemiology and new drugs for treatment. Eye 17, 893–905. https://doi.org/10.1038/sj.eye.6700563.

Shadoud, L., Almahmoud, I., Jarraud, S., Etienne, J., Larrat, S., Schwebel, C., Timsit, J.-F., Schneider, D., Maurin, M., 2015. Hidden selection of bacterial resistance to fluoroquinolones *in vivo*: the case of Legionella pneumophila and humans. EBioMedicine 2, 1179–1185. https://doi.org/10.1016/j.ebiom.2015.07.018.

Shaheen, M., Ashbolt, N.J., 2018. Free-living amoebae supporting intracellular growth may produce vesicle-bound respirable doses of *Legionella* within drinking water systems. Expo. Health 10, 201–209. https://doi.org/10.1007/s12403-017-0255-9.

Sivagnanam, S., Podczervinski, S., Butler-Wu, S.M., Hawkins, V., Stednick, Z., Helbert, L.A., Glover, W.A., Whimbey, E., Duchin, J., Cheng, G.S., Pergam, S.A., 2017. Legionnaires' disease in transplant recipients: a 15-year retrospective study in a tertiary referral center. Transpl. Infect. Dis. 19.

Skiada, A., Pavleas, I., Drogari-Apiranthitou, M., 2020. Epidemiology and diagnosis of mucormycosis: an update. J. Fungi (Basel) 6, 265. https://doi.org/10.3390/jof6040265.

Souli, M., Galani, I., Giamarellou, H., 2008. Emergence of extensively drug-resistant and pandrug-resistant Gram-negative bacilli in Europe. Eurosurveillance 13 (47), 19045. https://doi.org/10.2807/ese.13.47.19045-en.

Spencer, H., Banerjee, R., Wilson, G., Boswell, T., 2021. Cluster of invasive *Pseudomonas aeruginosa* infections in a neonatal intensive care unit. Antimicrob. Steward. Healthc. Epidemiol. 1, s74–s75. https://doi.org/10.1017/ash.2021.147.

Stout, J.E., Yu, V.L., 1997. Legionellosis. N. Engl. J. Med. 337, 682–687. https://doi.org/10.1056/NEJM199709043371006.

Stovall, S.H., Wisdom, C., McKamie, W., Ware, W., Dedman, H., Fiser, R.T., 2010. Nosocomial transmission of *Cupriavidus pauculus* during extracorporeal membrane oxygenation. ASAIO J. 56, 486–487. https://doi.org/10.1097/MAT.0b013e3181f0c80d.

Suenaga, H., Yamazoe, A., Hosoyama, A., Kimura, N., Hirose, J., Watanabe, T., Fujihara, H., Futagami, T., Goto, M., Furukawa, K., 2015. Draft genome sequence of the polychlorinated biphenyl-degrading bacterium *Cupriavidus basilensis* KF708 (NBRC 110671) isolated from biphenyl-contaminated soil. Genome Announc. 3. https://doi.org/10.1128/genomeA.00143-15. e00143-15.

Surman-Lee, S., Fields, B., Hornei, B., Ewig, S., Exner, M., Tartakovsky, I., Lajoie, L., Dangendorf, F., Bentham, R., Cabanes, P.A., Fourrier, P., Trouvet, T., Wallet, F., 2007. Legionella and the prevention of Legionellosis. In: Bartram, J., Chartier, Y., Lee, J.V., Pond, K., Surman-Lee, S. (Eds.), Ecology and Environmental Sources of *Legionella*. WHO.

Taylor, S.M., 2016. Impact of physical therapy for a patient with multiple limb amputations secondary to legionnaires' disease: a case report. J. Prosthet. Orthot. 28, 118–123. https://doi.org/10.1097/JPO.0000000000000096.

UKHSA, 2021a. *Pseudomonas aeruginosa* Bacteraemia: Annual Data. GOV.UK.

UKHSA, 2021b. MRSA, MSSA and Gram-negative Bacteraemia and CDI: Annual Report. GOV.UK.

Uzodi, A.S., Schears, G.J., Neal, J.R., Henry, N.K., 2014. *Cupriavidus pauculus* bacteremia in a child on extracorporeal membrane oxygenation. ASAIO J. 60, 740–741. https://doi.org/10.1097/MAT.0000000000000120.

Valdés-Corona, L.F., Maulen-Radovan, I., Videgaray-Ortega, F., 2021. *Cupriavidus pauculus* bacteremia related to parenteral nutrition. IDCases 24, e01072. https://doi.org/10.1016/j.idcr.2021.e01072.

van der Kooij, D., Veenendaal, H.R., Italiaander, R., 2020. Corroding copper and steel exposed to intermittently flowing tap water promote biofilm formation and growth of Legionella pneumophila. Water Res. 183, 115951. https://doi.org/10.1016/j.watres.2020.115951.

Vandamme, P., Coenye, T., 2004. Taxonomy of the genus *Cupriavidus*: a tale of lost and found. Int. J. Syst. Evol. Microbiol. 54, 2285–2289. https://doi.org/10.1099/ijs.0.63247-0.

Vander Elzen, K., Zhen, H., Shuman, E., Valyko, A., 2019. The hidden truth in the faucets: a quality improvement project and splash study of hospital sinks. In: American Journal of Infection Control, 46th Annual Conference Abstracts, APIC 2019, Philadelphia, PA 47, S26., https://doi.org/10.1016/j.ajic.2019.04.048.

Vaneechoutte, M., Kämpfer, P., De Baere, T., Falsen, E., Verschraegen, G., 2004. *Wautersia* gen. nov., a novel genus accommodating the phylogenetic lineage including *Ralstonia eutropha* and related species, and proposal of *Ralstonia [Pseudomonas] syzygii* (Roberts et al. 1990) comb. nov. Int. J. Syst. Evol. Microbiol. 54, 317–327. https://doi.org/10.1099/ijs.0.02754-0.

Vay, C., García, S., Alperovich, G., Almuzara, M., Lasala, M.B., Famiglietti, A., 2007. Bacteremia due to *Cupriavidus pauculus* (formerly CDC Group IVc-2) in a hemodialysis patient. Clin. Microbiol. Newsl. 29, 30–32. https://doi.org/10.1016/j.clinmicnews.2007.02.002.

Vincent, J.-L., Sakr, Y., Singer, M., Martin-Loeches, I., Machado, F.R., Marshall, J.C., Finfer, S., Pelosi, P., Brazzi, L., Aditianingsih, D., Timsit, J.-F., Du, B., Wittebole, X., Máca, J., Kannan, S., Gorordo-Delsol, L.A., De Waele, J.J., Mehta, Y., Bonten, M.J.M., Khanna, A.K., Kollef, M., Human, M., Angus, D.C., EPIC III Investigators, 2020. Prevalence and outcomes of infection among patients in intensive care units in 2017. JAMA 323, 1478–1487. https://doi.org/10.1001/jama.2020.2717.

Walker, J.T., Jhutty, A., Parks, S., Willis, C., Copley, V., Turton, J.F., Hoffman, P.N., Bennett, A.M., 2014. Investigation of healthcare-acquired infections associated with *Pseudomonas aeruginosa* biofilms in taps in neonatal units in Northern Ireland. J. Hosp. Infect. 86, 16–23. https://doi.org/10.1016/j.jhin.2013.10.003.

Waman, V.P., Vedithi, S.C., Thomas, S.E., Bannerman, B.P., Munir, A., Skwark, M.J., Malhotra, S., Blundell, T.L., 2019. Mycobacterial genomics and structural bioinformatics: opportunities and challenges in drug discovery. Emerg. Microbes Infect. 8, 109–118. https://doi.org/10.1080/22221751.2018.1561158.

Warris, A., Voss, A., Verweij, P.E., 2001. Hospital sources of *Aspergillus*: new routes of transmission? Rev. Iberoam. Micol. 18, 156–162.

WHO, 2017. Global Spending on Health: A World in Transition.

Wilks, S.A., Koerfer, V.V., Prieto, J.A., Fader, M., Keevil, C.W., 2021. Biofilm development on urinary catheters promotes the appearance of viable but nonculturable bacteria. mBio 12. https://doi.org/10.1128/mBio.03584-20. e03584-20.

Williams, B.J., Dehnbostel, J., Blackwell, T.S., 2010. *Pseudomonas aeruginosa*: host defence in lung diseases. Respirology 15, 1037–1056. https://doi.org/10.1111/j.1440-1843.2010.01819.x.

Winthrop, K.L., Marras, T.K., Adjemian, J., Zhang, H., Wang, P., Zhang, Q., 2020. Incidence and prevalence of nontuberculous mycobacterial lung disease in a large U.S. managed care health plan, 2008-2015. Ann. Am. Thorac. Soc. 17, 178–185. https://doi.org/10.1513/AnnalsATS.201804-236OC.

Wollenweber, H.W., 1935. Die fusarien: ihre beschreibung, schadwirkung und bekämpfung.

Yabuuchi, E., Kosako, Y., Yano, I., Hotta, H., Nishiuchi, Y., 1995. Transfer of two *Burkholderia* and an *Alcaligenes* species to *Ralstonia* gen. Nov.: proposal of *Ralstonia pickettii* (Ralston, Palleroni and Doudoroff 1973) comb. Nov., *Ralstonia solanacearum* (Smith 1896) comb. Nov. and *Ralstonia eutropha* (Davis 1969) comb. Nov. Microbiol. Immunol. 39, 897–904. https://doi.org/10.1111/j.1348-0421.1995.tb03275.x.

Yahya, R., Mushannen, A., 2019. *Cupriavidus pauculus* as an Emerging Pathogen: A Mini-review of Reported Incidents Associated With its Infection. [WWW Document]. URL: https://www.semanticscholar.org/paper/Cupriavidus-pauculus-as-an-Emerging-Pathogen-%3A-A-of-Yahya-Mushannen/a39ed20ce20bf674b76b9a9fbafcd332afaa8bab. (Accessed 23 May 2022).

Clinical surveillance of waterborne infections

27

Introduction

Modern healthcare facility water systems are becoming more and more complex, and the number of immunologically vulnerable patients is increasing to an extent that prevention of healthcare-associated waterborne infections needs to be a priority. Design, building, and commissioning of hospital buildings tend to result in large complex water systems that are colonized and contaminated by waterborne pathogens prior to the entry of vulnerable patients. Controlling the presence of these waterborne pathogens is extremely difficult, and as a consequence, healthcare facilities should conduct clinical surveillance of waterborne infections and environmental monitoring to understand and remediate sources of transmission promptly.

If it was possible to reliably detect all or even most waterborne transmission events, then deficiencies in the built environment or practices would be corrected. Unfortunately, this being far from the case allows poor design and practices to flourish. An effective backstop is lacking, but why should this be the case?

In 1967, in response to a leading article in the British Medical Journal, Joachim Kohn wrote to the journal stating *"your leading article is perhaps too cautious in saying that the evidence of infection from sinks and drains is not clearly established"* (Kohn, 1967). He went on to provide scientific data proving acquisition by patients of sensitive strains of *Pseudomonas aeruginosa* from water outlets and drains. Unfortunately, his views were ridiculed, and an unwritten microbiology folklore that organisms went from the patient to the sink and not vice versa remained in place for the next 45 years. Hence, it has taken many years and fatalities of patients, such as the death of four neonates from waterborne pseudomonas in 2012 to partially bring the medical profession to its senses (Anaissie et al., 2002; RQIA, 2012; Walker et al., 2014). While the microbiology folklore was no longer in vogue with water being accepted as a route of transmission, this did not produce the necessary reduction in the burden of ill-health. Much of this as we will see relates to the lack of sensitivity of clinical surveillance. For the mythology to have remained in place for 45 years (quite unbelievable) required an ally, a flawed surveillance system.

Flying is the safest form of travel. One of the reasons for this is that if something goes wrong, it is immediately apparent, often to the whole world. This results in an investigation and implementation of findings. Thus, the airline safety surveillance system is robust, extending even further to include near miss events.

While several healthcare facilities describe themselves as "centers of excellence," the numbers of preventable infections does not always reflect this statement. (van Buijtene and Foster, 2019) In Europe, over 4 million patients acquire a

Safe Water in Healthcare. https://doi.org/10.1016/B978-0-323-90492-6.00013-6

healthcare-associated infection (HCAI) every year with approximately 37,000 deaths. (PHE, 2016) In the United States, HCAI is the fifth leading cause of death in acute care hospitals. (Septimus et al., 2014) Detailed analysis has established that water-related investigations represented 22% of CDC HCAI, and reports have indicate that *P. aeruginosa* accounted for an estimated 1400 deaths per year. (Anaissie et al., 2002; Perkins et al., 2019) Extrapolating the US data to Europe would equate to 8140 deaths every year as a direct consequence of waterborne-associated HCAI. In what other industry would this number of deaths be acceptable?

UKHSA has published guidance on the different approaches to surveillance indicating that it is a core function to ensure that we have the right information available to us at the right time to inform public health decisions and actions (PHE, 2017).

Surveillance may be divided into electronic, laboratory, and clinical.

Clinical surveillance

The ultimate aim of clinical surveillance is to reduce healthcare associated infections and should consist of data collection, validation, analysis, interpretation, and feedback to the clinical teams.

Caution should be applied to only reacting to electronic data and the action list of microorganisms. Microbiology staff and clinicians need to be mindful of unusual microorganism that may occasionally be reported.

The conundrum with clinical surveillance and certain organisms is being able to distinguish the source being endogenous or exogenous. To put this into simpler terms, some patients will carry the organism as part of their normal bacterial flora into high-risk units—this is a term known as colonization. Many of the necessary interventions required for medical care put the patient at risk of infection, and this can either be with their own organisms (endogenous infection) or from organisms acquired on the unit (exogenous infection).

With *P. aeruginosa* colonization/infection in particular, it is necessary to distinguish between exogenous and endogenous acquisition as some patients are admitted to units already colonized with this organism.

With neonates, the situation is simpler. During the process of birth, the baby acquires its bacterial flora from the birth canal of the mother, and *P. aeruginosa* is rarely carried by pregnant mothers. Therefore, the detection of a single case of *P. aeruginosa* infection/colonization in neonates is enough to prompt an incident meeting (Buttery et al., 1998). One of the authors worked in a hospital that housed a regional neonatal intensive care unit. During the course of a year, only one baby was detected with *P. aeruginosa,* and this could be traced to a transfer from another unit where the baby was positive on admission.

With adults, the situation is complicated as colonization does occur, so patients may bring their own strain onto a unit. The problem then arises as to what is an acceptable level of colonization, beyond which acquisition on a unit should be suspected. No one has set a baseline for endogenous carriage. The ensuing problem that this causes is highlighted in the following two publications;

1. Tissot et al. described how the rates of *P. aeruginosa* rose from 32.2 to 44.7/1000 admissions in a burns unit through unrecognized environmental contamination of hydrotherapy equipment (Tissot et al., 2016). Despite the occurrence of unusual early-onset *P. aeruginosa* infections and conventional epidemiological typing not demonstrating an outbreak, the conundrum was whether the increase reflected natural variation in endogenous carriage or whether this was acquisition on the unit (exogenous). Eventually, a new typing system confirmed an outbreak. When hydrotherapy practices were reviewed, it was readily evident that there were major deficiencies leading to transmission of *P. aeruginosa*.
2. Halstead et al. used whole genome sequencing to compare water isolates of *P. aeruginosa* and patient isolates over a 16-week period in the augmented care units of four hospitals (Halstead et al., 2021). None of these hospitals suspected ongoing transmission of *P. aeruginosa* from water systems to patients. Results showed that in three out of four hospitals definitive ongoing transmission of *P. aeruginosa* was occurring from water outlets to patients.

The key to surveillance is being able to make a microbiological diagnosis, which requires the following steps:

1. **Clinical teams need to send the appropriate specimens**. For example, if a blood culture or respiratory sample is not sent, then the appropriate microbiological diagnosis may not be made.
2. **The laboratory needs to perform the correct test**. *Legionella pneumophila* is the greatest risk concern for patient safety. In France, 98% of culture-confirmed Legionnaires' disease cases are attributable to *L. pneumophila* (Campese et al., 2011). Similar statistics are also seen in travel-associated Legionnaires' disease cases, (ECDC, 2019) in Japan, (Amemura-Maekawa et al., 2018) and in US CDC outbreaks (CDC, 2021). An analysis of 10 years of culture diagnoses from a study of more than 40,000 patients across the EU (including the United Kingdom) found that even for healthcare-acquired Legionnaires' disease, 98% of the cases were caused by *L. pneumophila* (Beauté et al., 2020). In recent years, urinary antigen has become the mainstay for diagnosis of Legionnaires' disease and *Legionella pneumophila* serogroup 1. In England and Wales, nonpneumophila strains now account for less than 1% of clinical cases, and this low rate has been thought to be due to the predominant use of the urinary antigen test (PHE, 2018). However, clinical data from Denmark where that 80%–90% of Legionnaires' disease cases are diagnosed by polymerase chain reaction (PCR) have indicated that 93% of PCR-diagnosed cases were a result of infections caused by *L. pneumophila*, (including community acquired, hospital/nursing home-acquired, and travel within Denmark cases). Thirteen with only one of the 112 culture-verified cases confirmed as a nonpneumophila species. Nonetheless, these data do not lessen the importance of clinical surveillance in healthcare to identify nonpneumophila strains and that resources have to be implemented to identify the source to prevent further outbreaks. This has important implications within healthcare facilities. If a contamination of the water system is found with nonserogroup 1 Legionella, then alternative clinical diagnostic tests are required for their detection.
3. **The laboratory must be able to correctly identify the organism**—The advent of matrix-assisted laser desorption ionization-time-of-flight mass spectrometry (MALDI-TOF MS) has revolutionized the identification of organisms, especially nonfermenting organisms that may be found in water systems. The recent association and detection of *Cupriavidus pauculus* in water systems may well be related to this new technology (Inkster et al., 2022). Prior to the newer technology, organisms, which were biochemically inert, were often reported as

"nonfermenting environmental organism of dubious clinical significance." Inkster et al also reported inconsistent identification of *C. pauculus* by other identification platforms (Inkster et al., 2022). However, there are certain challenges in using MALDI-TOF MS for the identification of less frequently isolated environmental waterborne microorganisms (Popović et al., 2017).

4. **Assigning alert organisms**—an increasing number of organisms have been identified, which can be transmitted by water systems. The laboratory needs to ensure that organisms are reported and infection control needs to ensure that their infection control software will list these as alert organisms. These alert organisms would then automatically come to the attention of the infection control team—they do not have the resources or the need to review every positive isolate coming out the laboratory. Consideration, therefore, needs to be given as to when water should be tested. Excellent advice on the surveillance of waterborne microorganisms has been compiled by the National Services Scotland, who have compiled a check list (National Services Scotland, 2019).
5. **Recognizing there is an issue**—a single case of hospital-acquired Legionella is sufficient to hold an immediate incident meeting. The assumption that the first case is actually the index case should always be questioned. For example, in one hospital, which was specializing in immunocompromised patients, a nonserogroup 1 Legionella was detected in patient respiratory samples. This turned out to be the fourth case—the average time between cases was 4 months, and some patients who had attended this tertiary referral hospital had been diagnosed at their source hospital. Some waterborne outbreaks can take even longer (e.g., 2–3 years) to identify particularly where unusual microorganisms are involved. Such scenarios identify the challenges of surveillance and importance of retrospective surveillance over long time periods.

   Biofilms are ubiquitous in water systems, and it is likely that other potential water pathogens will also be represented in the biofilm (Burmølle et al., 2014). Once one waterborne pathogen is detected, a general sweep of the infection control database should be made for other organisms. In one hospital where they had two virtually identical intensive care units, on one they had three cases of *Stenotrophomonas maltophilia* over a period of weeks. They were uncertain of the significance but decided to do a trawl for other waterborne organisms. On the unit with the cases of *Stenotrophomonas maltophilia,* they found several other waterborne pathogens, none of which were present on the other ITU. This provided the impetus to investigate the water services and as has been shown elsewhere a range of common and rare waterborne microorganism can be associated with infections (Inkster et al., 2022; Perkins et al., 2019).

Polyclonality, an issue affecting both clinical and environmental samples, may be a hindrance to establishing a link between clinical specimens and the environment. This results in an incident either not being recognized or the link to water and wastewater systems not being identified. The traditional method of investigating person-to-person spread assumes a single clone. Environmental outbreaks are different—there may be several different clones/strains present in water and wastewater systems. Polyclonal outbreaks are recognized, but the traditional system of investigating outbreaks virtually precludes detection of polyclonal outbreaks. This is mostly due to when positive cultures are either obtained from patient or the environment. Unless there is a phenotypic difference between colonies, the assumption is that the one colony picked and sent for typing is representative of all the colonies. We know this not to be the case, but how common polyclonality is in environmental outbreaks has not been established.

Traditionally hospital-acquired cases are analyzed at ward level, i.e., is their spread within a ward. Wastewater systems link every area of the hospital so outbreaks can be hospital-wide (Breathnach et al., 2012). As a consequence, surveillance for these alert organisms needs to be hospital-wide not just by ward locality. Nucleus areas in hospitals are areas where patients may visit or temporarily reside, but then they return to or are sent to other wards during their stay. Failure to control an outbreak in a nucleus area may seed the hospital (Weinbren and Inkster, 2021). The move to manage patients in the outpatient setting or community provides further challenges to surveillance. For example, an increasing number of patients are being managed with long lines as outpatients. The longline provides a portal of entry for viruses of organisms including water/wastewater. Tracking down the link between water acquired infection and the area, this occurrence can be difficult, especially as the incubation period for some line infections can be long (i.e., NTM infections). Where teams are proactive, they will identify such outpatient settings and ensure staff understand the risks and have the correct processes in place.

With clinical surveillance, the lack of detected infections does not necessarily mean patient care is satisfactory. Too often an outbreak with a resistant organism highlights poor clinical practice, which should have been detected beforehand. The take-home message should be that clinical surveillance, especially with sensitive organism outbreaks, is far from perfect. As antimicrobial resistance becomes more commonplace, it is likely that even highly antibiotic-resistant strains will fail to attract the attention to spur an incident meeting.

The lack of an effective backstop in surveillance suggests that a more proactive approach needs to be taken in the design and use of water services in order to prevent transmission in the first instance.

Environmental surveillance

Environmental surveillance from a micro-budget perspective may be divided into water sampling and swabbing of environmental surfaces.

Water sampling

Water samples may be collected at various stages during the life of the building:

- commissioning
- routine/ad hoc environmental sampling.

It is important to remember that water sampling is not a control mechanism. Negative water samples, assuming that the collection of water and testing have been done correctly, only provide information about the quality of water at that point in time. There are numerous examples where people have focused on negative water samples to the exclusion of the control measures and of course discovered to their shock that the water samples suddenly become positive.

Commissioning is a particularly dangerous phase in the lifespan of a building. Poor planning and execution of the commissioning phase will facilitate overgrowth of biofilm within the water system, which once established may cause ongoing problems and costs for the lifetime of the occupation of the building.

The topic is so important that water sampling during the commissioning phase is dealt with under the section dealing with commissioning.

Routine environmental water sampling is indicated in the following circumstances:

Potable water quality for drinking—test for total viable counts, *Escherichia coli* and coliforms, *Enterococci*, TVC, and *Clostridium perfringens.*

The recommendations for routine surveillance of Legionella are laid out in HSG 274 (HSE, 2014):

- water systems treated with biocides where water is stored or distribution temperatures are reduced. Initial testing should be carried out monthly to provide early warning of loss of control. The frequency of testing should be reviewed and continued until such a time as there is confidence in the effectiveness of the regime;
- water systems where the control levels of the treatment regime, e.g., temperature or disinfectant concentrations, are not being consistently achieved. In addition to a thorough review of the system and treatment regimes, frequent testing, e.g., weekly, should be carried out to provide early warning of loss of control. Once the system is brought back under control as demonstrated by monitoring, the frequency of testing should be reviewed;

HTM O4-01 recommends 6-monthly routine testing for *P. aeruginosa* in augmented care areas (DHSC, 2016). The definition of augmented care areas is difficult to define precisely and will always need some local interpretation.

In terms of high-risk areas or where there is a population with increased susceptibility, augmented care was a definition that many people struggled to define (DHSC, 2016). High-risk settings in healthcare associated with water infections include hematology and oncology units, bone marrow and stem cell transplant units, neonatal, paediatric and adult ICUs, and other care areas where patients are severely immunocompromised through disease or treatment.

Ensuring the periphery of the water systems is not a risk to high-risk patients is much more difficult as control measures tend to be less effective in this area. Hence, the importance of collecting preflush specimens in high-risk patient areas, as it is the microbial overgrowth in the first run off, of water, which is sufficient to make the patient unwell. Another reason for testing in high-risk areas, is that there may be local issues to the water system, which are not going to be detected by overall system controls.

The interval of 6-month testing for *P. aeruginosa* was not based upon scientific evidence and is a timeline that should be altered by hospital surveillance teams depending on the contamination identified. Garvey et al. have shown re-contamination of outlets within a 6-month period (Garvey et al., 2016b). The lack of sensitivity of current surveillance systems especially with regard to sensitive organisms in effect means that there is no effective backstop in detecting transmission. There is therefore an argument that more frequent testing should be conducted on units to ascertain the risk specific to local circumstances. If samples return negative, then time between tests can be expanded.

Point-of-use or microbial retention filters can be a contentious control measure, but experience has shown this can be useful. Primary filter failure in quality branded filters is rare. However, poor practices (contamination of base of filter through inappropriate cleaning practices) or incorrect filter placement can produce pseudo-filter failure (Garvey et al., 2016a). Taking water samples after it has passed through a filter may serve to identify such failures and might be considered useful. If testing is to be undertaken, then testing for TVC is probably the best test.

Routine/ad hoc environmental sampling

Routine water sampling is discussed above and set out in guidance for *P. aeruginosa* and *Legionella*. It should be noted that these is only a guideline and that hospitals can increase the frequencies where an higher risk has been identified (DHSC, 2016; HSE, 2014). In addition, clinical surveillance may indicate that other waterborne organisms may be of interest.

Where routine water testing is undertaken, a total viable count (TVC) may be sufficient for organisms other than *P. aeruginosa* and *Legionella* to give an indication of the quality of the water. It would be advisable to type to species level during incidents/outbreaks investigations and particularly if there are clinical cases, even over long periods of time.

When taking water samples, it would be prudent to take pre- and postflush samples to determine if the microbial contamination is located at an outlet or upstream of the outlet component (DHSC, 2016).

Sampling and swabbing of water system components

During incidents or outbreaks, components of the water system may need to be swabbed. While tap outlets and shower heads may be easily accessible, other components such as EPDM hoses, thermostatic mixer valves, or drains may need plumbing components to be dismantled to determine the source of the contamination (O'Donnell et al., 2011; Walker et al., 2014, 2017).

The water safety group should ensure that no components of water systems or devices using water are wet tested during the manufacturing process or should water testing be required that adequate steps are taken to mitigate risk. Where there are concerns that components could be contaminated, especially in a new build project, then swabbing of a percentage of components might be useful as a quality assurance check. There are a wide range of water-related devices in the clinical environment, and many are discussed in other sections and chapters. It would be advisable to include as many of these water-related devices in the checklist for managing waterborne incidents and outbreaks.

Routine environmental sampling of drains is not routinely recommended as invariably biofilm and antibiotic resistant strains will be present. Sampling is therefore normally restricted to when there are concerns over transmission events.

Some hospitals have looked at screening wastewater for Carbapenemase-producing organisms. While this often identifies their presence, it is not necessarily linked to patient transmission events. Therefore, in the absence of suspected transmission, this would appear to be of little value. However, some surveillance studies have identified extensive transmission (Decraene et al., 2018).

With environmental surveillance, such as water sampling, a negative result does not necessarily mean the water system is being run correctly. It is not a control mechanism, and if the water system is not under control, that negative sample can rapidly change to a high count of a pathogen. Equally Infection prevention and control should be about preventing infections. Unfortunately, the current situation is that many infection control practitioners, through no fault of their own, have not been trained over the risks from water/wastewater systems, and therefore, a preventative approach is frequently lacking.

Actions associated with surveillance of healthcare water-associated infections

Criteria	Yes	No
Where surveillance has identified clinical cases have they been isolated or charted?		
Has a clinical assessment been carried out for patients in the locality?		
Has a retrospective review been carried out to identify microbiology linked cases?		
Have a range of water sources been identified for surveillance through water and or swab sampling, e.g., taps, showers, baths, water based equipment and drains?		
Have the isolates from the environmental sampling been identified to look at commonalities?		
Where links (both local at the ward level and across the hospital) have been identified between clinical and environmental isolates, has a water safety plan been implemented by the water safety group to prevent transmissions?		
Has a review been undertaken to control transmission from the water sources from which the isolates have been recovered?		
Where individual outlets have been identified as being positive for isolates, have remedial measures been implemented, e.g., microbial retention filters, review of the water system temperatures and flow rates or preferrable removal of that outlet.		

Criteria	Yes	No
Has an inspection been carried out of all point-of-use outlets to identify any design flaws, faults, or signs of deterioration/contamination/colonization/debris—this should include an assessment of water flow rate from outlets and corresponding impact points on basins/drains/shower trays to assess excess splash/spray?		
Where systemic contamination has been identified, has a remedial plan been devised and implemented?		
Has an assessment been carried out of the drain areas of hand wash stations and sink units as well drain traps (p-traps)?		
Has the provision of water-free care been considered (waterless oral care, use of disposable sponges or wash cloths), use of detergent wipes for cleaning reusable patient equipment (note that detergent wipes are not suitable for disinfection of reusable patient equipment), and use of sterile water/boiled water for preparation of antiseptic solutions and drinking?		
Has a clinical staff training and education program been implemented to refresh and update staff on the issues and microbiological risks to patients from water including high-risk patients, usage of hand wash stations and sinks		
Have the estates and facilities staff been trained in hygienic plumbing and how to set up hand wash stations, sink operation handles (where fitted) and outlets to ensure they are not directed into drains or create excessive splashing?		

Guidance for surveillance

HSG 274 Legionnaires' disease—Technical guidance Part 2: The control of legionella bacteria in hot and cold water systems Technical Guidance.
HTM 04-01: Safe water in healthcare premises.
National Services Scotland, 2019. Prevention and management of healthcare water-associated infection incidents/outbreaks.

Acknowledgment

We are grateful to Teresa Inkster for her insightful overview and comments on this chapter.

References

Amemura-Maekawa, J., Kura, F., Chida, K., Ohya, H., Kanatani, J., Isobe, J., Tanaka, S., Nakajima, H., Hiratsuka, T., Yoshino, S., Sakata, M., Murai, M., Ohnishi, M., Working Group for Legionella in Japan, 2018. *Legionella pneumophila* and other *Legionella* species isolated from Legionellosis patients in Japan between 2008 and 2016. Appl. Environ. Microbiol. 84.

Anaissie, E.J., Penzak, S.R., Dignani, M.C., 2002. The hospital water supply as a source of nosocomial infections: a plea for action. Arch. Intern. Med. 162, 1483–1492.

Beauté, J., Plachouras, D., Sandin, S., Giesecke, J., Sparén, P., 2020. Healthcare-Associated Legionnaires' Disease, Europe, 2008−2017. Emerg. Infect. Dis. 26, 2309–2318.

Breathnach, A.S., Cubbon, M.D., Karunaharan, R.N., Pope, C.F., Planche, T.D., 2012. Multidrug-resistant *Pseudomonas aeruginosa* outbreaks in two hospitals: association with contaminated hospital waste-water systems. J. Hosp. Infect. 82, 19–24. https://doi.org/10.1016/j.jhin.2012.06.007.

Burmølle, M., Ren, D., Bjarnsholt, T., Sørensen, S.J., 2014. Interactions in multispecies biofilms: do they actually matter? Trends Microbiol. 22, 84–91. https://doi.org/10.1016/j.tim.2013.12.004.

Buttery, J.P., Alabaster, S.J., Heine, R.G., Scott, S.M., Crutchfield, R.A., Bigham, A., Tabrizi, S.N., Garland, S.M., 1998. Multiresistant *Pseudomonas aeruginosa* outbreak in a pediatric oncology ward related to bath toys. Pediatr. Infect. Dis. J. 17, 509–513. https://doi.org/10.1097/00006454-199806000-00015.

Campese, C., Bitar, D., Jarraud, S., Maine, C., Forey, F., Etienne, J., Desenclos, J.C., Saura, C., Che, D., 2011. Progress in the surveillance and control of Legionella infection in France, 1998–2008. Int. J. Infect. Dis. 15, e30–e37. https://doi.org/10.1016/j.ijid.2010.09.007.

CDC, 2021. National Outbreak Reporting System (NORS) Dashboard.

Decraene, V., Phan, H.T.T., George, R., Wyllie, D.H., Akinremi, O., Aiken, Z., Cleary, P., Dodgson, A., Pankhurst, L., Crook, D.W., Lenney, C., Walker, A.S., Woodford, N., Sebra, R., Fath-Ordoubadi, F., Mathers, A.J., Seale, A.C., Guiver, M., McEwan, A., Watts, V., Welfare, W., Stoesser, N., Cawthorne, J., Group, the T.I, 2018. A large, refractory nosocomial outbreak of *Klebsiella pneumoniae* carbapenemase-producing *Escherichia coli* demonstrates carbapenemase gene outbreaks involving sink sites require novel approaches to infection control. Antimicrob. Agents Chemother. 62. https://doi.org/10.1128/AAC.01689-18.

DHSC, 2016. HTM 04-01: Safe water in healthcare premises.

ECDC, 2019. Legionnaires' Disease—Annual Epidemiological Report for 2019. European Centre for Disease Prevention and Control.

Garvey, Bradley, C.W., Jumaa, P., 2016a. The risks of contamination from tap end filters. J. Hosp. Infect. 94, 282–283. https://doi.org/10.1016/j.jhin.2016.08.006.

Garvey, M.I., Bradley, C.W., Tracey, J., Oppenheim, B., 2016b. Continued transmission of *Pseudomonas aeruginosa* from a wash hand basin tap in a critical care unit. J. Hosp. Infect. 94, 8–12. https://doi.org/10.1016/j.jhin.2016.05.004.

Halstead, F.D., Quick, J., Niebel, M.O., Garvey, M., Cumley, N., Smith, R., Neal, T., Roberts, P., Hardy, K., Shabir, S., Walker, J.T., Hawkey, P., Loman, N.J., 2021. *Pseudomonas aeruginosa* infection in augmented care: the molecular ecology and transmission dynamics in four large UK hospitals. J. Hosp. Infect. https://doi.org/10.1016/j.jhin.2021.01.020.

HSE, 2014. HSG 274 Legionnaires' Disease—Technical Guidance Part 2: The Control of Legionella Bacteria in Hot and Cold Water Systems Technical Guidance. http://www.hse.gov.uk/pubns/books/hsg274.htm. (Accessed 5 January 2021).

Inkster, T., Wilson, G., Black, J., Mallon, J., Connor, M., Weinbren, M., 2022. *Cupriavidus* spp. and other waterborne organisms in healthcare water systems across the UK. J. Hosp. Infect. 123, 80–86. https://doi.org/10.1016/j.jhin.2022.02.003.

Kohn, J., 1967. *Pseudomonas* infection in hospital. Br. Med. J. 4, 548. https://doi.org/10.1136/bmj.4.5578.548.

National Services Scotland, 2019. Prevention and Management of Healthcare Water-associated Infection Incidents/Outbreaks. https://www.nipcm.hps.scot.nhs.uk/media/1680/2019-08-water-incidents-info-sheet-v1.pdf.

O'Donnell, M.J., Boyle, M.A., Russell, R.J., Coleman, D.C., 2011. Management of dental unit waterline biofilms in the 21st century. Future Microbiol. 6, 1209–1226. https://doi.org/10.2217/fmb.11.104.

Perkins, K.M., Reddy, S.C., Fagan, R., Arduino, M.J., Perz, J.F., 2019. Investigation of healthcare infection risks from water-related organisms: summary of CDC consultations, 2014—2017. Infect. Control Hosp. Epidemiol. 40, 621–626. https://doi.org/10.1017/ice.2019.60.

PHE, 2016. A Point Prevalence Survey on Levels of Healthcare-Associated Infections (HAI) and Levels of Antimicrobial Use in Hospitals in England in 2016. GOV.UK.

PHE, 2017. Public Health England: Approach to Surveillance. GOV.UK.

PHE, 2018. Legionnaires' Disease in Residents of England and Wales—2016.

Popović, N.T., Kazazić, S.P., Strunjak-Perović, I., Čož-Rakovac, R., 2017. Differentiation of environmental aquatic bacterial isolates by MALDI-TOF MS. Environ. Res. 152, 7–16. https://doi.org/10.1016/j.envres.2016.09.020.

RQIA, 2012. Independent Review of Incidents of Pseudomonas aeruginosa Infection in Neonatal Units in Northern Ireland. https://www.rqia.org.uk/RQIA/files/ee/ee76f222-a576-459f-900c-411ab857fc3f.pdf.

Septimus, E., Weinstein, R.A., Perl, T.M., Goldmann, D.A., Yokoe, D.S., 2014. Approaches for preventing healthcare-associated infections: go long or go wide? Infect. Control Hosp. Epidemiol. 35, 797–801.

Tissot, F., Blanc, D.S., Basset, P., Zanetti, G., Berger, M.M., Que, Y.-A., Eggimann, P., Senn, L., 2016. New genotyping method discovers sustained nosocomial *Pseudomonas aeruginosa* outbreak in an intensive care burn unit. J. Hosp. Infect. 94, 2–7. https://doi.org/10.1016/j.jhin.2016.05.011.

van Buijtene, A., Foster, D., 2019. Does a hospital culture influence adherence to infection prevention and control and rates of healthcare associated infection? A literature review. J. Infect. Prev. 20, 5–17. https://doi.org/10.1177/1757177418805833.

Walker, J.T., Jhutty, A., Parks, S., Willis, C., Copley, V., Turton, J.F., Hoffman, P.N., Bennett, A.M., 2014. Investigation of healthcare-acquired infections associated with *Pseudomonas aeruginosa* biofilms in taps in neonatal units in Northern Ireland. J. Hosp. Infect. 86, 16–23. https://doi.org/10.1016/j.jhin.2013.10.003.

Walker, J., Moore, G., Collins, S., Parks, S., Garvey, M.I., Lamagni, T., Smith, G., Dawkin, L., Goldenberg, S., Chand, M., 2017. Microbiological problems and biofilms associated with *Mycobacterium chimaera* in heater–cooler units used for cardiopulmonary bypass. J. Hosp. Infect. 96, 209–220. https://doi.org/10.1016/j.jhin.2017.04.014.

Weinbren, M., Inkster, T., 2021. Role of the kitchen environment. Infect. Prev. Prac. 3, 1–3.

Role of the water safety group and water safety plans

28

Introduction

Water in healthcare premises is a significant source of healthcare-acquired infections especially in the very young, elderly, those with underlying conditions, and patients immunocompromised as a result of illness and/or treatment (Falkinham, 2020; Gómez-Gómez et al., 2020; Kaul et al., 2022). While historically infections from legionellae have been foremost in guidance, there is an increasingly wide range of other opportunistic pathogens including *Pseudomonas aeruginosa*, *Stenotrophomonas* spp., *Klebsiella pneumoniae,* nontuberculous *Mycobacteria* spp. *Cupriavidus paucu-lus,* and *Aspergillus* spp. These opportunistic pathogens have shown to be the cause of healthcare-acquired infections associated with exposure to water or aerosols derived from water. With increasing technological advances in laboratory diagnostics for example, Matrix-Assisted Laser Desorption Ionization-Time-of-Flight Mass Spectrometry (MALDI-ToF), the identification of microorganisms from traditionally difficult sources, such as water and other environmental reservoirs, has become much quicker and easier. As a result, it is likely that this list of waterborne opportunistic pathogens is going to increase at a frightening rate, and we need be able to ensure there are effective processes to ensure we keep up to speed with the relevant ecological niches, modes of exposure, and those most likely to be infected for each newly recognized pathogen.

Sources of opportunistic pathogens in healthcare buildings

Contrary to common belief, infections caused by waterborne pathogens (i.e., those associated with growth in biofilms in water systems and/or damp environments) are not just a result of poor estate management or poor design and build of water systems. Even when all measures have been taken to avoid contamination during the construction, installation, and commissioning of a new water systems, it only takes a number of different scenarios to result in contamination. For example, water systems can be contaminated through the following processes:

- Contamination of new plumbing components installed using unhygienic plumbing practices.
- Installation of new plumbing component or fittings, which has been contaminated during the manufacturing process.
- Equipment brought to site and connected into the plumbing network without due diligence checks to ensure it has not been contaminated when used previously.

Safe Water in Healthcare. https://doi.org/10.1016/B978-0-323-90492-6.00009-4

- Bathroom and showers that are no longer needed and now used as a storeroom resulting in contaminated stagnant water.
- Drinks discarded in clinical wash hand basins because it is nearer than the kitchen can provide nutrients, which encourage opportunistic pathogens to grow upwards toward the drain.
- Discarding patient waste, which may potentially contain antibiotic-resistant microorganisms, into clinical hand wash basins or kitchen sinks that will contaminate the drains and outlets.
- Filling water jugs in transplant ward kitchen sink with a spray insert, which has never been cleaned because it's the catering staff who are responsible for cleaning the kitchens, and they have not been educated on the risk of inserts in augmented care outlets.
- Placing water jugs on the base of kitchen sinks in hematology oncology leading to contamination of the base of the water jugs with *P. aeruginosa* from the drain as the staff had not been trained and educated in the risks (Fig. 1).
- Reuse of single use equipment that has been exposed to water and contaminated, for example, pan scrubbers used to clean the sink or rubber ducks used in pediatrics, all of which would grow harmful microorganisms and potentially harm patients.

All of these instances can result in the growth of waterborne pathogens in the stagnant pipework, fittings, outlets, and drains (Fig. 2) resulting in retrograde contamination, which can spread from that area through the distribution network carried by the flow to some distant basin and shower outlet or directly out to patients via utensils and personal equipment.

It is the high-risk patient, highly susceptible, transplant, or neutropenic patients who are most at risk when they, for example:

- turn on the tap and are exposed to a potentially high numbers of pathogenic bacteria.
- use the hand wash basin where antibiotic-resistant bacteria have colonized the drain, basin, and outlet.
- handle water jugs that have been filled in kitchen sinks where antibiotic microorganisms have contaminated and formed biofilms in the drain and on the sink surfaces.

Fig. 1 Drinking water jug in a hematology oncology kitchen sink placed in close proximity to the drain while being filled and then provided to highly susceptible patients.

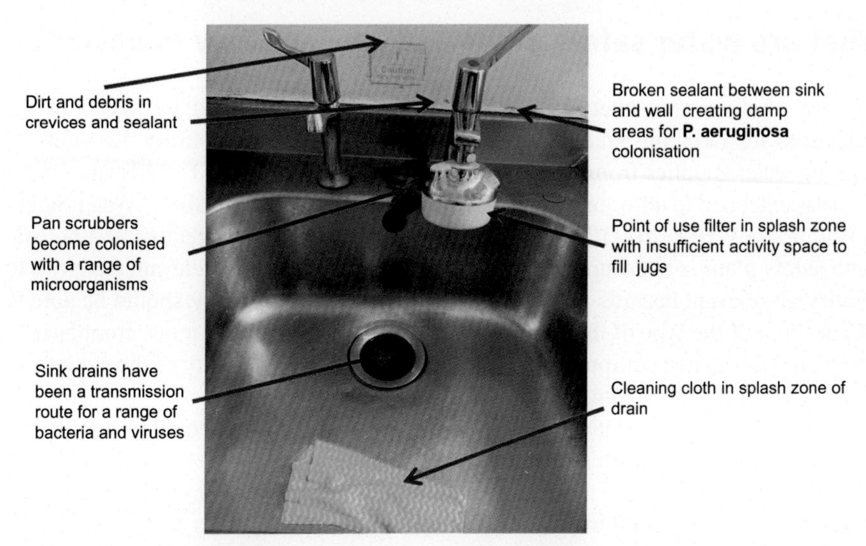

Fig. 2 Kitchen sink unit with several potential sources of waterborne pathogen exposure to patients.

It is important that all those involved in caring for highly susceptible patients directly or indirectly understand that hospital-acquired infections can be caused by exposure to opportunistic pathogens from various routes including:

- water itself,
- splashes and aerosols derived from water outlets (Allegra et al., 2016; Benoit et al., 2021),
- cross-contamination from outlets, drains, and damp areas of the fabric of the building (Hajar et al., 2019; Inkster and Weinbren, 2021),
- objects left in the splash zone of the sink and drains including personal equipment such as shaving brushes (Kim et al., 2020),
- contaminated cleaning cloths (Bergen et al., 2009).

The transmission of infections via such routes can result in increased morbidity and mortality, be difficult to diagnose, and may have limited or intolerable treatment options. The costs and consequences to affected patients, their families, and colleagues are difficult to estimate (Collier et al., 2021). Infections from waterborne pathogens can have significant consequences for healthcare providers with increased costs due to longer patient hospital stays; extended periods in augmented care settings such as intensive care and high-dependency units; increased requirements for interventions such as further surgery, wound debridement, and significantly increased costs of drugs. In addition, there is also a clear link to the presence of such opportunistic pathogens in healthcare water distribution systems, components, fittings, and associated sanitation systems to increasing the risk of transfer of antibiotic resistance between different bacterial populations and onward to patients and the community (Jamal et al., 2019; Tang et al., 2020). It is clear that the management of these risks is not just a matter of following guidance on how to engineer and manage water systems but needs a much more holistic approach including water safety plans.

What are water safety plans and why are they important?

The use of the water safety plan approach was first advocated by WHO in the Guidelines for Drinking Water Quality in 2004 as a means of ensuring the safety of drinking water supplies from water catchment to consumers (WHO and Team, 2004). This was extended to all water within buildings in the WHO publication "Water Safety In Buildings (WHO, 2011)." A key component for developing effective and holistic water safety plans is to ensure you have access to people with the relevant expertise to identify all relevant hazards associated with the source water. They should be able to take account of the type of use, materials, and properties of the systems, components, fitting, and associated equipment, which could impact on the quality of the stored and distributed water at the point of use (DHSC, 2016; HSE, 2014).

Holistic and bespoke water safety plans developed and implemented by a multidisciplinary team with the necessary skills and experience can have a significant impact on minimizing the risks of infection from waterborne pathogens. They would also avoid adverse other health impacts including those posed by water treatment chemicals, which may be toxic to dialysis patients or neonates, for example, or may result in a reduced life cycle of the plumbing infrastructure due to their adverse impact on the materials of the system.

Physical risks associated with water use should also be considered and mitigated including scalding as a result of hot water temperatures, inappropriate placing of water outlets and drains, which then pose indirect risks to patient and staff safety from cross-contamination due to splashing or misuse of sinks for wastewater disposal. A comprehensive water safety plan will also ensure there are policies and processes in place to ensure that those designing and engineering water systems take into account the absolute requirement to put patient safety at the forefront of their design (BSRIA, 2018; DHSC, 2016; HSE, 2014; RIBA, 2020).

What are the benefits of the water safety plan approach?

Ultimately, the water safety plan is there to save lives both in existing premises and those about to be built. The ongoing management of unsafe water systems poses a real burden on organizational resources. It's not just old, antiquated buildings that pose risks of waterborne infections there are many examples of unnecessary illness and deaths associated with poor design, installation, and commissioning of water systems in new hospitals.

Why do you need a water safety group?

Following outbreaks in Stafford and Queens Medical Centre in Nottingham, the focus of preventing waterborne infections in healthcare shifted to managing risks associated with causing Legionnaires disease, with the Duty Holder (generally the person or persons with overall responsibility for health and safety on site) required to appoint a "responsible person" (RP) to manage the delivery of safe water on a day-to-day basis (Brundtett, 1991; Colville et al., 1993). The responsible person was usually an engineer; often the Head or Deputy Head of Estates/Facilities (HSE, 2013). Good practice within healthcare premises was to include the management of *Legionella* on the infection prevention and control team (IPC) agenda; however, in many cases, especially as other more clinically relevant pressures emerged such as MRSA, *Clostridium difficile*, and VRE, the management of water systems often lowers priority until an untoward event happened such as a suspected case of Legionnaires' disease (Department of Health, 2005).

Following the neonatal deaths resulting from *P. aeruginosa* in Northern Ireland in 2012 (Walker et al., 2014; Wise, 2012). It was realized that not only was the current *Legionella* specific guidance for healthcare premises inadequate to protect patients from all forms of waterborne infections, but also there was no one single person that had all the skills necessary to be able to identify and manage all risks associated with the increasing range of recognized waterborne pathogens, which pose a risk of harm to highly susceptible patients.

A major difficulty when developing and implementing water safety plans for healthcare is that there are no internationally accepted levels of what is a safe number of each pathogen for each different patient groups, i.e., the setting of safe healthcare-based targets. Because of the lack of evidence for setting a safe microbiological level and the relatively low level of sensitivity for microbiological methods for the highest risk patients, the safest approach is to work on the precautionary principle and for the highest-risk patients, this may mean excluding water exposure and putting in place a clinical pathway, which includes when water can safely be reintroduced depending on the patient's susceptibility. Previous research where patients have been exposed only to water via a point of use filter and more recently in the Netherlands where water exposure was excluded for Intensive care units have shown that not only were *P. aeruginosa* infections reduced but all Gram-negative infections were reduced in these patient areas (Hopman et al., 2017; Parkinson et al., 2020).

Putting together the team

The first hurdle is to identify the skills needed to develop and implement a WSP, which should include all uses of water on site to manage the associated risks effectively. For a small nursing home with patients who care mainly for themselves, the infection control lead, site manager, engineer, cleaning supervisor with input as required from the water treatment provider, may be all that is needed. However, in a large healthcare hospital where there are many uses of water for patient diagnostics and treatment, the skills set and the number and experts required will need to reflect these complexities.

As you go through this book, we hope you will learn much about the skills needed to manage all risks from water use in healthcare, the intention is not that you do this is on your own and become a Guru of all things water. If that is the case, then you have missed the point. The complexities of ensuring hospital water systems and associated equipment are safe for all uses, and all users are far too complicated for one person, the authors of this book have many decades of experiences between us (yes some of us are that old!!), but not one of us would claim to have the all the answers. Remember the definition of an expert is that they know more and more about less and less! Water is used in healthcare for many purposes including drinking, food preparation, other domestic purposes, bathing and laundry, but also for diagnostic and treatment purposes.

There will be many different types of systems, equipment, and potential modes and routes of exposure, and there will be an unequivocal need to have a range of specific experts and expertise to cover the different uses. These uses may include hand hygiene, decontamination purposes (both of hospital equipment and personnel), for pharmacy preparations, hydrotherapy pools, for medical physics equipment and firefighting systems.

There is a need for input from:

- nursing and clinical teams on the wards who understand what water is used for;
- clinical teams to determine which patients are at increased risk of waterborne infections and how much protection is required;
- microbiologists and infection prevention and control specialists who understand all the potential modes of transmission, i.e., routes of exposure, for example, to advise on patient surveillance, microbiological methodology, and interpret sampling results;
- estates and facilities who understand how the water system operates and can advise on the inherent risks that have been identified as well as on control strategies within the water system;
- independent water treatment specialists will be required who have expertise on the pros and cons of chemical and nonchemical treatment regimes, their breakdown products, and benefits and contraindications on each type of water treatment including any adverse effects on the patients and materials in the water systems;
- responsible for finance and/or corporate health and safety, experts who can provide independent auditing services, review risk assessments, and check documentation (a fresh pair of eyes), and from each of the specialist service providers such as dialysis, aquatic physiotherapists, patient support staff responsible for cleaning and maintaining water assets, etc.;
- senior management (Board level) to provide a Governance, oversight, and take overall responsibility.

However, a point of caution, while it may be perceived that having lots of water safety group members is better, there actually needs to be a balance as the larger the numbers on a committee, the more difficult it is to get things done. The system that seems to work better is to have a core water safety group with input into the group from others (subgroups or other experts) as needed and to provide support to the core group. Task-based subgroups can also be used, for example, auditing of cleaning (e.g., ward sinks) and showers by infection prevention and control staff or cleaning supervisors, advising on actions, checks and monitoring needed for ward/unit movements, change or use, capital, and refurbishment projects with input from microbiologists and other experts when needed.

Understanding modes of infection

While legionellosis is primarily caused by inhalation and aspiration in healthcare settings, there are many other potential modes of infection, which the water safety group needs to consider when reviewing or carrying out risk assessments in augmented care settings including:

- **direct or indirect consumption**: from drinking contaminated water or ice (Anon, 1993; Bangsborg et al., 1995) **and/or consuming food irrigated or prepared using contaminated water**
- **by contact** with contaminated water, e.g., during bathing, irrigating or cleaning wounds, cooling burns, hydrotherapy pools, use of specialist equipment such as dental chairs, footbaths, and debridement baths
- **by indirect contact**: e.g., from washing hands in contaminated water, splash contamination from contaminated sinks, drains, etc., bath toys internally colonized with *P. aeruginosa* (Buttery et al., 1998; Döring et al., 1991)
- from use of contaminated mouthwashes, eye drops, etc.
- poor cleaning of medical equipment.
- contaminated cleaning equipment.
- **by inhalation of aerosols**, e.g., legionellae in the drift from evaporative cooling towers used for comfort cooling or refrigeration, etc., aerosols produced from distributed water such as when using showers, wash hand basins, and toilet flushing.
- from respiratory therapy equipment rinsed under contaminated water, e.g., nebulizers.
- indoor fountains, water features, water jets, irrigation systems, pressure washers, etc.
- **by aspiration (water going down "the wrong way")**, i.e., contaminated water unintentionally entering the lungs during drinking or sucking ice made from contaminated water, a particular problem for patients with swallowing difficulties, for example, following a stroke, recent surgery, patients with nasogastric tubes, and those with neurological conditions (Bencini et al., 2005)

System description and risk assessment

What is in place what is missing?

As the multidisciplinary water safety group is being formed, it is important that it is:

- formally appointed,
- the remit is agreed,
- there is governance and reporting to an Executive Board Member,
- resources are allocated to enable the group to meet and function on a regular basis—this may include administration support for organizing meetings, taking minutes, storage of data and records,
- gap analyses are carried out to identify where additional internal or external expertise is required,
- competency checks are carried out for internal and external staff to formally assess and approve staff are competent and that recurrent training is in place,
- training schedules implemented for those attending the water safety group (provide suitably qualified internal as well as external experts).

In the United Kingdom and many other parts of the world, it is a legal requirement to have *Legionella* risk assessments and management plans in place (HSE, 2013, 2014). If these are effective, then many of the engineering aspects to control the risks from other waterborne pathogens will also be in place.

Particularly in healthcare settings, where the population is likely to be at an increased risk of infection, the advantages of the water safety approach are that it requires a systematic and detailed assessment of each potential use of water, describing the system and ensuring there are accurate plans and a register of all associated assets.

For such purposes, an asset is each system, piece of equipment, and the relevant components and fittings that identifies the item, where it is located in the system, when cleaning, servicing, and maintenance should be carried out, and who is responsible for such work to ensure the asset is safe and fit for purpose. For each asset identified on the register, the relevant hazards should be identified and managed; hazards could include biological, chemical, or physical agents with the potential to cause adverse health effects.

The risk assessment should identify the likelihood of harm occurring, the frequency and the consequences of the harm that could be caused, so a priority list for action to manage the risks ideally using a multibarrier approach can be developed and implemented.

The water safety management scheme should include the identification of appropriate control measures whether they are filtration, point-of-use filters, chemical treatment, UV, temperature, or pH together with any validation and operational monitoring requirements. Depending on the complexity of the control strategy, the water safety group may have to consider bringing in external expertise to provide assurance that the chosen strategies will be effective and what monitoring systems should be implemented to ensure safe water is provided.

Water safety plan risk assessments

Risk assessments should be carried out by those who have the relevant experience to identify:

- whether there is appropriate governance, accountability, and responsibility and corporate oversight of water safety
- whether there are appropriate chains of communication in place
- if all those involved in water safety management, including support staff and contractors, have the training, competencies, and experience to carry out their tasks effectively and understand their role in keeping patients safe
- all potential hazards relevant to each system or associated equipment
- the environmental factors that can increase the risk of introducing a hazard or allowing hazards to increase to levels that could cause harm
- the level of risk for each identified hazard taking account of the exposed population
- the appropriate measures to control the identified risks to ensure that all water at each point of exposure is safe
- the relevant standard, guidelines, parameters, and targets relevant to each water use and that the targets are being consistently met
- the validation and verification required to ensure the controls are appropriate and continue to be effective
- the records and documentation are appropriate and support that there is an effective management plan and all risks are controlled as far as possible
- if there are adequate and effective audit/review, support, and training programs in place to maintain patient safety in the long term
- whether any improvements can be made to improve patient safety.

An important step in the water safety plan is ensuring that there is adequate documentation and communication both for normal operation of the systems and following incidents, e.g., when there have been failures in controls (e.g., temperature), in critical equipment (such as biocide dosing units) and subsequent cases of illness directly or indirectly associated with the water system (surveillance) (Bartram, 2007; BSI, 2020; DHSC, 2016; DWI, 2009; HSE, 2013, 2014; WHO, 2011).

A holistic WSP should also include a plan for dealing with predictable/foreseeable problems (e.g., the breakdown of a critical piece of equipment such as a pump, which can be predicted to occur sometime in the lifetime of a system) including the remedial actions to be taken in the event of such occurrences. There should also be a list, with contact details, of those to be informed if such an event occurs.

Can healthcare organizations afford not have a water safety group?

A barrier that needs government change of policy is the separation of the costs of capital projects from the ongoing operational budget. This would ensure that future costs of a poorly designed and engineered project do not cost more to the:

- organization in terms of the costs associated with remedial actions to correct the design and build failures and ongoing microbial control measures
- patients in additional costs for extended patient treatments, stay in hospital and support, and long-term increased risk of spreading antimicrobial resistance.

Training and education

All those involved in managing water safety risks from the cleaners up to the CEO need to be made aware of the risks posed by water. It is vital to ensure water safety is taken seriously by all those involved in the water safety group but also those who deal with water for patient care on a daily basis. There have been instances where hospital trusts have been resistance to, for example, educating cleaners so they understand why they clean the way they do and how important their front-line role is in infection prevention. A job description does not make you competent to understand water. Without a doubt, training and educating staff so that they are able to use their eyes and ears to identify unsafe engineering and clinical water practices including poor temperature control, poorly fitted elbow taps, unused outlets, sinks used inappropriately, blocked sinks can actually save lives. Once trained, members of the water safety group can then cascade training but only if resources and project planning provide the mechanisms.

Conclusions

An effective water safety plan needs to include a description of all systems and associated equipment as well as fittings, components (assets) to ensure all potential modes and sources of exposure are identified, risk assessments and monitoring are carried out to determine if water within the water system and at point of use is safe for all who might be exposed, i.e., meets the health-based targets and minimizes the risk to susceptible patients, staff, and visitors as far as possible.

A water safety plan developed and implemented specifically for a healthcare facility is the most effective and resilient way to ensure that all potential sources of infection are considered and prevention measures put in place to keep patients safe from all infections associated with water or damp environments, which takes account of the susceptibilities and vulnerabilities of the population. A holistic and effective water safety plan can also be used to provide governance (responsibility at the Board level), due diligence and that all water meets the applicable standards as required within regulations, standards, and guidance appropriate to its intended use.

References

Allegra, S., Leclerc, L., Massard, P.A., Girardot, F., Riffard, S., Pourchez, J., 2016. Characterization of aerosols containing *Legionella* generated upon nebulization. Sci. Rep. 6, 33998. https://doi.org/10.1038/srep33998.

Anon, 1993. Ice as a source of infection acquired in hospital. Commun. Dis. Rep. CDR Wkly 3, 241.

Bangsborg, J.M., Uldum, S., Jensen, J.S., Bruun, B.G., 1995. Nosocomial legionellosis in three heart-lung transplant patients: case reports and environmental observations. Eur. J. Clin. Microbiol. Infect. Dis. 14, 99–104. https://doi.org/10.1007/BF02111866.

Bartram, J., 2007. WHO Legionella and the Prevention of Legionellosis. https://www.who.int/water_sanitation_health/publications/legionella/en/.

Bencini, M., Yzerman, E.P.F., Koornstra, R.H.T., Nolte, C.C.M., den Boer, J.W., Bruin, J.P., 2005. A case of Legionnaires' disease caused by aspiration of ice water. Arch. Environ. Occup. Health 60, 302–306. https://doi.org/10.3200/AEOH.60.6.302-306.

Benoit, M.-È., Prévost, M., Succar, A., Charron, D., Déziel, E., Robert, E., Bédard, E., 2021. Faucet aerator design influences aerosol size distribution and microbial contamination level. Sci. Total Environ. 775, 145690. https://doi.org/10.1016/j.scitotenv.2021.145690.

Bergen, L.K., Meyer, M., Høg, M., Rubenhagen, B., Andersen, L.P., 2009. Spread of bacteria on surfaces when cleaning with microfibre cloths. J. Hosp. Infect. 71, 132–137. https://doi.org/10.1016/j.jhin.2008.10.025.

Brundtett, G.W., 1991. Outbreak of Legionnaires' disease at Stafford District General Hospital. Build. Serv. Eng. Res. Technol. 12, 53–64. https://doi.org/10.1177/014362449101200201.

BSI, 2020. BS 8680—Water Quality. Water Safety Plans. Code of practice. https://shop.bsigroup.com/ProductDetail?pid=000000000030364472. (Accessed 5 January 2021).

BSRIA, 2018. Soft Landings Framework. [WWW Document]. URL: https://www.bsria.com/uk/product/QnPd6n/soft_landings_framework_2018_bg_542018_a15d25e1/. (Accessed 9 August 2022).

Buttery, J.P., Alabaster, S.J., Heine, R.G., Scott, S.M., Crutchfield, R.A., Bigham, A., Tabrizi, S.N., Garland, S.M., 1998. Multiresistant *Pseudomonas aeruginosa* outbreak in a pediatric oncology ward related to bath toys. Pediatr. Infect. Dis. J. 17, 509–513. https://doi.org/10.1097/00006454-199806000-00015.

Collier, S.A., Deng, L., Adam, E.A., Benedict, K.M., Beshearse, E.M., Blackstock, A.J., Bruce, B.B., Derado, G., Edens, C., Fullerton, K.E., Gargano, J.W., Geissler, A.L., Hall, A.J., Havelaar, A.H., Hill, V.R., Hoekstra, R.M., Reddy, S.C., Scallan, E., Stokes, E.K., Yoder, J.S., Beach, M.J., 2021. Estimate of Burden and Direct Healthcare Cost of Infectious Waterborne Disease in the United States. Emerg. Infect. Dis. 27, 140–149. https://doi.org/10.3201/eid2701.190676.

Colville, A., Crowley, J., Dearden, D., Slack, R.C., Lee, J.V., 1993. Outbreak of Legionnaires' disease at University Hospital, Nottingham. Epidemiology, microbiology and control. Epidemiol. Infect. 110, 105–116. https://doi.org/10.1017/s0950268800050731.

Department of Health, 2005. Saving Lives: A Delivery Programme to Reduce Healthcare Associated Infection Including MRSA. [WWW Document]. URL: https://webarchive.nationalarchives.gov.uk/+/http://www.dh.gov.uk/en/Publicationsandstatistics/Publications/PublicationsPolicyAndGuidance/DH_4113889. (Accessed 2 April 2019).

DHSC, 2016. HTM 04-01: Safe Water in Healthcare Premises.

Döring, G., Ulrich, M., Müller, W., Bitzer, J., Schmidt-Koenig, L., Münst, L., Grupp, H., Wolz, C., Stern, M., Botzenhart, K., 1991. Generation of *Pseudomonas aeruginosa* aerosols during handwashing from contaminated sink drains, transmission to hands of hospital personnel, and its prevention by use of a new heating device. Zentralbl. Hyg. Umweltmed. 191, 494–505.

DWI, 2009. Drinking Water Safety: Guidance to Health and Water Professionals.

Falkinham, J.O., 2020. Living with Legionella and Other Waterborne Pathogens. Microorganisms 8, 2026. https://doi.org/10.3390/microorganisms8122026.

Gómez-Gómez, B., Volkow-Fernández, P., Cornejo-Juárez, P., 2020. Bloodstream infections caused by waterborne bacteria. Curr. Treat. Options Infect. Dis. 12, 332–348. https://doi.org/10.1007/s40506-020-00234-5.

Hajar, Z., Mana, T.S.C., Cadnum, J.L., Donskey, C.J., 2019. Dispersal of gram-negative bacilli from contaminated sink drains to cover gowns and hands during hand washing. Infect. Control Hosp. Epidemiol. 40, 460–462. https://doi.org/10.1017/ice.2019.25.

Hopman, J., Tostmann, A., Wertheim, H., Bos, M., Kolwijck, E., Akkermans, R., Sturm, P., Voss, A., Pickkers, P., vd Hoeven, H., 2017. Reduced rate of intensive care unit acquired Gram-negative bacilli after removal of sinks and introduction of 'water-free' patient care. Antimicrob. Resist. Infect. Control 6, 59. https://doi.org/10.1186/s13756-017-0213-0.

HSE, 2013. Legionnaires' Disease. The Control of Legionella Bacteria in Water Systems. Approved Code of Practice Legionnaires' Disease. ACOP. https://www.hse.gov.uk/pubns/books/l8.htm. (Accessed 5 January 2021).

HSE, 2014. HSG 274 Legionnaires' Disease—Technical Guidance Part 2: The Control of Legionella Bacteria in Hot and Cold Water Systems Technical Guidance. http://www.hse.gov.uk/pubns/books/hsg274.htm. (Accessed 5 January 2021).

Inkster, T., Weinbren, M., 2021. Water springing to life the fungal desert. J. Hosp. Infect. 111, 65–68. https://doi.org/10.1016/j.jhin.2021.02.015.

Jamal, A., Brown, K.A., Katz, K., Johnstone, J., Muller, M.P., Allen, V., Borgia, S., Boyd, D.A., Ciccotelli, W., Delibasic, K., Fisman, D., Leis, J., Li, A., Mataseje, L., Mehta, M., Mulvey, M., Ng, W., Pantelidis, R., Paterson, A., McGeer, A., 2019. Risk factors for contamination with carbapenemase-producing Enterobacteriales (CPE) in exposed hospital drains in Ontario, Canada. Open Forum Infect. Dis. 6, S441. https://doi.org/10.1093/ofid/ofz360.1091.

Kaul, C.M., Chan, J., Phillips, M.S., 2022. Mitigation of nontuberculous mycobacteria in hospital water: challenges for infection prevention. Curr. Opin. Infect. Dis. 35, 330–338. https://doi.org/10.1097/QCO.0000000000000844.

Kim, E.J., Park, W.B., Yoon, J.-K., Cho, W.-S., Kim, S.J., Oh, Y.R., Jun, K.I., Kang, C.K., Choe, P.G., Kim, J.-I., Choi, E.H., Oh, M.D., Kim, N.J., 2020. Outbreak investigation of *Serratia marcescens* neurosurgical site infections associated with a contaminated shaving razors. Antimicrob. Resist. Infect. Control 9, 64. https://doi.org/10.1186/s13756-020-00725-6.

Parkinson, J., Baron, J.L., Hall, B., Bos, H., Racine, P., Wagener, M.M., Stout, J.E., 2020. Point-of-use filters for prevention of health care–acquired Legionnaires' disease: field evaluation of a new filter product and literature review. Am. J. Infect. Control 48, 132–138. https://doi.org/10.1016/j.ajic.2019.09.006.

RIBA, 2020. The Royal Institute of British Architects (RIBA) Plan of Work. [WWW Document]. URL: https://www.architecture.com/knowledge-and-resources/resources-landing-page/riba-plan-of-work. (Accessed 8 August 2022).

Tang, L., Tadros, M., Matukas, L., Taggart, L., Muller, M., 2020. Sink and drain monitoring and decontamination protocol for carbapenemase-producing *Enterobacteriaceae* (CPE). Am. J. Infect. Control 48, S17. https://doi.org/10.1016/j.ajic.2020.06.132.

Walker, J.T., Jhutty, A., Parks, S., Willis, C., Copley, V., Turton, J.F., Hoffman, P.N., Bennett, A.M., 2014. Investigation of healthcare-acquired infections associated with *Pseudomonas aeruginosa* biofilms in taps in neonatal units in Northern Ireland. J. Hosp. Infect. 86, 16–23. https://doi.org/10.1016/j.jhin.2013.10.003.

WHO (Ed.), 2011. Water Safety in Buildings.

WHO, Water, Sanitation and Health Team, 2004. Guidelines for Drinking-water Quality. Vol. 1, Recommendations. World Health Organization.

Wise, J., 2012. Three babies die in pseudomonas outbreak at Belfast neonatal unit. BMJ 344, e592. https://doi.org/10.1136/bmj.e592.

Controlling the microbial quality of water systems

29

Introduction

Control strategies for water systems are conventionally considered as the application of physical and/or chemical technologies to prevent the proliferation of microorganisms after infections and outbreaks have occurred (Alangaden, 2011; Baron et al., 2020; Exner et al., 2005; Inkster et al., 2021; Kinsey et al., 2017). Yet retrospective action is like shutting the barn door after the horse has bolted, i.e., trying to manage an out-of-control system once the microorganisms have proliferated.

In industry, a hierarchy of control is a system that is used to eliminate or reduce exposure to risk in the workplace and should start from the top of the hierarchy down (CDC, 2022; HSE, 2022). In most circumstances, a combination of control measures often termed "the scheme of control" is chosen to effectively reduce the risk posed by a hazard and may include the following elements

- Elimination (removes the hazard at the source)
- Substitution (using a safer alternative to the source of the hazard)
- Engineering controls (reduce or prevent hazards from coming into contact with patients)
- Infection and prevention control measures to eliminate or reduce contact with harmful hazards
- Administrative controls (establish work practices that reduce the duration, frequency, or intensity of exposure to hazards)
- Personal protective equipment (equipment worn to minimize exposure to hazards)

While several healthcare facilities describe themselves as "centers of excellence," the numbers of preventable hospital acquired waterborne infections does not always reflect this statement. Analysis by Perkins et al. indicated that water-related investigations represented 22% of hospital-acquired infection consultations, and Anaissie et al reported that *Pseudomonas aeruginosa* accounted for 1400 deaths per year (Anaissie et al., 2002; Perkins et al., 2019). Extrapolating the United States Perkins et al data to Europe would equate to 8140 deaths per year as a direct consequence of waterborne-associated hospital-acquired infections. In what other workplace would this number of deaths be acceptable?

Should increasing rates of morbidity and mortality in high-risk patients be an inevitable collateral damage of failures to control waterborne infection in the built environment?

Control strategies

Instead of waiting for microbial problems and waterborne infections to manifest, what if there was another way?

Safe Water in Healthcare. https://doi.org/10.1016/B978-0-323-90492-6.00001-X

It is important to recognize that no one control strategy will be suitable for all situations and that once a building is colonized, it is unlikely to provide safe water for the rest of its life cycle. As such prevention and control should be the preferred option. The development of a water safety plan should include a multibarrier approach to provide safe water. So if one barrier fails, there is a back-up to protect patients vulnerable to infection in healthcare settings (BSI, 2020; WHO, 2017). What works in one building system does not always work in other buildings. There are many interacting factors that need to be considered when carrying out a risk assessment. These factors may include engineering, building occupation factors, types of water use, modes of exposure, vulnerability of those exposed, surrounding environment, and human behavior. It is imperative for system controls to remain effective that the water safety group has the experience and skills to ensure systems are risk assessed, maintenance and controls are appropriately applied and validated as well as ongoing verification by monitoring and microbial sampling of relevant parameters to ensure each water system and any water-associated equipment remains safe.

The model of active and latent failures, typically referred to as the Swiss cheese model of human error and accident causation, has become well known in most safety and healthcare circles (Fig. 1) (Stein and Heiss, 2015; Wiegmann et al., 2022). This model moves away from the single elements toward identifying holistic approaches where each cheese slice is a defensive layer in the process or system.

The holes represent opportunities for failures in the system including specification, design, build, commissioning, water supply quality, storage conditions, hot and cold water temperatures, stagnation, handwash stations, outlets, drains, environmental and clinical surveillance, education, signage, and unsafe acts. While one or two holes in the model may be an issue that can resolved, when increasing numbers of the holes in the defensive layers start to align, then the result is an increased potential for adverse catastrophic events resulting in harm to patients and sometimes also staff and visitors.

History teaches us many things, but often we do not actually learn from past lessons (Ayliffe, 2000; Hunter and Ensign, 1947; Kohn, 1967; Watt, 1946). To quote Florence

Fig. 1 Swiss cheese model of active and latent failures.

Nightingale in 1859, "The very first requirement in a hospital is that it should do the sick no harm." Indeed this is a fundamental statement for healthcare professionals as they undertake their roles to look after their patients to be the best of their ability with the basic premise of *do no harm*.

This section on control measures will take a holistic approach and very much like the Swiss cheese model will look at the many different aspects that are required if healthcare premises are to achieve a reduction in waterborne infections especially in high-risk patients. To paraphrase James Watt from 1945, "there was sufficient time... and ample opportunity for introduction and maintenance of contamination ...since their care was a case of everybody's job being nobody's job." Water is every healthcare professional's concern for they all use water on a daily basis in and around patients, some of whom may be high risk.

The Royal National Lifeboat Institute has a saying "Respect the Water." This is analogous to healthcare where we need everyone to understand the implications to patients when that respect is not applied.

Design of water systems in healthcare

Those that are involved in the design of water systems require education and training to understand all the aspects of water system design that could result in microbial colonization and growth in order to reduce the potential sources of exposure and number of transmission events. The more outlets, hand hygiene stations, showers, and the more complex the water system in the building, then the greater likelihood that harmful hazards, both microbial and chemical, can increase to levels that can cause harm. If these first principles are not understood and implemented, then it is likely that upon commissioning, water systems and associated equipment will already be colonized and a return to a safe system will be difficult and in most cases impossible, to achieve without a great deal of work and investment to replace colonized pipework and components. To understand the issues of water contamination, designers have to (i) work with specialist teams competent in understanding and managing risks in the built environment, ideally as part of a project-specific water safety group to ensure that waterborne risks are kept to a minimum; and (ii) and at the very beginning of the project all, i.e., the concept stage, all those involved in the project, in any capacity, should undertake training and education programs aimed at understanding the risks and how to deliver a safe water system especially for high-risk patients.

Such approaches require a complete rethink in how hospitals are designed and engineered and what designers and water safety groups need to take into consideration when designing water systems in healthcare.

Yet in reality, we are still enabling the design of water systems that encourage the growth of waterborne pathogens in buildings, and in effect we are establishing sick and potentially lethal buildings into which we are then putting high-risk patients (Jansz, 2011).

There are simple statements to keep in mind in terms of water system safety:

- Keep the cold water cold
- Keep the hot water hot
- Keep the water moving

Yet the complexity of water systems as they are currently designed results in heat loss from the hot water system, heat gain in the cold water system, and the water flows is slow or even not moving at all, particularly prior to low flow or rarely used outlets, regardless of the purpose of that outlet. Traditional designs of hot and cold water systems were to build large centralized system with long pipe runs with hot and cold and even steam and central heating pipes all in proximity. Therefore, the quality of water in a building will deteriorate due to poor design, inappropriate installation, inadequate commissioning and operation practices, and underuse of the water system so that water at the point of use can no longer be considered wholesome or safe.

Cold water supply

Cold water supplied to hospitals may be a mixture of water supplied by a water utility and a private water supply (usually a borehole on site) to ensure resilience of supply and reduce costs. However, supplied water entering the site and building must be wholesome and comply with relevant national drinking water regulations (DHSC, 2016; DWI, 2021; EU, 2020; Larson, 2020; UK Government, 2018). Wholesome water is water that complies with all the parameters in the relevant regulations, chemical and microbiological, and is fit to use for drinking, cooking, food preparation, or washing without any potential danger to human health (UK Government, 1991). Consequently, the water delivered to hospitals at the point of entry is considered wholesome and safe for the general population. The tests undertaken by regulators and water companies are stringent and set to detect particular types of bacteria under set conditions based on the absence of fecal indicators (Black et al., 1979; SCA, n.d.). Therefore, while water supplied at the point of entry to the healthcare establishment is not sterile, the microbial content is controlled through the above quality standards. It should be noted that the above tests do not assess for the presence of naturally occurring microbial populations associated with growth in biofilms such as *Legionella*, *Pseudomonas aeruginosa,* and nontuberculous Mycobacteria.

Cold water storage and distribution

The design guidance parameters for cold water storage and distribution are set out in Chapter 2 (cold water systems), but time and again, cold water storage tanks are too large, over capacity, inadequately insulated, have inadequate cross flow, and have poorly fitting lids such that there is little protection from ingress by birds and rodents. In addition, little thought is often given to access space for inspection, maintenance or cleaning, and disinfection, and they are often also placed in warm plant rooms or on roof tops without consideration for accessibility.

The basic concept of keeping cold water cold, e.g., the temperature at the outlet should be no more than 2°C above the incoming supply temperature. Cold water pipes are often placed in the same conduit as hot water pipes, and this lack of separation and inadequate insulation leads to heat gain. With the rise in incoming cold water due to climate change, above the target of 20°C into buildings, this will inevitably lead to the exponential growth of microorganisms. For all patients, microbial growth needs to be controlled, and designers need to compensate and redesign for such eventualities. Poor labeling of pipework is also identified as a significant factor in causing outbreaks of waterborne disease caused by inadvertent cross-connection of clean and dirty systems (WHO, 2017).

Flushing

Flushing of cold water systems in little used areas is necessary to keep the microbial control measures effective right up to the outlet. However, with the increasing risk of drought, the flow of this water to the drain will be costly and unacceptable. Some hospitals already use automatic flushing devices to keep water moving, especially in areas where access to patient accommodation is difficult, e.g., where nuclear medicine is given or intensive care units with severely immunocompromised patients. Their needs to be a balance of risks, and this likely to become more complex as hospitals try to achieve net zero targets, which means there must be even more emphasis on specialist training so those with the aim to improve sustainability targets do not consequentially increase the risk of harm to patients as well as staff and visitors. Additional information is provided on cold water recirculation systems later in this chapter.

Hot water system

Hot water systems need to be tightly controlled to ensure the hot water achieves the target temperature of 55°C within 1 min (DHSC, 2016). In well-plumbed buildings, this should take seconds rather than minutes. The design parameters and guidance documents, which are set out in Chapter 3 list the characteristics that are required to maintain microbial control. Generating water at the required temperature would seem a simple aspect of physics and engineering, yet maintaining the required temperatures through hot water systems seems to be too much of a challenge in most healthcare establishments. This all too common, lack of temperature control may be due to many factors, for example: designing large and complex reticulated systems as opposed to smaller more manageable localized systems. Another common problem encountered during investigations into cases of gross contamination in water systems is that higher specification fittings and/or components are often downgraded at the value engineering stage to reduce costs. The unintended consequences are that there is an increased susceptibility to microbial growth, increased corrosion, component failures, and components can be more difficult to clean, maintain, and disinfect. This inevitably results in additional long-term operational costs, but budget allocation means that this is not a consideration at the building design stage. Examples of poor design engineering include insufficient separation, insulation, long spurs with trace heating as opposed to returns (incorporating return loops increases the cost of pipework needed), faulty flow balancing valves, poor balancing, stagnation prior to outlets, undercapacity heating, overcapacity storage. Hot water system can be too complex to achieve such consistent temperatures. Nevertheless, maintaining hot water temperatures for microbial control should be achievable in a well-designed system, and modern real-time monitoring technologies are beginning to assist in that goal (Chapter 25).

Outlets including hand wash station, showers, and drinking water

Ensuring adequate hand hygiene has become a major issue for infection control specialists. In 2009, the World Health Organization identified that alcohol hand rubs should be the first option for hand hygiene (Maroldi et al., 2017). Yet guidelines for hand hygiene have major implications when designs incorporate a hand hygiene station next to every patient bed without regard for how often they will be used or potential for cross-contamination to the patient, their belongings and equipment. While hand hygiene is essential for safe patient care, an oversupply of hand hygiene stations has consequences for the transmission of waterborne infections. Microbial growth at outlets can occur through the water system or through retrograde contamination by contact with patients and/or staff. Poor installation of outlets can also increase the risk of cross-contamination from outlets post handwashing as all too often the operating levers on hand wash basins cannot be operated hands-free as intended as the elbow or hand-operating levers have not been fitted in the correct position. Touchless operation has major advantages that should be part of the design principle, and lessons should be learnt from some touchless systems, which due to the complexity of the design and/or the materials used in manufacture have resulted in an increased risk of microbial colonization (Charron et al., 2015; Hargreaves et al., 2001). Hand hygiene stations are frequently misused by staff and patients and are often sited too close to patients beds and furniture as well as clinical trolleys, for example, so there is a risk of splash contamination, Because of this risk of cross-contamination from splashing and drains, which poses a real transmission risk at distances of less 2 m, the advice in HTM 04-01 is "All preparation areas for aseptic procedures and drug preparation and any associated sterile equipment should not be located where they are at risk of splashing/ contamination from water outlets." In addition, clean utilities and other areas used for aseptic drug preparation, which will be given intravenously or via a central/Hickman line, then wash hand basins should be located outside the area as these present a risk of cross-contamination from the sink to such preparations. DH Guidance states these should be located outside areas used for pharmacy preparations (including drug infusions and TPN) because of the risk of cross-contamination from splashing and drains. HBN 14-01 states that "Sinks and drains should be excluded from areas where aseptic operations are carried out" (NHS England, n.d.).

Choice of plumbing fixtures and fitting

It has been noted that some designers prefer particular manufacturers' outlet fittings because of their aesthetic looks irrespective of performance and the ability to be easily disinfected, cleaned, and fitting point-of-use filters when required. However, each outlet/tap/basin should be risk assessed based on the manufacturer's testing and published research on the aspects of microbial control to provide risk reduction properties. This should include:

- In high-risk areas can the tap components including the spout and TMV (where fitted) be removed for cleaning, descaling, and disinfection.
- Where fitted point-of-use filters should allow (in augmented care) sufficient activity space.

- Tap components that reduce the potential for microbial growth in water.
- Materials and surfaces that will deter biofilm growth.
- Outlet design and its impact on flow.
- Complex nature of flow regulators (outlets) and their ability to harbor and generate biofilms.

  Contamination potential for manual hand touch-operated levers.
- Design and setup of hand wash stations to reduce contact points.
- Infrared sensors should be visible to reduce inadvertent contact contamination.
- Design of drain strainer area to reduce to splash.
- Splash and aerosol production when outlet is in use.
- Position of drain and potential for splashing and contamination of surrounding area with antibiotic-resistant strains.

  Hand wash basins should not provide surrounding space that can be used as a shelf.

  Hard-piped plumbing connections rather than the fitting of flexible tails.

The move to single room accommodation while having some benefits for patients also has the unintended consequence of significantly increasing the risk of waterborne infections for high-risk patients and an increased operational cost to ensure the ensuite outlets remain safe. Too little thought is given as to whether a patient is likely to use the outlets, especially when admitted for severe illness or major procedures, which leaves them unable to use the facilities for long periods of time. In surgical wards, for example, patients are often asked to shower before admission and go home the same day or the next day without having had a shower—hence, there is little or no turnover of the water. Following major surgery, the patient may be too unwell to shower. Therefore, risk assessments should consider placement of patients in rooms with a toilet and wash hand basin and full ensuites only in rooms where patients are likely to use them.

Outlets such as showers should be reduced to a minimum but where required should be designed with touchless control, attention should be given to ensure the fall of the floor flows to the drain to minimize the risk of standing water and the ease of cleanliness. Standing water may pose a risk of retrograde growth from the drains and transmission of *P. aeruginosa* and other opportunistic waterborne through splash contamination. Fixed showers are preferable to showers with flexible hoses, especially where ligature-resistant showers are required. If flexible shower hoses are required, the hoses should be tethered to prevent inadvertent insertion in water (toilets, pooled water on the floor) and subsequent back-contamination. Flexible shower hoses should be placed of the asset register to ensure that they are changed on a frequent basis.

Where shower rooms and debridement baths, for example, are designated as being essential, such as in burns units, then a risk assessment should determine whether additional precautions are needed such as point-of-use filtration via a sterilizing grade filter and the frequency of replacement. Some high-risk units where it is expected usage will be low, but the shower is still considered essential as single-use showers. Particular attention should be applied to the servicing, maintenance, cleaning, and monitoring as key aspects in the design as these units could play a critical role in transmission of waterborne pathogens.

When contamination of the water supply or outlet has been identified, the placement of a microbial retention filter is often the first line of defense. While these will be discussed later, the designer's job is to ensure that the design of the wash hand station is able to accommodate the fitting of such a filter. With few exceptions, point-of-use filters should not be viewed as a permanent solution to a water system that is not able to provide wholesome water. Too often the fitting of a filter results in a reduction of the activity space such that effective hand hygiene cannot be undertaken and a breach in the water fittings regulations as there is no accommodation for the air gap that is a reequipment.

Senior management must take responsibility for outbreaks associated with hand wash basins and drains. It is not acceptable for high-risk patients to knowingly be exposed to highly transmittable pathogenic microorganisms in the built environment (Decraene et al., 2018).

At a time when major new builds are in the planning process, perhaps it is time to consider the work undertaken by Joost Hopman and colleagues in The Netherlands where they have introduced patient-free water care (Hopman et al., 2017, 2018). This approach still requires hand hygiene stations but relies on removing the hand hygiene stations from close proximity to the patients where they are not required. Removing the majority of the hand hygiene stations not only reduces the exposure to high-risk patients but also removes the associated drains (Kenters et al., 2018). Obviously where hands are visually soiled, they need to be washed to remove the contaminating material as gelling will not clean them sufficiently.

Drains

Misuse of hand hygiene stations also includes the disposal of patient material, nutrient-rich water used for patient hygiene, used giving sets, including those used for antibiotic infusions and objects put into drains to the extent that the drain becomes partially blocked. The ability of some people to get items such as bottle teats, syringe covers, and nail cleaners into the p-trap ceases to amaze. It should not be surprising that the same antibiotic-resistant strains from drains/p-traps have also been isolated from patients (Aranega-Bou et al., 2018; De Geyter et al., 2021). The extent of the role of drains is only just starting to be recognized, and we have a duty care to control this route of transmission. Various strategies, with varying degrees of success, have been investigated for control of waterborne pathogens and clinical strains in the drain and p-trap.

The removal of infrequently used hand hygiene stations will also remove the associated contaminated drains. However, where the hand hygiene stations are retained, then temporary and transient treatment of the drains will only have a limited impact on the presence of the microorganisms (Tang et al., 2020). In addition, there will be drains associated with showers, scrub sinks, kitchen sinks, and sluice rooms, which will also be contaminated and will be risk-assessed to prevent transmission of waterborne pathogens via this route.

Risk mitigation must take a multidisciplinary holistic approach by those with the expertise and competence to do so, including training, servicing, replacement, and treatment. Those without the necessary skills and knowledge may increase the risk of spreading pathogens from the drains to the environment and personnel in the vicinity. Removal of drains then and associated plumbing needs to carried out with extreme caution to avoid disruption of biofilm and cross-contamination of adjacent areas and new plumbing (Aranega-Bou et al., 2018; Decraene et al., 2018; Kotay et al., 2020).

Method statements should be reviewed and approved by the multidisciplinary WSG with input from external expertise when needed. Drains should be added to the asset register to ensure that there are servicing and maintenance of the drain units to ensure they are clean and not full of debris and nutrients that will encourage growth. Education and training must enable staff to identify and provide the support to staff to report water supply and drainage-associated problems with the confidence that work will be carried out by competent personnel to resolve the problems that have been identified.

A wide range of treatment regimens have been proposed including flushing, heating, and biocide application (Bédard et al., 2021; De Geyter et al., 2017; Jamal et al., 2019; Jung et al., 2020; Parkes and Hota, 2018; Smolders et al., 2019; Tang et al., 2020). Technologies that are specifically designed for the continuous application of biocides are more likely to be successful at controlling the presence of pathogens in the sink drain (Coleman et al., 2020). In addition, commercial self-disinfecting sink trap systems are available that will thermally and chemically treat the drain trap (Medizinische Hygiene-Siphone BIOREC MoveoMed, Dresden, Germany) (Regev-Yochay et al., 2018; Wolf et al., 2014).

Examples of units caring for high-risk patients

Particular attention needs to be given to specialist units that care for high-risk patients, and renal units are used here as an example. During an average week of hemodialysis, patients will be exposed to 300–600L of water, which has the potential to expose patients to waterborne pathogens (Rao et al., 2009) (Chapter 29). Transmission of waterborne pathogens has caused a number of infections in dialysis patients and is avoidable if the renal unit water system has been designed appropriately from the outset (Thet et al., 2019; Ward, 2011). Adverse patient outcomes including fatalities have also occurred due to patient exposure to dialysate with water containing high levels of chemicals being used to decontaminate a water system (Ward, 2011). It is vital to ensure the water used to perform dialysis is safe, and designers need to be aware that separate water systems may be required for such units.

Endoscopy units

Sterile services departments are an example of a specialist department responsible for the decontamination of medical devices including endoscopes (Chapter 24) (Khalsa et al., 2014; Marek et al., 2014). Such department will require water of a high quality produced through a reverse osmosis unit for the final rinse water to ensure that all chemical pollutants are removed prior to the use of the endoscope on patients (NHS Scotland, 2019). This rinse water needs to be free of bacteria. As such designers need to be made aware of the requirements of departments to ensure that the water quality is sufficient to maintain the cleanliness and safety of such instruments. This will require designated water systems specifically for these departments and ensuring that the water supply to these departments is of a high quality when delivered to that department(s).

Pipe and components

Scientific research and outbreak investigations have demonstrated that all pipe materials will become contaminated with biofilm and some materials more so than others. Despite guidance and standards testing and development of "antibiofilm formation" components, all materials, including copper, will in time succumb to biofilm formation and proliferation of waterborne pathogens. This is because products are tested under certain laboratory conditions for a relatively short period of time, but once a product is incorporated into fittings and components, the conditions in water system are very different particularly where there is also biocide-mediated corrosion of metals and hardening of plastics. Water conditions, time, flow, water quality and chemistry, nutrients, and the presence of other microorganisms create a very different environment. Biofilms will also form on copper piping and the myriad of plastic and rubber components used in plumbing components, whether they have antimicrobial products incorporated or not (Moritz et al., 2010; Rogers et al., 1994; Waines et al., 2011; Walker et al., 2014). Therefore, it is imperative that designers of plumbing components and water systems understand the implication of these findings. In addition, it is clear that where integrity testing of water components is required to be carried out by the manufacturer, i.e., wet testing for leaks using water, that this practice can pose a risk of pathogen contamination. It is the manufacturer's responsibility to minimize the risk, but those specifying and purchasing products should ensure that products are certificated to state either water has not been used, i.e., they have been dry tested or the product has been disinfected by an independently validated method to ensure that components are safe to use in healthcare (this will include modular buildings).

The simple goals in healthcare water system design are to minimize water stagnation, ensure that there is minimum controlled heat loss from the hot water system and little or no temperature gain in the cold water system with regulated controlled flow and minimal nutrient contaminants added to the water systems from pipes and components.

New developments and technologies—Pipes

Cold water circulation is not necessarily a new concept and was developed by Kemper some years ago (Kemper, 2022). This technology enables the cold water to be circulated around the building to ensure that the temperature is maintained as close to the supply temperatures as possible.

Kemper has enabled retrofitting technology such that existing building can have cold water circulation installed by using a flow splitter to ensure that temperatures of <20°C can be maintained even in buildings.

Other new developments in piping include the "Eco-Duo" system, a revolutionary method of maintaining hot water temperatures throughout a building, and are being marketed by "Water Kinetics" (Water Kinetics, 2022). The concepts sound extremely simple in that it is pipe within a pipe. The supply flow of hot water passes through the outer layer and returns via the inner layer, and as such the piping provides its own insulation to maintain the hot water at the appropriate temperature for microbial control. In addition, special tap connectors enable recirculation right up to the outlet, re-

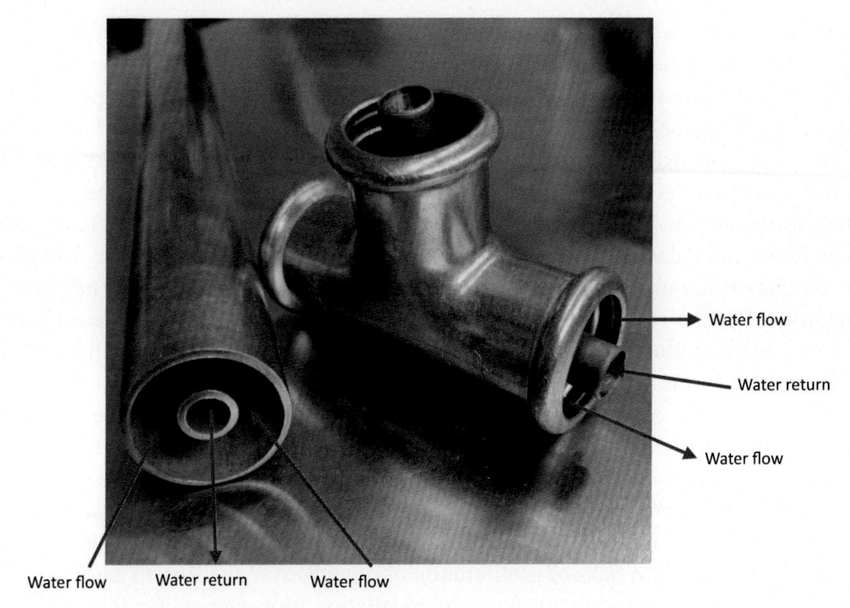

Fig. 2 Eco-Duo flow and return pipework for hot and cold water systems.

moving potential "dead legs." The manufacturers claim that the same system can also be used to recirculate cold water. All manufacturer's claims should be supported by peer-reviewed trial data, which should be reviewed by the water safety group (Fig. 2).

The manufacturers claim the potential advantages of this technology including:

- Keeping the water moving right up to the outlet—reducing stagnation.
- Maintaining the hot water at temperatures required for control of microbial growth across the whole system.
- Maintaining the cold water at supply temperatures, which reduces the proliferation of microorganisms.
- Reductions in energy use due to maintaining a constant temperature across the entire system compared with conventional water systems; hence, carbon reduction targets may be associated with this technology.
- Reductions in energy use due to the minimizing compared with conventional water systems; hence, carbon reduction targets may be associated with this technology.

Commissioning is often taken for granted or indeed the importance of the process is overlooked in terms of protecting high-risk patients (Chapter 2).There are many considerations that require awareness to ensure that all aspects of the design and installation serve to ensure that there is minimal introduction of contamination to the system. In respect of this, bringing outlets and pipes to site, the contractor/project team should ensure that this is completed in the cleanest and safest way possible to reduce the likelihood of contamination and biofilms.

Commissioning a water system is a part of the process where the new water system is brought into operation and applies to all component parts of a building water system including attached equipment (HSE, 2014).

However, such wholescale commissioning is rarely if ever carried out in a single session. Water systems are more likely to be commissioned on a piecemeal basis where only specific sections or floors are assessed through pressure testing to ensure systems are leakproof and operate to design specifications. This will entail repairing problems such as leaks and that the flow of hot and cold water systems is balanced. As this is carried out progressively across different floors as they are completed and across different parts of the hospital, it is not surprising that there are problems with water flows and balancing across different sections. Whereas in healthcare, installation of many water using components, specialized systems, and fitting will take place sometime after the circulating system has been filled the WSG need to ensure the system is filled as close to handover as possible, a biocide is added when filling takes place, and the levels are maintained. Water storage tanks are bypassed until full occupation, and there is an agreed plan for flushing and checking of biocide levels at distal ends of the filled system loops. Often hosepipes are used at this stage to take the flushed water to drain before fittings are installed without a residual biocide, these are likely to support microbial growth including of *Legionella* and *P. aeruginosa*.

Fitting out can occur long before the building is brought into full occupation, and it is critical to manage the risks of proliferation of organisms in the system and installed fittings and equipment during this phase of installation. However, with a flushing program in place at this point, the water system will not be used to full capacity, which will result in underuse of the system, stagnation, thermal transfer, and potential ingress of contamination resulting in the growth of microorganisms and biofilms.

Once a building is contaminated, it is very difficult if not impossible to remediate, resulting in an ongoing risk to patients, staff, and visitors and ongoing high costs for testing and disinfection or the installation of point-of-use filters to provide safe water.

HSG 274 recognizes that from a microbiological perspective, the period between filling the system and bringing it into normal use is potentially the most hazardous, and designers should be aware of these aspects (HSE, 2014). As such there needs to be a planned water sampling program, which the whole multidisciplinary team needs to agree. This will identify hygiene issues within the water system. Where microbial contamination has been identified, then the water safety group must ensure there are processes in place for appropriate remedial action following documented water safety group approved procedures so that at the handover stage, the water system is safe for patients.

In addition to this, it is also important to acknowledge that when commissioning of a new system is taking place, the commissioning team will require filling point or a connection point from an existing building, which will have been identified by the organization for which the build is being completed. The project WSG should have reviewed and agreed the processes and method statements for system filling based on the stage risk assessment and what and when this needs to be independently observed by the client's representative. Water used for filling should be from a potable drinking water supply and not a quick fill via a fire supply or other nonpotable source. When connecting into an existing system, which could potentially already be contaminated, the water delivered through connection could be introducing contaminated water into the newly commissioned and hopefully uncontaminated system. As such it is imperative that the water supplied from the connection point to the new system is safe even for the most vulnerable patients and that the water from the connection point should be sampled and tested to ensure that it meets the water regulations and is pathogen-free. In addition, the commissioning sampling plan and monitoring program should be included at the design and specification stage and agreed by the WSG between the trust and the contractor, and a timeline and criteria for handover agreed.

Modular construction

Modular construction provides some exciting moves forward in terms of the potential for safe building processes with clean room type conditions for construction and installation of whole wards/units and departments, for example. It would be a real disaster if these were then connected into a contaminated supply point. WSGs should ensure there are processes in place to ensure the connection is carried out by an approved contractor with a WIAPS qualification, which means they are trained and safe to work on a mains water connection. It is highly recommended that this stage is observed by a competent client representative.

Furthermore, the increase in modular construction on a smaller scale, such as slot in hand wash stations, shower units, also highlights a critical area where increased risk to patients is present. Companies who build modular units offsite that come prepared for installation often come with flexible hoses attached and currently have no clear process for managing and testing of the attached fixtures and pipework to ensure that they are not microbially contaminated. As with smaller components and fittings, it is likely that these will be leak tested following manufacture, those responsible for specifying and purchasing should ensure they are leak tested with air or disinfected flowing testing using a specific process, which has been independently validated. Following offsite testing and commissioning of modular units, a certificate of cleanliness should be available from the manufacturer.

There is no doubt that the water safety group, including the authorized engineers (water) and infection prevention and control specialists, needs to be aware of these limitations and needs to ensure they are involved in all stages of the design, build, and commissioning, and appropriate risk assessments are carried out at each stage and any gaps in safe water management addressed before progression to the next stage. There should be audited accounts that these processes have been carried out adequately with evidence that the water system is safe, including for high-risk patients at each stage including flow rates and pressures, biocide and temperatures monitoring, microbiological surveillance, and disinfection certification where required to demonstrate the wholesomeness of the water from the mains supply to each and every single outlet. This needs careful consideration to ensure that all parties are aware of their specific responsibilities and maintain good communication and collaboration throughout the building process, such as the construction teams being informed of what is an acceptable standard for microbiological quality at handover, so parameters are understood.

HSG 274 indicates that a water safety group is formed to undertake the commissioning (HSE, 2014). This is not usually the case as the water safety group is there to ensure there are processes documented within the water safety plan to ensure systems are designed, built, and commissioned safely with competent oversight from a client representative expert. The WSG should have sufficient expertise to audit the data and information from the commissioning team to ensure that the water delivered at the outlets is safe for it intended use and users. A set plan of works should be agreed by the water safety group including, for example, the schedule for filling the system and a systematic sampling plan to ensure that the water delivered is safe for patients. Handover should not take place if the water safety group does not get adequate assurance that the water system (and other services of course) does not pose a risk to the intended occupants.

It is imperative that during the commissioning stage, there is documented evidence that records the ongoing work, the reasons behind the decisions with an action log that is followed up by the WSG at each stage review. Such evidence should be signed off by the dedicated team and those who have been given responsibility for the built environment on the Executive team. The involvement of an authorized engineer for water may also provide valuable input to the processes.

Governance

Executive trust top management is legally accountable and responsible for health and safety on site, including water safety to prevent harm including from the transmission of waterborne infections.

The water safety group should be empowered and have the required skills and resources to manage all risks related to the water storage and use on site and its impact on patients, staff, and visitors including those vulnerable to harm from associated hazards whether of a biological, chemical, radiological, or physical nature. As individual members of staff and as a multidisciplinary group, they have a duty of care to assess the information passed to them from the design, commissioning, build, servicing, and maintenance teams. These assessments and their findings should be passed to the Trust Executive or top management who has responsibility against a set of legal requirements and governance standards. HTM 04-01 (England) provides examples of the HCAI code of practice and regulatory standards that influence and facilitate safe healthcare estates and facilities and for which Trust executives should be responsible (DHSC, 2016).

There should be a reportable mechanism by which the trust executive can assess the impact of waterborne infections over time. Only by enabling the water safety group to report near misses, abnormal microbiological results (failures), surveillance data, and waterborne transmission through an escalation process will the trust be able to understand and effectively determine the cost to patients, their families, and to the trust.

The Trust Executive must take responsibility for the overall build, installation, commissioning, delivery, and operation of a water system that supplies wholesome water that does not present a danger to high-risk patients.

Training and education

In a study by Maroldi et al., healthcare professionals believed that when undertaking hand hygiene, there was a higher efficacy of water and soap compared with alcohol.

Maroldi et al. discussed that this is outdated thinking, as the World Health Organization identified in 2009, has implemented a worldwide campaign recommending the use of alcohol hand rubs as the first option for hand hygiene in 2009 (Maroldi et al., 2017). Staff need to be aware that microorganisms such as *Clostridium difficile* will not be controlled using alcohol gel, and so hand washing in those circumstances must be available (Speight et al., 2011; Tarrant et al., 2018). This is an example where healthcare professionals may not be receiving sufficient in-service training and education for infection prevention or indeed are not following the training they have been given. For example, despite a recognition of the importance of hand washing in reducing transmission of microorganisms, compliance by health professionals is not always followed and protective equipment not always used appropriately (Pittet, 2000). As a consequence, if staff are not following the basic guidelines of hand hygiene, it is hardly surprising when they do not adhere to other infection prevention and control principles when it comes to the use of water in and around patients. It is not surprising then there needs to be a continuous training and dissemination of evidence-based knowledge backed by scientific evidence to combat the spread of waterborne infections in the healthcare setting and ensure water is being used safely in healthcare environment.

Furthermore, the lack of specific training for housekeeping/domestic workers has been of concern. Providing a supported learning environment with awareness workshops so all staff understand not just how to carry out a process but why they need to follow a specific schedule and the consequences to patient health of their role is a key to improved compliance with following specific methods. The simple a process is to follow, the more consistently it will be followed. Giving knowledge is empowering, and we have found that giving the basics of the bugs, which cause waterborne infections, is motivating, and they will question when they see poor practices, which should be encouraged, patient support staff can be the eyes and ears of infection control and estates staff if encouraged to so, faulty taps, items left in the splash zone, leaks, and unused outlets are all identified and reported much sooner if they are told what to look for. It is important though that these reports are acted on in a timely way to keep the motivation alive. Some hospitals use complicated cloth folding techniques for cleaning clinical sinks and patient handwash basins presumably as a cost-saving measure, it is highly predictable that mistakes will occur, it only takes a moment interruption or loss of concentration to forget where one is in the process and for cross-contamination to be carried from drains to sink handles. In high-risk areas, the use of multiple cloths is a far safe option, when looking at the bigger picture preventing just one case of HAI infection could be a lot less expensive than the costs to the patient, extra treatment, and longer stays as well as the associated costs of investigations, legal costs, compensation, meetings, report writing, and remedial actions not taking into account the loss of moral because a patient has been harmed while in the hospital's care.

Others who are in need of specific training in terms of hygiene practices include the estates and facilities staff, contractors, plumbers, supervisors, samplers, and engineers in fact anyone who has an influence on water quality including external service providers and auditors. It is critical that all staff are included in training that emphasizes the role of using aseptic technique and personal hygiene in preventing contamination of water fittings and the water system during building refurbishment and maintenance.

This would include the use of clean decontaminated tools, overalls, and decontamination of replacement parts before inserting them in the system as well as safe removal of potentially contaminated components they have removed especially in high-risk areas. A permit to work system can ensure that access in only permitted to those who have the necessary training and understand the risks to patients from poor practice.

Blocked drains are an important aspect of plumbing work in hospitals and so the training and understanding of the requirement to manage the risk from cross-contamination by good technique when removing contaminated fittings, components and pipework, good hand hygiene, and personal protective equipment to protect patients cannot be over emphasized.

Clinical staff

Infection, prevention, and control staff seem to be responsible for managing many facets of environmental contamination, including the water system, dealing with clinical infections, and tracking down transmission routes. They are often the go-to staff when problems arise or the water safety group is asked questions. Yet this highly specialized team of staff cannot be everywhere in the hospital, often do not having training in identifying and managing risks from the built environment or be responsible for water systems that are failing whether it is due to design, maintenance, servicing, or repair. These staff may also not be familiar with the engineering aspect of water system nor the waterborne microorganisms causing infection.

Estates and facilities

The estates and facilities staff have traditionally been deemed the responsible team when *Legionella*-associated infections occur. This is because legionellae are able to grow and proliferate within the water system itself, and national guidance until recently has focused on preventing Legionnaires' disease. As a result, the Estates team is often the first port of call when there is a case of waterborne infection and an assumption that it is the engineering at fault. Yet this is not always the case as *L. pneumophila* as other waterborne pathogens, *P. aeruginosa* and nontuberculous Mycobacteria, for example, can colonize shower heads and taps, i.e., at the periphery of the system instead of the perception of systemic colonization or at distal points such as the tank. In more recent times, the periphery has become more important as contamination has been shown to occur not only from the water system per se but also from retrograde sources external from the water systems, for example, from patient waste or even the drains (Jung et al., 2020).

Professor Kevin Kerr was instrumental in demonstrating that the route from the ward environment was fundamentally important in the facilitation of transmission of waterborne infections including *P. aeruginosa* (Kerr and Snelling, 2009). While Estates teams have a good knowledge of the engineering aspects, they also need to understand how the water system and their engineering mitigations and actions impact on the control of waterborne infections, the different modes of transmission of infection as well as the roles of others such as clinical, the role of infection prevention, and how staff and patients interact with the system. The duty holder for health and safety has a responsibility to ensure that all members of staff including estates and facilities staff are trained and have the skills necessary to carry out their role competently and have an understanding of the consequences to patient safety if there are failures. The Trust should also ensure that staff are competent to recognize and act on issues and problems appropriate to their role, which may impact on high-risk patients and escalate these to the water safety group in an appropriate timescale.

Water safety group

For all members being part of a multidisciplinary water safety group improves learning and understanding of aspects that are not part of their routine day job and a better understanding of what is realistic and achievable in terms of water system management (Chapter 28). It also builds resilience as members learn from each other for the future.

There has been much written about water safety groups in recent years (DHSC, 2016). There is no doubt that a functioning water safety group is essential for the protection of high-risk patients, staff, and visitors. However, many of those staff who are appointed to the water safety group are doing so on top of many other responsibilities. This is a particular issue where these staff feel that they are not responsible for water issues. Yet to quote Watt, "since their care was a case of everybody's job being nobody's job." Basically, everybody uses water in some way or another in the near patient environment, but everybody considers that it is everybody else's responsibility to ensure that the water is safe. Everyone needs to be concerned about water and ensure that it is safe for use and does not present a transmission route to vulnerable patients.

The water safety group may have a number of subgroups. For example, it may be prudent to introduce a water monitoring subgroup that reviews out-of-range data and can escalate these problems to the main water safety group. In addition, this may assist with the allocation of finances for strategic sampling and testing in specific areas or across the site and the implementation of mitigation strategies where increased risks have been identified.

Competence

The delivery of water systems requires a wide range of personnel with specific skills. Competence can be defined as a combination of training, skills, experience, and knowledge that a person has and their ability to apply them to perform a task safely.

It is clear that each member of staff has a very important role to play to ensure that the water is wholesome and to function effectively, the team must have individuals who understand specific issues, for example, architects and design engineers' drawings, specifications for materials and components, and the impact of poor design on risks to patients. While the role of the water safety group is described in guidance and elsewhere in this book, it is clear that a water safety group cannot contain all the skills required to deliver a hospital water system. Therefore, the Executive trust must take responsibility to ensure that where gaps are identified that staff with the required competencies are recruited not just for the completion and handover of the water system but the ongoing delivery of a safe water system for the lifetime of the building.

There also must be an understanding of finance and procurement departments in the impact of decisions made in relation to patient safety, for example, ensuring that equipment that is water-related needs to have maintenance and servicing contracts.

Identifying these competencies should be carried out at the project planning stage. There are too many roles to list and identify in this control section, but it is imperative that competency of staff is identified as a key delivery indicator of the whole project and that these competences are assessed on an ongoing basis at annual appraisals. Too often staff are identified to fulfil a competency gap without having the necessary background and skills to undertake the role, never mind being given the opportunity and resource to obtain the education and training in the built environment.

Human behavior

Human behavior is a complex area where we try to understand why individuals act in the way that they do. Regardless of education, training, and assessment, there are circumstances when we as humans will undertake tasks because that is the way we have always done them.

There are many examples of poor behavior related to water system use. One example is the use of hand hygiene stations. The sole purpose of a clinical hand wash station should be for hand hygiene and nothing else. However, Grabowski et al. found that hand washing accounted for only 4% of visits, i.e., 96% of visits were for the wrong purpose. (Grabowski et al., 2018) This included using the hand wash basin as a shelf, washing equipment, and disposal of fluids. Washing instrument trays in tap water resulted in contamination with *Stenotrophomonas maltophilia.* When the tray was used for drug preparation, either staff hands or equipment transferred the organism into a Hickman line resulting in a line infection. Controlling human behavior is a difficult endeavor, but without the necessary training, the findings of Grabowski et al. and others should not come as a surprise. (Grabowski et al., 2018) A hand wash station is often seen as a place of safety as it is linked to hand washing, a procedure that healthcare staff (as well as patients and the public) view as being the most effective means of preventing cross-infection so the perception is that the water delivered from it is clean and safe.

The role of designers and installers is more important than ever when it comes to controlling the misuse of hand hygiene stations. Manual elbow-operated taps are often placed in such a position that it is difficult or impossible to use them without using "contaminated" hands in the process. Clearly, this is a case where designers need to work with teams where the competences are able to identify the evidential transmission routes. While the components of hand-free outlets have themselves been identified as causing transmission of infections, manufacturers need to take responsibility for designing equipment that minimized the routes of exposure and transmission (Charron et al., 2015).

As such designers and manufacturers must recognize that transmission routes associated with human behavior can and should be designed out of the built environment, and that this applies to multiple aspects of water systems and involve all the staff whose actions may otherwise result in contamination through inappropriate human behavior.

Yet human behavior can also be strong driver in delivering a water system that does not result in the transmission of waterborne infections. There are number of other sectors where staff and the public are educated and informed to identify, recognize, and report risks and hazards.

As indicated earlier, the Royal National Lifeboat Institute have a saying "Respect the Water," which is related to preventing additional deaths while people are undertaking activities in and around water.

British Rail has a saying "see it, say it, sorted," which although related to identifying terrorist activity can be used to alert the British Transport Police to any issue where lives may be threatened or be at risk.

Such campaigns are aimed at changing behavior and raising awareness of issues and can easily be translated to empower healthcare staff to raise issues that impact on contamination of water systems including poor flow, infrequently used outlets, visually unclean outlets, blocked hand wash basins, blocked toilets, splashing from outlets and basins, to name but a few.

Portable wash hand basins

Where the water outlet (e.g., hand wash basin taps) supply has been found to be microbiologically unacceptable for high-risk patients, then the water safety group may consider the temporary installation of portable wash hand basins; however, these should be:

- risk assessed to determine that they have been stored appropriately, drained and with no stagnant water and are appropriate to deliver safe water for the area in which they are to be installed.
- monitored and sampled plan to ensure safe water is delivered from the outlets.
- placed on the water asset list.
- managed and maintained by suitable and competent staff.
- installed following manufacturer's instruction with a daily maintenance/decontamination programme.
- supported by contracts to ensure that the units are serviced and fit for purpose.
- seen as a temporary measure until the water system is treated and safer water is available at outlets.
- commissioned and decommissioned through procedures agreed by the WSG.

Monitoring and surveillance

The way water systems are monitored still seems to be in the dark ages. Current guidance for monitoring water temperature is based on manual monitoring, which means that there are inherent limitations in this type of approach based on human behavior and accessibility of monitoring points. Monthly checks of calorifier flow and return temperatures and hot water temperatures at sentinel and outlets and return legs on subordinate loops are recommended, as are annual cold water storage and monthly temperature checks at sentinel cold water outlets. These manual checks when carried out reflect the performance of the water system at a particular point in time and while accurate at that moment, that is as long as the equipment used to monitor them has been calibrated, they may be significantly different only a short time later.

Many healthcare buildings incorporate Building Management Systems to monitor a range of parameters including stored hot and cold water temperatures and calorifier flow and return temperatures. However, these readings are only taken at specific points in the system and are insufficient, even with manual temperature measurements to understand whether temperature control is being achieved throughout the system. Basically, there is little or no knowledge as to how the whole water system is actually performing.

Remote monitoring

In recent years, novel approaches have been developed to measure temperature measurements at multiple strategic points across water systems with minimal human intervention. By deploying temperature sensors attached to, for example, cold water tanks, calorifiers, the hot and cold water supplies through to the water outlets and showers automatic temperature measurements can be taken as frequently as required to gain an understanding of the system in both time of peak and low use, for example. In effect, such in-depth reporting enables a round-the-clock risk assessment of hot water regimens (Gavaldà et al., 2019).

Such artificial technology can be applied across whole systems and specific sentinel points to generate algorithms where the generation of data is able to identify high risks across the water system based on temperatures, usage, microbiological data, and patient infections. With data stored in a secure cloud systems means it is accessible to all with the appropriate permission even when not on site.

In terms of outlets, the gathering and analysis of accurate data can identify whether outlets are being used frequently or whether they present risk to patients because of lack of use and microbial growth within the outlet and supply pipework. Therefore, monitoring systems can be used to identify infrequently used outlets (taps and showers) and with the incorporation of automated flushing devices, the risks of water stagnation can be reduced (Macková and Peráčková, 2021). However, allowing water to be run to the drain would seem counterintuitive with sustainability targets. Alternative cold circulation systems have been devised and with the greater requirement on sustainability and carbon targets. These systems may become more popular as they retain the lower temperatures required for the control of microbial growth and would negate the requirement for flushing and as such reduced wasting such a precious resource (Macková and Peráčková, 2021).

In time, these automatically generated data will assist in the reduction of waterborne transmission by providing an understanding of the temperature, use and flow of water, biofilm measurement, and the risks that would otherwise be inherent where manual temperature monitoring is recorded. Such data will enable designers to understand exactly what is happening in a water system but also to design out redundancy in terms of water provision storage and underused outlets.

Surveillance of water and patients

More work must be carried out to link monitoring to microbiological results of the water system and clinical surveillance of infections. Each organization's water safety plan should include a monitoring and sampling plan for each water system right from the design stage through to build, commissioning and for the lifetime of the building based on risk assessment and agreed by the water safety group. Only through such a strategy can there be an understanding of the risks associated with the performance of each system.

Yet it is ironic that routine monitoring is not currently mandated, even in Scotland following relatively recent incidents (HPS, 2018).

Microbial monitoring during the commissioning stage has a vital role to play in demonstrating the general health of the water system.

In terms of clinical surveillance, a wide range of microorganisms have been associated with incidents and outbreak in healthcare buildings, and it is not feasible to monitor for all these microorganisms (National Services Scotland, 2019). However, a list of alert microorganisms enables those in the water safety team to start to build a picture of clinical infections over time and to determine whether control strategies have actually been effective. Surveillance may have to be undertaken over a number of years to generate sufficient evidence to realize that infrequent infections are being transmitted, and these data need to be reported to the responsible members of the Trust Executive.

Transmission of waterborne infections may involve the water system, water-based equipment, and contaminated water-based products, and so the group needs to be aware of the different nuances and routes of transmission to high-risk patients.

Each trust should have a policy that deals with surveillance, sampling strategies and defines water-associated incidents and outbreaks and triggers for action and reporting procedures to the trust executives on the board (National Services Scotland, 2019).

Where microorganisms on the alert list have been isolated from patients, then the water safety group needs to investigate the source of the microorganisms and what control measures are going to be implemented.

For example, *L. pneumophila* is considered the greatest risk concern for patient safety as it accounts for 98% of culture-confirmed Legionnaires' disease cases (Amemura-Maekawa et al., 2018; Beauté et al., 2020; Campese et al., 2011; CDC, 2021; ECDC, 2019). The urinary antigen test has become the mainstay for diagnosis of Legionnaires' disease and as it only detects *L. pneumophila* serogroup 1, the low rate of nonpneumophila strains (<1.5%) in clinical cases has been thought to be due to the predominant use of the urinary antigen test (PHE, 2018). However, clinical data using polymerase chain reaction have also demonstrated that 93% of PCR-diagnosed cases are a result of *L. pneumophila* (Uldum and Helbig, 2001). Nonetheless, these data do not lessen the importance of clinical surveillance in healthcare to identify non-pneumophila strains and that resources have to be implemented to identify the source to prevent further outbreaks. This has important implications within healthcare facilities. If a contamination of the water system is found with nonserogroup 1 Legionella, then alternative clinical diagnostic tests are required for their detection.

In addition, as discussed in Chapter 27, polyclonality may be a hindrance to establishing a link between clinical specimens and the environment. This results in an incident either not being recognized or the link to water and wastewater systems not being identified as the traditional method of investigating person-to-person spread assumes a single clone. In environmental outbreaks, there may be several different clones/strains present in water and wastewater systems (Halstead et al., 2021; Walker et al., 2014). Unless there is a phenotypic difference between colonies, the assumption is that the one colony picked and sent for typing is representative of all the colonies. Such issues need to be recognized and addressed by the water safety group.

New developments and technologies in hand wash stations

Relatively new technologies have also been developed for (i) the design of clinical wash hand basins and (ii) the automatic flow, temperatures, and microbial biofilm in hand hygiene stations and are being implemented in hospitals (Baillie, 2020) (Fig. 3).

These touch-free hand wash stations deliver soap, water, and hand gel automatically via a concealed tap spout and delivery tubes, hence reducing the potential for hand contact and potential retrograde contamination. Water safety groups should review manufacturer's data including peer-reviewed publication by independent groups. The water is delivered via microbial retention filters to ensure that microorganisms are not delivered via the water supply. Wireless sensors monitor the hot and cold water supply, flow, temperature, and biofilm accumulation in the tap outlet. The waste system has an onboard detection system that detects when any other fluids are poured down the waste and will produce alerts that such inappropriate use of the hand wash basin has taken place. Artificial intelligence assists in addressing human behavior issues as the use of RFID tags enables individual users to be identified, for example, tracking their handwashing habits. While individual reporting mechanisms may not be favorable among staff, this is an approach that can be used to address persistent and behaviors that are dangerous to high-risk patients.

These hand washing basins have been designed with a tubular glass section angled at 45 degrees into which the hands are inserted. As the hands are being washed, the tubular glass section contains splashes and aerosols, and any residual water flows down into the waste outlet and trap. The manufacturer's claim that the waste outlet and trap

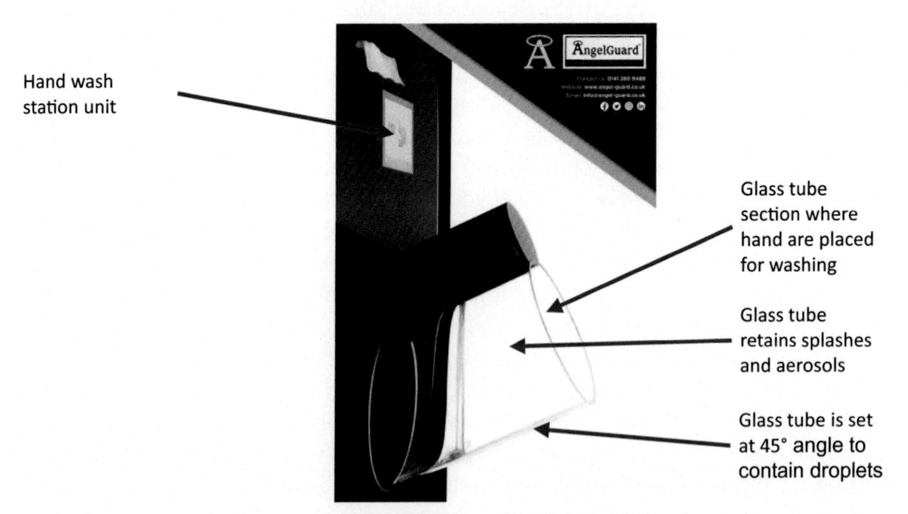

Fig. 3 Angel Guard wash hand station to contain splashes and aerosols.

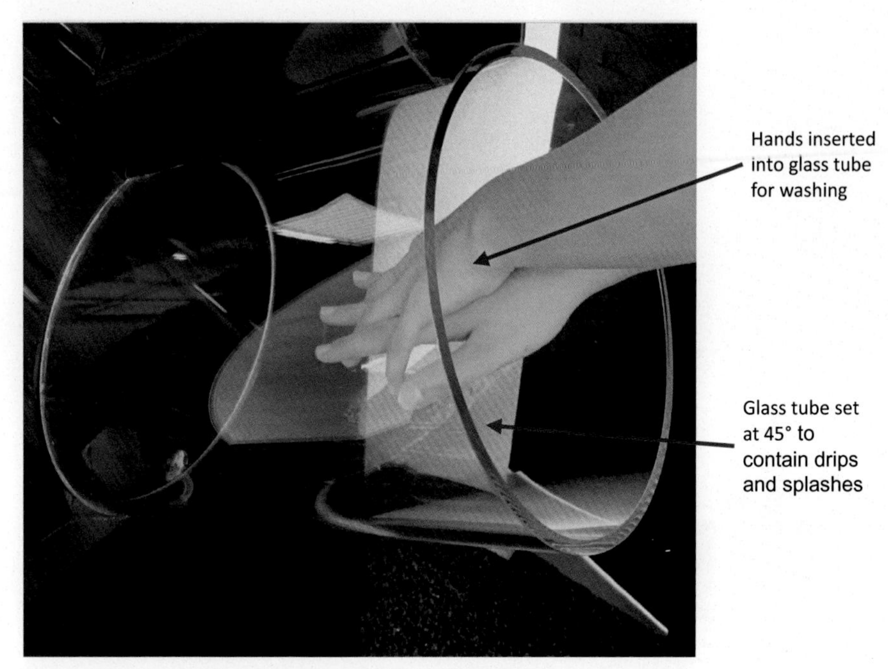

Fig. 4 Splash and aerosol containment wash hand basin.

incorporate "a special device," which "eliminates trap contents from coming back up into the basin." Hence, these devices could assist in the reduction of transmission of microorganisms being disseminated into the surrounding ward. Such newly developed systems will still require cleaning to ensure that areas, including the glass tube, are hygienically clean (Fig. 4).

Algorithms have been developed based on risk assessments that issue alerts to designated persons based on a number of criteria to enable targeted deployment of counter measures to reduce or combat the identified risk with the data securely stored in an offsite cloud system (Fig. 5).

As such these monitoring systems can enable on-site accurate measurements of designated parameters to identify and categorize levels of risk. There is the potential for these systems to replace the standard methods of manually recording temperatures and use of water outlets due to the artificial intelligence and reduction of hand wash basin acquired infection. As the water is only run when needed to wet and rinse the hand, then water savings will contribute to sustainable carbon reduction targets.

Remedial actions

All of the subject areas described above are control and remedial strategies although where followed they should result in a water system and a competent staff that would lead to a reduction in the transmission of waterborne infections. In an ideal world, a zero-tolerance approach should be adopted such that we are able to control transmission of waterborne microorganisms to a much greater extent than we currently do.

Fig. 5 Graphical representation of flow, biofilm, and temperature recording.

However, it is likely that hospital-acquired infections as a result of water transmission will continue to occur.

When infections occur, which are shown to be associated with exposure to water sources, based on the water system, the causative microorganism, route of exposure, and/or equipment involved, the water safety group and associated experts will need to decide on a range of remedial actions. More often than not, this would involve applying a range of different approaches and technologies related to the source of infection. It is important that where remedial action is required, the actions are sustainable and not akin to placing a band aid onto the contaminated system rather than treating and control the cause of the contamination even though some control strategies may be very costly and long term. However, the cost to patients, their infections, fatalities, and the impact on families is a cost far greater than that which will be borne by the hospital in bringing about these control remedies.

It should not be assumed that specialist disinfection expertise is available within the trust. It is a legal responsibility in the United Kingdom that those employing contractors and consultants should take all reasonable efforts to ensure the competency of those they employ. The water safety plan should have processes in place to ensure contractors providing water treatment options and equipment are sought who have independent validation of their product being effective preferably in peer-reviewed and published studies. Reliance on any manufacturer and supplier validation data is not demonstrating due diligence.

Where investigation has identified that local outlet fixtures and fittings are the source of microbial contamination, then then these will have to be decontaminated in situ, removed and decontaminated or replaced with new units (Garvey et al., 2017). Microbial monitoring should then be carried out to assess that the water is safe for high-risk patients.

The water safety plan should include documented procedures for the steps to take following the identification of cases of contamination. These should detail what investigations should be carried out including the extent of sampling to assess how far back in the system the contamination is and what other outlets in the vicinity are fed by the supply water by the pipework that has been identified as being contaminated. The investigation needs to differentiate between systemic localized colonization of the supply to that area where the local contamination has been identified or if the contamination is the result of retrograde contamination as a result of human behavior.

Disinfection of the sections of the water system or the entire water system may involve a range or combinations of technologies in the short and longer term. There are pros and cons associated with all of these, which need to be discussed and the strategy agreed by the water safety group.

Short-term strategies for microbiological control of the system may include:

- superheating/pasteurization (≥60°C) and continuous prolonged flushing (Gavaldà et al., 2019; Unterberg et al., 2021) carries with it a high risk of scalding so is often carried out at night with associated increased staff costs as the water needs to be pulled through each outlet. Calorifiers may not be able to produce enough hot water at this temperature. There is the potential for increasing the risk in the cold with water system with the increased heat in the hot pipework being transferred to the cold.
- hyperchlorination (Baron et al., 2020) may be effective in the short term but has the consequences of shortening the life cycle of the system components and pipework. Successive treatments may be necessary.
- point-of-use filtration (Parkinson et al., 2020) provides immediate protection but may involve altering outlet fittings to provide a fitting, which does not leak, does not breach the air gap required by the water fitting regulations to prevent backflow (in the United Kingdom) and has sufficient activity space (WRAS, 1999).

Point-of-use filters

- Point-of-use filters are often the first line of defense when contaminated water is identified at outlets as they instantly provide safe water and are in most cases considered a short-term measure (DHSC, 2016). Yet it is not unusual for filters to be installed on an almost permanent basis, without the underlying contamination being identified or even controlled. Whenever the decision is taken to install filter, the criteria for its safe removal should also be established. The instalment of point-of-use filters must be balanced against the ongoing costs and the underlying reasons for the continued microbial contamination, it must be stressed that filters are not a fit and forget policy in terms of the underlying contamination.
- filter fitting should be checked to ensure that it is not leaking.
- the longer the filter is in place, the more likelihood that the external fittings will become contaminated (Garvey et al., 2016).

- wash hand basins must have been designed and chosen located correctly in the first place to ensure that there is sufficient activity space in the basins.
- each filter must be part of the water fittings asset list to ensure that the service life of each filter is not breached.
- water sampling is still undertaken of the water outlet, with the filter in place, to ensure that s safe water is still being delivered.
- that the manufacturer's instructions are followed in terms of servicing, maintenance and hygiene.
- Investigation to the system must include the whole system including above and below the unit where there is an outlet contamination, it is very often forgotten that the building is a whole, and the water system is connected, therefore the contamination may be present in another area adjacent to the initial positive for that unit.
- cleaning staff have an approved protocol for how to clean outlets with filters and report if they see anything untoward such as leakage around the filter or their removal.

Short-term management may or may not achieve control of waterborne pathogens for the duration of the application of the technology. However, where conditions exist in the water system for growth, then as soon as the technology has been removed then microorganisms will proliferate.

Where the water system per se has been identified as failing and microbial control cannot be achieved through short-term treatments, longer-term control measures may include the use of (EPA, 2016; HSE, 2014);

- chlorine (Orsi et al., 2014)
- monochloramine (Lytle et al., 2021)
- chlorine dioxide (Vincenti et al., 2019)
- copper silver ionization (Cloutman-Green et al., 2019)
- silver hydrogen peroxide (Casini et al., 2017)
- Ozone (Carlson et al., 2020)
- ultraviolet light disinfection (Buse et al., 2022).

Chlorine

Chlorine is widely used to disinfect water supplies, and most mains water supplies will contain a low level chlorine residual (0.1–0.5 mg/L) at the entry point to the premises. However, this chlorine concentration may not be sufficient to inhibit legionella growth in the water systems. As such supplementary dosing with further chlorine may improve the control of legionella; however, it is recognized that chlorine is less effective than some other oxidizing biocides at penetrating biofilm. Chlorine efficacy is determined by the concentration, contact time, pH, temperature, presence of organic matter, and the extent of colonization of the water system.

There is a health-based guideline maximum concentration of 5.0 mg/L for total chlorine as a residual disinfectant in drinking water (Ref. WHO). However, at concentrations greater than 1.0 mg/L, the water would be unpalatable and may lead to excessive.

Maintaining a free chlorine residual of 0.5–1.0 mg/L at an outlet in a relatively clean water system will reduce biofilm in the preceding pipework and aid the control of legionella. Flushing is an important strategy as regularly flushing of the outlets will maintain a chlorine residual that can significantly control microbial contamination in pipework leading to little used outlets.

Routine inspection and maintenance of the dosing system should be carried and recorded:

Weekly—check the system operation and chemical stocks in the reservoir;

Monthly—the concentration of free chlorine at the sentinel taps should be measured to be 0.5–1.0 mg/L, and the chlorine concentration altered to the required residual at the sentinel sample points where required;

Annually—the chlorine concentration should be monitored at representative outlets throughout the distribution system with a target concentration of at least 0.5 mg/L free chlorine.

Monochloramine

The primary use of monochloramine (NH_2Cl) in water systems is to maintain a disinfectant residual in the distribution system and has been used for over 100 years and has been used more recently, particularly in the United States. Monochloramine can be formed by first adding chlorine, then ammonia, or vice versa. Often ammonia is added after chlorine has acted as a primary disinfectant for a period of time, and the resulting monochloramine is used as a residual disinfectant (EPA, 2016). The normal dosage rate for monochloramine is considered to be between 1.0 and 4.0 mg/L with 1 to 2 mg/L considered as an effective concentration for the control of planktonic cells and biofilm.

Monochloramine is effective for controlling bacterial regrowth and as is able to penetrate and controlling biofilms; however, excess ammonia can cause biofilm growth (LeChevallier et al., 1988; USEPA, 1999a). Monochloramine is more specific than chlorine, which reacts with a wider array of compounds. For similar chlorine concentrations, monochloramine has been shown to penetrate biofilm 170 times faster than free chlorine (Lee et al., 2011).

Marchesi et al. (2012) used monochloramine (1.5 and 3.0 mg/L) in an Italian hospital to treat a hot water system using continuously applied monochloramine (Marchesi et al., 2012). Prior to the trial, 97% of samples were positive for Legionella, and after dosing only 13% of samples were positive. Consequently, the authors concluded that, based on this study, continuous injection of monochloramine in a hospital hot water system has potential to control Legionella (Marchesi et al., 2012).

Potential water quality issues associated with monochloramine include corrosion, formation of disinfection by-products, and nitrification. Chloramine has also been demonstrated to result in the degradation of rubber and plastic components in a water system (Kirmeyer et al., 2004).

For most systems, the following checks should be undertaken and recorded:

Weekly—check the system operation and chemical stocks in the reservoir;

Monthly—test treated water for monochloramine at an outlet close to the point of dosing to verify the dosage rate and conversion yield;

Monthly—measure the concentration of monochloramine at the sentinel taps and adjust the dosage to the required residual at the sentinel sample points.

Chlorine dioxide

Chlorine dioxide is an oxidizing biocide/disinfectant that has been effective at controlling both legionella and biofilm growth in hot and cold water systems. Chlorine dioxide is not affected by the pH or hardness of the water.

Use of chlorine dioxide as a legionella control strategy is subject to BS EN 1267, and national conditions of use require that the combined concentration of chlorine dioxide, chlorite, and chlorate in the drinking water does not exceed 0.5 mg/L as chlorine dioxide (BS EN 12671, 2016).

Achieving and maintaining a chlorine dioxide residual (as total oxidant) of 0.1–0.5 mg/L at an outlet is usually sufficient to control legionella in the preceding pipework, although in a heavily colonized system, higher residuals may be necessary. Where a system is heavily colonized, then there will be a significant chlorine dioxide demand, and it may be time lag before a stable chlorine dioxide residual is established at the extremities of the system. As such a super disinfection with chlorine dioxide (20–50 mg/L) may be necessary, but this should only be undertaken following a detailed risk assessment, and great care should be taken to ensure the system is flushed thoroughly after cleaning.

It should be noted that excessive levels of chlorine dioxide should be avoided since they can encourage the corrosion of copper and steel pipework and high levels of chlorine dioxide can degrade certain types of polyethylene pipework particularly at elevated temperatures.

For most systems, the following checks should be undertaken and recorded:

Weekly—check the system operation and chemical stocks in the reservoir;

Monthly—test treated water for both chlorine dioxide and total oxidant/chlorite at an outlet close to the point of injection to verify the dosage rate and conversion yield;

Monthly—measure the concentration of chlorine dioxide at the sentinel taps where it should be at least 0.1 mg/L and adjust the dosage to the required residual at the sentinel sample points;

Annually—assess the chlorine dioxide and total oxidant/chlorite concentration at a representative selection of outlets throughout the distribution system to ensure the concentration is at least 0.1 mg/L chlorine dioxide.

Copper and silver ionization

Ionization is the term given to the electrolytic generation of copper and silver ions providing a continuous release of ions in water, which are generated by passing a low electrical current between copper and silver electrodes or copper and silver alloy electrodes. Copper and silver ionization has been shown to be effective at controlling legionella and can penetrate and control established biofilms.

Equipment manufacturers generally recommend copper (0.2–0.8 mg/L) and silver (0.02–0.08 mg/L) ion concentrations to control legionella. The Water Supply (Water Quality) Regulations 2001 set a standard for copper of 2 mg/L, which must not be exceeded (GOV.UK, 2001). Values of more than 0.2 mg/L copper and more than 0.02 mg/L silver are recommended at outlets to ensure effective control of legionella.

Ionization systems should be installed on the incoming mains supply before water storage treating both hot and cold water systems and after the softening system to avoid removal of some of the copper and silver ions by the water softening system resins. In hard water areas, a specific electrode evaluation and descaling procedure should be part of the program as the natural hardness will deposit on the copper and silver electrodes and reduce ionization efficiency. It should also be noted that the ionization process is pH-sensitive, and dosing levels may need increasing for pH levels greater than 7.6.

Routine inspection and maintenance should be undertaken to ensure appropriate ion concentrations are being achieved at outlets and any remedial action should be taken when necessary and recorded:

Weekly—check concentration of copper and silver ions in the water supply and install equipment capable of proportional dosing relative to flow;

Monthly—check copper and silver ion concentrations at sentinel outlets;

Annually—check the measurement of copper and silver ion concentrations at representative taps selected on a rotational basis once each year and assess the condition and cleanliness of the electrodes and the pH of the water supply.

Silver-stabilized hydrogen peroxide

Silver-stabilized hydrogen peroxide has been used for the control of legionella in water systems. Silver hydrogen peroxide solution is injected into the water system. As with any water treatment programme it should be validated to ensure it is effective in controlling legionella. Silver hydrogen peroxide should not be used in water systems supplying dialysis units.

Ozone

Ozone is an oxidizing disinfectant used in drinking water treatment (USEPA, 1999b, 2007) It is generated on-site as a gas using either air or liquid oxygen and is pumped (dissolved) into the water phase. Molecular ozone (O_3) is unstable when dissolved in water and decomposes to hydroxyl radical, which is a stronger and typically more reactive oxidizing agent than molecular ozone, which then decomposes quickly during water treatment.

Ozone can be used to oxidize iron, manganese, taste, and odor. It works by oxidizing organic matter into smaller molecules. However, ozone decays so quickly that it cannot maintain a disinfectant residual in the water system.

Ozone disinfection is a relatively complex process. Operational and maintenance demands are significantly greater than those for chlorine and chloramines (EPA, 2016).

UV disinfection

UV disinfection has been used extensively for the inactivation of pathogens in water systems.

There is not residual. Consequently, pathogens present in the water system downstream of a UV lamp, additional disinfection technologies will be required.

UV is only effective at inactivating pathogens in the water that flow through the UV lamp, and as such the impact on the waterborne bacteria is only effective at the point of application as there is no residual.

UV light systems have been developed that also achieve microbial control of the water prior to the discharge point of outlets. Though, it should be taken into consideration that the UV will not prevent retrograde contamination occurring at the exterior of the outlet.

UV reactors need to be maintained to remain effective as the lamps can be fouled by iron, manganese, calcium carbonate, or other deposits that decrease UV output. Lamps and other reactor components also need to be replaced periodically in order to maintain treatment effectiveness. Fouling of the UV lamps decreases efficacy of the UV disinfection. Liu et al. (1995) added filters to prevent scaling on UV lamps installed near the point of use in a hospital's cold and hot water systems (Liu et al., 1995). After treatment with superheat/flush and shock chlorination, and installation of filters to remove particles that foul the UV lamps, the UV intensity of the lamps remained at 100% throughout the trail, and the showers remained Legionella-free for a period of three months.

Care needs to be taken over installation of UV disinfection as the systems require routine maintenance such as cleaning and replacement of the lamps. Most smaller UV lamps will typically be rated for 8000–12,000 h of operation (1 year of continuous operation equals 8736 h).

The above list of products is not exhaustive, and a number of review, books, guidelines, and regulation documents provide far more details (Baron et al., 2020; Carlson et al., 2020; EPA, 2016; HSE, 2014; National Academies of Sciences et al., 2019). Of course, all of these technologies and strategies have been through numerous European Standard tests and trials that demonstrate that they can be effective against a range microorganisms under particular conditions in healthcare water systems. However, there is a big difference between a European Standard test, a laboratory test, laboratory models of water systems and actual water systems.

Disinfection technologies may fail to control the presence of waterborne microorganisms or may results in negative impacts corrosion, calcification of water, damage to plastics, rubbers, and plumbing components (Muzzi et al., 2020). In addition, patients will need protection from exposure to either increased temperatures or from the disinfection products or by-products, which during commission may have a prolonged presence in the system.

As discussed above, renal units require particular attention as the exposure of patients to disinfectants in renal units may lead to long-term illness and fatalities (DHSC, 2013; Rao et al., 2009).

Quite often, the underlying reason for failure may be that the products were not applied appropriately. The disinfection product needs to be in contact with the microorganisms and biofilms for a particular period of time, known as the "contact time" and needs to reach all the pipes, components, and parts that are contaminated. One off application of biocides is unlikely to achieve such stringent requirements especially when every single outlet has to be treated and flushed to achieve the target biocide concentration for the required time period. Biofilms, as discussed in the ecology section and demonstrated in numerous investigations, can be tolerant to biocides and as such will readily contaminate a supposedly safe water system.

The permutations of water types, chemicals, pipes, components, contact time, temperatures, microorganisms, and biofilms can be challenging to any control strategy and time and again such supposedly fool proof remedial measures fail. This can either be through to inexperience of the company/team contracted to undertake the remediation, lack of understanding the water system size, lack of schematics, number of and type of outlets, water chemistry, or the extent of contamination.

Similar technology as described above for monitoring temperatures needs to be applied to disinfection and biocides with a range of sensors placed through the water system to ensure that the appropriate concentrations are held for sufficient contact times and that this occurs across every pipe section and outlet. As discussed above, the requirements of intensive labor recording designated outlets manually using external biocide monitors should be designated to history.

While some disinfection companies may advise that the hot water temperature can be decreased to flows of less than 60°C while using their products, the decision needs to be taken at corporate level based on a corporate risk assessment with the trust executive understanding all the potential consequences (HSE, 2014). The trust executive needs to understand there is no single panacea for safe water management, and there is no single product that can compensate for a badly designed and poorly maintained water system. In hospitals where there are large system complicated systems and it is inevitable that not all outlets will reach target temperatures all of the time, a biocide should be seen as part of the multiple barrier approach as advocated by WHO (Guidelines for drinking water quality 4th edition, water safety in buildings, BS 8680). At some point, it is likely that the control measures may fail, the biocide is the back-up for when the temperature control fails. If the control temperatures cannot be achieved at the outlet for whatever reason, and if the water temperature is below 55°C, then it is likely that microbial pathogens will proliferate putting patients, staff, and visitors at risk.

Summary

Water systems should be managed so they are safe for all uses and all users. This should be considered at the design stage with a design water safety plan, which considers the safe management of all systems and associated equipment, potential uses taking account of how each system (and associated equipment) will be used, maintained, cleaned, and disinfected. The susceptibility to harm from all potential hazards associated with water including biological, chemical, physical, and where relevant radiological hazards. Water delivered to healthcare buildings while safe for the general population may not be safe for those more susceptible to infections particularly those who are immunocompromised as a result of their illness or treatment. For some patients exposure to distributed water may be harmful safe, and some uses of water may need additional treatment to ensure it is safe for its intended use. Even when water is safe at the point of entry for most patients conditions within the system or intermittent usage may mean that by the time it is delivered at the point of use, it poses a risk not only to patients but staff and visitors too.

Control strategies should not be considered as a bolt on technology that is applied to a water system once harm, such as an outbreak occurs, due to the growth of waterborne opportunistic pathogens. There are multiple aspects of control to consider to ensure safe water systems in healthcare buildings including the human behavior of the staff that work in the built environment. However, designers should be designing out water system risks, taking account of how water is to be used to ensure that at handover the building and its water systems do not pose a risk to patient health irrespective of the susceptibility of those exposed. This includes ensuring the design of the wastewater system is given equal attention. The management and oversight of the build and commissioning stages are also important in maintaining a hygienic water system, and this must be managed such that at each point of use, safe water is delivered to the outlets and medical equipment that uses water. Where independently validated as being effective new technologies should be adopted following a robust and documented assessment by the multidisciplinary WSG. The use of artificial intelligence for remote monitoring will start to ensure that the parameters set by guidance documents are actually achieved across the whole water network and not just at the outlets. Microbial surveillance (both environmental and clinical) is a strategy that is able to determine whether these control strategies that are being implemented are actually working. There is sufficient evidence in the public domain of the danger from water systems to patients especially those more susceptible because of illness or treatment. The knowledge already exists, and there are a wide range of guidance documents and peer-reviewed publications. While day-to-day responsibility can be delegated to the WSG, the overall responsibility of preventing transmission of waterborne infections in healthcare remains as the legal responsibility of the Duty holder, usually the CEO and Board of Directors. For this reason, it is essential that they and all other staff involved in ensuring the safety of patients, staff, and visitors are trained and competent to carry out their role in keeping water safe for all uses and all users on site.

Risk assessing criteria for controlling microorganisms in water systems

Criteria	Yes	No
Is there a holistic WSP based on risk assessments to BS 8580—parts 1 and 2 and does it consider all potential hazards, hazardous events, and all systems and equipment to which patients, staff, and visitors could be exposed?		
Has a gap analysis been carried out (BS 8680) to consider what is present and what is missing to ensure the safe management of all water on-site, including aspects of governance and supporting programs, e.g., training, audit, monitoring, sampling, and surveillance?		
Has competence of all healthcare staff in the clinical area as well as cleaners and those in the estates and facilities departments been demonstrated, recorded ongoing refresher courses to ensure that they understand the risks from water?		

Criteria	Yes	No
Are there processes within the WSP to ensure all external contractors/ consultants are competent including design teams, building contractors, subcontractors, service providers?		
Is there sufficient review and sign-off at each gateway stage of a project for both capital and refurbishment projects, by the designated project water safety group (design/build/commissioning) backed by evidential expertise and sampling results, where relevant, to ensure that water is safe for designated patient groups and that alert organisms are absent from the water?		
Are the as-fitted drawings accurate and up to date and any reflect major changes and managed accordingly to ensure that changes are recorded?		
Has a senior member of the executive/a director been nominated to accept responsibility water safety?		
Has the role of human behavior been addressed and training given as required across all relevant disciplines?		
Has appropriate training been provided on wash hand station use and abuse including to patients and visitors in high-risk wards/units?		
Has sufficient training been provided on the transmission risks of waterborne infections?		
Are staff empowered to act as whistle-blowers where they have identified water-associated activities that endanger patients?		
Is there sufficient and robust monitoring technology (preferably remote) to risk assess the thermal and where appropriate chemical control and usage of water to reduce the risk of microbial proliferation?		
Is there appropriate surveillance of clinical infections associated with waterborne microorganisms that issues alerts when numbers are exceeded and systems are place to investigate and carry out environmental investigations?		
Has an alert list of waterborne pathogens been identified?		
is there an up-to-date asset register, which includes all relevant components (including pumps, outlets, showers, TMVs, equipment, and drains of the water system and appropriate risk assessment, servicing and maintenance, and control regimens been identified?		
Have preventative control strategies been designed from the outset prior to occupation?		
Where control strategies have been predesigned, can they actually be implemented, for example, can hand washing be carried out once point-of-use filters have been fitted to hand wash basis?		
Has the expertise required to implement, monitor, and audit control strategies been identified?		
Where companies are being employed to implement control strategies, do they have the competence to understand what they need to do and how to achieve microbial control of the water system?		
Has the provision of water-free patient care been considered to reduce waterborne infections in high-risk wards/units?		
Have water risk assessments been undertaken on all systems and equipment and when there are changes that affect their validity?		

Continued

Criteria	Yes	No
Where appropriate can single-use equipment be used and that where possible, equipment is single patient use?		
Where equipment that has been in contact with patients is reused has an appropriate validated decontamination procedure been implemented?		
Has a clinical risk assessment been undertaken of the susceptibility of patients to waterborne pathogens?		
Is there an environmental sampling strategy in the WSP, including for following the identification of cases or an outbreak—what samples, how they will be taken, microorganisms to be tested, what will be done with the results?		
Are all historical water sampling results available for comparison to the current outbreak?		
Where outbreaks have been identified have all relevant staff including senior management been informed, risk assessments and control schemes reviewed and updated, and where appropriate retraining implemented and competences reviewed?		
Where control strategies have been implemented, has the water system been reviewed to identify ongoing weaknesses (faults, flaws, debris, biofilm, temperatures, and flow problems) and gaps in the control strategies that resulted in the microbial growth?		

Guidance documents for controlling waterborne pathogens in water systems

HSG 274 Legionnaires' disease—Technical guidance Part 2: The control of legionella bacteria in hot and cold water systems (HSE, 2014).

HTM 04-01 Safe water in healthcare. UK (DHSC, 2016).

References

Alangaden, G.J., 2011. Nosocomial fungal infections: epidemiology, infection control, and prevention. Infect. Dis. Clin. N. Am. 25, 201–225. https://doi.org/10.1016/j.idc.2010.11.003.

Amemura-Maekawa, J., Kura, F., Chida, K., Ohya, H., Kanatani, J., Isobe, J., Tanaka, S., Nakajima, H., Hiratsuka, T., Yoshino, S., Sakata, M., Murai, M., Ohnishi, M., Working Group for Legionella in Japan, 2018. *Legionella pneumophila* and other *Legionella* species isolated from Legionellosis patients in Japan between 2008 and 2016. Appl. Environ. Microbiol. 84.

Anaissie, E.J., Penzak, S.R., Dignani, M.C., 2002. The hospital water supply as a source of nosocomial infections: a plea for action. Arch. Intern. Med. 162, 1483–1492.

Aranega-Bou, P., George, R.P., Verlander, N.Q., Paton, S., Bennett, A., Moore, G., Aiken, Z., Akinremi, O., Ali, A., Cawthorne, J., Cleary, P., Crook, D.W., Decraene, V., Dodgson, A., Doumith, M., Ellington, M., Eyre, D.W., George, R.P., Grimshaw, J., Guiver, M., Hill, R., Hopkins, K., Jones, R., Lenney, C., Mathers, A.J., McEwan, A., Moore, G., Neilson, M., Neilson, S., Peto, T.E.A., Phan, H.T.T., Regan, M., Seale, A.C., Stoesser, N., Turner-Gardner, J., Watts, V., Walker, J., Sarah Walker, A., Wyllie, D., Welfare, W., Woodford, N., 2018. Carbapenem-resistant *Enterobacteriaceae* dispersal from sinks is linked to drain position and drainage ratesin a laboratory model system. J. Hosp. Infect. https://doi.org/10.1016/j.jhin.2018.12.007.

Ayliffe, G.A., 2000. Evidence-based practices in infection control. Br. J. Infect. Control. 1, 5–9. https://doi.org/10.1177/175717740000100402.

Baillie, J., 2020. Radical washbasin and new 'pipe within a pipe' system. Health Estates J. 74 (2), 1–6.

Baron, J.L., Morris, L., Stout, J.E., 2020. Control of Legionella in hospital potable water systems. In: Decontamination in Hospitals and Healthcare. Elsevier, pp. 71–100.

Beauté, J., Plachouras, D., Sandin, S., Giesecke, J., Sparén, P., 2020. Healthcare-associated Legionnaires' disease, Europe, 2008–2017. Emerg. Infect. Dis. 26, 2309–2318.

Bédard, E., Benoit, M.-È., Bourdin, T., Charron, D., DeLisle, G., Daraiche, S., Gravel, S., Robert, E., Constant, P., Déziel, E., Quach, C., Prévost, M., 2021. Implementation of a low-cost method to reduce bacterial load in patient-room sink drains. Antimicrobial Stewardship Healthc Epidemiol. 1, s21–s22. https://doi.org/10.1017/ash.2021.39.

Black, H.J., Holt, E.J., Kitson, K., Maloney, M.H., Phillips, D., 1979. Contaminated hospital water supplies. Br. Med. J. 1, 1564–1565.

BS EN 12671, E, 2016. BS EN 12671:2016 Chemicals used for treatment of water intended for human consumption. Chlorine dioxide generated in situ https://www.en-standard.eu.

BSI, 2020. BS 8680—Water Quality. Water Safety Plans. Code of practice. https://shop.bsigroup.com/ProductDetail?pid=000000000030364472. (Accessed 5 January 2021).

Buse, H.Y., Hall, J.S., Hunter, G.L., Goodrich, J.A., 2022. Differences in UV-C LED inactivation of *Legionella pneumophila* serogroups in drinking water. Microorganisms 10, 352. https://doi.org/10.3390/microorganisms10020352.

Campese, C., Bitar, D., Jarraud, S., Maine, C., Forey, F., Etienne, J., Desenclos, J.C., Saura, C., Che, D., 2011. Progress in the surveillance and control of Legionella infection in France, 1998–2008. Int. J. Infect. Dis. 15, e30–e37. https://doi.org/10.1016/j.ijid.2010.09.007.

Carlson, K.M., Boczek, L.A., Chae, S., Ryu, H., 2020. Legionellosis and recent advances in technologies for Legionella control in premise plumbing systems: A review. Water 12, 676. https://doi.org/10.3390/w12030676.

Casini, B., Aquino, F., Totaro, M., Miccoli, M., Galli, I., Manfredini, L., Giustarini, C., Costa, A.L., Tuvo, B., Valentini, P., Privitera, G., Baggiani, A., 2017. Application of hydrogen peroxide as an innovative method of treatment for Legionella control in a hospital water network. Pathogens 6, 15. https://doi.org/10.3390/pathogens6020015.

CDC, 2021. National Outbreak Reporting System (NORS) Dashboard.

CDC, 2022. NIOSH Hierarchy of Controls | NIOSH | CDC. [WWW Document]. URL: https://www.cdc.gov/niosh/topics/hierarchy/default.html. (Accessed 18 August 2022).

Charron, D., Bédard, E., Lalancette, C., Laferrière, C., Prévost, M., 2015. Impact of electronic faucets and water quality on the occurrence of *Pseudomonas aeruginosa* in water: a multi-hospital study. Infect. Control Hosp. Epidemiol. 36, 311–319. https://doi.org/10.1017/ice.2014.46.

Cloutman-Green, E., Barbosa, V.L., Jimenez, D., Wong, D., Dunn, H., Needham, B., Ciric, L., Hartley, J.C., 2019. Controlling, *Legionella pneumophila* in water systems at reduced hot water temperatures with copper and silver ionization. Am. J. Infect. Control 47, 761–766. https://doi.org/10.1016/j.ajic.2018.12.005.

Coleman, D.C., Deasy, E.C., Moloney, E.M., Swan, J.S., O'Donnell, M.J., 2020. 7—Decontamination of hand washbasins and traps in hospitals. In: Walker, J. (Ed.), Decontamination in Hospitals and Healthcare, second ed. Woodhead Publishing, pp. 135–161, https://doi.org/10.1016/B978-0-08-102565-9.00007-8. Woodhead Publishing Series in Biomaterials.

De Geyter, D., Blommaert, L., Verbraeken, N., Sevenois, M., Huyghens, L., Martini, H., Covens, L., Piérard, D., Wybo, I., 2017. The sink as a potential source of transmission of carbapenemase-producing *Enterobacteriaceae* in the intensive care unit. Antimicrob. Resist. Infect. Control 6, 24. https://doi.org/10.1186/s13756-017-0182-3.

De Geyter, D., Vanstokstraeten, R., Crombé, F., Tommassen, J., Wybo, I., Piérard, D., 2021. Sink drains as reservoirs of VIM-2 metallo-β-lactamase-producing *Pseudomonas aeruginosa* in a Belgian intensive care unit: relation to patients investigated by whole-genome sequencing. J. Hosp. Infect. 115, 75–82. https://doi.org/10.1016/j.jhin.2021.05.010.

Decraene, V., Phan, H.T.T., George, R., Wyllie, D.H., Akinremi, O., Aiken, Z., Cleary, P., Dodgson, A., Pankhurst, L., Crook, D.W., Lenney, C., Walker, A.S., Woodford, N., Sebra, R., Fath-Ordoubadi, F., Mathers, A.J., Seale, A.C., Guiver, M., McEwan, A., Watts, V., Welfare, W., Stoesser, N., Cawthorne, J., Group, the T.I, 2018. A large, refractory nosocomial outbreak of *Klebsiella pneumoniae* carbapenemase-producing *Escherichia coli* demonstrates carbapenemase gene outbreaks involving sink sites require novel approaches to infection control. Antimicrob. Agents Chemother. 62. https://doi.org/10.1128/AAC.01689-18.

DHSC, 2013. Health Building Note 07-02: Main renal unit.

DHSC, 2016. HTM 04-01: Safe Water in Healthcare Premises.

DWI, 2021. Legislation, Drinking Water Inspectorate. DWI.

ECDC, 2019. Legionnaires' Disease—Annual Epidemiological Report for 2019. European Centre for Disease Prevention and Control.

EPA, 2016. Technologies for Legionella Control in Premise Plumbing Systems: Scientific Literature Review. https://www.epa.gov/ground-water-and-drinking-water/technologies-legionella-control-premise-plumbing-systems. 139.

EU, 2020. Directive (EU) 2020/2184 of the European Parliament and of the Council of 16 December 2020 on the Quality of Water Intended for Human Consumption. https://eur-lex.europa.eu/eli/dir/2020/2184/oj.

Exner, M., Kramer, A., Lajoie, L., Gebel, J., Engelhart, S., Hartemann, P., 2005. Prevention and control of health care–associated waterborne infections in health care facilities. Am. J. Infect. Control 33, S26–S40. https://doi.org/10.1016/j.ajic.2005.04.002.

Garvey, Bradley, C.W., Jumaa, P., 2016. The risks of contamination from tap end filters. J. Hosp. Infect. 94, 282–283. https://doi.org/10.1016/j.jhin.2016.08.006.

Garvey, M.I., Bradley, C.W., Wilkinson, M.A.C., Bradley, C., Holden, E., 2017. Engineering waterborne *Pseudomonas aeruginosa* out of a critical care unit. Int. J. Hyg. Environ. Health 220, 1014–1019. https://doi.org/10.1016/j.ijheh.2017.05.011.

Gavaldà, L., Garcia-Nuñez, M., Quero, S., Gutierrez-Milla, C., Sabrià, M., 2019. Role of hot water temperature and water system use on Legionella control in a tertiary hospital: an 8-year longitudinal study. Water Res. 149, 460–466. https://doi.org/10.1016/j.watres.2018.11.032.

GOV.UK, 2001. The Water Supply (Water Quality) Regulations 2001.

Grabowski, M., Lobo, J.M., Gunnell, B., Enfield, K., Carpenter, R., Barnes, L., Mathers, A.J., 2018. Characterizations of handwashing sink activities in a single hospital medical intensive care unit. J. Hosp. Infect. 100, e115–e122.

Halstead, F.D., Quick, J., Niebel, M.O., Garvey, M., Cumley, N., Smith, R., Neal, T., Roberts, P., Hardy, K., Shabir, S., Walker, J.T., Hawkey, P., Loman, N.J., 2021. *Pseudomonas aeruginosa* infection in augmented care: the molecular ecology and transmission dynamics in four large UK hospitals. J. Hosp. Infect. https://doi.org/10.1016/j.jhin.2021.01.020.

Hargreaves, J., Shireley, L., Hansen, S., Bren, V., Fillipi, G., Lacher, C., Esslinger, V., Watne, T., 2001. Bacterial contamination associated with electronic faucets: a new risk for healthcare facilities. Infect. Control Hosp. Epidemiol. 22, 202–205. https://doi.org/10.1086/501889.

Hopman, J., Tostmann, A., Wertheim, H., Bos, M., Kolwijck, E., Akkermans, R., Sturm, P., Voss, A., Pickkers, P., vd Hoeven, H., 2017. Reduced rate of intensive care unit acquired Gram-negative bacilli after removal of sinks and introduction of 'water-free' patient care. Antimicrob. Resist. Infect. Control 6, 59. https://doi.org/10.1186/s13756-017-0213-0.

Hopman, J., Donskey, C.J., Boszczowski, I., Alfa, M.J., 2018. Multisite evaluation of environmental cleanliness of high-touch surfaces in intensive care unit patient rooms. Am. J. Infect. Control 46, 1198–1200. https://doi.org/10.1016/j.ajic.2018.03.031.

HPS, 2018. Summary of incident and findings of the NHS Greater Glasgow and Clyde: Queen Elizabeth University Hospital/Royal Hospital for Children water contamination incident and recommendations for NHS Scotland. [WWW Document]. URL: https://www.gov.scot/binaries/content/documents/govscot/publications/factsheet/2019/02/qe-university-hospital-royal-hospital-children-water-incident/documents/queen-elizabeth-university-hospital-royal-hospital-for-chidren-water-contamination-incident-hps-report/queen-elizabeth-university-hospital-royal-hospital-for-chidren-water-contamination-incident-hps-report/govscot%3Adocument. (Accessed 17 April 2019).

HSE, 2014. HSG 274 Legionnaires' Disease—Technical Guidance Part 2: The Control of Legionella Bacteria in Hot and Cold Water Systems Technical Guidance. http://www.hse.gov.uk/pubns/books/hsg274.htm. (Accessed 5 January 2021).

HSE, 2022. Hierarchy of Control; What it is & How it Works—HSEWatch.

Hunter, C.A., Ensign, P.R., 1947. An Epidemic of Diarrhoea in a New-born Nursery Caused by *Pseudomonas aeruginosa*. 4.

Inkster, T., Peters, C., Wafer, T., Holloway, D., Makin, T., 2021. Investigation and control of an outbreak due to a contaminated hospital water system, identified following a rare case of *Cupriavidus pauculus* bacteraemia. J. Hosp. Infect. https://doi.org/10.1016/j.jhin.2021.02.001.

Jamal, A., Brown, K.A., Katz, K., Johnstone, J., Muller, M.P., Allen, V., Borgia, S., Boyd, D.A., Ciccotelli, W., Delibasic, K., Fisman, D., Leis, J., Li, A., Mataseje, L., Mehta, M., Mulvey, M., Ng, W., Pantelidis, R., Paterson, A., McGeer, A., 2019. Risk factors for contamination with carbapenemase-producing *Enterobacteriales* (CPE) in exposed hospital drains in Ontario, Canada. Open Forum Infect. Dis. 6, S441. https://doi.org/10.1093/ofid/ofz360.1091.

Jansz, J., 2011. Theories and knowledge about sick building syndrome. In: Abdul-Wahab, S.A. (Ed.), Sick Building Syndrome: In Public Buildings and Workplaces. Springer, Berlin, Heidelberg, pp. 25–58, https://doi.org/10.1007/978-3-642-17919-8_2.

Jung, J., Choi, H.-S., Lee, J.-Y., Ryu, S.H., Kim, S.-K., Hong, M.J., Kwak, S.H., Kim, H.J., Lee, M.-S., Sung, H., Kim, M.-N., Kim, S.-H., 2020. Outbreak of carbapenemase-producing *Enterobacteriaceae* associated with a contaminated water dispenser and sink drains in the cardiology units of a Korean hospital. J. Hosp. Infect. 104, 476–483. https://doi.org/10.1016/j.jhin.2019.11.015.

Kemper, 2022. Cold Water Circulation System. https://www.kemper-uk.com/building-technology/product-information/khs-drinking-water-hygiene/cold-water-circulation-khs-coolflow/?L=0.

Kenters, N., Gottlieb, T., Hopman, J., Mehtar, S., Schweizer, M.L., Tartari, E., Huijskens, E.G.W., Voss, A., 2018. An international survey of cleaning and disinfection practices in the healthcare environment. J. Hosp. Infect. 100, 236–241. https://doi.org/10.1016/j.jhin.2018.05.008.

Kerr, K.G., Snelling, A.M., 2009. *Pseudomonas aeruginosa*: a formidable and ever-present adversary. J. Hosp. Infect. Proc. Lancet Conf. Healthc.-Assoc. Infect. 73, 338–344. https://doi.org/10.1016/j.jhin.2009.04.020.

Khalsa, K., Smith, A., Morrison, P., Shaw, D., Peat, M., Howard, P., Hamilton, K., Stewart, A., 2014. Contamination of a purified water system by *Aspergillus fumigatus* in a new endoscopy reprocessing unit. Am. J. Infect. Control 42, 1337–1339. https://doi.org/10.1016/j.ajic.2014.08.008.

Kinsey, C.B., Koirala, S., Solomon, B., Rosenberg, J., Robinson, B.F., Neri, A., Halpin, A.L., Arduino, M.J., Moulton-Meissner, H., Noble-Wang, J., Chea, N., Gould, C.V., 2017. *Pseudomonas aeruginosa* outbreak in a neonatal intensive care unit attributed to hospital tap water. Infect. Control Hosp. Epidemiol. 38, 801–808. https://doi.org/10.1017/ice.2017.87.

Kirmeyer, G.K., et al., 2004. Optimizing Chloramine Treatment. second ed., The Water Research Foundation.

Kohn, J., 1967. *Pseudomonas* infection in hospital. Br. Med. J. 4, 548. https://doi.org/10.1136/bmj.4.5578.548.

Kotay, S.M., Parikh, H.I., Barry, K., Gweon, H.S., Guilford, W., Carroll, J., Mathers, A.J., 2020. Nutrients influence the dynamics of *Klebsiella pneumoniae* carbapenemase producing enterobacterales in transplanted hospital sinks. Water Res. 176, 115707. https://doi.org/10.1016/j.watres.2020.115707.

Larson, R., 2020. Water law and the response to COVID-19. Water Int. 45, 716–721. https://doi.org/10.1080/02508060.2020.1835422.

LeChevallier, M.W., Cawthon, C.D., Lee, R.G., 1988. Inactivation of biofilm bacteria. Appl. Environ. Microbiol. 54, 2492–2499. https://doi.org/10.1128/aem.54.10.2492-2499.1988.

Lee, W.Y., Wahman, D.G., Bishop, P.L., Pressman, J.G., 2011. Free chlorine and monochloramine application to nitrifying biofilm: comparison of biofilm penetration, activity, and viability. Environ. Sci. Technol. 45 (4), 1412–1419. https://doi.org/10.1021/es1035305 (Epub 2011 Jan 12).

Liu, Z., Stout, J.E., Tedesco, L., Boldin, M., Hwang, C., Yu, V.L., 1995. Efficacy of ultraviolet light in preventing Legionella colonization of a hospital water distribution system. Water Res. 29, 2275–2280. https://doi.org/10.1016/0043-1354(95)00048-P.

Lytle, D.A., Pfaller, S., Muhlen, C., Struewing, I., Triantafyllidou, S., White, C., Hayes, S., King, D., Lu, J., 2021. A comprehensive evaluation of monochloramine disinfection on water quality, Legionella and other important microorganisms in a hospital. Water Res. 189, 116656. https://doi.org/10.1016/j.watres.2020.116656.

Macková, D., Peráčková, J., 2021. Ensuring the required potable water temperature in water pipeline inside buildings. Period. Polytech. Mech. Eng. 65, 345–353. https://doi.org/10.3311/PPme.18205.

Marchesi, I., Cencetti, S., Marchegiano, P., Frezza, G., Borella, P., Bargellini, A., 2012. Control of Legionella contamination in a hospital water distribution system by monochloramine. Am. J. Infect. Control 40, 279–281. https://doi.org/10.1016/j.ajic.2011.03.008.

Marek, A., Smith, A., Peat, M., Connell, A., Gillespie, I., Morrison, P., Hamilton, A., Shaw, D., Stewart, A., Hamilton, K., Smith, I., Mead, A., Howard, P., Ingle, D., 2014. Endoscopy supply water and final rinse testing: five years of experience. J. Hosp. Infect. 88, 207–212. https://doi.org/10.1016/j.jhin.2014.09.004.

Maroldi, M.A.C., Felix, A.M.d.S., Dias, A.A.L., Kawagoe, J.Y., Padoveze, M.C., Ferreira, S.A., Zem-Mascarenhas, S.H., Timmons, S., Figueiredo, R.M., 2017. Adherence to precautions for preventing the transmission of microorganisms in primary health care: a qualitative study. BMC Nurs. 16, 49. https://doi.org/10.1186/s12912-017-0245-z.

Moritz, M.M., Flemming, H.-C., Wingender, J., 2010. Integration of *Pseudomonas aeruginosa* and *Legionella pneumophila* in drinking water biofilms grown on domestic plumbing materials. Int. J. Hyg. Environ. Health 213, 190–197. https://doi.org/10.1016/j.ijheh.2010.05.003. First Ph.D. student seminar on "Water & Health" organized within the "Cannes Water Symposium", Cannes, July 2009.

Muzzi, A., Cutti, S., Bonadeo, E., Lodola, L., Monzillo, V., Corbella, M., Scudeller, L., Novelli, V., Marena, C., 2020. Prevention of nosocomial legionellosis by best water management: comparison of three decontamination methods. J. Hosp. Infect. 105, 766–772. https://doi.org/10.1016/j.jhin.2020.05.002.

National Academies of Sciences, Engineering, and Medicine, Health and Medicine Division, Division on Earth and Life Studies, Board on Population Health and Public Health Practice, Board on Life Sciences, Water Science and Technology Board, Committee on Management of Legionella in Water Systems, 2019. Regulations and Guidelines on Legionella Control in Water Systems, Management of Legionella in Water Systems. National Academies Press (US).

National Services Scotland, 2019. Prevention and Management of Healthcare Water-associated Infection Incidents/Outbreaks. https://www.nipcm.hps.scot.nhs.uk/media/1680/2019-08-water-incidents-info-sheet-v1.pdf.

NHS England, n.d. HBN Medicines Management Health Building Note 14-01.

NHS Scotland, 2019. Guidance for the Interpretation and Clinical Management of Endoscopy Final Rinse Water. V1.0. National Services Scotland.

Orsi, G.B., Vitali, M., Marinelli, L., Ciorba, V., Tufi, D., Del Cimmuto, A., Ursillo, P., Fabiani, M., De Santis, S., Protano, C., Marzuillo, C., De Giusti, M., 2014. Legionella control in the water system of antiquated hospital buildings by shock and continuous hyperchlorination: 5years experience. BMC Infect. Dis. 14, 394. https://doi.org/10.1186/1471-2334-14-394.

Parkes, L.O., Hota, S.S., 2018. Sink-related outbreaks and mitigation strategies in healthcare facilities. Curr. Infect. Dis. Rep. 20, 42. https://doi.org/10.1007/s11908-018-0648-3.

Parkinson, J., Baron, J.L., Hall, B., Bos, H., Racine, P., Wagener, M.M., Stout, J.E., 2020. Point-of-use filters for prevention of health care–acquired Legionnaires' disease: Field evaluation of a new filter product and literature review. Am. J. Infect. Control 48, 132–138. https://doi.org/10.1016/j.ajic.2019.09.006.

Perkins, K.M., Reddy, S.C., Fagan, R., Arduino, M.J., Perz, J.F., 2019. Investigation of healthcare infection risks from water-related organisms: Summary of CDC consultations, 2014–2017. Infect. Control Hosp. Epidemiol. 40, 621–626. https://doi.org/10.1017/ice.2019.60.

PHE, 2018. Legionnaires' Disease in Residents of England and Wales–2016.

Pittet, D., 2000. Improving compliance with hand hygiene in hospitals. Infect. Control Hosp. Epidemiol. 21, 381–386. https://doi.org/10.1086/501777.

Rao, C.Y., Pachucki, C., Cali, S., Santhiraj, M., Krankoski, K.L.K., Noble-Wang, J.A., Leehey, D., Popli, S., Brandt, M.E., Lindsley, M.D., Fridkin, S.K., Arduino, M.J., 2009. Contaminated product water as the source of *Phialemonium curvatum* bloodstream infection among patients undergoing hemodialysis. Infect. Control Hosp. Epidemiol. 30, 840–847. https://doi.org/10.1086/605324.

Regev-Yochay, G., Smollan, G., Tal, I., Zade, N.P., Haviv, Y., Nudelman, V., Gal-Mor, O., Jaber, H., Zimlichman, E., Keller, N., Rahav, G., 2018. Sink traps as the source of transmission of OXA-48–producing *Serratia marcescens* in an intensive care unit. Infect. Control Hosp. Epidemiol. 39, 1307–1315. https://doi.org/10.1017/ice.2018.235.

Rogers, J., Dowsett, A.B., Dennis, P.J., Lee, J.V., Keevil, C.W., 1994. Influence of plumbing materials on biofilm formation and growth of *Legionella pneumophila* in potable water systems. Appl. Environ. Microbiol. 60, 1842–1851. https://doi.org/10.1128/aem.60.6.1842-1851.1994.

SCA, n.d. Water Quality, Epidemiology and Public Health https://standingcommitteeofanalysts.co.uk/microbiology-working-group/.

Smolders, D., Hendriks, B., Rogiers, P., Mul, M., Gordts, B., 2019. Acetic acid as a decontamination method for ICU sink drains colonized by carbapenemase-producing *Enterobacteriaceae* and its effect on CPE infections. J. Hosp. Infect. 102, 82–88. https://doi.org/10.1016/j.jhin.2018.12.009.

Speight, S., Moy, A., Macken, S., Chitnis, R., Hoffman, P.N., Davies, A., Bennett, A., Walker, J.T., 2011. Evaluation of the sporicidal activity of different chemical disinfectants used in hospitals against *Clostridium difficile*. J. Hosp. Infect. 79, 18–22. https://doi.org/10.1016/j.jhin.2011.05.016.

Stein, J.E., Heiss, K., 2015. The Swiss cheese model of adverse event occurrence—Closing the holes. In: Seminars in Pediatric Surgery, SI: Improving Pediatric Surgery Quality and Outcomes in the 21st Century 24, 278–282., https://doi.org/10.1053/j.sempedsurg.2015.08.003.

Tang, L., Tadros, M., Matukas, L., Taggart, L., Muller, M., 2020. Sink and drain monitoring and decontamination protocol for carbapenemase-producing *Enterobacteriaceae* (CPE). Am. J. Infect. Control 48, S17. https://doi.org/10.1016/j.ajic.2020.06.132.

Tarrant, J., Jenkins, R.O., Laird, K.T., 2018. From ward to washer: the survival of *Clostridium difficile* spores on hospital bed sheets through a commercial UK NHS healthcare laundry process. Infect. Control Hosp. Epidemiol. 39, 1406–1411. https://doi.org/10.1017/ice.2018.255.

Thet, K., Pelobello, M.L.F., Das, M., Alhaji, M.M., Chong, V.H., Khalil, M.A.M., Chinniah, T., Tan, J., 2019. Outbreak of nonfermentative Gram-negative bacteria (*Ralstonia pickettii* and *Stenotrophomonas maltophilia*) in a hemodialysis center. Hemodial. Int. 23, E83–E89. https://doi.org/10.1111/hdi.12722.

UK Government, 1991. Water Industry Act 1991.

UK Government, 2018. The Private Water Supplies (England) Regulations 2016.

Uldum, S.A., Helbig, J.H., 2001. *Legionella* serogroup and subgroup distribution among patients with Legionnaires' disease in Denmark. In: Legionella. John Wiley & Sons, Ltd, pp. 200–203, https://doi.org/10.1128/9781555817985.ch36.

Unterberg, M., Rahmel, T., Kissinger, T., Petermichl, C., Bpsmanns, M., Niebius, M., Schulze, C., Jochum, H.-P., Parohl, N., Adamzik, M., Nowak, H., 2021. Legionella contamination of a cold-water supplying system in a German university hospital—assessment of the superheat and flush method for disinfection. J. Prev. Med. Hyg. 62, E751–E758. https://doi.org/10.15167/2421-4248/jpmh2021.62.3.1944.

USEPA, 1999a. Alternative Disinfectants and Oxidents—Guidance Manual. https://nepis.epa.gov/Exe/ZyNET.exe/2000229L.TXT?ZyActionD=ZyDocument&Client=EPA&Index=1995+Thru+1999&Docs=&Query=&Time=&EndTime=&SearchMethod=1&TocRestrict=n&Toc=&TocEntry=&QField=&QFieldYear=&QFieldMonth=&QFieldDay=&IntQFieldOp=0&ExtQFieldOp=0&XmlQuery=&File=D%3A%5Czyfiles%5CIndex%20Data%5C95thru99%5CTxt%5C00000015%5C2000229L.txt&User=ANONYMOUS&Password=anonymous&SortMethod=h%7C-&MaximumDocuments=1&FuzzyDegree=0&ImageQuality=r75g8/r75g8/x150y150g16/i425&Display=hpfr&DefSeekPage=x&SearchBack=ZyActionL&Back=ZyActionS&BackDesc=Results%20page&MaximumPages=1&ZyEntry=1&SeekPage=x&ZyPURL.

USEPA, 1999b. Microbial Disinfection Byproduct Rules—Simultaneous Compliance Guidance Manual.

USEPA, 2007. Simultaneous Compliance Guidance Manual For The Long Term 2 And Stage 2 Dbp Rules. https://ldh.la.gov/assets/oph/Center-EH/engineering/SDWP/SCGM_LT2_S2.pdf.

Vincenti, S., de Waure, C., Raponi, M., Teleman, A.A., Boninti, F., Bruno, S., Boccia, S., Damiani, G., Laurenti, P., 2019. Environmental surveillance of *Legionella* spp. colonization in the water system of a large academic hospital: Analysis of the four-year results on the effectiveness of the chlorine dioxide disinfection method. Sci. Total Environ. 657, 248–253. https://doi.org/10.1016/j.scitotenv.2018.12.036.

Waines, P.L., Moate, R., Moody, A.J., Allen, M., Bradley, G., 2011. The effect of material choice on biofilm formation in a model warm water distribution system. Biofouling 27, 1161–1174. https://doi.org/10.1080/08927014.2011.636807.

Walker, J.T., Jhutty, A., Parks, S., Willis, C., Copley, V., Turton, J.F., Hoffman, P.N., Bennett, A.M., 2014. Investigation of healthcare-acquired infections associated with *Pseudomonas aeruginosa* biofilms in taps in neonatal units in Northern Ireland. J. Hosp. Infect. 86, 16–23. https://doi.org/10.1016/j.jhin.2013.10.003.

Ward, R.A., 2011. Avoiding toxicity from water-borne contaminants in hemodialysis: new challenges in an era of increased demand for water. Adv. Chronic Kidney Dis. 18, 207–213. https://doi.org/10.1053/j.ackd.2011.01.007.

Water Kinetics, 2022. EcoDuo. https://www.water-kinetics.co.uk/.

Watt, J., 1946. Practical implications of the epidemiology of the diarrheal diseases of the newborn. Obstet. Gynecol. Surv. 1, 214. https://doi.org/10.1097/00006254-194604000-00067.

WHO, 2017. Guidelines for Drinking-Water Quality, 4th edition, Incorporating the 1st Addendum.

Wiegmann, D.A., Wood, L.J., Cohen, T.N., Shappell, S.A., 2022. Understanding the "Swiss cheese model" and its application to patient safety. J. Patient Saf. 18, 119–123. https://doi.org/10.1097/PTS.0000000000000810.

Wolf, I., Bergervoet, P.W.M., Sebens, F.W., van den Oever, H.L.A., Savelkoul, P.H.M., van der Zwet, W.C., 2014. The sink as a correctable source of extended-spectrum β-lactamase contamination for patients in the intensive care unit. J. Hosp. Infect. 87, 126–130. https://doi.org/10.1016/j.jhin.2014.02.013.

WRAS, 1999. The Water Supply (Water Fittings) Regulations 1999.

Index

Note: Page numbers followed by *f* indicate figures and *t* indicate tables.

L

M

Printed in the United States
by Baker & Taylor Publisher Services